The Application of Machine Learning in Geotechnical Engineering

The Application of Machine Learning in Geotechnical Engineering

Editor

Wei Gao

MDPI

Basel • Beijing • Wuhan • Barcelona • Belgrade • Novi Sad • Cluj • Manchester

Editor
Wei Gao
Hohai University
Nanjing
China

Editorial Office
MDPI AG
Grosspeteranlage 5
4052 Basel, Switzerland

This is a reprint of articles from the Special Issue published online in the open access journal *Applied Sciences* (ISSN 2076-3417) (available at: https://www.mdpi.com/journal/applsci/special_issues/WE8Y1M34P1).

For citation purposes, cite each article independently as indicated on the article page online and as indicated below:

Lastname, A.A.; Lastname, B.B. Article Title. *Journal Name* **Year**, *Volume Number*, Page Range.

ISBN 978-3-7258-2247-8 (Hbk)
ISBN 978-3-7258-2248-5 (PDF)
doi.org/10.3390/books978-3-7258-2248-5

Contents

About the Editor

Wei Gao

Wei Gao is a Professor at the College of Civil and Transportation Engineering of Hohai University, Nanjing, P.R. China. He is also a Professor at the Key Laboratory of the Ministry of Education for Geomechanics and Embankment Engineering. He received both his B.S. and M.S. degrees from the China University of Mining and Technology, Xuzhou, P.R. China, and received his Ph.D. from the Logistical Engineering University, Chongqing, P.R. China. From 2002 to 2005, he completed a postdoctoral fellowship at the Institute of Rock and Soil Mechanics, Chinese Academy of Sciences, Wuhan, P.R. China, and then worked as an Associate Researcher in the same institute from 2005 to 2009. From 2009 to 2013, he worked as a Professor at the School of Civil and Architectural Engineering, Wuhan University, Wuhan, P.R. China. At present, he is the Distinguished Professor of Jiangsu and The Leading Talent of Jiangsu Province. His research interests include the stability of geotechnical engineering and the application of intelligent methods in civil engineering. He has published more than 100 journal papers and five monographs.

applied sciences

MDPI

Editorial

The Application of Machine Learning in Geotechnical Engineering

Wei Gao

Key Laboratory of Ministry of Education for Geomechanics and Embankment Engineering, College of Civil and Transportation Engineering, Hohai University, Nanjing 210098, China; wgaowh@163.com

1. Introduction

Geotechnical engineering is civil engineering constructed in rock and soil and includes three main types: underground, foundation, and slope engineering. Because rock and soil are typical natural geological bodies, their mechanical properties and internal structures are very complex [1]. They are characterized by complicated physical, mechanical, and chemical behaviors, and their structure is a three-phased system which changes depending on water content and environmental conditions in a highly non-linear manner. Moreover, rock and soil are anisotropic and heterogeneous by nature owing to differences in their origins and formation processes. Therefore, most geotechnical engineering problems involve the coupling of multiple fields and multiple phases; generally, they cannot be solved easily. Moreover, unsafe geotechnical engineering will bring about serious disasters, such as landslides and surface subsidence, which cannot be solved well by traditional (e.g., theoretical, numerical, and experimental) methods. However, the development of artificial intelligence has supported the development of better solutions to geotechnical engineering problems. Machine learning methods have been applied widely to this end and have gained prominence in current research trends [2,3]. The first study to use machine learning in geotechnical engineering was conducted by Stanford in 1983 [4] and addressed the potential applications of expert systems in the domain of slope engineering specifically. From that moment on, machine learning methods have been gaining increasingly widespread use in the field of geotechnical engineering. Because geomechanics are the foundation of geotechnical engineering, nowadays, there are numerous studies on the applications of machine learning methods in geomechanics. These include two main aspects: geotechnical parameters and geotechnical constitutive models [5,6]. Numerous studies have also explored using machine learning methods in the three main types of geotechnical engineering: underground engineering [7,8], foundation engineering [9], and slope engineering [10]. Moreover, machine learning methods have been used to solve specific geotechnical engineering problems, including soil–structure interactions [11] and seepage of dams [12]. In addition, machine learning methods have been employed to predict geotechnical engineering disasters, such as ground surface settlements [13], landslides [14], rock bursts [15], etc. Currently, machine learning methods have become prevalent in the development of boring machines for underground engineering: the construction of integrated management systems for tunnel boring machine operations [16], the safety prediction of shield tunnel construction [17], the prediction of shield attitudes [18], and main drive torque estimation in shield tunnelling [19] are some of the areas in which they have proven useful. Therefore, nowadays, machine learning methods have been applied in most, if not all, fields of geotechnical engineering and geomechanics. Almost all types of machine learning methods have been employed in this field; examples include expert systems, fuzzy systems, artificial neural networks, deep learning methods, swarm intelligence, evolutionary algorithms, big data analysis, biological computation, nature-inspired computation, support vector machine, and Gaussian process regression. Out of these, the use of artificial neural networks has been the most extensive.

Citation: Gao, W. The Application of Machine Learning in Geotechnical Engineering. *Appl. Sci.* **2024**, *14*, 4712. https://doi.org/10.3390/app14114712

Received: 5 May 2024
Accepted: 27 May 2024
Published: 30 May 2024

This Special Issue presents new applications of machine learning methods in the field of geotechnical engineering, from planning and design to construction. It contains 19 articles, which I will briefly describe in the next section. It is not the purpose of this Editorial to elaborate on each of the papers, but rather to encourage the reader to explore them on their own.

2. An Overview of the Published Papers

In the study by Elsawy et al. (contribution 1), the undrained shear strength of sensitive alluvial soft clay located in northern areas of the Nile River, Egypt, is determined using the machine learning approach. This analysis is key to assessing the stability of foundation engineering in this type of soil. Generally, the main method to determine the undrained shear strength of soil is through laboratory tests. However, this is costly and time-consuming. Moreover, extracting undisturbed samples of sensitive clay from construction sites it extremely difficult. Therefore, based on a dataset of 111 geotechnical testing points from laboratory and field tests, Elsawy et al. use several machine learning models (including linear regression, Gaussian process regression, regression trees, ensembles of regression trees, and support vector machine) to determine the undrained shear strength of sensitive alluvial soft clay according to the soil's key features, which include water content, liquid limit, dry unit weight, plasticity index, consistency index, void ratio, specific gravity, and pocket penetration shear. The performance of each machine learning model is assessed through its coefficient of determination. The results show that the most accurate model is the support vector regression model.

The study by Zheng et al. (contribution 2) is focused on the back analysis of surrounding rock parameters of underground engineering. For the construction of large-span arch cover metro stations, the determination of the surrounding rock parameters is very important. In this paper, to obtain the surrounding rock parameters of Shikui Road station, located on Dalian Metro Line Five, an intelligent back analysis method, called the Gaussian process differential evolution co-optimization algorithm (GP-DE algorithm), is proposed. In this method, based on the data obtained by the FLAC3D finite element model using the orthogonal scheme for numerical results, the relationship between monitoring data (displacement and stress) and the surrounding rock parameters is constructed by the Gaussian process model. Based on data measured in real-life engineering, through the optimization of the differential evolution algorithm, suitable surrounding rock parameters can be obtained. Finally, by using the forward calculation of FLAC3D, the accuracy of the inversion parameters is verified. The results show that the forward calculation results obtained using the inversed parameters are in good agreement with those obtained in real life, and the accuracy of this back analysis method is very high.

The third article in this Special Issue is also focused on the back analysis of geomaterial mechanical parameters of underground engineering. In this study, a novel back analysis method combining a reduced-order model (ROM) and grey wolf optimization (GWO) has been proposed. In this method, the ROM is adopted to construct a surrogated model between the response of the underground structure and the surrounding rock parameters based on a numerical model, and GWO is used to optimize the surrounding rock parameters based on the ROM. The proposed method is applied to determine the surrounding rock parameters of two tunnels. The results show that the obtained surrounding rock parameters are in excellent agreement with the actual parameters, and the deformation results computed based on those parameters are consistent with the theoretical deformation results.

In the fourth article in this collection, the synthetic aperture radar (SAR) and optical image registration are studied. The SAR and optical images collect rich spectral information for ground objects, but their qualities are seriously affected by atmospheric attenuation and weather conditions. Therefore, improving the quality of these images is crucial to improving geotechnical engineering surveys. In this study, a novel method for SAR and optical image registration is proposed. In the new method of feature point extraction, phase consistency intensity screening and scale space grid division are combined to obtain stable and uniform feature points from images, and in feature description, the extended phase consistency method is applied to calculate the gradient amplitude and direction of the images. The experimental results showed the superior matching performance of the new method compared with current state-of-the-art methods.

The study by Zhan et al. (contribution 5) is also focused on the back analysis of surrounding rock parameters of underground engineering. In this study, based on data from a numerical model established based on a fluid–solid coupling finite element model, a surrogated model between the response of the surrounding rock (displacement and pore water pressure distribution) and surrounding rock parameters is constructed by the Gaussian process. According to the measured displacement and pore water pressure of the surrounding rock, this surrogated model searches for the suitable rock parameters using differential evolution. Therefore, the study proposes a new intelligent back analysis method for fluid–solid coupling of surrounding rock in tunnels in water-rich areas. Finally, this new back analysis method is verified by comparing the inversion parameters with the ones measured in real-life.

In the next paper, by Li and Dias (contribution 6), the assessment of rock elasticity modulus is conducted using four hybrid random forest models. The study proposes a data-driven method based on the hybrid random forest for the determination of rock elasticity modulus, which is vital for rock engineering design and cannot be solved using traditional methods such as experimental analysis and empirical formulas. In this new method, four metaheuristic optimization algorithms (backtracking search optimization algorithm, multi-verse optimizer, golden eagle optimizer, and poor and rich optimization algorithm) are used to optimize the random forest, thus enabling the construction of four hybrid random forest models. Based on the collected database consisting of 120 rock samples, the hybrid random forest models are applied to predict the rock elasticity modulus according to four factors (porosity, P-wave velocity, Schmidt hammer rebound number, and point load index). Moreover, the performance of the four hybrid random forest models is evaluated according to four indices (root-mean-square error, mean absolute error, determination coefficient, and Willmott's index). The results show that the prediction accuracy of the hybrid random forest model based on the poor and rich optimization algorithm is the best, and that porosity is the most important factor.

The study by Yang et al. (contribution 7) is also focused on estimating the undrained shear strength of clay. In this study, based on the collected database of 202 Finnish clay samples, a CatBoost–Bayesian hybrid model adaptively coupled with modified theoretical equations is constructed to determine the undrained shear strength of clay according to 11 main parameters (organic content, clay content, void ratio, natural water content, liquid limit, plastic limit, effective overburden pressure, preconsolidation pressure, overconsolidation ratio, compression index, and sensitivity). The CatBoost–Bayesian hybrid model, in which the Bayesian optimization algorithm is used to optimize the CatBoost algorithm, is employed to obtain the feature importance level of the 11 parameters and is adaptively coupled with the theoretical equation of undrained shear strength derived from the modified Cambridge model. The constructed model is verified using the experimental samples of Finnish clay. The results indicate that the established model can successfully estimate the undrained shear strength of isotropically consolidated clays.

The next study, by Lee et al. (contribution 8), uses machine learning for the prediction of ground subsidence risk in urban areas in Korea. Because ground subsidence in urban areas, caused by damage to underground utilities, can cause serious disasters, Lee et al. construct a machine learning-based ground subsidence risk prediction model based on the collected attribute information and historical ground subsidence data on six types of underground utility lines (water supply, sewage, power, gas, heating, and communication). Firstly, the target area is divided into a square grid from which the attribute information and historical ground subsidence data are extracted. Twenty-four datasets are developed, including single-type attribute information, merged by six types of underground utility lines, and three risk levels, categorized from the number of ground subsidence occurrences. Then, based on the datasets, three machine learning models (random forest, extreme gradient boosting, and light gradient boosting machine) are applied to classify ground subsidence risk levels. The results show that the new method can successfully predict risk levels in the target region.

The study by Chala and Ray (contribution 9) is focused on comparing four machine learning-based soil classification methods using cone penetration test data. In this study, the four machine learning methods include artificial neural network, random forest, support vector machine, and decision trees. The database used is collected from the International Society for Soil Mechanics and Geotechnical Engineering (ISSMGE) database and includes 232 cone penetration test (CPT) datasets, and soil is classified according to Robertson's soil behavioral types. Moreover, the quantitative metrics used to evaluate the machine learning methods include overall accuracy, sensitivity, precision, F1_score, and confusion matrices. The results show that all machine learning methods can accurately classify most soils, and most evaluation metrics of the four machine learning methods indicate high scores. Moreover, the support vector machine and random forest have outstanding performance on both majority and minority soil classes and can be applied for rapid and accurate soil classification.

The article by Lendo-Siwicka et al. (contribution 10) is focused on the determination of the damping ratio of clay soils. In this study, to ease the difficulty of determining the properties of soils, which are affected by many complex factors, a new method to determine the value of the clay soil damping ratio using an artificial neural network model is proposed. Based on a dataset consisting of 1227 examples from the testing 15 soil samples in the Resonance Column, the damping ratio of clay soils affected by seven factors (shear strain, normalized shear modulus, liquid limit, mean effective stress, silt content, plasticity limit, and clay content) is predicted. Moreover, a comparison of the new artificial neural network model with empirical formulas is conducted. The results show that the new method boasts a much higher prediction accuracy and a more flexible application.

The next study, by Cheng et al. (contribution 11), is focused on the prediction of undrained bearing capacity of skirted foundation in spatially variable soils. The bearing capacity of the skirted foundation, widely used in offshore and subsea engineering, is seriously affected by variabilities in soil undrained shear strength. To predict its uniaxial bearing capacity factors under pure horizontal and moment loads, an efficient machine learning method using a two-dimensional convolutional neural network is proposed. In this new method, datasets with 600 samples from the random finite element numerical simulation are used; the input is a random field data matrix for the soil domain in a numerical model, and the output is the corresponding bearing capacity factor. The results show that the prediction performance of the new method is satisfactory in terms of the variation coefficients and the fluctuation scale in two directions, with a high determination coefficient and low root-mean-square error.

The study by Daghistani and Abuel-Naga (contribution 12) is also focused on the prediction of soil behavior using machine learning methods. In this study, based on a dataset from 1068 tests on sand (including microscopy, direct shear, oedometer, and specific gravity tests), a machine learning model for evaluating the influence of sand particle morphology on shear strength is constructed. Two machine learning methods—multiple linear regression and random forest regression—are applied and compared. The features of sand particle morphology considered by the two machine learning models include mean particle size, uniformity coefficient, curvature coefficient, dry density, normal stress, and particle regularity. The results show that the prediction accuracy of both models is very high compared to the experimental results, and the most important factor affecting the shear strength of sand is mean particle size.

The next study, by Chala and Ray (contribution 13), is focused on the prediction of soil shear wave velocity, which is an essential parameter in evaluating the seismic response in foundation engineering. In this study, based on a collected dataset with

1000 cone penetration test data, four machine learning methods (random forest, support vector machine, decision trees, and extreme gradient boosting) are employed to predict soil shear wave velocity according to four features: cone tip resistance, sleeve friction, friction ratio, and soil depth. Moreover, to improve the performance of the four machine learning methods, their hyperparameters are optimized using Bayesian optimization with the k-fold cross-validation method. Eight metrics (root-mean-square error, mean absolute error, mean absolute percentage error, coefficient of determination, performance index, scatter index, uncertainty analysis at 95% confidence level, and proportion of samples that fall within ±10% deviation from the predicted values compared with the target value) are used to evaluate the performance of the proposed machine learning methods. The results show that the performance of the random forest is the best, achieving the highest accuracy and the lowest level of errors.

The study by Ma et al. (contribution 14) focuses on the application of the multi-objective optimization method for the optimization of the foundation pit dewatering scheme, which is very important for the safety of foundation engineering. Using the foundation pit dewatering theory and the multi-objective optimization algorithm of the non-dominated sorting genetic algorithm (NSGA-II), the authors optimize the foundation pit dewatering scheme for a foundation pit dewatering project at an inverted siphon section of the Xixiayuan canal head. This multi-objective optimization has three objectives, which are minimum total cost of dewatering, minimum amount of land subsidence caused by dewatering, and maximum drawdown of water level in the center of the foundation pit. Using this new method, a Pareto-optimal solution set with uniform distribution is obtained, and the optimization scheme is applied to the solution set to produce multiple feasible schemes for real dewatering. The results show that the obtained foundation pit dewatering scheme meets the requirements for water level and settlement control.

The article by Almasoudi et al. (contribution 15) focuses on the effects of dry density and moisture content on kaolin–brass interfacial shear adhesion. In this study, to evaluate the interface shear adhesion between compacted kaolin clay and a metallic surface, a new testing method is proposed. Compacted kaolin clay specimens with various energy levels and moisture contents are used to determine the interface shear adhesion strength between reconstituted kaolin clay and a metallic surface. The results show that to provide the highest density and divide the compaction curve into dry and wet sides, the optimum moisture content is 30%, and as the clay's dry density increases, the interface shear adhesion strength increases too. Moreover, as the moisture content rises on the wet side of the compaction curve, the strength decreases significantly.

The next article, by Abed et al. (contribution 16), focuses on the accurate estimation of soil compaction parameters. In this study, based on a collected dataset with 226 entries, the multivariate adaptive regression splines model algorithm is used to predict essential soil compaction parameters, including optimum water content and maximum dry density, according to six factors (liquid limit, plastic limit, compaction energy, sand content, fines content, and gravel content). The hyperparameter of the multivariate adaptive regression splines model is searched for using the grid search approach with cross-validation strategies. To evaluate the performance of the proposed machine learning method, three metrics (root-mean-square error, mean absolute error, and coefficient of determination) are applied. The results show that the performance of the proposed model is excellent, with a high coefficient of determination and a low root-mean-square error and mean absolute error. Thus, the model's robustness and reliability in predicting soil compaction parameters are all very high.

The following article, by Gajan (contribution 17), predicts the acceleration amplification ratio of rocking shallow foundations. In this study, based on a dataset from 140 rocking foundation experiments comprising a total of 9 series of centrifuge and shaking table experiments, the maximum acceleration transmitted to structures on rocking shallow foundations during earthquake loading is predicted by various machine learning models (including artificial neural network, k-nearest neighbors regression, support vector regression, random forest, adaptive boosting regression, and gradient boosting regression models) according to three non-dimensional rocking system capacity parameters (critical contact area ratio of the rocking foundation, slenderness ratio of the rocking structure, and rocking coefficient of the soil–foundation structure system) and two earthquake loading demand parameters (peak horizontal ground acceleration of earthquake shaking and arias intensity of earthquake). Here, the acceleration amplification ratio is defined as the maximum acceleration at the gravity center of a structure by the peak ground acceleration of the earthquake. To evaluate the performance of the machine learning models, two metrics (mean absolute percentage error and mean absolute error) are applied. The results show that the artificial neural network model is the most accurate and most consistent.

The study by Yang et al. (contribution 18) focuses on the prediction of the advanced rate of dual-mode shield tunneling in complex strata. In this study, based on the geological and on-site monitoring parameters of dual-mode shield tunneling collected from the left tunnel of Shenzhen Metro Line 13 (China), the advanced rate of shield tunneling is predicted using a long short-term memory recurrent neural network. To this end, the influence factors of advanced rate of shield tunneling, including tunneling parameters, shield tunneling mode, and strata parameters, are used. Moreover, the original data are preprocessed by the isolation forest algorithm and the improved mean filtering algorithm to obtain steady-state phase parameters. Meanwhile, the hyperparameters of the long short-term memory recurrent neural network are optimized using the particle swarm optimization, genetic algorithm, differential evolution, and Bayesian optimization algorithms. The performance of the optimized long short-term memory recurrent neural network is evaluated using the evaluation metrics of mean absolute error, root-mean-square error, and mean absolute percentage error, according to which the Bayesian optimization–long short-term memory recurrent neural network achieves superior performance. Finally, by combining the dropout algorithm and five-fold time series cross-validation with the best model, a multi-algorithm-optimized recurrent neural network model for tunneling speed prediction is constructed. The results show that the new prediction model has high prediction accuracy and operational efficiency under different excavation modes.

The last paper, by Gao et al. (contribution 19), focuses on the prediction of utility tunnel performance in soft foundations during operation periods. In this study, based on a total of 15,376 data collected from field tests on utility tunnel engineering in Suqian City, Jiangsu Province, China, utility tunnel performance in soft foundations during operation periods, represented by the main structure responses (displacement and stress), is predicted using deep learning according to five main disturbance factors (four vehicle operating load parameters and one operating time parameter). The deep belief network is applied to treat big data. To improve the network's performance and optimize its hyperparameters, the whale optimization algorithm is applied, resulting in the construction of a new deep learning model. To evaluate the prediction accuracy of the proposed model, three evaluation indexes (root-mean-square error, mean absolute error, and correlation coefficient) are applied. The results show that the new deep learning model can accurately predict the performance of utility tunnels during operation periods, with suitable applicability.

3. Conclusions

The papers collated in this Special Issue encompass the applications of machine learning methods across almost all geotechnical engineering disciplines, including underground and foundation engineering and cover a variety of stages, including planning and design, construction, and operation. This indicates that machine learning can be used across all stages of geotechnical engineering—that is, its applications can cover the full geotechnical engineering lifecycle. Moreover, machine learning methods can prove useful in the field of geomechanics, e.g., in the evaluation of the physical and mechanical parameters of geomaterials (including soil and rock). In addition, this collection covers numerous machine learning methods, including regression methods (Gaussian process regression, regression trees, ensembles of regression trees, support vector machine, reduced-order model, random forest, extreme gradient boosting, light gradient boosting machine, artificial neural network, decision tree, multivariate adaptive regression splines model, k-nearest neighbors regression, adaptive boosting regression, and CatBoost algorithm); intelligent optimization methods, from swarm intelligence and evolutionary algorithms to nature-inspired computation (differential evolution algorithm, grey wolf optimization, backtracking search optimization algorithm, multi-verse optimizer, golden eagle optimizer, poor and rich optimization algorithm, Bayesian optimization algorithm, non-dominated sorting genetic algorithm (NSGA-II), particle swarm optimization, genetic algorithm, and whale optimization algorithm); and even some deep learning methods (two-dimensional convolutional neural network, long short-term memory recurrent neural network, and deep belief network). Many of the studies presented here construct data-driven models with the aid of machine learning methods. For these studies, the key is to obtain a suitable dataset. Various methods are employed to achieve this goal, including field tests, laboratory tests, numerical simulation, previous studies, etc. Some of those datasets can be called big data.

In terms of the subjects covered by this Special Issue, the application of machine learning methods in geomechanics emerges as the main discussion topic, with eight papers. In seven of these, the research object is obtaining the properties of soil, as this can easily be achieved by tests. Of those seven papers, three focus on estimating mechanical parameters, another three on estimating physical parameters, and only one on soil classification. The final paper in this group of eight analyzes the mechanical parameters of rock. In these studies, information about the property parameters of geomaterials generally comes from laboratory tests.

The next main topic in this Special Issue is the application of machine learning methods in underground engineering, and there are six papers in this field. Among them, the most popular research object is the construction stage of engineering, which is the focus of four papers. Three of these are on the back analysis of tunnels and one on ground subsidence risks for underground utilities. Engineering during the design stage is the research object of one paper, which determines shield tunneling parameters. Finally, there is one paper whose research object is engineering during the operation period. It analyzes utility tunnel performance in soft foundations affected by operation factors.

The third major topic in this Special Issue is the application of machine learning methods in foundation engineering, which is discussed in four papers. The majority are concerned with engineering during the construction stage, with three papers covering three different research objects—the undrained bearing capacity of skirted foundations, the interaction between soil and structure, and the dynamic response of rock foundations. In the remaining paper, the research object is engineering during the design period. In this study, the optimization of a foundation pit dewatering design is conducted.

The last topic in this Special Issue is the application of machine learning methods in the planning of geotechnical engineering, with only one paper concentrating on this area. This paper focuses on the identification of geophysical prospecting information.

From the above analysis, it is evident that the geomechanics subjects in this compilation of articles are almost exclusively the parameters of geomaterials, and that one main geomechanics subject is lacking, namely the constitutive model of geomaterials. Moreover, in terms of geotechnical engineering subjects, a major area, namely slope engineering, is equally missing from this Special Issue. On top of the subjects mentioned above, machine learning methods can also be applied in the study of engineering disasters related to geotechnical engineering. Therefore, the papers in this Special Issue are only some of the new developments in the application of machine learning in geotechnical engineering, and should be seen not only as new results, but also as starting points, inviting readers to conduct future studies on the themes explored here.

As a final note, I would like to highlight the particularity that all papers in this Special Issue are on the construction of data-driven models, which is the main application of machine learning methods in geotechnical engineering. However, datasets obtained in geotechnical engineering, no matter what method is used (including field tests, laboratory tests, and numerical simulations), are not complete datasets, and their number is very limited; that is, these datasets only describe very limited, partial properties of real engineering. Therefore, purely data-driven models will run into essential problems. However, the properties of real engineering can partly be described by mechanical models on the soil and rock mechanics. Therefore, in order to apply machine learning in geotechnical engineering, mechanical models should be included as well, as is the case in physics-based machine learning [20]. Moreover, to obtain more big datasets, the development of modern monitoring technology is another way that will promote the application of machine learning in geotechnical engineering.

Conflicts of Interest: The author declares no conflicts of interest.

List of Contributions:

1. Elsawy, M.B.D.; Alsharekh, M.F.; Shaban, M. Modeling Undrained Shear Strength of Sensitive Alluvial Soft Clay Using Machine Learning Approach. *Appl. Sci.* **2022**, *12*, 10177. https://doi.org/10.3390/app121910177.
2. Zheng, F.; Jiang, A.; Guo, X.; Min, Q.; Yin, Q. Back Analysis of Surrounding Rock Parameters of Large-Span Arch Cover Station Based on GP-DE Algorithm. *Appl. Sci.* **2022**, *12*, 12590. https://doi.org/10.3390/app122412590.
3. Zhao, L.; Liu, X.; Zang, X.; Zhao, H. Back Analysis of Geotechnical Engineering Based on Data-Driven Model and Grey Wolf Optimization. *Appl. Sci.* **2022**, *12*, 12595. https://doi.org/10.3390/app122412595.
4. Zhang, W.; Zhao, Y. SAR and Optical Image Registration Based on Uniform Feature Points Extraction and Consistency Gradient Calculation. *Appl. Sci.* **2023**, *13*, 1238. https://doi.org/10.3390/app13031238.
5. Zhan, T.; Guo, X.; Jiang, T.; Jiang, A. Intelligent Feedback Analysis of Fluid–Solid Coupling of Surrounding Rock of Tunnel in Water-Rich Areas. *Appl. Sci.* **2023**, *13*, 1479. https://doi.org/10.3390/app13031479.
6. Li, C.; Dias, D. Assessment of the Rock Elasticity Modulus Using Four Hybrid RF Models: A Combination of Data-Driven and Soft Techniques. *Appl. Sci.* **2023**, *13*, 2373. https://doi.org/10.3390/app13042373.
7. Yang, H.; Liu, Z.; Li, Y.; Wei, H.; Huang, N. CatBoost–Bayesian Hybrid Model Adaptively Coupled with Modified Theoretical Equations for Estimating the Undrained Shear Strength of Clay. *Appl. Sci.* **2023**, *13*, 5418. https://doi.org/10.3390/app13095418.
8. Lee, S.; Kang, J.; Kim, J. Prediction Modeling of Ground Subsidence Risk Based on Machine Learning Using the Attribute Information of Underground Utilities in Urban Areas in Korea. *Appl. Sci.* **2023**, *13*, 5566. https://doi.org/10.3390/app13095566.
9. Chala, A.T.; Ray, R. Assessing the Performance of Machine Learning Algorithms for Soil Classification Using Cone Penetration Test Data. *Appl. Sci.* **2023**, *13*, 5758. https://doi.org/10.3390/app13095758.
10. Lendo-Siwicka, M.; Zabłocka, K.; Soból, E.; Markiewicz, A.; Wrzesiński, G. Application of an Artificial Neural Network (ANN) Model to Determine the Value of the Damping Ratio (D) of Clay Soils. *Appl. Sci.* **2023**, *13*, 6224. https://doi.org/10.3390/app13106224.
11. Cheng, H.; Zhang, H.; Liu, Z.; Wu, Y. Prediction of Undrained Bearing Capacity of Skirted Foundation in Spatially Variable Soils Based on Convolutional Neural Network. *Appl. Sci.* **2023**, *13*, 6624. https://doi.org/10.3390/app13116624.
12. Daghistani, F.; Abuel-Naga, H. Evaluating the Influence of Sand Particle Morphology on Shear Strength: A Comparison of Experimental and Machine Learning Approaches. *Appl. Sci.* **2023**, *13*, 8160. https://doi.org/10.3390/app13148160.
13. Chala, A.T.; Ray, R.P. Machine Learning Techniques for Soil Characterization Using Cone Penetration Test Data. *Appl. Sci.* **2023**, *13*, 8286. https://doi.org/10.3390/app13148286.
14. Ma, Z.; Wang, J.; Zhao, Y.; Li, B.; Wei, Y. Research on Multi-Objective Optimization Model of Foundation Pit Dewatering Based on NSGA-II Algorithm. *Appl. Sci.* **2023**, *13*, 10865. https://doi.org/10.3390/app131910865.
15. Almasoudi, R.; Abuel-Naga, H.; Daghistani, F. Effects of Dry Density and Moisture Content on the Kaolin–Brass Interfacial Shear Adhesion. *Appl. Sci.* **2023**, *13*, 11191. https://doi.org/10.3390/app132011191.

16. Abed, M.S.; Kadhim, F.J.; Almusawi, J.K.; Imran, H.; Bernardo, L.F.A.; Henedy, S.N. Utilizing Multivariate Adaptive Regression Splines (MARS) for Precise Estimation of Soil Compaction Parameters. *Appl. Sci.* **2023**, *13*, 11634. https://doi.org/10.3390/app132111634.
17. Gajan, S. Prediction of Acceleration Amplification Ratio of Rocking Foundations Using Machine Learning and Deep Learning Models. *Appl. Sci.* **2023**, *13*, 12791. https://doi.org/10.3390/app132312791.
18. Yang, T.; Wen, T.; Huang, X.; Liu, B.; Shi, H.; Liu, S.; Peng, X.; Sheng, G. Predicting Model of Dual-Mode Shield Tunneling Parameters in Complex Ground Using Recurrent Neural Networks and Multiple Optimization Algorithms. *Appl. Sci.* **2024**, *14*, 581. https://doi.org/10.3390/app14020581.
19. Gao, W.; Ge, S.; Gao, Y.; Yuan, S. Prediction of Utility Tunnel Performance in a Soft Foundation during an Operation Period Based on Deep Learning. *Appl. Sci.* **2024**, *14*, 2334. https://doi.org/10.3390/app14062334.

References

1. Stead, D.; Wolter, A. A critical review of rock slope failure mechanisms: The importance of structural geology. *J. Struct. Geol.* **2015**, *74*, 1–23. [CrossRef]
2. Ebid, A.M. 35 Years of (AI) in Geotechnical Engineering: State of the Art. *Geotech. Geol. Eng.* **2021**, *39*, 637–690. [CrossRef]
3. Baghbani, A.; Choudhury, T.; Costa, S.; Reiner, J. Application of artificial intelligence in geotechnical engineering: A state-of-the-art review. *Earth-Sci. Rev.* **2022**, *228*, 103991. [CrossRef]
4. Stanford, G. Potential Applications of Expert Systems in Geotechnical Engineering. Master's Thesis, Carnegie-Mellon University, Pittsburgh, PA, USA, 1983.
5. Puri, N.; Prasad, H.D.; Jain, A. Prediction of Geotechnical Parameters Using Machine Learning Techniques. *Procedia Comput. Sci.* **2018**, *125*, 509–517. [CrossRef]
6. Gao, W. A comprehensive review on identification of the geomaterial constitutive model using the computational intelligence method. *Adv. Eng. Inform.* **2018**, *38*, 420–440. [CrossRef]
7. Liu, L.; Song, Z.; Li, X. Artificial intelligence in tunnel construction: A comprehensive review of hotspots and frontier topics. *Geohazard Mech.* **2024**, *2*, 1–12. [CrossRef]
8. Wang, X.; Lu, H.; Wei, X.; Wei, G.; Behbahani, S.S.; Iseley, T. Application of Artificial Neural Network in Tunnel Engineering: A Systematic Review. *IEEE Access* **2020**, *8*, 119527. [CrossRef]
9. Shahin, M.A. State-of-the-art review of some artificial intelligence applications in pile foundations. *Geosci. Front.* **2016**, *7*, 33–44. [CrossRef]
10. Gao, W.; Ge, S. A comprehensive review of slope stability analysis based on artificial intelligence methods. *Expert Syst. Appl.* **2024**, *239*, 122400. [CrossRef]
11. Jong, S.C.; Ong, D.E.L.; Oh, E. State-of-the-art review of geotechnical-driven artificial intelligence techniques in underground soil-structure interaction. *Tunn. Undergr. Sp. Tech.* **2021**, *113*, 103946. [CrossRef]
12. Beiranvand, B.; Rajaee, T. Application of artificial intelligence-based single and hybrid models in predicting seepage and pore water pressure of dams: A state-of-the-art review. *Adv. Eng. Softw.* **2022**, *173*, 103268. [CrossRef]
13. Niu, G.; He, X.; Xu, H.; Dai, S. Tunnelling-induced ground surface settlement: A comprehensive review with particular attention to artificial intelligence technologies. *Nat. Hazards Res.* **2024**, *4*, 148–168. [CrossRef]
14. Merghadi, A.; Yunus, A.P.; Dou, J.; Whiteley, J.; Thaipham, B.; Bui, D.T.; Avtar, R.; Abderrahmaneet, B. Machine Learning Methods for Landslide Susceptibility Studies: A Comparative Overview of Algorithm Performance. *Earth-Sci. Rev.* **2020**, *207*, 103225. [CrossRef]
15. Basnet, P.M.S.; Mahtab, S.; Jin, A. A comprehensive review of intelligent machine learning based predicting methods in long-term and short-term rock burst prediction. *Tunn. Undergr. Sp. Tech.* **2023**, *142*, 105434. [CrossRef]
16. Loy-Benitez, J.; Song, M.K.; Choi, Y.H.; Lee, J.; Lee, S.S. Breaking new ground: Opportunities and challenges in tunnel boring machine operations with integrated management systems and artificial intelligence. *Automat. Constr.* **2024**, *158*, 105199. [CrossRef]
17. Ge, S.; Gao, W.; Cui, S.; Chen, X.; Wang, S. Safety prediction of shield tunnel construction using deep belief network and whale optimization algorithm. *Automat. Constr.* **2022**, *142*, 104488. [CrossRef]
18. Wang, L.; Pan, Q.; Wang, S. Data-driven predictions of shield attitudes using Bayesian machine learning. *Comput. Geotech.* **2024**, *166*, 106002. [CrossRef]
19. Glab, K.; Wehrmeyer, G.; Thewes, M.; Broere, W. Predictive machine learning in earth pressure balanced tunnelling for main drive torque estimation of tunnel boring machines. *Tunn. Undergr. Sp. Tech.* **2024**, *146*, 105642. [CrossRef]
20. Vadyala, S.R.; Betgeri, S.N.; Matthews, J.C.; Matthews, E. A review of physics-based machine learning in civil engineering. *Results Eng.* **2022**, *13*, 100316. [CrossRef]

applied sciences

MDPI

Article

Modeling Undrained Shear Strength of Sensitive Alluvial Soft Clay Using Machine Learning Approach

Mohamed B. D. Elsawy [1,2], Mohammed F. Alsharekh [3,*] and Mahmoud Shaban [3,4]

1 Department of Civil Engineering, Faculty of Engineering, Geotechnical Engineering and Foundations at University of Tabuk, Tabuk 71491, Saudi Arabia
2 Department of Civil Engineering, Faculty of Engineering, Aswan University, Aswan 81542, Egypt
3 Department of Electrical Engineering, College of Engineering, Qassim University, Unaizah 56452, Saudi Arabia
4 Department of Electrical Engineering, Faculty of Engineering, Aswan University, Aswan 81542, Egypt
* Correspondence: m.alsharekh@qu.edu.sa

Abstract: Soft soils are commonly located in many regions near seas, oceans, and rivers all over the world. These regions are vital and attractive for population and governments development. Soft soil is classified as problematic soil owing to sustaining low shear strength and high settlement under structures. Constructing structures and/or infrastructures on soft soil is a considerable risk that needs great attention from structural engineers. The bearing capacity of structure foundations on soft soil depends mainly on their undrained shear strength. This soil feature strongly influences the selection of appropriate soil improvement methods. However, determining undrained shear strength is very difficult, costly, and time-consuming, especially for sensitive clay. Consequently, extracting undisturbed samples of sensitive clay faces several difficulties on construction sites. In this research, accurate field-tested data were fed to advanced machine learning models to predict the undrained shear strength of the sensitive clay to save hard effort, time, repeated laboratory testing, and costs. In this context, a dataset of 111 geotechnical testing points were collected based on laboratory and field examinations of the soil's key features. These features included the water content, liquid limit, dry unit weight, plasticity index, consistency index, void ratio, specific gravity, and pocket penetration shear. Several machine learning algorithms were adopted to provide the soft clay modeling, including the linear, Gaussian process regression, ensemble and regression trees, and the support vector regression. The coefficient of determination was mainly used to assess the performance of each predictive model. The achieved results revealed that the support vector regression model attained the most accurate prediction for soil undrained shear strength. These outcomes lay the groundwork for evaluating soil shear strength characteristics in a practical, fast, and low-cost way.

Keywords: soft clay; undrained shear strength; machine learning; ICT; predictive modeling

Citation: Elsawy, M.B.D.; Alsharekh, M.F.; Shaban, M. Modeling Undrained Shear Strength of Sensitive Alluvial Soft Clay Using Machine Learning Approach. *Appl. Sci.* **2022**, *12*, 10177. https://doi.org/10.3390/app121910177

Academic Editor: Wei Gao

Received: 10 September 2022
Accepted: 4 October 2022
Published: 10 October 2022

1. Introduction

Patches of soft soil are distributed throughout many vital regions near oceans, seas, and rivers. These regions are essential for human activities and the development of various structures such as roads, bridges, embankments, buildings, railroads, tunnels, etc., and new cities are also constructed on this type of soil. The problems presented by soft soil are high compressibility and low shear strength. Moreover, structure settlement continues for long periods. The failure of structures mainly occurs due to a lack of soft soil shear strength. Soft soil is one of the problematic soils encountered all over the world. Consequently, the "soft soil" [1], is defined as soil that typically exhibits an undrained shear strength (USS) in the range of 20 to 40 kPa, while the very soft soil possesses USS values of less than 20 kPa. The performance of soft soil depends not only on water content, but also on its structure. Generally, the soft soil remains stiff in the dry state until it is subjected to a drastic increase in its water content, at which it converts to weak and soft soils [2]. The saturated soft soil causes several problems for structures on it, such as low bearing capacity and long-term excessive settlement [3]. Constructing heavy structures on native soft clay soil is a difficult task. The geotechnical design depends mainly on the (short-term) undrained shear strength of the soft soil, as it has a smaller value compared to that of the (long-term) drained shear strength.

Hence, the undrained shear strength of the soft soil is the most important parameter to be determined when calculating its bearing capacity. The selection of suitable improvement methods for the sensitive clay is also mainly dependent on the accurate values of its undrained shear strength. However, there is a difficulty associated with extracting and sampling the soft soil, especially the very sensitive soft clay. The sensitive clay usually has an undrained shear strength in the laboratory smaller than that in the field due to the effect of disturbance on the sample. Since the design of structure foundations depends on the undrained shear strength of the sensitive clay, it must be carefully determined and evaluated from the results of both field and laboratory tests.

The precision of the laboratory tests, such as direct shear, unconfined compression, or undrained triaxial compression, significantly depend on the quality of the collected sensitive, undisturbed clay samples [2,4], and additionally rely on the thickness and the friction of the sharp edges of the circular samplers with soil. On the other hand, field test results (field vane test and piezocone cone penetration test—CPTU) are also influenced by selection techniques [5,6]. In 2021, Ayadat [4] stated that the field vane shear test is more accurate than the Swedish cone shear test.

Due to the difficulty of determining the undrained shear strength of the sensitive clay, empirical equations are utilized based on clay properties, like clay activity and Atterberg limits [4,7–11], water content, plasticity index, and over consolidation ratio [12]. However, there is a shortage of past studies that investigate the functional form and the influential soil properties of the existing empirical correlations. Moreover, Mataic et al. [13] stated that pre-consolidation pressures can differ significantly owing to soil disturbances during sampling, which occasionally causes unpredictable values of the pre-consolidation pressures. Subsequently, empirical correlations include only a few soil parameters despite there being greater soil parameter influences on the undrained shear strength of the soft soil.

Recently, machine learning (ML) models have been widely utilized to predict the same important soil parameters that are costly, time-consuming, and require great effort to determine in the laboratory or the field. Generally, machine learning has been recently used by geotechnical engineers in the prediction of skin friction of driven piles [14], the bearing capacity of shallow foundations [15], or the shear strength of soil [16–18]. ML models are rarely used to estimate undrained shear strength of soft soil based on different geotechnical properties [12,19,20]. In 2022, Tran et al. [6] only utilized ML models in the prediction of the undrained shear strength for marine-sensitive soft soil, with a coefficient of correlation of 0.715.

Predictions of undrained shear strength of the alluvial sensitive soft clay using ML advanced modeling have not yet been thoroughly studied by geotechnical researchers. Therefore, the main objective of the current research was to introduce an accurate ML model to predict the undrained shear strength of sensitive alluvial soft clay using many important soft clay soil properties. The ML models have been formulated using a realistic dataset. The advantage of the ML model is to make an accurate prediction for the undrained shear strength of alluvial sensitive soft clay, in order to avoid disturbance difficulty during its extracting and sampling. Moreover, handling and performing shear strength tests on very soft and soft soil, either in the field or in the laboratory, typically comes with high costs, is time-consuming, and requires hard work and effort from geotechnical staff. ML prediction of the undrained shear strength of sensitive clay is expected to introduce significant savings in the above-mentioned factors. The ML models predict undrained shear strength based on soft clay essential properties, which can be easily determined in the laboratory. Soft soil properties, acting as predicting features, include moisture content, specific gravity, void ratio, dry unit weight, liquid limit, plasticity and consistency indexes, and pocket penetration shear. In this context, the ML model has been formulated using precise realistic datasets (features) for the sensitive soft clay samples collected from the delta region of Egypt, located in northern areas of the Nile River. Generally, alluvial soft clay regions are very crowded all over the world, with various structures like roads, embankments, bridges, railways, tunnels, airports, buildings, etc. The accuracy of the ML models is evaluated by the coefficient of correlation method. An accurate ML prediction of the undrained shear strength will achieve sustainably for the structures on non-treated or treated sensitive soft clay by gathering between its long-term high performance and economic side. The used undrained shear strength values of the sensitive clay were determined from the results of precise field vane shear tests. Collecting realistic and feature-enriched datasets, along with the selection of the appropriate algorithm, led to a substantial improvement in the prediction accuracy of the model.

2. Materials and Methods

The soft clay is located in northern areas of the Nile River, in Egypt. The sensitive clay was formed in alluvial sediments. For the ML model study, 111 soft clay samples were collected from the northern Nile delta. The data set included geotechnical features of natural water content (*W*), dry unit weight (γ_d), liquid limit (*LL*), plasticity index (*PI*), consistency index (*CI*), void ratio (*e*), specific gravity (*Gs*), and pocket penetration shear (q_p). The utilized features were easily determined or calculated from the geotechnical laboratory tests. The performed experimental tests were natural water content, specific gravity, bulk density, liquid and plastic limits, and Pocket penetration, according to ASTM numbers D 2216, D 854, D1556, D 4318, and WK27337, respectively. The latter tests are traditional geotechnical tests, and are also more cost-effective and less time-consuming.

Moreover, the adopted dataset was selected because the included soil features significantly influence the undrained shear strength of the studied soft clay. For example, increasing the natural water content causes a decrement in the cohesive force between soil particles, which leads to a weakening of soil consistency and reduces shear strength. Increasing dry density leads to a minimizing of the pores between soil particles and void ratio, causing an effective increment in the undrained shear strength. The void ratio of soil not only affects its permeability, but it also significantly influences the undrained shear strength of the soil. The increments of the liquid limit, plasticity index, and consistency index offer strong evidence for the existence of clay content. Consequently, increasing clay content in soft soil directly increases its undrained shear strength. Specific gravity is considered the main factor in predicting shear strength. Pocket penetration shear value represents an important indication of the undrained shear strength of the saturated cohesive soil value, which has a significant correlation with the undrained shear strength of the sensitive soft clay.

It is known that determining undrained shear strength, either in the field or in the laboratory, is highly difficult. Moreover, several problems were faced in extracting undisturbed samples of sensitive clay. According to the unified soil classification system, the sensitive clay soil was categorized as a high plasticity clay (CH). The classification of CH in soft soil was a result of the high values of the plasticity index. The liquid limit and plasticity index ranged from 46% to 103%, and from 20% to 69%, respectively. The natural water content ranged between 40% to 71%, as illustrated in Figure 1 and Table 1. The undrained shear strength of the clay was determined by field vane shear tests (ASTM D2573-08) [21]. The vane test was suitable for the studied soil, having saturation conditions and a soft consistency. The field van shear equipment is shown in Figure 2. The results of the undrained shear strength stated that the sensitive clay ranged from very soft to soft soil (5 kPa–45 kPa). The sensitivity values of the clay were between 2.5 and 28.6, as depicted in Figure 3. Clay sensitivity was the ratio of the undrained shear strength in the field to that in the laboratory, for the same soil. According to the sensitivity classification of Das and Sobhan (2014) [20], the current study soil included low, medium, and high sensitivity clay. To overcome the difficulties presented in extracting samples of sensitive clay and determining its undrained shear strength, as well as saving time and reducing costs, accurate ML models were utilized.

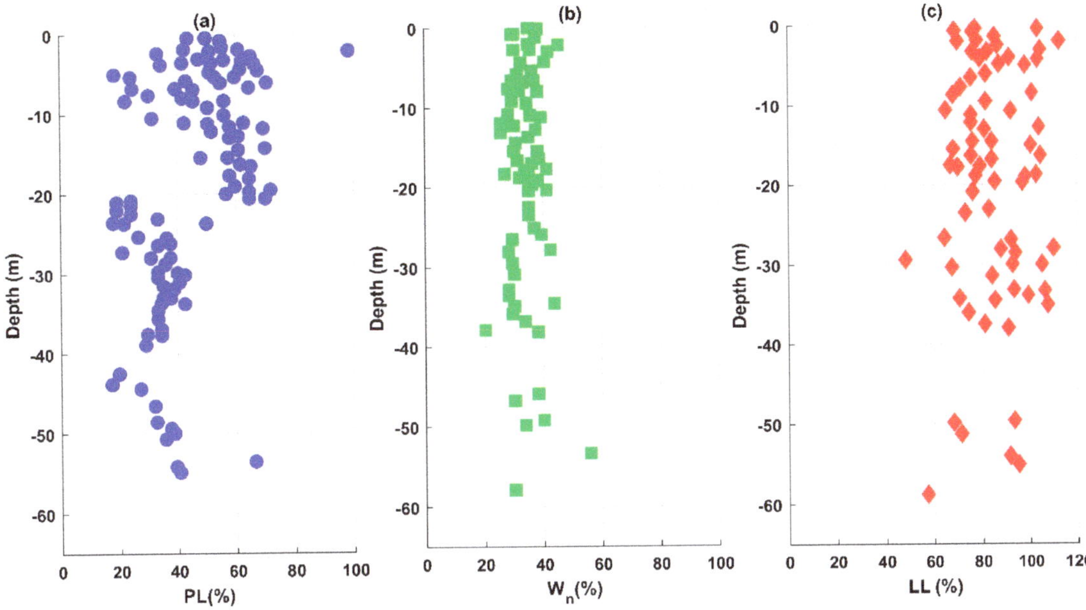

Figure 1. Plot of (**a**) PL, (**b**) W_n, and (**c**) LL versus depth.

Table 1. Statistical analysis of the dataset of soft clay features.

Soft Clay Feature (Unit)	Minimum	Maximum	Mean	Standard Deviation
γ_d (kN/m^3)	8.926	12.40	10.49	0.769
W_n (%)	40.00	71.00	54.80	6.834
LL (%)	46.00	103.0	75.58	12.77
PI (%)	20.00	69.00	43.07	9.587
CI	0.060	1.150	0.491	0.243
e	1.149	2.017	1.586	0.191
Qp (kPa)	10.00	80.0	38.19	15.75
Gs	2.680	2.740	2.711	0.0135
USS (kPa)	5.00	45.00	20.815	8.199

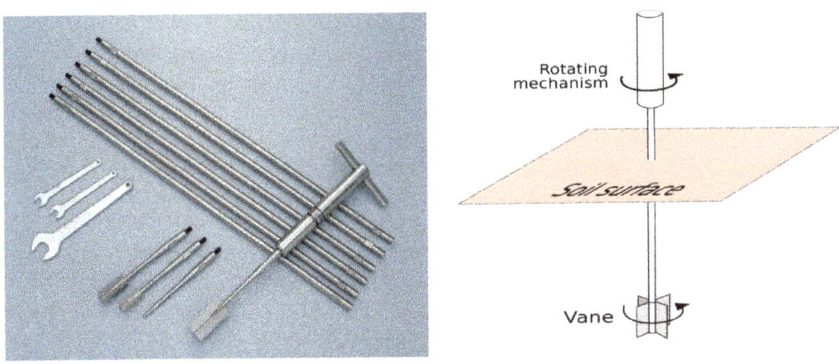

Figure 2. Field vane shear equipment.

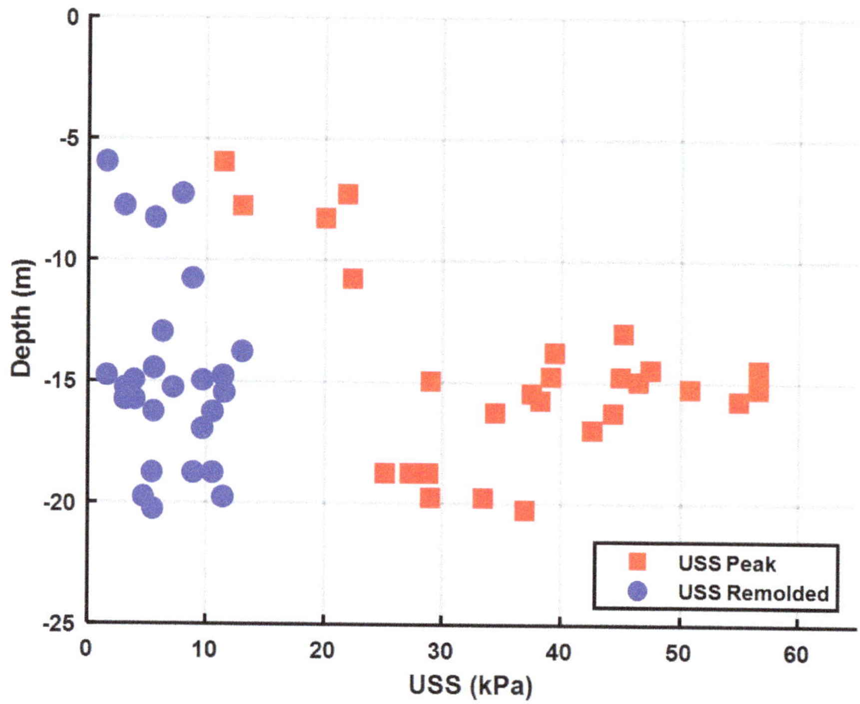

Figure 3. Sensitivity of the soft soil.

In this study, we considered eight input features to create an ML predictive modeling of the soft clay soil. The datasets were initially preprocessed to eliminate any outliers that were commonly observed in realistic datasets. To eliminate the impact of incompatible data scales on model training time, and therefore make the computations much faster, the data was normalized using the following equation:

$$x_n = \frac{x - x_{min}}{x_{max} - x_{min}} \tag{1}$$

where x refers to the original data, x_n refers to the normalized data, x_{min}, and x_{max} signify the minimum and maximum values detected in the dataset, respectively. The modified output data was simply retrieved in its original form, after model training, utilizing the formula:

$$x = x_n(x_{max} - x_{min}) + x_{min} \tag{2}$$

We used a statistical procedure known as feature selection in order to identify the most important features that have a significant influence on model predictions. The procedure comprises a correlation matrix (CM) alongside a principal component analysis (PCA). Following that, various supervised ML algorithms including linear regression, Gaussian process regression, regression trees, ensembles of regression trees, and support vector machine (SVM) approaches were adopted in the training phase.

The SVM approach [22] is widely used for ML modeling of classification or regression problems. With a few minor exceptions, the support vector machine regression (SVR) utilizes the same concepts as the SVM for classification [23–26]. In the case of regression, a margin of tolerance (ε) is specified as a rough approximation to the SVM that the issue would have already requested. However, there is a more problematic reason: the algorithm is more intricate and, therefore, it must be considered. The SVM algorithm operates by using the largest margin to identify the most optimum hyperplane that splits data into many classes. We used the regression function *f(x)* to optimally approximate the supplied training dataset $\{(x_i\ y_i)\}_i^N$, where x_n is a multivariate collection of N observations with observed response values y_n. For the simplest case, the function *f(x)* is expressed as [22]:

$$f(x) = wx + b \tag{3}$$

The optimized values of w and b can be acquired by minimizing the following expression [22]:

$$\min \frac{1}{2}\|W\|^2 + C\sum_{i=1}^{N}(\xi_i + \xi_i^*) \tag{4}$$

Subject to:

$$\begin{cases} y_i - wx_i - b \le \varepsilon + \xi_i \\ wx_i + b - y_i \le \varepsilon + \xi_i^* \\ \xi_i,\ \xi_i^* \ge 0 \end{cases} \tag{5}$$

where ε is an ε-insensitive tube indicating the error tolerance, and C is a compromise between the empirical error and the general term. The regression function may be stated as follows, using Lagrangian multipliers and optimum constraints [22]:

$$y = \sum_{i=1}^{N}(\alpha_i + \alpha_i^*)K(x_i, x) + b \tag{6}$$

where $K(x_i, x_j)$ is the kernel function. Examples of the most famous kernel functions are the linear, polynomial, sigmoidal, Gaussian, and radial basis functions. The latter is expressed as [22]:

$$K(x_i, x_j) = \exp\left(-\frac{\|x_i - x_j\|^2}{2\sigma^2}\right) \tag{7}$$

where $\|x_i - x_j\|$ denotes the Euclidean distance between the two feature vectors, and σ is the spread of the kernel function's distribution.

To evaluate and verify ML models, several assessment indices are commonly utilized. In this study, the following measures were used to evaluate the model prediction performance:

(1) Residuals, r_i, which characterizes the error for each data point. Most regression metrics are typically derived using r_i given by:

$$r_i = (y_i - \hat{y}_i)^2 \tag{8}$$

where y_i denotes the measured or original data, while $\hat{y}_i$ denotes the predicted data obtained by the model.

(2) Root mean square error (RMSE), which is computed by:

$$RMSE = \sqrt{\frac{1}{N}\sum_{i=1}^{N}(y_i - \hat{y}_i)^2} \tag{9}$$

(3) Furthermore, the coefficient of determination, R^2, is frequently used to assess and compare various regression models and is calculated by:

$$R^2 = 1 - \frac{\sum_{i=1}^{N}(y_i - \hat{y}_i)^2}{\sum_{i=1}^{N}(y_i - \overline{y}_i)^2} \tag{10}$$

where $\overline{y}_i$ denotes the mean values of the predicted output.

Following the choice of the most optimal regression modeling, hyperparameters were optimized (tuned) using further optimization techniques. In this regard, model hyperparameters are special configurations that are externally set to the model, using optimization procedures, and their values cannot be estimated from data. As a result, for correctly configured model hyperparameters, the objective function, known as the loss function, is minimized. In this process, for each training iteration, the mean squared error (MSE) statistics are obtained and computed using:

$$MSE = \frac{1}{N}\sum_{i=1}^{N}(y_i - \hat{y}_i)^2 \quad (11)$$

After optimizing the model hyperparameters, they are deployed and can be employed to predict soft clay undrained shear strength. The framework of the modeling procedure is further illustrated as shown in Figure 4.

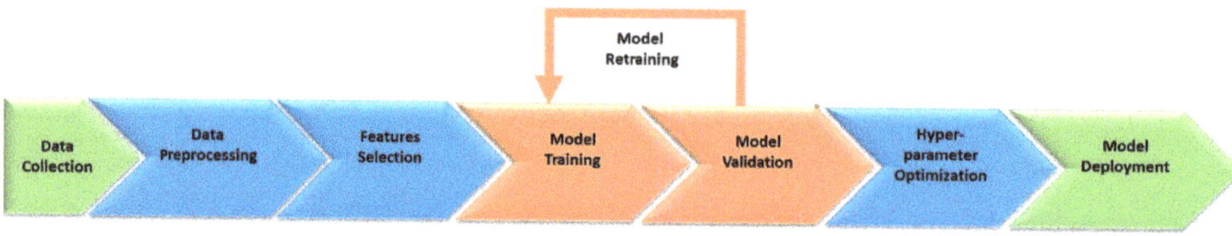

Figure 4. Framework of soft clay ML modeling.

3. Results and Discussion

For the examined soft clay soil, descriptive statistics of its geotechnical parameters, as well as the linear correlation chart, are illustrated in Figure 5. The studied clay soils exhibited a wide range of dry density (8.93–12.4 kN/m^3), initial void ratio (1.149–2.017), specific gravity (2.68–2.74), consistency index (0.06–1.15), and pocket penetration shear (10–80 KPa,). Table 1 summarizes the statistical analysis of the dataset.

The increments in the sensitive clay initial properties, like dry unit-weight, liquid limit, plasticity index, and consistency index, caused an increase in the values of the undrained shear strength, as illustrated in Figure 5. The undrained shear strength expressed the cohesion shear strength parameter of the normally consolidated clayey soil. The dry density increase led to additional inter-cohesion between clay particles. Consequently, the increment of liquid limit, plasticity, and consistency indexes was an indication of the high values of clay contents in the studied sensitive soil. The higher value of clay contents caused higher values of cohesion (undrained shear strength). The later results were found also by [7]. On the other hand, the water content and the void ratio had a negative influence on the USS of the sensitive soil. This is because the excessive voids and water in the soil logically decreased the shear resistance of the soil. Finally, the specific gravity had approximately no impact on the soil USS values, as depicted in Figure 5.

Data partitioning using a procedure known as cross-validation (CV) is essentially required for accurate modeling. Using this technique, the training dataset was randomly partitioned into two groups: the training set and the validation set. The training set was used in the learning phase and was typically the largest dataset and was mainly utilized to obtain the parameters of the model under development, while the validation set was basically used to tune the model hyperparameters. CV assisted data partitioning into a set of folds (k-folds). This process was run either during training or when estimating the average test error across all folds. Subsequently, this technique guards against overfitting better than others; however, it requires numerous fits, and consequently, it is suitable for small and medium-sized datasets. In this modeling approach, we used a CV partitioning of 5 folds. Sequentially, several ML techniques, including linear regression, regression trees, Gaussian process regression, the ensemble of trees, and support vector regression, were trained and validated to attain their optimum hyperparameters. For the SVR, the most important hyperparameters were the box constraint, denoted by the parameter C, which was a positive numeric variable that determined the penalty imposed on samples that fell beyond the epsilon edges (ε) and assisted in the prevention of overfitting or satisfying model regularization. In addition, the selection of the kernel function, as well as its optimized kernel scale, plays an important role in scaling the input features. The model performance is further evaluated using a new dataset, namely the testing set. This set is used to assess the model's performance and guarantees that it can generalize effectively to new and unknown data points. The evaluation metrics for RMSE and R^2 for both the training and testing sets are listed in Table 2. Among the adopted algorithms, the fine Gaussian SVR gave the best-fit results. The RMSE of the training and testing sets of the optimized model were 2.57 and 1.65 kPa, respectively. The model hyperparameters, including box constraint, ε, kernel function, and kernel scale of the optimized model based on SVR, are outlined in Table 3. The attained values of the determination coefficient, R^2-Score, were 0.90 and 0.96, respectively. These results indicate that the formulated model is reliable and accurate enough to be employed for forecasting the USS of soft clay soil.

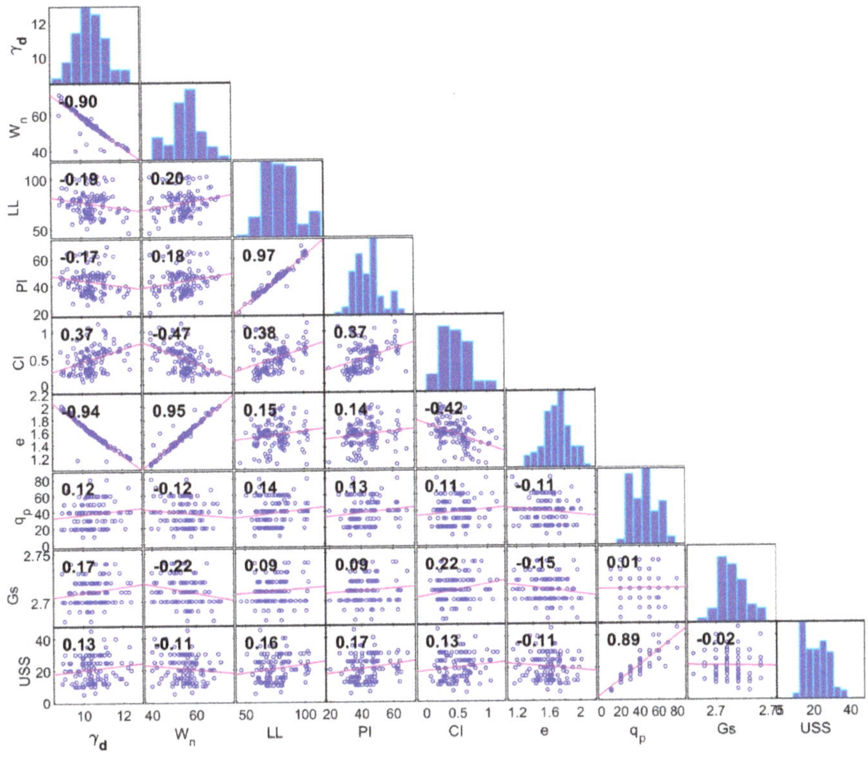

Figure 5. Correlation chart of soft clay features.

Table 2. Evaluation metrics of soft clay ML models for both the training and testing sets.

	Training Set		Testing Set	
ML Algorithm	**RMSE (kPa)**	R^2	**RMSE (kPa)**	R^2
Linear regression	3.65	0.80	2.89	0.87
Regression trees	3.66	0.80	2.68	0.89
Gaussian process regression	3.17	0.85	3.17	0.85
Ensemble of trees	3.79	0.79	2.79	0.88
Fine Gaussian SVR	2.57	0.90	1.65	0.96

Table 3. List of hyperparameters of the optimized model.

C	ε	Kernel Function	Kernel Scale
11.8255	0.02199	Gaussian	8.6869

Figure 6 demonstrates assessment plots of the predicated undrained shear strength against the actual one for the modeled dataset. As depicted in the figure, most values of the measured samples lie close to the line of equality with the predicted values. Furthermore, Figure 7 represents the correlation between the actual and the predicted data as regression curves. As can be observed, the model's predictability is good, especially for the testing dataset. As a result, the use of an SVR model to forecast soil USS is achievable with high accuracy and low error.

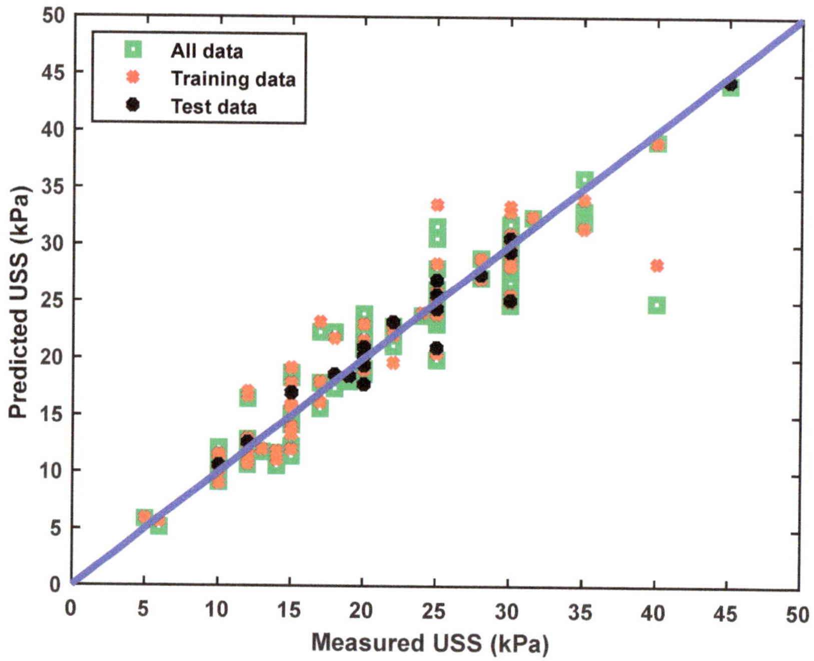

Figure 6. Comparison results of the trained model, predicted versus actual output data of vane-tested USS. The straight line denotes to line of equality.

Based on the above results and analysis, accurate ML prediction of the undrained shear strength of the alluvial sensitive soft clay has an important and positive impact on the field of geotechnical engineering. The current ML model achieved more accuracy compared with other previous studies [6,19]. Precise determination of USS for the sensitive clay is a challenging task, being time-consuming, requiring great effort, and incurring high financial costs. The input data of the basic soil properties effectively and directly influence the USS values with positive or negative impacts. The selected features in the present model are essential properties of the studied soil that can be easily determined in a short time. The collected realistic and feature-enriched datasets, as well as the selection of the appropriate ML algorithm, resulted in a significant improvement in the prediction accuracy of the presented model. The accurate ML prediction of USS of the alluvial sensitive clay contributes toward solving many problems, such as saving time, effort, and costs. This is achieved by reducing the number of samples and tests that are essentially required using traditional methods.

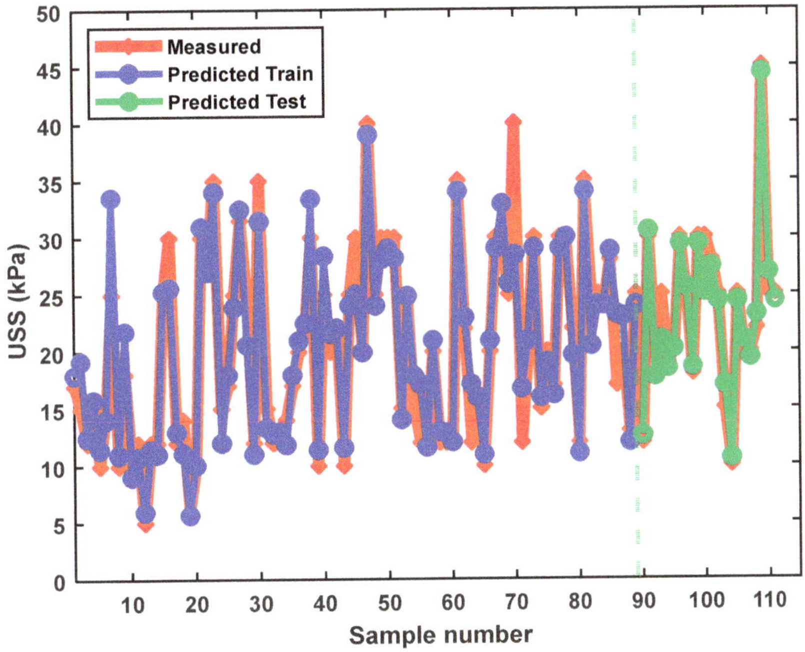

Figure 7. Soft clay USS plotted against sample number of the measured, predicted, and tested dataset.

4. Conclusions

This study presents ML predictive modeling of the alluvial sensitive soft clay soil undrained shear strength, an essential parameter, and intensively investigated topic in geotechnical engineering. The creation of an accurate machine learning model facilities to achieve a significant cost reduction in field testing. In this regard, data is essential in machine learning challenges in order to build an accurate and reliable prediction model. One aspect of this investigation was the collection of extra data comprising many important features of the soft clay soil. Collecting realistic and rich datasets along with the selection of the appropriate algorithm leads to substantial improvement of prediction accuracy. The dataset that was utilized to forecast soil undrained shear strength has eight input variables, namely the moisture content, specific gravity, void ratio, dry unit weight, liquid limit, plasticity and consistency indexes, and Pocket penetration shear. Several ML approaches were trained including linear regression, regression trees, Gaussian process regression, the ensemble of trees, and SVR. Among those, the fine Gaussian SVR accurately exhibited the best fit model. Many evaluation metrics of the model prediction performance were computed including MSE, RMSE, and R^2 metrics. The latter value for the investigated soft clay dataset was evaluated to be 0.96 for the testing dataset. The attained results confirm the model reliability to accurately forecast the sensitive alluvial soft clay vane-test undrained sheer strength. This accurate prediction of ML model has a great positive impact in the geotechnical and structural engineering field saving hard work, high costs, and long time.

Author Contributions: Conceptualization, M.B.D.E. and M.S.; Data curation, M.B.D.E.; Formal analysis, M.B.D.E. and M.S.; Investigation, M.F.A. and M.S.; Methodology, M.B.D.E., M.F.A. and M.S.; Project administration, M.F.A.; Software, M.S.; Validation, M.F.A.; Writing—original draft, M.S.; Writing—review and editing, M.F.A. All authors have read and agreed to the published version of the manuscript.

Funding: This research received no external funding.

Institutional Review Board Statement: Not applicable.

Informed Consent Statement: Not applicable.

Data Availability Statement: Not applicable.

Acknowledgments: The authors would like to thank the Deanship of Scientific Research, Qassim University for funding the publication of this study.

Conflicts of Interest: The authors declare no conflict of interest.

References

1. Institution, B.S. *Code of Practice for Design of Buildings and Structures for Agriculture. Appendices: Legislation, Technical Data and References*; British Standards Institution: London, UK, 1986.
2. Firoozi, A.A.; Olgun, C.G.; Firoozi, A.A.; Baghini, M.S. Fundamentals of soil stabilization. *Int. J. Geo.-Eng.* **2017**, *8*, 26. [CrossRef]
3. Ali, O.K.; Abbas, H.O. Performance assessment of screw piles embedded in soft clay. *Civ. Eng. J.* **2019**, *5*, 1788–1798. [CrossRef]
4. Ayadat, T. Determination of the Undrained Shear Strength of Sensitive Clay Using Some Laboratory Soil Data. *Stud. Eng. Technol.* **2021**, *8*, 14–27. [CrossRef]
5. Ly, H.-B.; Pham, B.T. Prediction of shear strength of soil using direct shear test and support vector machine model. *Open Constr. Build. Technol. J.* **2020**, 14.
6. Tran, Q.A.; Ho, L.S.; Le, H.V.; Prakash, I.; Pham, B.T. Estimation of the undrained shear strength of sensitive clays using optimized inference intelligence system. *Neural Comput. Appl.* **2022**, *34*, 7835–7849. [CrossRef]
7. Pham, B.T.; Hoang, T.-A.; Nguyen, D.-M.; Bui, D.T. Prediction of shear strength of soft soil using machine learning methods. *Catena* **2018**, *166*, 181–191. [CrossRef]
8. Hansbo, S. *New Approach to the Determination of the Shear Strength of Clay by the Fall-Cone Test*; Royal Swedish Geotechnical Institute: Stockholm, Sweden, 1957.
9. Chandler, R.J. *The In-Situ Measurement of the Undrained Shear Strength of Clays Using the Field Vane*; ASTM International: West Conshohocken, PA, USA, 1988.
10. D'Ignazio, M.; Phoon, K.-K.; Tan, S.A.; Länsivaara, T.T. Correlations for undrained shear strength of Finnish soft clays. *Can. Geotech. J.* **2016**, *53*, 1628–1645. [CrossRef]
11. El-Bosraty, A.H.; Ebid, A.M.; Fayed, A.L. Estimation of the undrained shear strength of east Port-Said clay using the genetic programming. *Ain Shams Eng. J.* **2020**, *11*, 961–969. [CrossRef]
12. Ng, I.-T.; Yuen, K.-V.; Dong, L. Estimation of undrained shear strength in moderately OC clays based on field vane test data. *Acta Geotech.* **2017**, *12*, 145–156. [CrossRef]
13. Mataic, I.; Wang, D.; Korkiala-Tanttu, L. Effect of destructuration on the compressibility of Perniö clay in incremental loading oedometer tests. *Int. J. Geomech.* **2016**, *16*, 04015016. [CrossRef]
14. Samui, P. Prediction of friction capacity of driven piles in clay using the support vector machine. *Can. Geotech. J.* **2008**, *45*, 288–295. [CrossRef]
15. Kuo, Y.; Jaksa, M.; Lyamin, A.; Kaggwa, W. ANN-based model for predicting the bearing capacity of strip footing on multi-layered cohesive soil. *Comput. Geotech.* **2009**, *36*, 503–516. [CrossRef]
16. Padmini, D.; Ilamparuthi, K.; Sudheer, K. Ultimate bearing capacity prediction of shallow foundations on cohesionless soils using neurofuzzy models. *Comput. Geotech.* **2008**, *35*, 33–46. [CrossRef]
17. Kanungo, D.; Sharma, S.; Pain, A. Artificial Neural Network (ANN) and Regression Tree (CART) applications for the indirect estimation of unsaturated soil shear strength parameters. *Front. Earth Sci.* **2014**, *8*, 439–456. [CrossRef]
18. Zhang, W.; Wu, C.; Zhong, H.; Li, Y.; Wang, L. Prediction of undrained shear strength using extreme gradient boosting and random forest based on Bayesian optimization. *Geosci. Front.* **2021**, *12*, 469–477. [CrossRef]
19. Pham, B.T.; Qi, C.; Ho, L.S.; Nguyen-Thoi, T.; Al-Ansari, N.; Nguyen, M.D.; Nguyen, H.D.; Ly, H.-B.; Le, H.V.; Prakash, I. A novel hybrid soft computing model using random forest and particle swarm optimization for estimation of undrained shear strength of soil. *Sustainability* **2020**, *12*, 2218. [CrossRef]
20. Das, B.M. *Principles of Geotechnical Engineering*; Cengage Learning: Belmont, CA, USA, 2021.
21. International, A. Standard Guide for Application of Continuous Process Verification to Pharmaceutical and Biopharmaceutical Manufacturing. Available online: https://www.astm.org/e2537-16.html (accessed on 9 May 2022).
22. Vapnik, V. *The Nature of Statistical Learning Theory*; Springer Science & Business Media: New York, NY, USA, 1999.
23. Cherkassky, V.; Ma, Y. Practical selection of SVM parameters and noise estimation for SVM regression. *Neural Netw.* **2014**, *17*, 113–126. [CrossRef]
24. Ahmad, A.S.; Hassan, M.Y.; Abdullah, M.P.; Rahman, H.A.; Hussin, F.; Abdullah, H.; Saidur, R. A review on applications of ANN and SVM for building electrical energy consumption forecasting. *Renew. Sustain. Energy Rev.* **2014**, *33*, 102–109. [CrossRef]
25. Blanco, V.; Japón, A.; Puerto, J. A mathematical programming approach to SVM-based classification with label noise. *Comput. Ind. Eng.* **2022**, *172*, 108611. [CrossRef]
26. Ahmed, N.I.; Nasrin, F. Reducing Error Rate for Eye-Tracking System by Applying SVM. In *Machine Intelligence and Data Science Applications*; Springer: New York, NY, USA, 2022; pp. 35–47.

applied sciences

MDPI

Article

Back Analysis of Surrounding Rock Parameters of Large-Span Arch Cover Station Based on GP-DE Algorithm

Fu Zheng [1], Annan Jiang [1,*], Xinping Guo [1], Qinghua Min [2] and Qingfeng Yin [2]

[1] Institute of Road and Bridge Engineering, Dalian Maritime University, Dalian 116026, China
[2] China Construction Transportation Construction Group Co., Ltd., Beijing 100010, China
* Correspondence: jiangannan@163.com; Tel.: +86-0411-84724258

Abstract: Due to the characteristics of soil–rock composites and large-span arches, the surrounding rock parameters of stations are difficult to obtain accurately under soft upper and hard lower geological conditions when the arch cover method is used to carry out the construction of a large-span underground excavation station. To optimize the design of stations and guide the next step of construction, an intelligent inverse analysis method, the Gaussian process differential evolution co-optimization algorithm (GP-DE algorithm), is proposed for the arch cover method for station construction. Taking the Shikui Road station of the Dalian Metro Line Five as the engineering background, the finite element model of FLAC3D is established. By combining the measured data of the sensor and the monitoring data obtained using the orthogonal scheme, this algorithm is used for the joint back analysis of displacement stress and the accuracy of the inversion parameters is verified by forwarding the calculation for FLAC3D. By using the obtained surrounding rock parameters, the demolition length of the center diaphragm to the Shikui Road station is optimized. Under different numbers of training samples, the inversion effect of the GP-DE algorithm and the other three common back-analysis algorithms is compared and analyzed. Finally, based on the iteration rate and convergence effect, the value range of the differential evolution algorithm parameters F and CR is given. The results show that the forward calculation results of the parameters obtained from the back analysis are in good agreement with the actual values, and the accuracy of the back-analysis results is high.

Keywords: arch-cover method; back analysis; Gaussian process; differential evolution algorithm; orthogonal design

Citation: Zheng, F.; Jiang, A.; Guo, X.; Min, Q.; Yin, Q. Back Analysis of Surrounding Rock Parameters of Large-Span Arch Cover Station Based on GP-DE Algorithm. *Appl. Sci.* **2022**, *12*, 12590. https://doi.org/10.3390/app122412590

Academic Editor: Wei Gao

Received: 1 November 2022
Accepted: 29 November 2022
Published: 8 December 2022

1. Introduction

In the construction of subway stations under complex geological conditions, the arch cover method is widely used because of its outstanding advantages, such as less impact on the surrounding environment, fewer procedures, high efficiency, and safe and reliable construction [1]. In construction using the arch cover method, due to the complexity and heterogeneity of the rock mass structure, it is difficult to accurately describe the failure mechanism of the rock mass, and it is also difficult to give accurate rock mass parameters in the project [2]. Because of the large span, the geological conditions of the crossing are more complex, so the accuracy of the rock stratum information in the exploration process is difficult to guarantee, and some parameters can only give the corresponding interval range, which cannot provide a certain reference value for the input parameters of the numerical simulation. Therefore, the back-analysis method combined with the field-measured data becomes one of the most effective methods to accurately obtain rock mechanical parameters [3,4].

At present, the back-analysis method is widely used. In the study of the back analysis, Kavanagh first put forward the method of the back analysis of displacement in 1971. According to the amount of deformations after tunnel excavation, the mechanical parameters of rock mass were calculated with the finite element method [5]. Gioda calculated some mechanical parameters of rock mass by monitoring information on displacement [6]. Gao proposed a new neural network back-analysis method based on the black hole algorithm. Based on the measured converging displacement, the physical parameters of the surrounding rock of two deep roadways in the Huainan coal mine of China were back-calculated [7]. Su used FLAC3D to initialize the stochastic mechanical parameters, and, then, based on the differential evolution algorithm, the displacement difference between the numerical value and the in situ measurement value was regarded as the fitness value for the identification of rock mechanical parameters [8]. Zhao proposed an inverse analysis method for identifying rock mass parameters around tunnels using the ELM extreme learning machine model optimized with particle swarm optimization (PSO), taking into account displacement loss and spatial effects [9]. Jiang and others, through a differential evolution algorithm, combined the automatic acquisition system with a numerical simulation to conduct a displacement stress joint back analysis of surrounding rock parameters in tunnel engineering [10]. Jin and others inversed the joint surface parameters of the rock mass of the Baihetan hydropower station by combining the discrete element method and neural network, applying the results to the model, and comparing them with the measured values [11]. Wang and others conducted a back analysis on surrounding rock parameters by compiling an intelligent displacement back-analysis program based on a differential evolution algorithm and using tunnel convergence and arch crown settlement displacement values [12]. Huang and others used a genetic algorithm to invert the surrounding rock parameters of a highway tunnel and then took the obtained parameters into the numerical model for a forward analysis. The difference between the results and the measured values was small [13]. In previous studies, some algorithms were mainly used to back-analyze the rock parameters surrounding a tunnel in combination with the actual site displacement, but only relying on the displacement for the back analysis, which has certain limitations, and there have only been a few intelligent back-analysis studies for the stations constructed using the arch cover method. In practical projects, especially under soft upper and hard lower geological conditions, the unique arch cover construction method of large span concealed excavation makes it difficult to predict the mechanical parameters of the surrounding rock. Therefore, an accurate prediction of surrounding rock parameters has a certain guiding role for the station constructed with the arch cover method.

Table 1 shows the application research of the Gaussian process machine learning algorithm. After 2009, the application of the Gaussian process in civil engineering increased, mainly in the time series prediction of rock and soil mass deformation during the construction process. There are only a few studies on the algorithm being used for the inversion of rock and soil mechanical parameters. The Gaussian process for machine learning can better solve the problem of supervised learning. Two well-known algorithms, the neural network (ANN) and support vector machine (SVM), do not handle some aspects well. For example, the neural network has certain limitations in dealing with small-sample problems, and the support vector machine is also difficult to use for determining its kernel function and loss function. With the continuous development of machine learning, the GP algorithm has become a focus point in the field of international machine learning, and has been successfully applied to regression classification. Based on the Bayesian learning theory, this algorithm has significant advantages in dealing with small samples, nonlinearity, and other problems, and the GP algorithm has flexible nonparametric inference, adaptive parameter acquisition, and a simple implementation process [14].

This paper takes the Shikui Road Station of the Dalian Metro as the engineering background. First, according to 25 groups of orthogonal design schemes, the forward calculation is carried out through the established three-dimensional FLAC3D numerical model. After the results are obtained, a sensitivity analysis is carried out on the results, the key parameters are found, and 25 groups of displacement stress results are normalized. Then, the Gaussian process differential evolution (GP-DE) algorithm is used to train the 25 groups of normalized orthogonal design results and the measured displacement and stress are taken as the input samples for the inversion to obtain accurate surrounding rock parameters, which intend to provide a theoretical basis and reference value for the optimization design of the station and the guidance of the next step of construction in arch cover construction.

Table 1. Application research of Gaussian process machine learning algorithm.

Authors	Research Introduction	Year
Carl Edward Rasmussen and Christopher K. I. Williams [15]	Gauss process machine learning algorithm, proposed in 1994, can better solve the problem of supervised learning	1994
Sofiane Brahim-Belhouari et al. [16]	Gaussian process is used in prediction of non-stationary time series, and compared with a neural network, it is proved that Gaussian process is more superior	2004
Hyun-OmlKimet al. [17]	Use Gaussian process to train human characteristics as samples for gender recognition	2006
Su Guoshao et al. [18,19]	Application of Gaussian process to time series prediction of slope deformation and foundation pit displacement	2007
Fyfe, Colin et al. [20]	Three analysis methods based on Gaussian process are proposed and relevant analysis is carried out	2008
Liu Kaiyun et al. [21]	Analysis of slope deformation time series using Gaussian process and genetic algorithm	2009
Liu Kaiyun et al. [22]	Parameter inversion of elasto-plastic model in geotechnical engineering by introducing Gaussian process regression algorithm	2011
He Peng et al. [23]	Prediction of the deformation convergence value of the surrounding rock of the working face based on Gaussian process	2017
Zhang Yunpeng et al. [24]	Establishment of prediction model for large deformation of tunnel surrounding rock using Gaussian process	2018

The remainder of this article is organized as follows. Section 2 introduces the geological uncertainty in the construction of a large-span arch cover station. Section 3 introduces the station parameter identification method based on the GP-DE arch cover method. Section 4 introduces the application of this method in engineering. Section 5 optimizes and analyzes the project's temporary support removal scheme according to the determined parameters. Section 6 discusses the inversion effect of GP-DE and the influence of differential evolution parameters on the back-analysis effect. Finally, Section 7 concludes.

2. The Problem of Geological Uncertainty in the Large-Span Arch Cover Station Construction

The geotechnical environment has a great influence on underground engineering because geotechnical mass is not only the generating medium of underground engineering, but also the source of load [25]. In particular, the core idea of the arch and cover method construction is to make use of the bearing capacity and stability of the underlying surrounding rocks, as the rock mass bears and disperses the load to different degrees, with multiple identities as well as functions [26,27]. Rock mechanical parameters are one of the most important factors affecting the mechanical characteristics of tunnel construction, together with the access point for disaster control and structure design [6]. However, the limitations of measurement technologies and equipment, especially the incomprehensibility and spatial variations in the geological environment, bring great challenges to the identification of mechanical parameters of surrounding rocks. Therefore, a reasonable determination of physical and mechanical parameters of rocks as well as soil mass has a great influence on the design and construction of underground engineering.

To ensure the safety of station construction through the arch and cover method, it is necessary to design or adjust the construction scheme, while an optimization analysis of the excavation methods, temporary support, and demolition measures are necessary [28,29]. One of the ways to optimize the rationality of the scheme is to use the numerical simulation of the site [27,30], as is shown in Figure 1. To ensure the effect and accuracy of the numerical model, it is necessary to ensure the accuracy of the formation parameters in the simulation process.

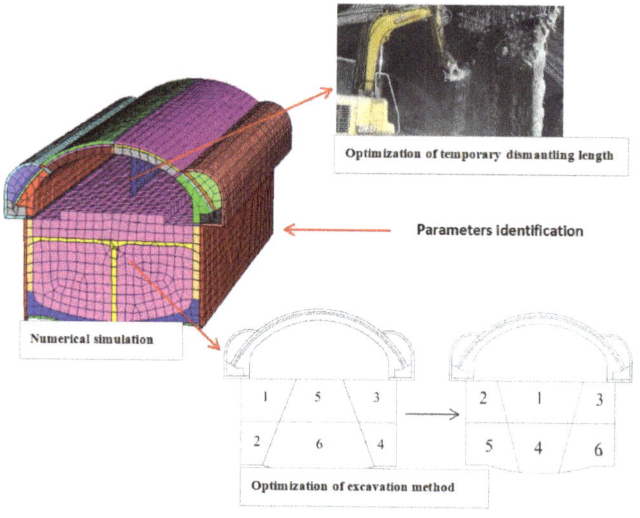

Figure 1. Relationship among parameter identification, numerical simulation and construction scheme optimization.

3. Parameter Identification Methods for Stations through Arch Cover Method Based on GP-DE

In the excavation process of a station through the arch cover method, if the joint reverse analysis of the displacement stress is used, we can more comprehensively consider the relationship between the changes of station surrounding rocks and parameters, and the accuracy of the inversion results should be higher. Combined with the automatic monitoring data and a reasonable optimization model, the optimal mechanical parameters of the surrounding rocks are obtained through the reverse analysis method, and then the actual deformation law of surrounding rocks is found, so as to provide a scientific basis for the subsequent optimization design of the station [31]. How to minimize the number of function evaluations in the optimization process while achieving the global optimal solution becomes the key. To this end, a joint inverse analysis of displacement stress based on the GP-DE algorithm is used.

3.1. The GP Respond Surface Optimized by DE

GP algorithm is mainly determined by the mean function and covariance function. Consider regression model [16,20,32].

This is example 1 of an equation:

$$y = f(x) + \varepsilon \tag{1}$$

where x is the input variation; y is the observed value; Noise $\varepsilon \sim N(0, {\sigma_n}^2)$. The expectation and variance of the predicted value y' can be obtained from the Bayesian posterior probability formula:

This is example 1 of an equation:

$$\mu'_y = C(x', X)\left[C(X, X) + {\sigma_n}^2 I_n\right]^{-1} y \tag{2}$$

$$\sigma_{y'}^2 = C(x', x') - C(x', X)\left[C(X, X) + \sigma_n^2 I_n\right]^{-1} C(X, x') \tag{3}$$

In GP, the covariance function is equivalent to the kernel function, and the squared exponential covariance function is the commonly used covariance function.

$$C(x_i, x_j) = \sigma_f^2 \exp\left[-\frac{1}{2}(x_i - x_j)^T M(x_i - x_j)\right] + \sigma_n^2 \sigma^{ij} \tag{4}$$

where: σ_f^2 is the signal variance of the kernel function, $M = diag(I^{-2})$ is the diagonal matrix of the hyperparameter, I is the variance scale, and σ^{ij} is the Kronecker symbol.

$\theta = \left\{M, \sigma_f^2, \sigma_n^3\right\}$ is made the hyperparameter, and the optimal hyperparameter can be obtained using the log function maximum likelihood method. The partial derivatives of the negative log-likelihood function and the hyperparameter θ are:

$$L(\theta) = \frac{1}{2} y^T C^{-1} y + \frac{1}{2}\lg|C| + \frac{n}{2}\lg 2\pi \tag{5}$$

$$\frac{\partial L(\theta)}{\partial \theta_i} = \frac{1}{2} tr\left[\left(\alpha\alpha^T - C^{-1}\right)\frac{\partial C}{\partial \theta_i}\right] \tag{6}$$

The DE algorithm is a differential evolution algorithm, which is also a population-based evolutionary algorithm. In the DE algorithm, the initial population is generated first, all the new individuals in the initial population are selected as the same parent probability, and better individuals are selected as the next generation between new and parent individuals. If the population generation is G and the number is N_P, the population vector of generation G is represented as $x_i(G), i = 1, 2, 3 \cdots, N_P$. Each vector individual contains D components. The basic process of the DE algorithm is as follows [8,33].

3.1.1. Generate Initial Population

In D-dimensional space, N_P chromosomes satisfying the upper and lower bounds of independent variables are randomly generated. The formula is:

$$x_{ij}(0) = rand_{ij}(0,1)(x^U{}_{ij} - x^L{}_{ij}) + x^L{}_{ij}$$
$$(i = 1, 2, \cdots, N_P,\ j = 1, 2, \cdots, D) \tag{7}$$

where: $x^U{}_{ij}$, $x^L{}_{ij}$ are the upper and lower bounds of the jth component in the ith vector; $rand_{ij}(0,1)$ is a random number between (0,1).

3.1.2. Mutation Operation

During evolution, the variable of the difference between any two target vector individuals is superimposed on a third vector individual according to a certain law, so that it becomes a brand-new variable, which is the variation vector. Through mutation operation, not only the searchability of the population is enhanced, but the diversity of the population is also maintained. For the target vector of generation G, the jth component of the variation vector is:

$$u_{ij}(G+1) = x_{r1j}(G) + F(x_{r2j}(G) - x_{r3j}(G)) \tag{8}$$

where: $r1$, $r2$, and $r3$ are arbitrary and mutually different integers in [1, N_P]; F is the scaling factor, which plays the role of adjusting the step amplitude of vector difference.

3.1.3. Cross Operation

The target vector $x_i(G)$ and the variation vector $x_i(G)$ are hybridized according to Formula (4) to generate a new test vector $u_i(G+1)$.

$$u_i(G+1) = \begin{cases} v_i(G+1),\ r_j \le CR \parallel j = n_i \\ x_i(G+1),\ r_j > CR \,\&\, j \ne n_i \end{cases} \tag{9}$$

where $r_j \in [0,1]$ is a random number, which corresponds to the jth component of the vector; $CR \in [0,1]$ is the cross probability, and a new individual is randomly generated by probability; and n_i is an integer randomly selected in $1, 2, \cdots, D$, that is, one or more components of the variation vector shall be ensured to be applied by the test vector.

3.1.4. Select Action

The greedy search method is selected for the selection operation. The fitness function is used to compare each objective vector of DE with the test vector. If the value of the objective function corresponding to the test vector is small, the test vector is selected; if not, it is retained.

The Gaussian process machine learning algorithm is used as the GP-DE algorithm to establish the nonlinear mapping relationship among the learning samples. Taking the optimal hyperparameters of the GP as the population, the hyperparameters in the GP are optimized through the mutation cross-selection operation based on the DE algorithm, so as to optimize the GP model and predict an output value closer to the optimal solution. Using the square difference between the optimal output value and the control value predicted by the GP model as the fitness function, the prediction ability of the GP model is improved by optimizing the sample population to approach the optimal solution, so that the predicted target vector is closer to the optimal solution. This cycle is iterated until the global optimal solution of each optimal hyperparameter is found.

3.2. *The Parameter Identification Flowchart*

According to the meaning of the model parameters, the stress and displacement are taken as the objective function to control the target value as well as to reach the minimum objective function.

$$\left.\begin{array}{l} \min F(x) = \min E(x) + \min D(x) \\ \min E(x_1, x_2, \cdots, x_n) = \frac{1}{m}\sum\limits_{i=1}^{m} \left[Y_i^0 - Y_i\right]^2 \\ \min D(x_1, x_2, \cdots, x_n) = \frac{1}{l}\sum\limits_{i=1}^{i} \left[S_i^0 - S_i\right]^2 \\ x_i^a \le x_i \le x_i^b (i = 1, 2, \cdots, n) \end{array}\right\} \quad (10)$$

where $\min E(x)$ is a function of the displacement control value; $\min D(x)$ is the stress control value; Y_i^0 and S_i^0 are the measured deformation and stress values of the surrounding rock; Y_i and S_i are implicit functions with the surrounding rock parameters as independent variables, which need to be solved by numerical simulation; x_i is a parameter; n is the number of parameters; x_i^a and x_i^b are the upper and lower limits of the parameter x_i; and m is the number of observation values.

The specific flow of displacement stress joint back analysis of surrounding rock parameters based on GP-DE is as follows:

1. Orthogonal design of surrounding rock parameters includes: the tensile strength T, Poisson's ratio μ, cohesion c, and internal friction angle φ. Different values of the five surrounding rock parameters of elastic modulus E are taken within a certain range for an orthogonal design, and an orthogonal scheme is formed.
2. Second item, sample generation: FLAC^{3D} is used to establish a numerical model, and the orthogonal scheme is used to simulate the model. The results of the orthogonal scheme are normalized after they are obtained, and the processed results are test samples, which are regarded as GP learning training samples and prediction test samples, and the influential factors obtained through a sensitivity analysis are used as the variables to be inverted.
3. The following method is used to establish a nonlinear mapping relationship model of surrounding rock parameters: use GP technology to establish the nonlinear mapping relationship, use superparameters to carry out GP learning and prediction on the test samples, and evaluate the fitness of the obtained results obtained. Then, repeat the operation of mutation, crossover, selection, GP prediction, and fitness evaluation on

the test samples. When the objective function reaches the preset value, the GP-DE algorithm has been optimized.

4. Optimization of surrounding rock parameters: first of all, take the objective function of Formula (10) as the fitness function of the DE algorithm, generate a random group of parameters in the solution space as the initial population, and repeat the operation of mutation, crossover, selection, GP prediction, and fitness evaluation. The surrounding rock parameters that meet the minimum fitness are the back-analysis parameters. The detailed process is shown in Figure 2.

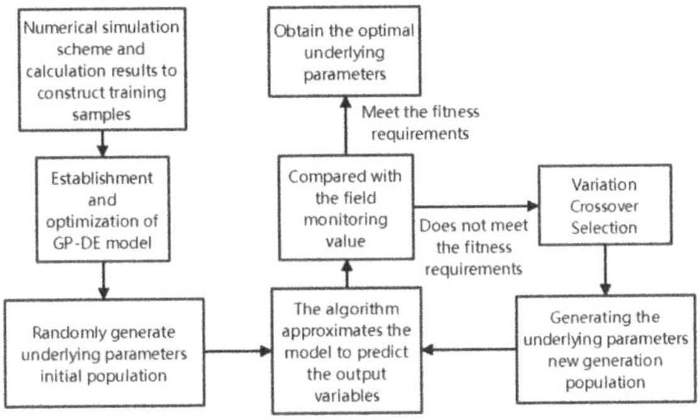

Figure 2. Flow chart of back-analysis method based on GP-DE algorithm.

Due to the different dimensions of displacement and stress in the results obtained based on the orthogonal design, a direct comparison and calculation will make the final results inaccurate, so they cannot be compared. To make the results accurate and reliable, normalization can be a good choice. After normalization processing, the data of different dimensions can be transformed into dimensionless data for subsequent processing. The normalization formula is as follows:

$$r_i = (k_i - k_{i\min})/(k_{i\max} - k_{i\min}) \quad (i = 1, 2, \cdots, 25) \tag{11}$$

where: k_i is the calculated value of the ith output in the sample, $k_{i\min}$ is the minimum value of the output in the orthogonal scheme, $k_{i\max}$ is the maximum value of the output in the orthogonal scheme, and the normalized value of r_i is $r_i \in [0, 1]$.

4. Engineering Application

4.1. Engineering Overview

Shikui Road Station of Dalian Metro Line 5 is located in the urban area of Dalian, which is surrounded by residential and commercial areas, with numerous underground pipelines. The main body of the station is located directly below the road. The overall terrain of the site is a gentle slope, whose geomorphic unit belongs to denuded low hills. The soil in the upper part of the station is plain fill, and the lower part is mainly constructed with moderately to strongly weathered quartzite. The excavation diameter of the tunnel face of the station is about 21 m, and the soil thickness is 11.9~19.4 m. The stratum through which the station passes mainly consists of moderately weathered quartzite, which is a class-IV surrounding rock with a low compressibility foundation, a high bearing capacity of bearing stratums, a uniform foundation, and a large buried depth. The project is a large-span underground excavation project, where the arch cover method is adopted for construction. During construction, a small pilot tunnel is excavated first; then, a middle pilot tunnel is excavated; then, the side walls of the small pilot tunnel are removed; and finally, and the arch cover is formed. During excavation, it is easy to cause collapse along the fracture structural plane, and the excavation depth is large, making it easy to cause soil deformation. The longitudinal geological profile and the station cross-section are shown in Figure 3. The project location is Shikui Road Station of Dalian Metro Line 5 in Dalian, Liaoning Province, China. The project location is shown in Figure 4.

4.2. Construction Procedure of Arch Cover Method

The construction sequence of symmetry is mainly adopted in the excavation of the Shikui Road arch cover pilot tunnel. First of all, a side pilot tunnel is excavated, then follows a middle pilot tunnel, and a supporting structure is timely constructed to prevent excessive settlement. Necessary monitoring measures shall also be taken during the construction to adjust the excavation footage and support scheme according to the monitoring data. The construction scheme is shown in Table 2. The specific construction process is as follows:

Table 2. Schematic diagram of arch cover method construction process.

Construction Steps	Diagrammatic Sketch	Construction Steps	Diagrammatic Sketch
1	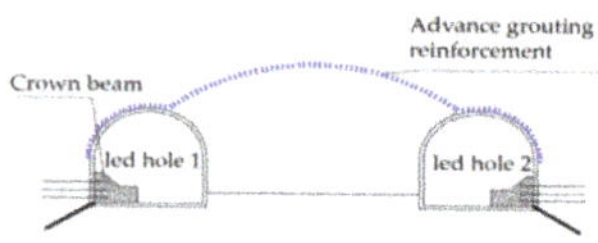	2	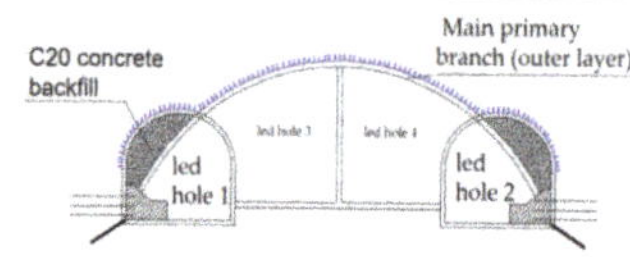
3	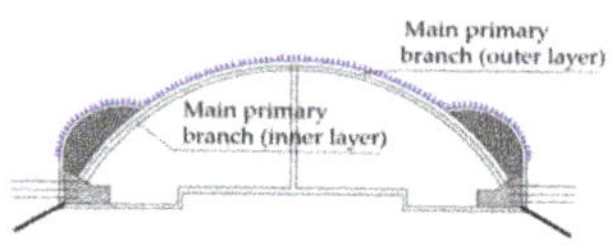	4	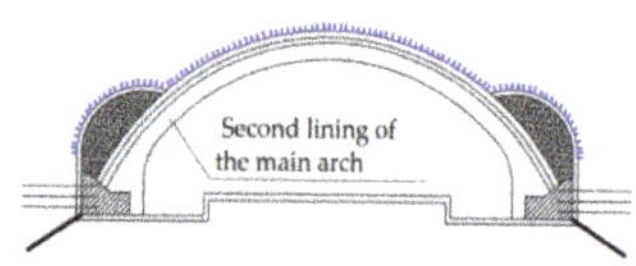
5	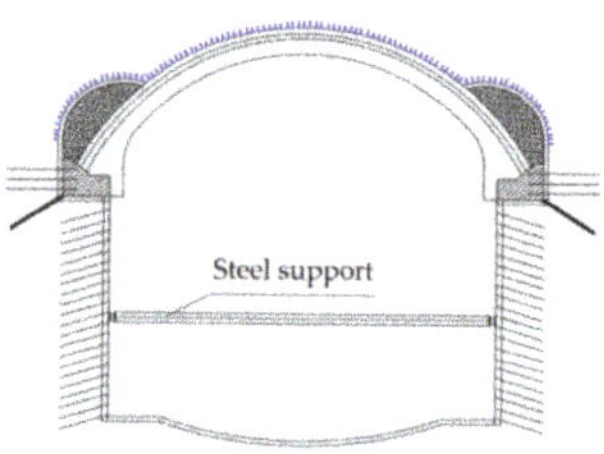	6	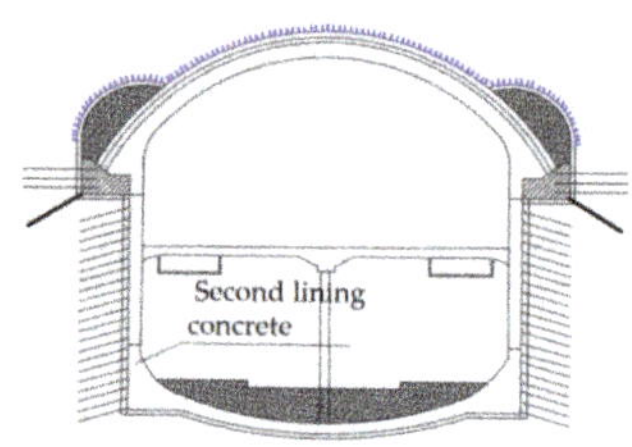

(1) The side pilot tunnels 1 and 2 of the main body are grouted and reinforced with advanced small conduits, which are then excavated and supported in time. The mortar and foot lock anchor bolts are set at the arch foot, and a crown beam is set when the side pilot tunnel is through.
(2) Advanced small tremie grouting reinforcement is carried out for pilot tunnel No. 3 and No. 4 in the main body, and then excavation is carried out, support is to be provided in time, and plain concrete is backfilled in the empty part of the side pilot tunnel.
(3) The side walls of side pilot tunnel No. 1 and No. 2 are dismantled, and secondary primary support is provided at the same time.
(4) Advanced small tremie grouting reinforcement is carried out for pilot tunnel No. 3 and No. 4 in the main body; then, excavation is carried out, support is to be provided in time, and plain concrete is backfilled in the empty part of the side pilot tunnel.
(5) The side walls of side pilot tunnel No. 1 and No. 2 are dismantled, and secondary primary support is provided at the same time.
(6) According to the measured data, the middle partition is removed at a certain distance, a waterproof layer is laid on the bottom of the crown beam, and a secondary lining structure is constructed in time.

(7) The lower soil mass is excavated layer by layer, and the side walls are initially supported, and at the same time, the side wall is provided with anchor cables and steel supports. After reaching the tunnel bottom elevation, a bottom cushion is constructed.
(8) The waterproof layer is paved, the secondary lining structure and medium plate beam column are constructed, and then the internal structure of the station is constructed.

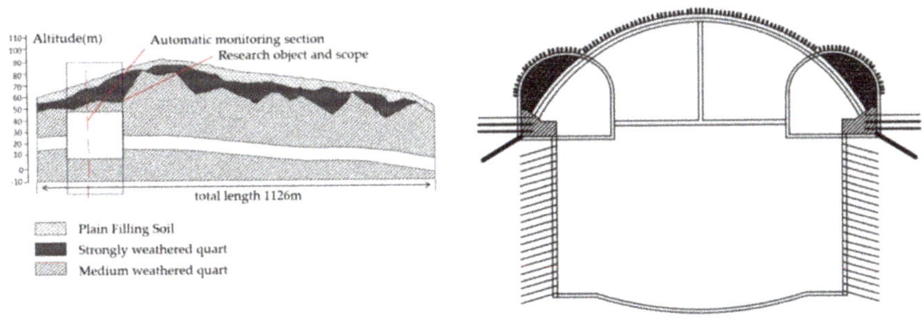

(**a**) (**b**)

Figure 3. The longitudinal geological profile and the station cross-section: (**a**) the longitudinal geological profile; (**b**) the station cross-section.

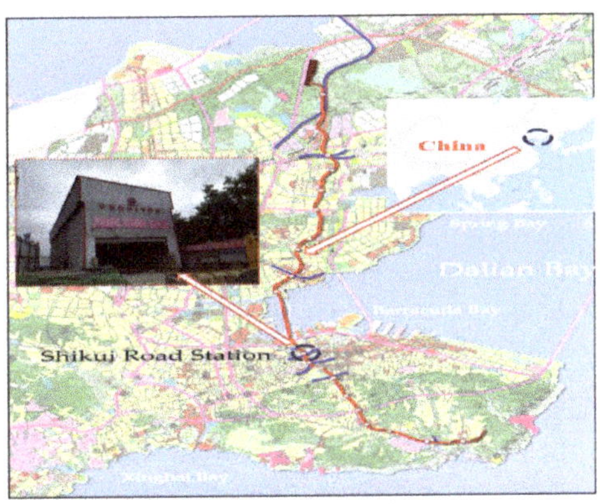

Figure 4. The project location map.

4.3. Numerical Simulation Model

A numerical calculation model is established using FLAC^{3D} finite difference software. In order to eliminate the influence of the boundary effect, the left and right boundary of the model are taken as more than two times the station span, the lower boundary is taken as two times the station span, the upper boundary is taken vertically to the surface, and the depth direction of the three-dimensional model is taken as 50 m, with an overall model size of 100 m $\times$ 60 m $\times$ 74.2 m, whose size scale is simulated according to the data given in the geological survey report. The model includes 200,853 nodes and 126,890 units. Ansys 15.0 software is used to establish and grid the model, as well as densify the grid of main structures and surrounding rock areas. The numerical calculation and analysis are completed using FLAC^{3D}3.0. The Mohr–Coulomb yield criterion is adopted in the analysis process of the model. A linear elastic model is used to simulate the primary support, secondary lining, backfill concrete, crown beam, and other structures. The boundary around the model is a normal constraint, the bottom boundary is a 3D fixed constraint, and the upper one is free. The input geotechnical parameters are based on Table 3. The numerical calculation model calculation model is shown in Figure 5. Note: the "e" in the figure represents "10", and the number after it is his exponent. For example, 2.5000e-003 represents 2.5000×10^{-003}.

Table 3. Basic mechanical parameters of surrounding rock.

Surrounding Rock	Thickness/m	*E*/GPa	μ	γ/kN·m^3	*c*/MPa	φ/°
Plain filling soil	3.6	0.008	0.4	17	0.01	15
Strongly weathered quartzite	2	0.05	0.35	23	0.08	30
Medium weathered quartzite	68.6	1.3~6	0.3~0.35	26.5	0.2~0.7	27~39

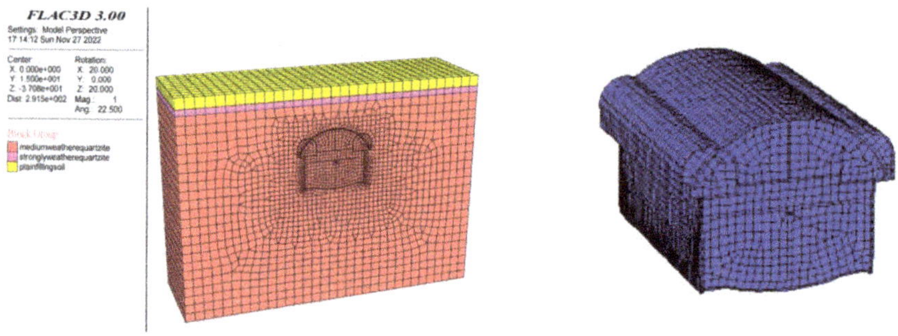

(a) (b)

Figure 5. Numerical calculation model: (**a**) overall model; (**b**) model of the station.

The Mohr–Coulomb yield criterion is adopted to the station. The first layer is constructed with plain fill, the second layer consists of strongly weathered quartzite, and the third layer is made of moderately weathered quartzite (parameter layer to be inverted). The basic mechanical parameters are shown in Table 3 regarding field investigation and investigation data.

4.4. Monitoring and Measurement

According to the basic requirements in the construction specifications of the Shikui Road Station, the construction methods and surrounding rock characteristics of Shikui Road Station were monitored, including changes in the stress and settlement of the arch cover during the construction period. To obtain a better inversion effect, it was necessary to select appropriate monitoring information for inversion. Therefore, an automatic monitoring section was set at the arch cover of the station, and an earth pressure box as well as a single-point displacement meter was used to monitor the stress and displacement changes in the arch cover in real time.

The earth pressure box (TY-1, TY-2, and TY-3) and single-point displacement meter (DW-1, DW-2, and DW-3) were arranged as shown in Figure 6.

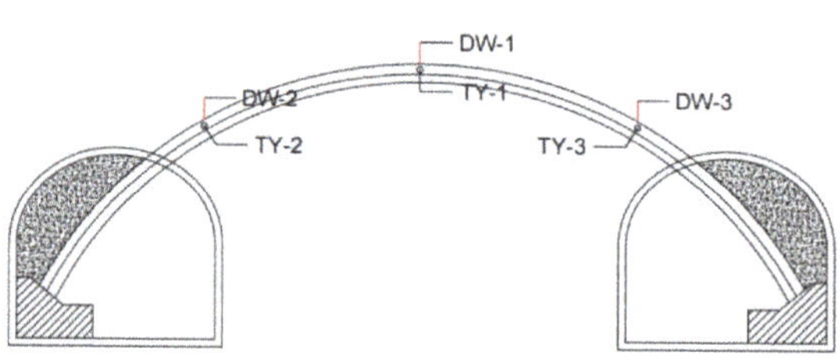

Figure 6. Automated sensor placement location.

The actual monitoring data and the calculation results were compared and analyzed. Figures 7 and 8 show the actual monitoring data of the earth pressure box and the single-point displacement meter.

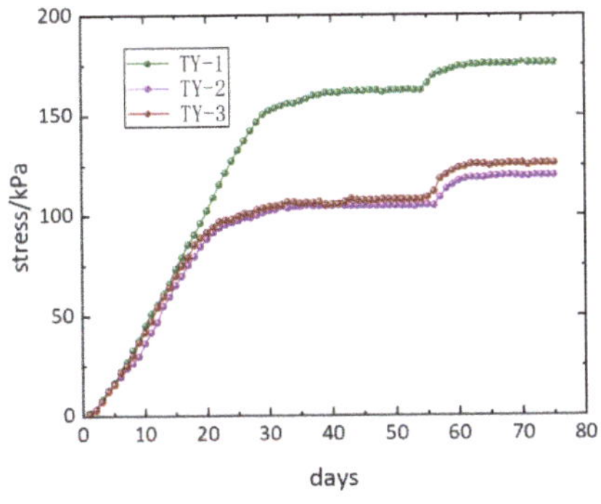

Figure 7. Field monitoring data curve of earth pressure box.

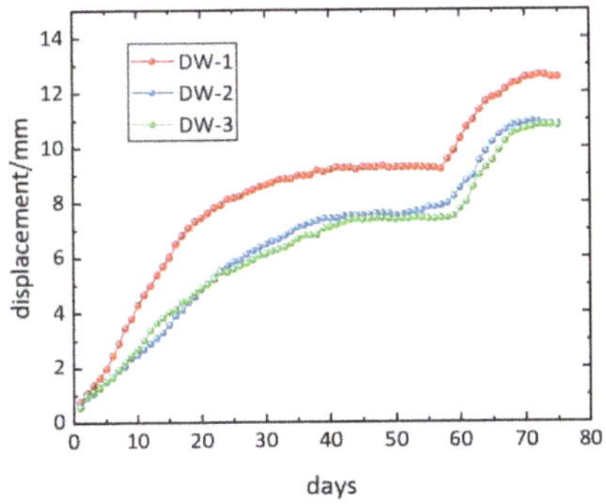

Figure 8. Field monitoring data curve of displacement measuring point.

4.5. Orthogonal Design

Orthogonal design is a multi-factor and multi-level test method, through which representative points can be selected for test design, thus reducing the difficulties caused by a large number of tests and the facilitation of efficient tests [34].

In the back analysis, the stability and displacement settlement of the station mostly depend on the surrounding rock parameters, and the influence of different surrounding rock parameters on the failure and deformation of rock mass is different; at the same time, if there are many inversion parameters, the authenticity of parameters obtained from the back analysis will also be affected. Therefore, the parameters of surrounding rocks that have a great impact on the station were obtained through a sensitivity analysis for back analysis.

An orthogonal design scheme was adopted, the parameter values were divided into five grades according to the determined parameter range, and then the values of different grades of each parameter were reasonably combined according to the orthogonal scheme; that is, 25 combinations were formed according to the L25 (5^5) orthogonal table (as is shown in Table 4). According to the field investigation and survey data, 25 combinations of five test parameters at five levels were designed. The parameter and value range: tensile strength t = 0.07~0.19 MPa, Poisson's ratio μ = 0.3~0.35, cohesion c = 0.2~0.7 MPa, internal friction angle φ = 27°~39°, and elastic modulus E = 1.3~6 GPa. The parameters of each scheme were substituted into a three-dimensional finite element model for forward calculation, and the three points (as is shown in Figure 7), where the sensors were located at the station arch cover, were, respectively, selected to obtain the corresponding displacement and stress, whose results are shown in Table 4.

Table 4. Parametric orthogonal experiment scheme.

Factor	*E*/GPa	*μ*	*c*/MPa	*φ*/°	*t*/MPa	DW-1/mm	DW-2/mm	DW-3/mm	TY-1/kPa	TY-2/kPa	TY-3/kPa
test1	1.30	0.30	0.20	27	0.07	13.64	11.67	11.69	234.95	174.19	169.67
test2	2.48	0.31	0.32	30	0.07	7.01	5.88	5.86	149.88	136.15	99.83
test3	3.65	0.32	0.45	33	0.07	4.87	4.09	4.05	133.20	116.90	96.41
test4	4.83	0.33	0.57	36	0.07	3.79	3.2	3.16	128.27	116.34	88.09
test5	6.00	0.35	0.70	39	0.07	3.11	2.63	2.59	125.51	120.03	87.13
test6	4.83	0.30	0.32	33	0.10	5.18	4.36	4.33	107.73	78.63	43.08
test7	6..00	0.31	0.45	36	0.10	3.1	2.59	2.56	109.42	115.60	81.28
test8	1.30	0.32	0.57	39	0.10	11.84	9.97	9.87	172.40	132.75	99.45
test9	2.48	0.33	0.70	27	0.10	6.92	5.79	5.74	133.63	117.19	83.77
test10	3.65	0.35	0.20	30	0.10	6.73	5.96	5.78	189.64	138.41	151.20
test11	2.48	0.30	0.45	39	0.13	8.86	7.47	7.36	138.23	65.41	24.61
test12	3.65	0.31	0.57	27	0.13	4.95	4.13	4.09	109.75	109.84	70.04
test13	4.83	0.32	0.70	30	0.13	3.83	3.19	3.16	102.77	103.65	69.79
test14	6.00	0.33	0.20	33	0.13	4.14	3.69	3.59	180.82	126.39	145.73
test15	1.30	0.35	0.32	36	0.13	14.43	12.29	12.27	190.41	85.35	52.84
test16	6.00	0.30	0.57	30	0.16	3.15	2.63	2.6	89.68	99.91	55.69
test17	1.30	0.31	0.70	33	0.16	11.96	10.06	9.97	163.99	114.17	90.04
test18	2.48	0.32	0.20	36	0.16	7.58	6.61	6.59	213.93	142.20	173.48
test19	3.65	0.33	0.32	39	0.16	5.15	4.34	4.35	122.05	111.87	96.55
test20	4.83	0.35	0.45	27	0.16	3.93	3.28	3.26	98.24	90.76	65.44
test21	3.65	0.30	0.70	36	0.19	5.05	4.22	4.17	101.81	93.96	60.39
test22	4.83	0.31	0.20	39	0.19	4.22	3.71	3.68	205.64	141.33	174.70
test23	6.00	0.32	0.32	27	0.19	3.44	2.92	2.93	112.19	117.19	101.08
test24	1.30	0.33	0.45	30	0.19	12.24	10.32	10.29	165.71	113.95	77.52
test25	2.48	0.35	0.57	33	0.19	13.64	11.67	11.69	121.43	106.48	62.03

In the back analysis, the mechanical parameters had a great relationship with the stability of the surrounding rocks. The failure and deformation of rock mass were different with different rock mass parameters. Moreover, the authenticity of the parameters obtained from back analysis decrease with the increase in the number of inversion parameters [35]. Therefore, sensitivity analysis was used in this paper to obtain the factors that have a greater impact on the stability of the station for an inverse analysis, and a range analysis was carried out on the settlement displacement value of the arch crown and arch waist. See Table 5 for the range value of DW-1, DW-2, and DW-3 as well as the order of their impact on the displacement.

Table 5. Numerical calculation results (Cont).

Factor	*E*/GPa	*μ*	*c*/MPa	*φ*/°	*t*/MPa
range (DW-1)	7.970	1.892	1.150	1.216	1.184
order (DW-1)	1	2	5	3	4
range (DW-2)	7.973	1.890	1.150	1.216	1.183
order (DW-2)	1	2	5	3	4
range (DW-3)	9.434	2.118	1.302	1.382	1.364
order (DW-3)	1	2	5	3	4
Range mean	8.856	2.033	1.229	1.320	1.292
0rder	1	2	5	3	4

In Tables 4 and 5, DW-2 is the settlement displacement value at the left arch waist of the arch cover; DW-3 is that at its right arch waist; DW-1 is that at its arch crown; TY-1, TY-2, and TY-3 are the total stress at these three points of the arch cover of the station; and the range, order as well as average range of DW-1, DW-2, and DW-3 are the range value of the settlement displacement at the corresponding measuring points of at the arch cover, the primary and secondary relationship of the influence of various parameters, as well as the average range value.

As can be seen from Table 3, the order of sensitive factors affecting the three points at the arch cover is elastic modulus E, Poisson's ratio μ, internal friction angle φ, tensile strength t, and cohesion force c. Elastic modulus has the greatest influence, followed by Poisson's ratio and internal friction angle. Therefore, the first three factors with a great influence on the sensitivity factor analysis are selected for the analysis in this paper.

4.6. Parameter Identification Results

The back analysis of displacement and stress can be used to further solve not only the problem of the low accuracy, but also the problem of the low authenticity caused by the increase in back-analysis parameters. At the same time, real changes in the surrounding rocks of the station can be fully considered in the joint analysis of displacement and stress, so that it is more real and effective.

According to the actual monitoring, the settlement displacement of each point of the surrounding rocks was DW-1 = 12.53 mm, DW-2 = 10.75 mm, DW-3 = 10.71 mm, TY-1 = 176.97 kPa, TY-2 = 125.8 kPa, and TY-3 = 120.55 kPa. Taking the measured displacement and stress values as the control values while applying MATLAB to conduct a joint back analysis on displacement and stress, the initial parameters setting of differential evolution are shown in Table 6.

Table 6. Initial parameters of differential evolution.

Initial Parameter Type	Number of Optimization Variables	Number of Population	Cross Factor CR	Variation Factor F	Maximum Evolution Algebra	Population Size NP
numerical value	5	100	0.7	0.8	200	100

It was found that when the cross factor was CR = 0.7 and F = 0.8, the optimal mechanical combination parameter of moderately weathered quartzite obtained through inversion was E = 1.3 GPa, μ = 0.33, and φ = 35°, and the inversion results of surrounding rock parameters are shown in Table 7. The results obtained from the inversion were substituted into the finite-element software for a forward calculation, the results obtained from which were compared with the monitoring values at the monitoring section of Shikui Road Station and the single displacement back analysis. The results are shown in Table 8.

Table 7. Inversion results of surrounding rock parameters.

Factor	E/GPa	μ	φ/°
Parameter value	1.3	0.33	35

Table 8. Comparison of inversion calculation results and actual measurement results.

Measuring Point	Actual Monitoring Value	Displacement Back Analysis		Joint Back Analysis of Displacement and Stress	
		Calculated Value	Relative Error/%	Calculated Value	Relative Error/%
DW-1/mm	12.53	11.86	5.35	12.58	0.40
DW-2/mm	10.75	10.18	5.30	10.84	0.84
DW-3/mm	10.71	10.15	5.23	10.81	0.93
TY-1/kPa	176.97	—	—	180.20	1.83
TY-2/kPa	125.8	—	—	130.35	3.62
TY-3/kPa	120.55	—	—	124.79	3.52

The back analysis of displacement and stress can be used to further solve not only the problem of the low accuracy, but also the problem of the low authenticity caused by the increase in back-analysis parameters. At the same time, the joint analysis of displacement and stress real changes in the surrounding rocks of the station can be fully considered, so that it is more real and effective.

According to the data in Table 4, only when the settlement value of the displacement at the arch cover of the station was used as the control value for the back analysis of surrounding rock parameters could the maximum relative error be 5.35%, while when the displacement and stress value at the arch cover of the station was combined for the back analysis, the maximum error was only 3.62%. The inversion effect was improved by 32.3% through a joint back analysis of displacement and stress compared with the single back analysis of displacement. This method further illustrates the applicability of the joint back analysis of displacement stress using the GP-DE algorithm through the arch and cover method. The surrounding rock parameters obtained from the joint back analysis can be applied to the problem of later scheme optimization, thus providing a reliable reference for the subsequent calculations.

5. Optimization Analysis of Temporary Dismantling Based on Determined Parameters

To analyze the stress characteristics of the center diaphragm wall of Shikui Road Station during the excavation process, the parameters obtained from the joint back analysis of displacement and stress based on the GP-DE algorithm were used as calculation parameters in the removal process of the center diaphragm wall support of the station. By analyzing the ground settlement and vault settlement of the station, the optimal length of the removal of its center diaphragm wall with class-IV surrounding rocks was studied. The overall model was consistent with Figure 9. A numerical simulation was carried out with FLAC3D 3.0 software. The stratum and constitutive as well as physical and mechanical parameters were consistent with Table 1. The center diaphragm wall was established with solid units. After the center diaphragm wall was removed, the two-layer primary support structure and the secondary lining structure were constructed immediately. The monitoring section was set as 3 m in the longitudinal direction of the station, and 35 monitoring points of the ground settlement were set in its transverse direction. The ground and vault settlement under the length of 6 m, 12 m, 18 m and 24 m were analyzed, respectively, to determine the optimal removal length of the center diaphragm wall of the station.

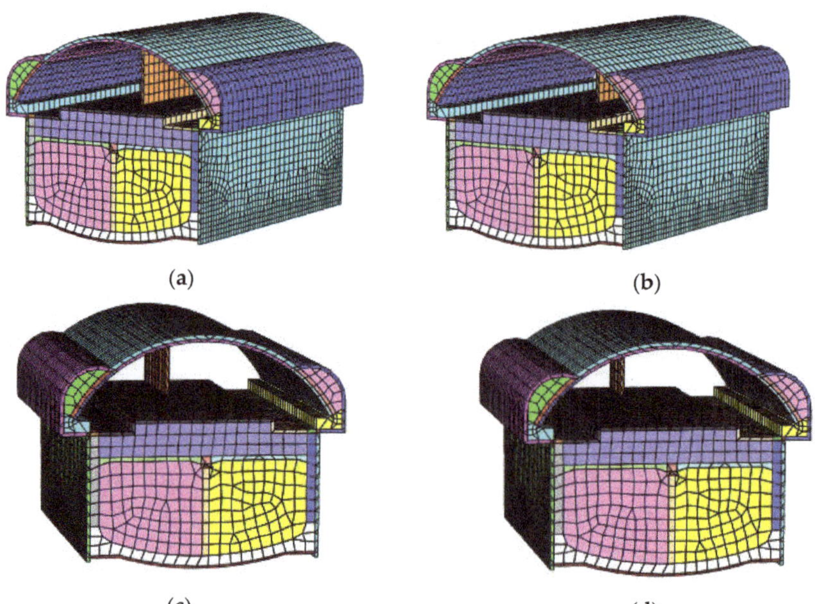

(a) (b) (c) (d)

Figure 9. Numerical model of demolition of partition walls in different length: (**a**) dismantle 6 m; (**b**) dismantle 12 m; (**c**) dismantle 18 m; (**d**) dismantle 24 m.

5.1. Surface Subsidence Analysis

To better analyze the suitable distance of the removal length of the center diaphragm wall in the station, the optimization was carried out based on the length range provided by the design institute. Four lengths were selected to analyze the removal length of the center diaphragm wall, so as to obtain a more appropriate removal length of the center diaphragm wall. The ground settlement under four schemes, namely 6 m, 12 m, 18 m, and 24 m, was compared, respectively, to determine the optimal removal length of the center diaphragm wall. The surface settlement curve is shown in Figure 10.

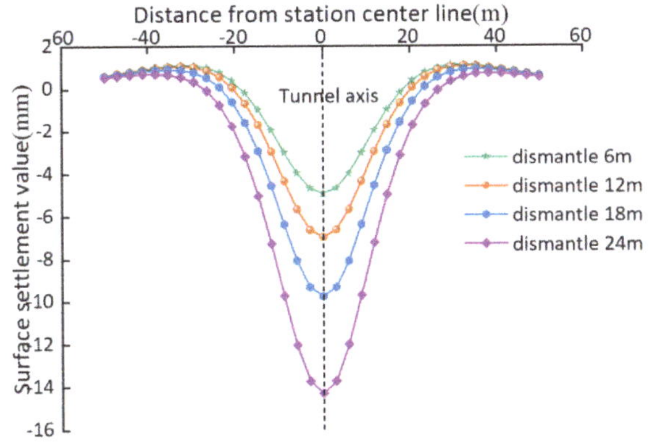

Figure 10. Surface settlement curve.

It can be seen from the surface settlement curve that, during the removal of the center diaphragm, the surface was symmetrically distributed to both sides with the center diaphragm as the center, and the surface settlement was distributed in a "V" shape. All the maximum surface settlement values were on the tunnel axis. When the removal length of the center diaphragm of the station was 6 m, the maximum surface settlement value was 4.9 mm, which was 6.65 mm, 9.75 mm and 14.31 mm when the removal length of the station center diaphragm was 12 m, 18 m and 24 m, respectively. It could be found that, with the increase in the removal length of the center diaphragm, the maximum surface settlement gradually increased. When the center diaphragm was removed for 6 m, the minimum surface settlement occurred, and when it was removed for 24 m, the maximum surface settlement occurred. This is because the stress conversion of the main pilot tunnel of the station after excavation is relatively stable when the removal length of the center diaphragm is small, while when the removal length of the center diaphragm is too long, the stress conversion of the main pilot tunnel of the station after excavation is severe, which may lead to the instability of the structure, thus causing a greater settlement of the ground surface. It can be seen from the surface settlement curve that the removal length of the center diaphragm had a great impact on the surface settlement, which is not enough to be judged only by the surface settlement, so a further analysis is needed.

5.2. Settlement Analysis of Arch Crown

During the construction and removal of the center diaphragm wall of the station, the value of the arch crown settlement is a monitoring item that cannot be ignored. The monitoring of the arch crown settlement plays an important role in the safety of the tunnel. If the value of the arch crown settlement exceeds the control range, it may bring great dangers to the tunnel. Therefore, the monitoring of the arch crown settlement should be focused on. The demolition length of the center diaphragm wall can be further determined by the crown settlement through the value of the crown settlement of different demolition lengths of the center diaphragm wall. The Figure 11 shows the vertical displacement cloud diagram of the demolition length of each center diaphragm wall. Note: the "e" in the figure represents "10", and the number after it is his exponent. For example, 2.5000e-003 represents 2.5000×10^{-003}.

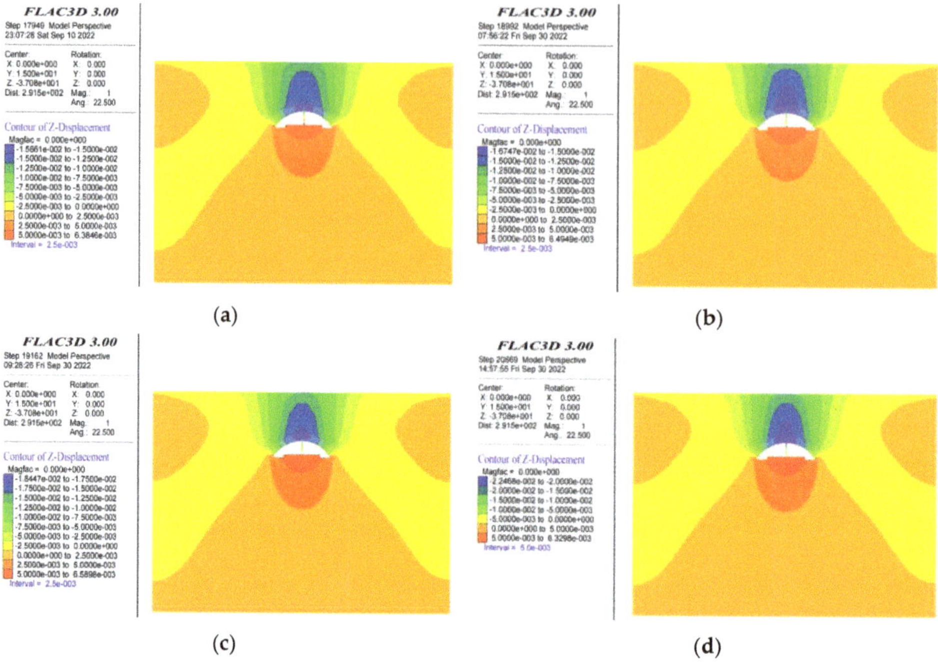

Figure 11. Vertical displacement cloud diagram of each demolished length: (**a**) dismantle 6 m; (**b**) dismantle 12 m; (**c**) dismantle 18 m; (**d**) dismantle 24 m.

It can be seen from the vertical displacement program of the center diaphragm of different removal lengths in Figure 10 that different removal lengths had different effects on the settlement of the station vault. When the removal length of the center diaphragm was 6 m, 12 m, 18 m and 24 m, the settlement of the station vault was 15.57 mm, 16.43 mm, 18.12 mm and 21.87 mm, respectively. It could be found that, with the increase in the removal length of the center diaphragm, the settlement value of the station vault showed a significant increase. When the removal length of the center diaphragm was 24 m, the settlement value of the vault was the largest. According to the field monitoring and data, the control value of the station vault settlement was 20 mm, so when the removal length of the center diaphragm was 24 m, the control value was exceeded. If this is the removal length, the stability of the station will be weakened, which may affect the safety of the whole station.

It can be seen from the surface settlement curve and the dome top settlement cloud diagram that after a comprehensive comparison of the surface settlement value and the dome top settlement value, when the removal length of the center diaphragm wall at the arch cover of the station was 18 m, the value of surface settlement and dome top settlement was within the control value range. Through the removal of the center diaphragm wall with this length, not only the budget cost for the removal of the center diaphragm wall of the station can be reduced, but also the safe construction of the tunnel can also be ensured.

6. Discussion

6.1. Effect Analysis of Back Analysis of GP-DE Model

To verify the algorithm of GP-DE, the displacement stress in the construction process through the large-span arch cover method and numerical model simulation was used to analyze the inversion effect of the algorithm on the mechanical parameters of the upper soft stratum and the lower hard stratum. To further analyze the applicability of this algorithm, the methods including GP-DE, BP, LSSVM and GP commonly used in back analyses were used to compare and analyze the inversion effect of the mechanical parameters of the upper and lower hard layer through the large-span arch cover method for different training samples with an identical numerical model and mechanical parameter input. The Figure 12 shows the relative error of monitoring point values of different strategies.

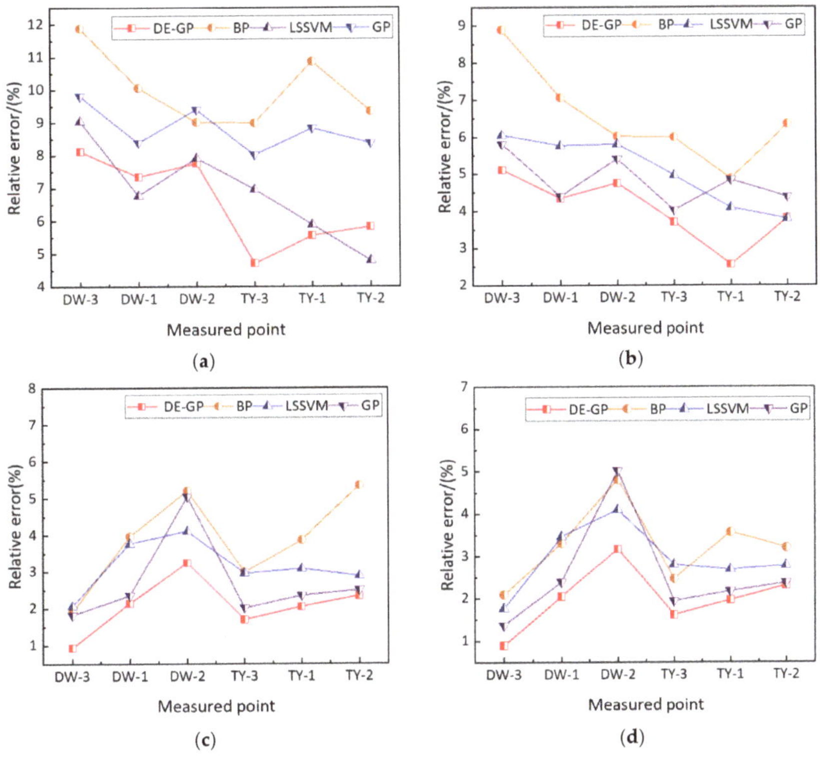

Figure 12. Relative error of monitoring point values of different strategies: (**a**) 5 samples; (**b**) 10 samples; (**c**) 25 samples; (**d**) 41 samples.

Training sets with 5, 10, 25 and 41 samples were used to train the above algorithms. A line chart of relative errors was drawn corresponding to the mechanical effect at each measurement point based on different training sets and different inversion algorithms, as is shown in the figure. Samples were obtained through numerical simulation. The rules for selecting samples from different sample sets were as follows: a number of samples (5 and 10 samples) were randomly selected from the table of Sample L25(5^5), respectively, and the training sample set 41 was the orthogonal scheme L25(5^5) plus L16(4^5).

When the number of samples was 5, there were fewer training samples at this time. It can be found from the figure that LSSVM and GP-DE inversion was significantly better than the BP neural network and GP, which shows that GP-DE and LSSVM can also have certain inversion performance of rock formation parameters when the sample size is small. BP and GP in this case, especially BP, are more likely to fall into a local solution state. When the number of trainings increased to 10, the GP inversion effect was significantly improved, and the GP-DE effect was the best. The BP neural network could easily fall into local solutions, and some relative errors were large.

When the number of training samples increased to 25, the results of the BP neural network had a better effect than LSSVM, so BP needs a sufficient number of training samples, which can better give play to the advantages of the algorithm. An advantage of LSSVM is that it can have a good effect when the number of training samples is small, which is suggested to be selected when the number of samples is small. When the number of training samples reached 41, little improvement could be found in the inversion results compared with 25, and the inversion results of the selected algorithm were at a better level. Compared with the other three algorithms, both the inversion results and the convergence speed had a good performance. The overall inversion effect of the GP-DE algorithm was excellent, which also had a good performance with a small sample size.

6.2. Influence of Differential Evolution Parameters on Optimization Results

In DE, variation factor F, cross factor CR, and population size NP, different strategies may affect the convergence rate. To obtain better calculation results, the control variable method was used in this paper to analyze the impact of key parameters such as DE, variation coefficient F, and cross coefficient CR on the DE algorithm under the above conditions with optimized parameters, 25 training samples, and the mechanical parameter input dimension of the numerical model; thus, the appropriate value range of variation coefficient F and cross coefficient CR was determined.

F represents the single search contraction range, and CR represents the probability of cross-recombination between newly generated mutants and individuals in the original population, which is used to enhance the diversity of the population. F was fixed as 0.3, 0.5, 0.7, and 0.9, respectively, and the convergence of CR that was between 0.3 and 0.9 was compared and analyzed, as is shown in Figure 13.

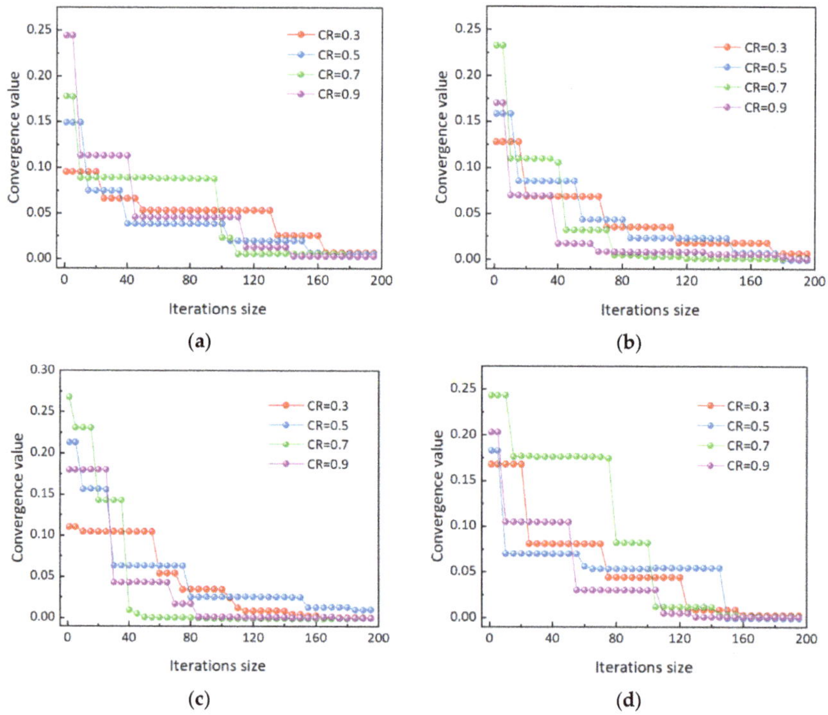

Figure 13. Training convergence curve under different F and CR values: (**a**) F = 0.3; (**b**) F = 0.5; (**c**) F = 0.7; (**d**) F = 0.9.

From the perspective of the overall iterative effect, when CR was 0.7 and 0.9, the iteration effect and rate were relatively better. When the CR value was 0.3 and 0.5, the iteration rate was high in the beginning part, but slowed down significantly in the later part. When F was 0.5 and 0.7, the overall iteration rate was higher than 0.3 and 0.9. Based on the iteration rate and convergence effect, it is suggested that the parameters and the value of CR be selected, respectively, from 0.5 to 0.7 and from 0.7 to 0.9.

7. Conclusions

The GP-DE back-analysis method was used to back-analyze the rock strata parameters of the station constructed using the arch cover method. The main conclusions are as follows:

1. The advantages of the GP-DE (Gaussian process differential evolution co-optimization) algorithm in establishing a nonlinear mapping relationship and its strong optimization ability are used to back-analyze the surrounding rock parameters of the power station. The calculation results showed that the error between the forward calculation of inversion parameters and the measured values was only 3.62%, which can meet the needs of the project to guide the next step of construction.
2. The current content is "based on the parameters inversed by the GP-DE algorithm"; it was studied that the optimal length for removing the station middle partition (including Class-IV surrounding rocks) was 18m, which can not only reduce the budget cost for removing the station middle partition, but also ensure the safety and efficiency of tunnel construction.
3. With different training samples, the inversion effect of the mechanical parameters of the upper soft rocks and the lower hard rocks through the long-span arch cover method was compared with that through the GP-DE, BP, LSSVM, and GP methods. Compared with the other three algorithms, the inversion effect of the GP-DE algorithm was generally good, which performed well with a small-sample training amount.
4. For the key parameters of DE, the variation coefficient F, and the cross-coefficient CR, the control variable method was used to analyze the impact of these two parameters on the DE algorithm. Based on the iteration rate and convergence effect, it is recommended that the selection range of parameter F be 0.5~0.7 and the value of CR be 0.7~0.9.

Author Contributions: Conceptualization, F.Z. and A.J.; methodology, F.Z.; software, X.G.; validation, F.Z.; formal analysis, F.Z.; investigation, Q.M., Q.Y. and F.Z.; resources, A.J., Q.M. and Q.Y.; data curation, X.G., Q.M. and Q.Y.; writing—original draft preparation, F.Z. and X.G.; writing—review and editing, A.J.; visualization, A.J.; supervision, A.J.; project administration, A.J.; funding acquisition, A.J. All authors have read and agreed to the published version of the manuscript.

Funding: This research was funded by the National Natural Science Foundation of China, grant number 52078093; LiaoNing Revitalization Talents Program, grant number XLYC1905015; and Cultivation Program for the Excellent Doctoral Dissertation of Dalian Maritime University, grant number 2022YBPY009.

Institutional Review Board Statement: Not applicable.

Informed Consent Statement: Informed consent was obtained from all subjects involved in the study.

Conflicts of Interest: The authors declare no conflict of interest.

References

1. Zhong, G. Arch-Cover Method: A New Subsurface Excavation Construction for Metro Stations. *J. Technol.* **2012**, *15*, 145–148.
2. Xu, J.; Ni, Y. Displacement ratio dichotomy back analysis of surrounding rock-initial support system of weathered rock tunnel. *Arab. J. Geosci.* **2019**, *12*, 181. [CrossRef]
3. Wang, K.P.; Jian, D.L.; Liu, D.H. The Parameter Back Analysis of Rock Mass Based on Orthogonal Design. In Proceedings of the 2014 7th International Conference on Intelligent Computation Technology and Automation, Changsha, China, 25–26 October 2014; pp. 809–813.

4. Feng, X.T.; Zhou, H.; Li, X.J.; Chen, B.R. Integrated Intelligent Feedback Analysis of Rock Mechanics and Engineering Problems and Its Applications. *Chin. J. Rock Mech. Eng.* **2007**, *26*, 1737–1744.
5. Sakurai, S.; Takeuchi, K. Back Analysis of Measured Displacements of Tunnels. *Rock Mech. Rock Eng.* **1983**, *16*, 173–180. [CrossRef]
6. Gioda, G. Some Remarks on Back Analysis and Characterization Problems. In Proceedings of the 5th International Conference on Numerical Methods in Geomechanics, Nagoya, Japan, 1–5 April 1985; Volume 1, pp. 47–61.
7. Gao, W.; Chen, D.L.; Dai, S.; Dai, X. Back analysis for mechanical parameters of surrounding rock for underground roadways based on new neural network. *Eng. Comput.* **2017**, *34*, 25–36. [CrossRef]
8. Su, G.S. A New Intelligent Back-Analysis Method Based on Differential Evolution Algorithm. In Proceedings of the 2019 International Joint Conference on Computational Sciences and Optimization, Sanya, China, 24–26 April 2009; pp. 212–214.
9. Zhao, Y.; Feng, S.J. Back analysis of surrounding rock parameters of tunnel considering displacement loss and space effect. *Bull. Eng. Geol. Environ.* **2021**, *80*, 5675–5692. [CrossRef]
10. Jiang, Z.B.; Jiang, A.N.; Hu, X.F.; Liu, Y.H.; Chen, W.; Liu, G.Y. Displacement-Stress Based Multi-Information Intelligent Back Analysis in Highway Tunnels. *Mod. Tunn. Technol.* **2017**, *54*, 145–152.
11. Jin, C.Y.; Feng, X.T.; Zhang, C.S. Application of neural network to back analysis of mechanical parameters of columnar joint basalt. *J. Hydroelectr. Eng.* **2010**, *29*, 234–238.
12. Wang, J.X.; Jiang, A.N. The analysis of monitoring data and intelligent inversion of parameters of Dalian subway tunnel. *China Civ. Eng. J.* **2011**, *44*, 135–138.
13. Huang, C.T.; Wang, Y.; Zhang, Y.J. Application of displacement back analysis method in determination of surrounding rock parameters of highway tunnels. *Yangtze River* **2017**, *48*, 138–140.
14. Xu, C.; Liu, B.G.; Liu, K.Y.; Guo, J.Q. Intelligent model for time series analysis and prediction of landslide displacement based on particle swarm optimization Gaussian process regression coupling algorithm. *Geotech. Mech.* **2011**, *6*, 1670–1675.
15. Rasmussen, C.E.; Williams, C.K.I. *Gaussian Processes for Machine Learning*; MIT Press: Cambridge, MA, USA, 2006.
16. Brahim-Belhouari, S.; Bermak, A. Gaussian Process for Nonstationary Timeseries Prediction. *Comput. Stat. Data Anal.* **2004**, *47*, 705–712. [CrossRef]
17. Kim, H.C.; Kim, D.J.; Ghahramani, Z.; Bang, S.Y. Appearance-Based Gender Classification with Gaussian Processes. *Pattern Recognit. Lett.* **2006**, *27*, 618–626. [CrossRef]
18. Guo, S.S.; Liu, B.Y.; Yong, C.S. Gaussian Process for Non-Linear Displacement TimeSeries Prediction of Landslide. *J. China Univ. Geosci.* **2007**, *18*, 212–219.
19. Su, G.S.; Song, Y.C.; Yan, L.B. A new method for predicting the effect of rock blasting. *J. Rock Mech. Eng.* **2007**, *1*, 3509–3514.
20. Fyfe, C.; Leen, G.; Lai, P.L. Gaussian processes for Canonical Correlation Analysis. *Neurocomputing* **2008**, *71*, 3077–3088. [CrossRef]
21. Liu, K.Y. Intelligent model for time series analysis of nonlinear deformation of slope based on genetic combined kernel function Gaussian process regression algorithm. *J. Geotech. Mech. Eng.* **2009**, *10*, 2128–2134.
22. Liu, K.Y.; Fang, Y.; Liu, B.G. Parameter inversion of elastoplastic model of tunnel engineering based on evolutionary Gaussian process regression algorithm. *J. Geotech. Eng.* **2011**, *6*, 884–889.
23. He, P.; Li, S.C.; Li, L.P.; Zhang, Q.Q.; Xu, Z.H. Tunnel surrounding rock deformation response prediction and dynamic change permission mechanism based on data mining. *J. Rock Mech. Eng.* **2017**, *12*, 2940–2953.
24. Zhang, Y.P.; Li, L.P.; He, P.; Qin, C.S. Large deformation Gaussian process regression prediction model of tunnel surrounding rock and its engineering application. *Sci. Technol. Eng.* **2018**, *1*, 122–127.
25. Dai, S.L.; Lv, Y.L. Rockmass Stability Analysis of YunYang's Arch Dam Left Shoulder. *Adv. Mater. Res.* **2012**, *518–523*, 4405–4408. [CrossRef]
26. Li, W.; Bai, J.W.; Li, K.X.; Zhang, S.J. Experimental Analysis of Deformation Mechanics and Stability of a Shallow-Buried Large-Span Hard Rock Metro Station. *Adv. Civ. Eng.* **2020**, *2020*, 4031306. [CrossRef]
27. Wang, J.B.; Huo, Q.; Song, Z.P.; Zhang, Y.W. Study on adaptability of primary support arch-cover method for large-span embedded tunnels in the upper-soft lower-hard stratum. *Adv. Mech. Eng.* **2019**, *11*, 1687814018825375. [CrossRef]
28. Liu, T.; Huang, Y.L.; Lei, G. Analysis of the Optimum Support Dismantling Scheme for Shallow Bored Subway Stations in a Soil-Rock Compound Stratum. *Mod. Tunn. Technol.* **2015**, *52*, 131–137.
29. Liu, X.; Jiang, A.; Guo, X.; Lu, H. Effect of Excavation Blasting in the arch-cover method on Adjacent Existing Pipelines in a Subway Station. *Appl. Sci.* **2022**, *12*, 1529. [CrossRef]
30. Guo, X.P.; Jiang, A.N. Study on the stability of a large-span subway station constructed by combining with the shaft and arch-cover method. *Tunn. Undergr. Space Technol.* **2022**, *127*, 104582. [CrossRef]
31. Jiang, A.N. Optimizing generator socket construction schemes of Shuibuya Underground Powerhouse based on intelligent back analysis. *Rock Soil Mech.* **2008**, *29*, 1372–1376.
32. Zhang, Y.; Su, G.S.; Liu, B.C.; Li, T.B. A novel displacement back analysis method considering the displacement loss for underground rock mass engineering. *Tunn. Undergr. Space Technol.* **2020**, *95*, 103141. [CrossRef]
33. Tian, M.L.; Han, L.J.; Meng, Q.B. Nonlinear Regression Analysis for Deep Rock Mass Parameters of the Hoek-Brown Failure Criterion Based on the Differential Evolution. *KSCE J. Civ. Eng.* **2021**, *25*, 3160–3171. [CrossRef]

34. Tian, H.M.; Chen, W.Z.; Yang, D.S.; Dai, Y.H.; Yang, J.P. Application of the orthogonal design method in geotechnical parameter back analysis for underground structures. *Bull. Eng. Geol. Environ.* **2015**, *75*, 239–249. [CrossRef]
35. Jiang, Q.; Feng, X.T.; Su, G.S.; Chen, G.Q. Intelligent Back Analysis of Rock Mass Parameters for Large Underground Caverns Under High Earth Stress Based on Edz and Increment Displacement. *Chin. J. Rock Mech. Eng.* **2007**, *26*, 2654–2662.

applied sciences

MDPI

Article

Back Analysis of Geotechnical Engineering Based on Data-Driven Model and Grey Wolf Optimization

Lihong Zhao [1], Xinyi Liu [2], Xiaoyu Zang [2] and Hongbo Zhao [2,*]

[1] School of Fine Art, Shandong University of Technology, Zibo 255000, China
[2] School of Civil and Architectural Engineering, Shandong University of Technology, Zibo 255000, China
* Correspondence: hbzhao@sdut.edu.cn

Abstract: Geomaterial mechanical parameters are critical to implementing construction design and evaluating stability through feedback analysis in geotechnical engineering. The back analysis is widely utilized to identify and calibrate the geomaterial mechanical properties in geotechnical engineering. This study developed a novel back-analysis framework by combining a reduced-order model (ROM), grey wolf optimization (GWO), and numerical technology. The ROM was adopted to evaluate the response of the geotechnical structure based on a numerical model. GWO was used to search and identify the geomaterials properties based on the ROM. The developed back analysis framework was applied to a circular tunnel and a practical tunnel for determining the mechanical property of the surrounding rock mass. The results showed that the ROM could be an excellent surrogated model and replaced it with the numerical model. The obtained geomaterial properties were in excellent agreement with the actual properties. The deformation behavior captured by the developed framework was consistent with the theoretical solution in a circular rock tunnel. The developed framework provides a practical, accurate, and convenient approach for calibrating the geomaterial properties based on field monitoring data in practical geotechnical engineering applications.

Keywords: geomaterial; tunnel; back analysis; reduced-order model; grey wolf optimization

Citation: Zhao, L.; Liu, X.; Zang, X.; Zhao, H. Back Analysis of Geotechnical Engineering Based on Data-Driven Model and Grey Wolf Optimization. *Appl. Sci.* **2022**, *12*, 12595. https://doi.org/10.3390/app122412595

Academic Editor: Wei Gao

Received: 21 September 2022
Accepted: 27 November 2022
Published: 8 December 2022

1. Introduction

The geomaterial property is critical to guiding construction and assuring safety during the feedback analysis and dynamic design for geotechnical engineering [1–3]. The back analysis based on field monitoring data has been widely used in geotechnical engineering. The back analysis aims to obtain the geomaterial properties by minimizing the objective function, which represents the response difference of geotechnical structure between the monitored field and the predicted by the physical model, based on the optimization technology. The core of the back analysis technique is the optimization technique and physical model. Generally, the numerical method is widely used to calculate the structure response field in practical geotechnical engineering [4,5]. However, a numerical method is time-consuming due to the repetitive computing in back analysis, especially for the practical large-scale geotechnical structure. Meanwhile, optimal techniques often have problems in back analysis, such as high dimensionality, complexity, and local minima. In order to solve the above two problems, various surrogate models and optimal technology have been proposed in the back analysis [2,3,6]. In this study, the reduced order model (ROM) and the grey wolf algorithm (GWO) were combined to improve the practicability of back analysis.

Last century, displacement back analysis was proposed by Sakurai and Takeuchi to identify the rock mechanical parameters based on the monitoring information of the tunnel [7]. However, the numerical method is time-consuming and limits the engineering application of back analysis. In order to overcome the limitations of the numerical method, various machine learning-based surrogate models were a focus of attention to approximate the response of the geotechnical structure in the past decades [8–11]. The neural network method was utilized to construct the intelligent displacement back analysis model for identifying the mechanical property of the surrounding rock mass [12–15]. The support vector machine and the relevance vector machine were selected to build a displacement back analysis model to recognize the geomaterials parameters [16–18]. Machine learning provides an excellent tool for predicting the structural response and is selected as the surrogate model in the geotechnical back analysis. However, some limits of artificial intelligence, such as overfitting, trapping local minimum, etc., hinder the practical application in back analysis for geotechnical engineering. Meanwhile, the traditional surrogate model does not reflect the physical mechanism of geological engineering and only obtains information about the discrete response at some monitoring points. Therefore, obtaining the total response fields using this classical approach is challenging due to the learning efficiency. The ROM was developed that contains some knowledge about the engineering structure under consideration and can overcome the limits of machine learning and the traditional surrogate model. So, this study adopted the ROM to establish the surrogate model for back analysis.

In order to acquire the appropriate geomaterial property, various intelligent optimization methods, such as genetic algorithm [16,19,20], artificial bee colony [18], particle swarm optimization [21,22], etc., have been widely utilized to seek the geomechanical property in the back analysis. Due to the complexity of geomaterials, trapping the local minimum solution is the main drawback of the back analysis method, which hinders its application in practical engineering. Grey wolf optimization (GWO) is an efficient metaheuristic method developed recently [23]. The GWO algorithm is suitable for solving nonlinear and complex problems due to its simple concept, small number of adjustable parameters, fast convergence, and strong global optimization capacity. It only considers the function evaluation and does not need the derivative information, which is suitable for black box global optimization problems. So, GWO was selected as an optimal technology for the back analysis in this study.

This study proposed a novel back analysis approach by combining numerical models, ROM, and GWO to identify the geomaterial mechanical property in geotechnical engineering. ROM was utilized to construct a surrogate model to approximate and capture the response of the geotechnical structure for replacing the numerical model in the back analysis. The GWO algorithm was regarded as an optimal technology for seeking the unknown geomaterial property based on the idea of the back analysis. The developed framework was applied to a circular tunnel and an actual tunnel project. The remainder of this study is stated as follows. First, ROM and GWO algorithms are briefly introduced in Sections 2 and 3, respectively. Section 4 introduces the main ideas and procedures of the developed back analysis in detail. In Section 5, a circular tunnel and an actual tunnel, i.e., the experimental tunnel in the Goupitan Water conservancy project, China, are used to verify and investigate the developed back analysis framework. Finally, some conclusions are drawn in Section 6.

2. Reduced-Order Model

2.1. Constructing the ROM Model

The ROM was used to predict the system response using a low-order model based on numerical methods and the proper orthogonal decomposition in engineering. For any x_i, $i = 1, 2, \ldots, I$, and θ_j, $j = 1, 2, \ldots, J$, the proper orthogonal decomposition was utilized to obtain the following equation [24].

$$\widetilde{u}^h(x_i,\theta_j) = \sum_{k=1}^{K} \beta_k(\theta_j)\varphi^k(x_i) + \widetilde{g}(x_i,\theta_j) \quad (1)$$

where $\widetilde{u}^h$ denots the solution of field variables for geotechnical structure, θ_j and x_i denote the parameter and design variables of a geotechnical model, φ and β note the unknown coefficient of ROM, and $\widetilde{g}(x,\theta)$ is an extended boundary condition in the entire domain.

$$\widetilde{g}(x,\theta) = \begin{cases} g(x,\theta) \; on \; \partial\Omega \\ 0 \; elsewhere \end{cases} \tag{2}$$

Equation (1) can be rewritten as follows:

$$\widetilde{u}^h = \varphi\beta + \widetilde{g} \tag{3}$$

By utilizing the Latin hypercube sampling (LHS), the set of design variables θ_j, $j = 1,2,\ldots,J$. was constructed for determining the unknown coefficient φ. Then, the corresponding discrete solutions (snapshots) of the numerical model $w_j = u^h(\theta_j) - \widetilde{g}(\theta_j)$, $j = 1,2,\ldots,J$, were acquired based on numerical techniques such as finite element, discrete element, boundary element, etc. The spatial Gram matrix by $\mathbf{M}^x$ can be obtained as follows:

$$M_{ij}^x = (w_i \cdot w_j),\; i,j = 1,2,\ldots,J \tag{4}$$

where $(w_i \cdot w_j)$ notes the inner product between w_i and w_j.

The descending order of the positive eigenvalues of $\mathbf{M}^x$ is listed in the following form.

$$\lambda_1 \geq \lambda_2 \geq \ldots \geq \lambda_J \geq 0 \tag{5}$$

The first K eigenfunctions $\varphi^k(x)$, $k = 1,2,\ldots K$ corresponds to the first K eigenvalues, providing the orthogonal principal direction of snapshots. If $r^k = \left(r_j^k\right)_{j=i,i,\ldots J}$ denotes the kth eigenvector of $\mathbf{M}^x$, then its dual kth eigenfunctions $\varphi^k(x)$ can be determined according to the following form.

$$\varphi^k(x) = \sum_{j=1}^{K} r_j^k w_j(x) \tag{6}$$

where K notes the basis size of the proper orthogonal decomposition and can be solved in the following inequation.

$$\frac{\sum_{i=1}^{K}\lambda_i}{\sum_{i=1}^{J}\lambda_i} > k \tag{7}$$

where k is the user-specified tolerance and equals 0.9999 in this study.

The following penalized minimization problem can solve the unknown coefficient β.

$$\min_{\beta_j \in R^K} \|u^{h,j} - \varphi\beta_j - \widetilde{g}_j\|^2 + \mu\|\beta_j\|^2 \tag{8}$$

β_j can be solved by the following normal equation.

$$\left(\varphi^T\varphi + \mu I_K\right)\beta_j = \varphi^T\left(u^{h,j} - \widetilde{g}_j\right),\; j = 1,2,\ldots,J \tag{9}$$

where μ notes a small regularization parameter.

2.2. Predicting the Field Variables

In order to determine the field variables for unknown variables θ and x, the radial basis function (RBF) was adopted to expand the coefficient $\beta_k(\theta)$ in the following form.

$$\beta_k(\theta) = \sum_{j=1}^{J} \alpha_{jk}\psi\left(\frac{|\theta - \theta_j|}{\sigma}\right) \tag{10}$$

For any $\theta_{j'}$, $j' = 1, 2, \ldots, J$, Equation (10) can be presented as follows.

$$\sum_{j=1}^{J} \alpha_{jk} \psi \left(\frac{\left| \theta_{j'} - \theta_j \right|}{\sigma} \right) = \beta_{kj'} \quad (11)$$

where $\beta_{kj'}$ are determined by Equation (9). The above equations can be rewritten in the compacted form:

$$A\alpha_k = \beta_k \quad (12)$$

The following equation can solve the unknown coefficient α_k.

$$\left(A^T A + \mu I_J \right) \alpha_k = A^T \beta_k, \ k = 1, 2, \ldots, K \quad (13)$$

2.3. Procedure of the ROM Model

This study developed a ROM by combining the numerical method (including finite element, discrete element and boundary element, etc.) and the proper orthogonal decomposition. Using the LHS, the set of design variables was constructed for the snapshot. Then the numerical method was adopted to calculate the corresponding solution of design variable in the above set. Based on the obtained snapshots, the proper orthogonal decomposition basis vector and its coefficient were acquired using the proper orthogonal decomposition algorithm for geotechnical engineering. In order to acquire the unknown field of the new design variable, RBF functions were adopted to expand the coefficient of the proper orthogonal decomposition basis and then to determine the coefficients of the orthogonal decomposition ROM. The unknown field variables corresponding to the new design were determined using the ROM. Figure 1 shows the main flowchart and procedure of the ROM. In what follows, the procedure of the ROM model is presented in detail.

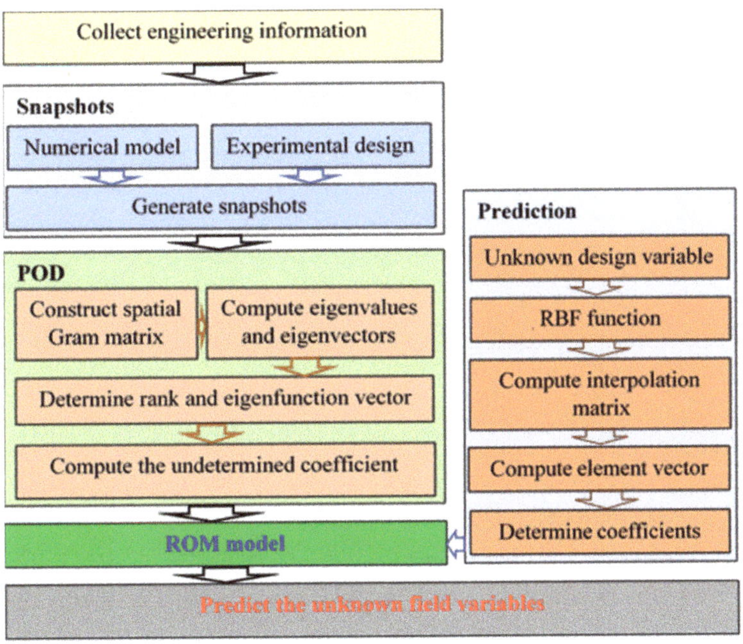

Figure 1. The main part of the ROM model.

Step 1: Collect the data of the geotechnical engineering, including project property, geostress, boundary conditions, etc.;

Step 2: Establish the numerical model (FEM) based on the above engineering information;
Step 3: Construct the design variables set $\boldsymbol{\theta}$ for the numerical model using LHS;
Step 4: Calculate the field variables w_i (displacement or stress field) at space domain $\mathbf{X}$ using a numerical method for each design variable. Collect all the field variables and acquire the snapshots;
Step 5: Build the spatial Gram matrix $\mathbf{M}^x$ based on the above snapshots;
Step 6: Solve the eigenvalues $\boldsymbol{\lambda}$ and eigenvectors $\boldsymbol{r}$ based on the spatial Gram matrix;
Step 7: Determine the rank number K of $\mathbf{M}^x$ and the first K eigenfunction vector $\boldsymbol{\varphi}$;
Step 8: Determine the undetermined coefficient β based on eigenfunction vector $\boldsymbol{\varphi}$ and snapshots;
Step 9: To a new design variable θ, construct element ϕ based on the design variables $\boldsymbol{\theta}$ generated by LHS using the RBF function;
Step 10: Determine the interpolation matrix $\mathbf{A}$ of elements ϕ;
Step 11: Determine the vector of element $\boldsymbol{\alpha}$ using the penalized linear systems;
Step 12: Solve the coefficients $\beta(\theta)$ based on the RBF function;
Step 13: Calculate the unknown field variables $\tilde{u}^h(\theta)$ based on coefficients $\beta(\theta)$ and eigenfunction vector $\boldsymbol{\varphi}$ using the ROM.

3. Grey Wolf Optimization (GWO)

Grey Wolf Optimization (GWO) is a heuristic optimization strategy inspired by the social hierarchy and hunting techniques of grey wolves. GWO mimics the leadership hierarchy and hunting technology of grey wolves. The hierarchy of the grey wolves is divided into four levels (alpha (α), beta (β), delta (δ), and omega (ω)), which present the optimal solution, the suboptimal solution, the third optimal solution, and the remaining candidate solutions, respectively [23]. There are three main stages of grey Wolf hunting, namely searching, encircling, and attacking prey. The GWO is the mathematical model of the hunting strategy and social hierarchy of grey wolf. A grey wolf can determine the prey by randomly changing its position based on the GWO algorithm. The hunting process is guided by α, β, δ, and ω according to the above three kinds of wolves in GWO.

In the encircling phase, the encircling behavior of wolves can be expressed in the following mathematical model:

$$\vec{D} = \left| \vec{D} \cdot \vec{X}_p(t) - \vec{X}(t) \right| \tag{14}$$

$$\vec{X}(t+1) = \vec{X}_p(t) - \vec{A} \cdot \vec{D} \tag{15}$$

where t denotes the step of the iteration, $\vec{D}$ denotes the searching vector, $\vec{X}$ and $\vec{X}_p$ are the vector and denote the position of a grey wolf and the prey, respectively. $\vec{A}$ and $\vec{C}$ denote the coefficient vectors and can be determined according to the following form:

$$\vec{A} = 2\vec{a} \cdot \vec{r}_1 - \vec{a} \tag{16}$$

$$\vec{C} = 2\vec{r}_2 \tag{17}$$

where $\vec{r}_1$ and $\vec{r}_2$ denote the vector and selected randomly in the range of zero to unity, the component $\vec{a}$ decreases linearly from 2 to 0 with the iterations.

In the hunting phase, the locations of other search agents (including omega) were updated according to alpha, beta, and delta knowledge based on the following equations.

$$\vec{D}_\alpha = \left| \vec{C}_1 \cdot \vec{X}_\alpha - \vec{X} \right| \tag{18}$$

$$\vec{D}_\beta = \left|\vec{C}_2 \cdot \vec{X}_\beta - \vec{X}\right| \tag{19}$$

$$\vec{D}_\delta = \left|\vec{C}_3 \cdot \vec{X}_\delta - \vec{X}\right| \tag{20}$$

$$\vec{X}_1 = \vec{X}_\alpha - \vec{A}_1 \cdot \vec{D}_\alpha \tag{21}$$

$$\vec{X}_2 = \vec{X}_\beta - \vec{A}_2 \cdot \vec{D}_\beta \tag{22}$$

$$\vec{X}_3 = \vec{X}_\delta - \vec{A}_3 \cdot \vec{D}_\delta \tag{23}$$

$$\vec{X}(t+1) = \frac{\vec{X}_1 + \vec{X}_2 + \vec{X}_3}{3} \tag{24}$$

where the subscripts α, β, and δ represent the alpha wolf, beta wolf, and delta wolf, respectively.

In the attacking prey phase, the final attack is determined by decreasing the $\vec{a}$ from 2 to 0 with the iterations $\vec{A}$ selected randomly in the range $[-2\vec{a}, 2\vec{a}]$. $\vec{A}$ will decrease with the reduction to the $\vec{a}$ and force the wolfs to approach the prey while $\left|\vec{A}\right|$ is less than 1.

In search of prey, grey wolves follow the leader, dispersing from one another in search of prey and gathering to attack. The number of wolves N_w and the generation N_G are the two essential parameters of GWO. Each generation represents the decision movement of a wolf. The number of wolves represents the function computational times in each generation. Figure 2 shows the flowchart of GWO and the main procedure. A detailed introduction of the GWO can be found in the literature [23].

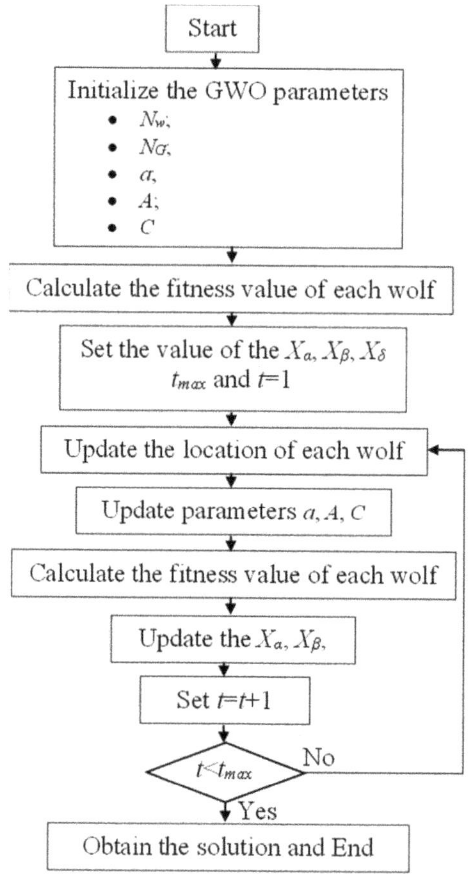

Figure 2. The flowchart of GWO.

4. ROM-Based Back Analysis Using GWO

Back analysis technology has been commonly utilized to identify the geomaterial property in geotechnical engineering. The numerical method and the optimal technology are the two critical elements of back analysis. This study developed a ROM-based back analysis combining numerical methods, ROM, and GWO. The ROM model was utilized to predict the nonlinear response of the geotechnical structure based on the numerical method. GWO was selected as an optimal technology to seek the geomaterial properties.

4.1. Back Analysis

In the 1970s, back analysis was proposed to identify the rock mass properties in rock engineering [25]. The back analysis provides a simple but effective way to identify the geomaterial properties based on the field data and numerical analysis. It also provides a helpful tool for guiding the dynamic design, reinforcement of surrounding rock mass, and safe construction of geotechnical engineering. Figure 3 shows the main parts of the back analysis and its basic idea. Field measurements provide basic information for back analysis. The physical model is the heart of the back analysis. Due to the complex geological conditions, it is not easy to determine the closed-form solution of geotechnical engineering. Meanwhile, optimal technology is essential to back analysis due to multi-extremum and multi-constrained optimization problems. This study adopted the ROM model to capture the physical model. GWO was selected as the optimal technology due to its excellent global optimizing capability.

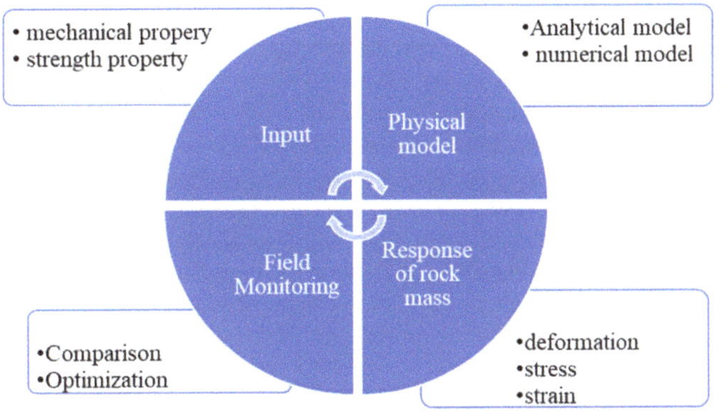

Figure 3. Schematics of back analysis.

4.2. ROM-Based Surrogate Model

The physical model characterizes the nonlinear implicit function mapping between geomaterial properties and their response during construction. In this study, a ROM-based physical model was established to capture the nonlinear mapping function between the geomaterial properties (Deformation modulus, Poisson's ratio, strength property, and in-situ stress) and corresponding structural response (displacement, stress, strain, etc.). The following equations define the physical model *ROM*(**X**).

$$ROM(\boldsymbol{X}) : R^N \rightarrow R^Q, \tag{25}$$

$$\boldsymbol{Y} = ROM\,(\boldsymbol{X}), \tag{26}$$

where $\boldsymbol{X} = (x_1, x_2, \ldots, x_N)$ is a vector and x_i ($i = 1, 2, \ldots, N$) is the ith geomaterial properties and $\boldsymbol{Y} = (y_1, y_2, \ldots, y_N)$ is a Q dimension vector and denotes the response induced by construction.

Some known training samples are necessary for establishing the surrogate model *ROM*(**X**) of the physical model in the back analysis. It is necessary to obtain the training samples for ROM based on a numerical method and the design of experiment.

4.3. Objection Function

This study constructed the objective function based on the geotechnical structure response difference between the field value and ROM prediction. The objection function forced the optimal technology to seek the optimal variable. The following root means square defines the objective function:

$$fitness = \sqrt{\sum_{i=1}^{n} \left(y_{pi} - y_i\right)^2 / n} \tag{27}$$

where n denotes the number of monitoring points, y_i and y_{pi} denote the predicted by ROM and monitoring response of the geotechnical structure in ith measurement point.

4.4. Procedure of the Developed Framework

This study developed a novel back-analysis framework combining ROM, GWO, and numerical technology. The ROM was adopted to capture the nonlinear mapping between the geomaterial properties and the corresponding response during excavation in combination with the ROM and numerical model. The experimental design was adopted to construct the combination of the unknown properties, and then the numerical method was utilized to calculate the structural response at each combination. The snapshots consist of a combination of the unknown parameters and the corresponding response. ROM was built based on the above snapshots. GWO is an optimal approach to seeking geomaterial properties based on the ROM. The detailed procedures of the proposed method are listed as follows (as seen in Figure 4):

Step 1: Collect the engineering data, such as the unknown (need to determine by back analysis) and known geomaterial mechanical and physical properties, boundary conditions, and the range of unknown geomaterial properties;
Step 2: Generate the combination of the unknown properties based on experimental design and calculate the structural response at each training sample. The snapshots consist of the combination of the unknown parameters and the corresponding response;
Step 3: Based on the determined snapshots, generate the ROM to capture the nonlinear function mapping between the geomaterial properties and the corresponding structural response in geotechnical engineering;
Step 4: Establish the objective function and call the GWO to seek the geomaterial properties based on the monitored data during the construction.

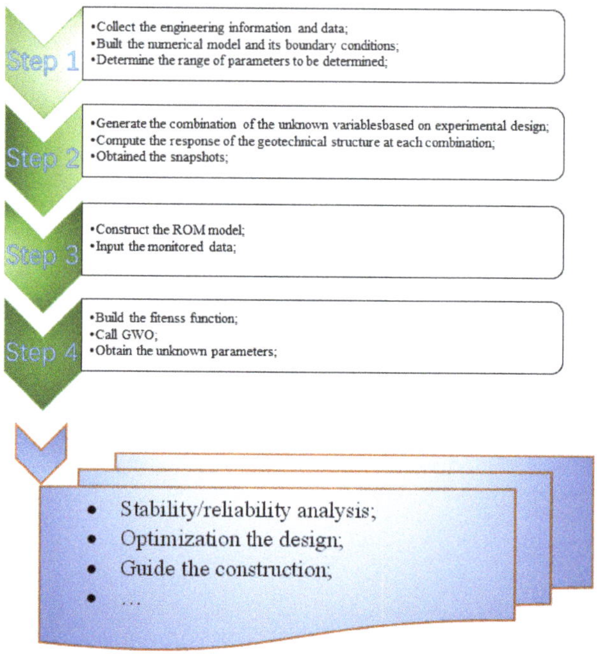

Figure 4. Flowchart of the proposed back analysis framework.

5. Numerical Example and Application

5.1. Numerical Example

A circular tunnel is excavated in a continuous, homogeneous, and isotropic rock mass. The hydrostatic far-field stress p_0 and uniform support pressure p_i are shown in Figure 5. When support pressure p_i is not enough to meet critical pressure p_{cr}, a plastic zone will exist. The values of p_{cr} could be computed as follows:

$$p_{cr} = \frac{2p_0 - \sigma_c}{k+1} \tag{28}$$

where σ_c notes the uniaxial compression strength. It could be obtained using the following equation.

$$\sigma_c = \frac{c(k-1)}{\tan\varphi} \tag{29}$$

where φ and c denote the cohesion and the friction angle, respectively. k could be determined as follows:

$$k = \frac{1 + \sin\varphi}{1 - \sin\varphi} \tag{30}$$

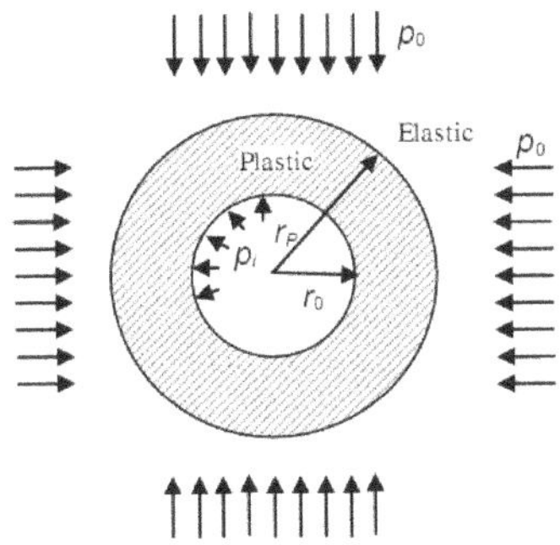

Figure 5. Numerical example—a circular tunnel.

According to the Mohr–Coulomb criterion, Duncan (1993) analyzed and inferred the inward displacement of tunnel wall u_{ip} and the plastic zone radius r_p [26].

$$\frac{r_p}{r_0} = \left[\frac{2(p_0 + s)}{(k+1)(p_i + s)}\right]^{\frac{1}{k-1}} \tag{31}$$

$$\frac{u_{ip}}{r_0} = \left[\frac{(1+\mu)}{E}\right]\left[2(1-\mu)(p_0 - p_{cr})(\frac{r_p}{r_0})^2 - 2(1-2\mu)(p_0 - p_i)\right] \tag{32}$$

where E notes the elastic modulus and μ notes Poisson's ratio. The values are computed as follows:

$$s = \frac{\sigma_c}{k-1} \tag{33}$$

In this study, back analysis was utilized to identify the far-field stress $p_{0,}$ cohesion c, and friction angle φ based on the deformation of the surrounding rock mass in the tunnel. Five horizontal direction monitoring points were placed to record the deformation of surrounding rock mass in the circular tunnel. The distance between the center of the tunnel and the five monitored points are 1.0 m, 1.2 m, 1.6 m, 1.8 m, and 2.0 m, respectively. The displacements of monitored points could be calculated using the above formula (Equation (32)). The tunnel radius is 1.0 m. The value of far-field stress $p_{0,}$ cohesion c, and friction angle φ are 32 MPa, 6.8 GPa, 3.2 MPa, and 32°, respectively. The displacements of 5 monitored points were calculated by the analytical solution and adopted as field measurement to back-calculate the unknown parameters of the surrounding rock mass using the proposed method. The snapshots were constructed and generated based on the experimental design and a numerical method.

Once the snapshots are obtained, the surrogate model could be established according to the ROM algorithm. Figure 6 shows the calculated displacement comparison between the ROM and the analytical solution (Equation (32)). The predicted displacement using the ROM surrogate model is in good agreement with the analytical solution. It shows that the ROM captured well the nonlinear function mapping between unknown properties and the tunnel deformation. The ROM-based surrogate model provides a feasible way to replace the analytical solution in back analysis.

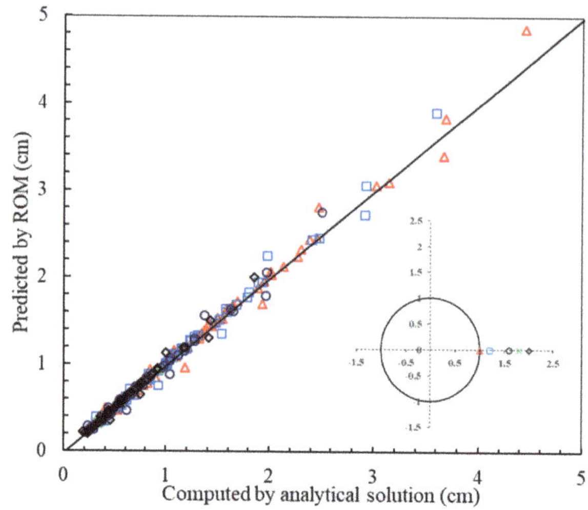

Figure 6. Deformation comparison between computed by the analytical solution and the ROM prediction.

ROM-based back analysis was utilized to identify the unknown properties of the surrounding rock mass using the above ROM surrogate model and to predict displacement in the tunnel. The far-field stress p_0, cohesion c, and friction angle φ are 32.01 MPa, 6.73 GPa, 3.40 MPa, and 31.03°, respectively (Table 1). The relative error is −0.03%, 1.02%, −6.25% and 3.03%, respectively. The maximum relative error is less than 7%. It shows that the identified parameters agree with the actual parameters using the proposed back analysis framework. The displacements comparison between the predicted by back analysis, ROM surrogate model, and calculated by the analytical by actual parameters are shown in Figure 7. Figure 8 shows the stress and displacement of the surrounding rock mass in the tunnel calculated based on the different methods. The results show that the proposed framework can be utilized to identify the mechanical property of the surrounding rock mass in the tunnel.

Table 1. The results and comparison.

	Actual	This Study	Relative Error (%)
p_0/MPa	32.00	32.01	−0.03
E/MPa	6800.00	6730.87	1.02
c/MPa	3.20	3.40	−6.25
φ/°	32.00	31.03	3.03

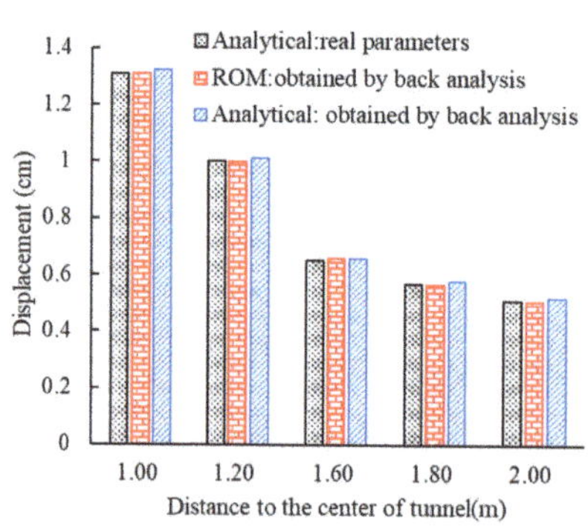

Figure 7. Displacement comparison of the different method at monitored point.

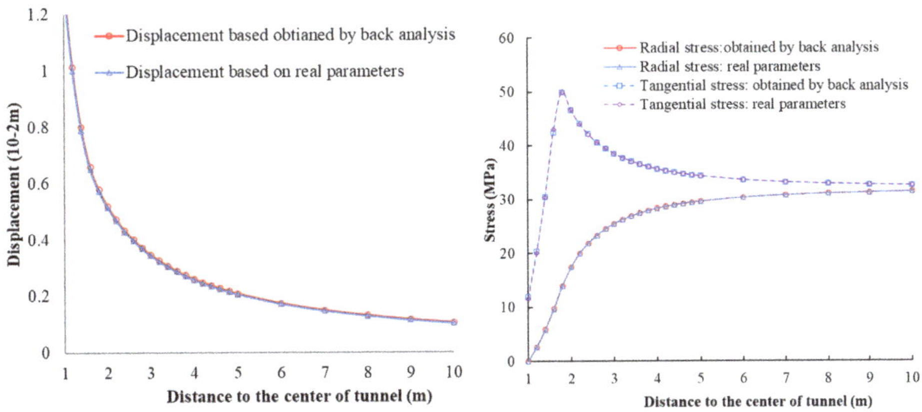

(**a**) Displacement (**b**) Stress

Figure 8. Displacement and stress of surrounding rock mass and their comparison.

Figure 9 shows the convergence of the unknown mechanical property of the surrounding rock mass using the developed back analysis method. The convergence property of the developed back analysis is shown in Figure 10. The unknown property can converge to the final value quickly. The developed method has excellent convergence and global optimization performance.

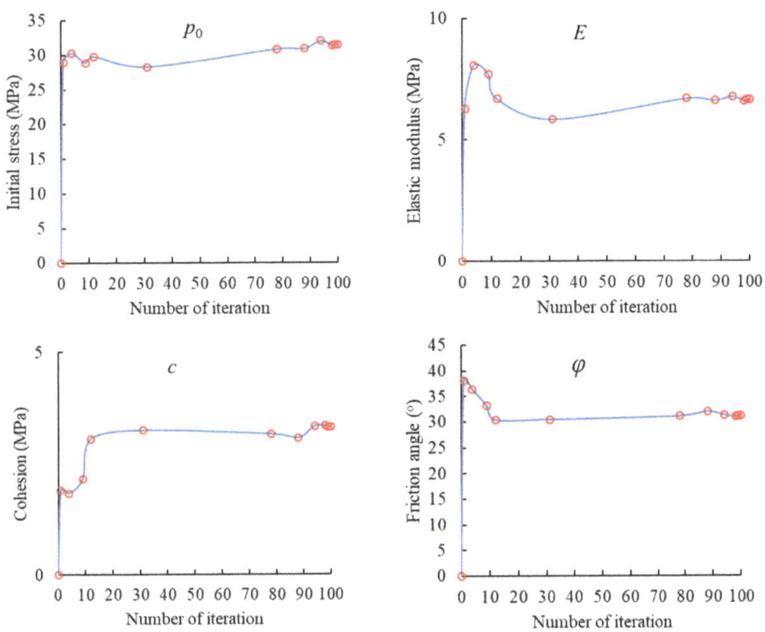

Figure 9. Convergence process of the unknown parameters.

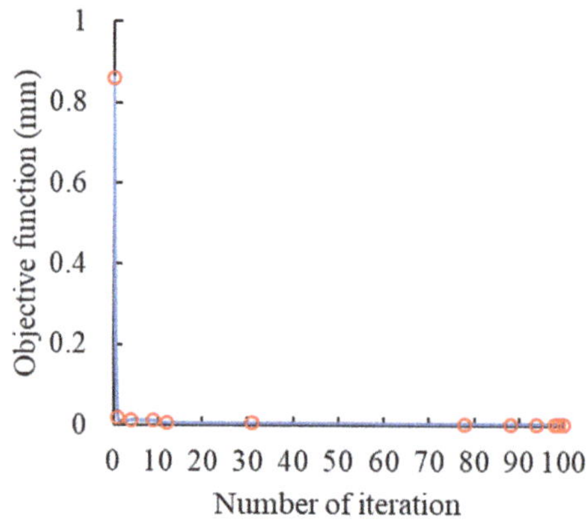

Figure 10. The convergence process of GWO.

5.2. Application: Goupitan Experimental Tunnel

Goupitan Hydropower Station is a landmark west-to-east transmission project located on the Wujiang River in Guizhou Province, China [27]. According to the preliminary design of the underground powerhouse, the right bank tailwater tunnel and the construction diversion tunnel pass through the soft clay rock mass. An experiment tunnel with a buried depth of 70 m was excavated to understand the rheological characteristics of clay rock. The tunnel size was 2 m in width and 2 m in height, respectively. Some monitoring points are set up to obtain the deformation of surrounding rock during tunnel excavation. Rock stratum S_{2h}^{1-2} and S_{2h}^{1-1} are located approximately 3 m below and 30 m above the tunnel, respectively (Figure 11). The 4# and 6# borehole of 7 m depth is set at the position of 11.6 m in the tunnel, where the 4# borehole is horizontal, the 5# borehole is 45° inclined, and the 6# borehole is vertical. Five monitoring points numbered 1–5 were arranged at a depth of 0, 1, 2, 4, and 6 m (Figure 11). Table 2 lists the deformation of the 3-day, 5-day, and 11-day at each monitoring point.

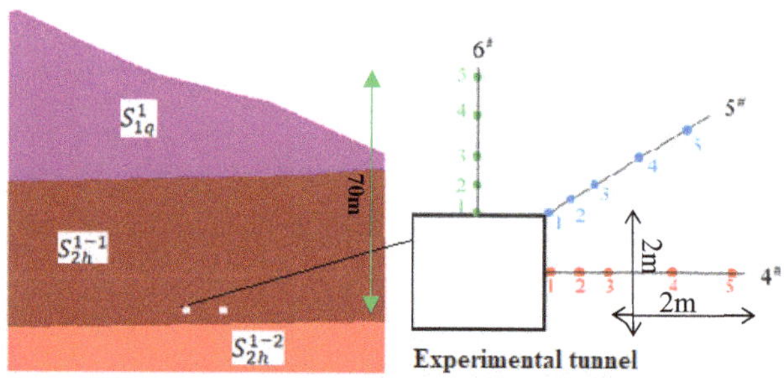

Figure 11. Goupitan experimental tunnel and monitored points.

Table 2. The monitoring displacement in the monitored borehole.

Time (Day)	Displacement (mm)	
	4#	6#
3	2.558	1.778
5	3.789	2.377
11	4.531	2.685

The developed back analysis framework was utilized to identify the rheological properties of the rock mass based on the 3rd, 5th, and 11th day monitored displacements at 4# and 6# boreholes. For the developed back analysis, a physical model is critical to identify the rheological mechanical properties of the rock mass. In this study, the rheological properties of the rock mass include the shear modulus G_1 and viscosity η_1 for the Kelvin model, the shear modulus G_2 and viscosity η_2 for the Maxwell model. In geotechnical engineering, it is not easy to determine an analytical solution for use. Although the numerical method is commonly used to understand the rheological mechanism and deformations behavior of the rock mass, it is time-consuming in practical and large-scale geotechnical engineering.

In this study, a ROM surrogate model, which replaced the numerical method, was used to improve the efficiency of back analysis. A uniform design was utilized to construct a group of 42 samples, and Fast Lagrangian Analysis of Continua (FLAC) software was utilized to solve the displacement of the tunnel wall in the rock mass. Rock masses S_{2h}^{1-1} and S_{2h}^{1-2} were regarded as Burger's material [28,29], and their rheological mechanical properties were identified by the developed method. Table 3 lists the range of unknown rheological mechanical parameters based on laboratory tests and a field survey. Figure 12 indicates the predicted displacement comparison by the ROM and the numerical model along 4# and 6# boreholes, respectively. The calculated deformations by the ROM were found very close to the deformation computed by the numerical solution. It proves that the ROM surrogate model characterized well the nonlinear mapping relationship between unknown rheological properties and deformation of the tunnel wall in the rock mass. Hence, it could replace the numerical method in the back analysis.

Table 3. Ranges of unknown rheological properties.

Clay-Green Clay Rock S_{2h}^{1-1}				Purple Clay Rock S_{2h}^{1-2}			
$G_1{}^h$ (GPa)	$G_2{}^h$ (GPa)	$\eta_2{}^h$ (GPa·d)	$\eta_1{}^h$ (GPa·d)	$G_1{}^z$ (GPa)	$G_2{}^z$ (GPa)	$\eta_2{}^z$ (GPa·d)	$\eta_1{}^z$ (10^3 GPa·d)
0.5–4.5	0.1–3.5	0.1–3.5	15–35	1–15	5–20	1–15	1.5–4.5

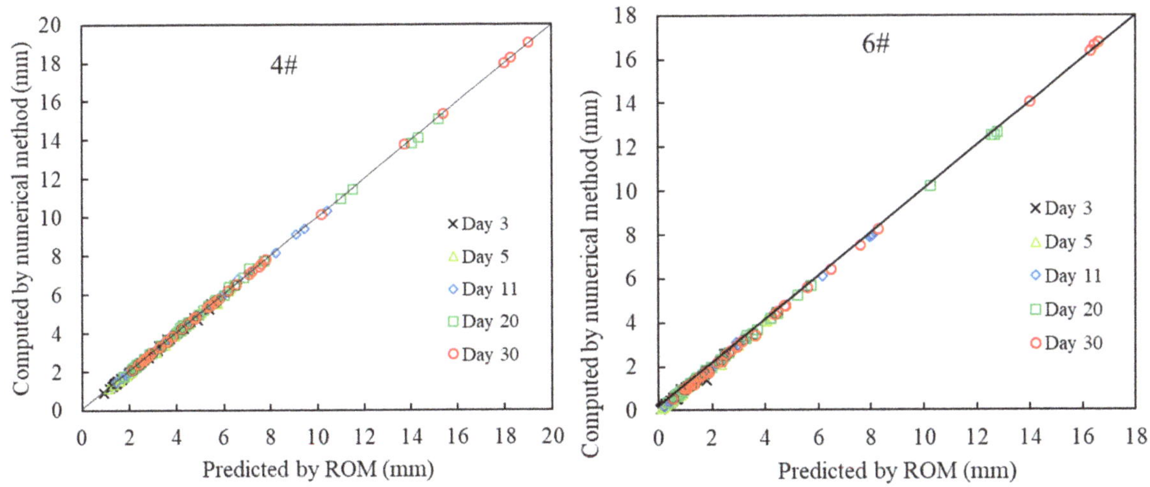

(**a**) 4# borehole (**b**) 6# borehole

Figure 12. Displacement comparison of monitoring borehole.

According to the back analysis theory and the proposed method, the rheological parameters of S_{2h}^{1-1} and S_{2h}^{1-2} are determined based on the displacement of the surrounding rock mass on the 3rd, 5th and 11th days of the 4# and 6# boreholes (Table 4). The developed method identified the rheological mechanical properties in a reasonable way. Compared with other methods, the rheological mechanical properties obtained by the proposed framework are closer to the actual properties of rock mass, and the rheological properties could be determined dynamically by rationally using the on-site displacement data monitored during construction. The developed back analysis framework costs approximately 73.65 s in PC with Intel(R) Xeon(R) Gold 5218 CPU @ 2.30 GHz &2.29 GHz to obtain the rheological properties. However, it takes approximately 2 min for a single tunnel stability analysis using the numerical simulation method. It is obvious that the ROM-based surrogate model could dramatically improve the efficiency of back analysis.

Table 4. Obtained rheological parameters.

Number of Monitored Day		3rd, 5th and 11th
Clay-green clay rock $S_{2h}{}^{1-1}$	$G_1{}^h$ (GPa)	1.39
	$G_2{}^h$ (GPa)	0.20
	$\eta_2{}^h$ (GPa·d)	0.12
	$\eta_1{}^h$ (GPa·d)	35.00
Purple clay rock $S_{2h}{}^{1-2}$	$G_1{}^z$ (GPa)	1.00
	$G_2{}^z$ (GPa)	20.00
	$\eta_2{}^z$ (GPa·d)	8.68
	$\eta_1{}^z$ (10^3 GPa·d)	1.96

It is critical to identify and understand the deformation and failure mechanism of the surrounding rock mass. The developed back analysis identifies the rheological properties based on the displacements monitored. The deformation of the rock mass was investigated based on the rheological properties identified by the developed back analysis framework using the ROM and GWO. The ROM then predicted the displacements of the monitored borehole. Their comparisons are shown in Figure 13 in which it is evident that the displacement determined by the ROM is in good agreement with the monitored displacement during the excavation. It also shows that the identified rheological properties well characterized the rheological behavior of the rock mass during the tunnel construction. This confirms that the developed method can be used for determining rheological properties and evaluating the time-dependent behavior of the rock mass. The rheological properties obtained by the developed method can be used for stability analysis, design, and safety construction during excavation in rock engineering. In addition, the predicted deformation by ROM diverges from the monitoring data on the 34th day due to the complexity of the construction site, which brings errors and uncertainty to the monitoring data. With the increasing monitoring data, the rheological properties of the surrounding rock mass will be updated dynamically to capture the trend.

The relationship between the rheological property (clay-green clay rock) and the number of iterations is shown in Figure 14 for the GWO. The variation process objective function is plotted in Figure 15. The results are similar to the above numerical example. GWO could seek the appropriate mechanical property of the rock mass quickly. It proved again that the GWO is an excellent optimal technology and has a good performance of global optimization.

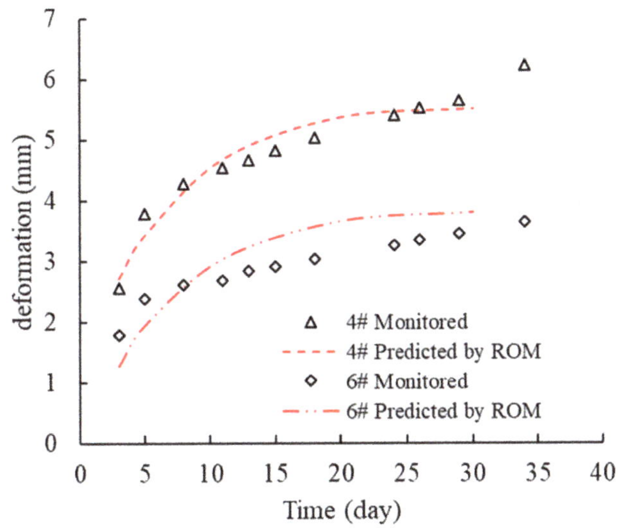

Figure 13. The monitored and predicted deformation based on obtained parameters and their comparison.

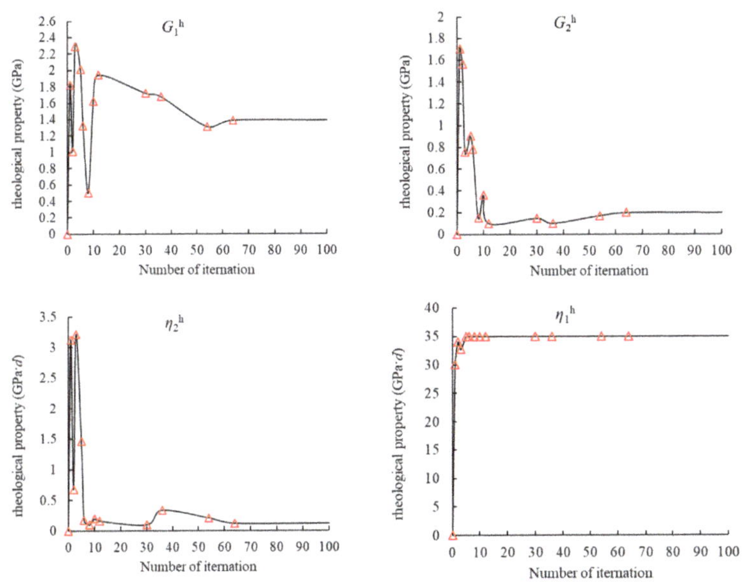

Figure 14. Variation process of the rheological property for clay-green clay rock.

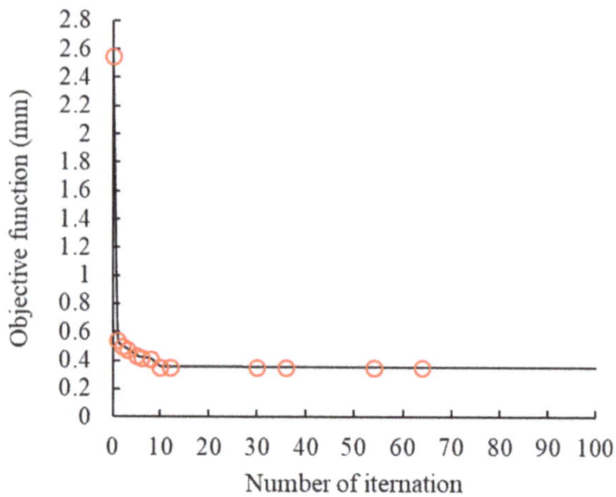

Figure 15. Convergence process of the developed framework.

6. Conclusions

This study developed a novel back analysis framework to identify the geomaterial property by combining a numerical model, ROM, and GWO. The ROM was utilized to establish the surrogate model for capturing deformation during excavation in geotechnical engineering. The numerical method was adopted to construct snapshots for the ROM based on the design of the experiment. The geomaterial properties were identified based on the monitored displacement data by the developed framework. Meanwhile, GWO was selected as an optimal technology for back analysis due to its global optimization performance. The developed back analysis framework was illustrated successfully by a numerical example and Goupitan experimental tunnel. The geomaterial properties identified by this study were compared with the ROM prediction. The predicted displacement of the surrounding rock mass was also close to the actual monitored displacement. The results showed that the developed back analysis framework provides a convenient, practical, and accurate way to understand the geomaterial properties based on the monitored response during construction.

(1) The ROM model was utilized to construct a low-order surrogate model for capturing the response-induced excavation in geotechnical engineering and replacing the numerical model in the back analysis. It is critical to practical engineering due to the difficulties in obtaining the analytical solution for geotechnical engineering;
(2) Back analysis is a scientific and practical tool widely used in geotechnical engineering. The numerical model and optimal technology are the two critical components of back analysis. The developed back analysis framework takes full advantage of the merits of ROM and GWO and provides a feasible way for determining the property of the surrounding rock mass in geotechnical engineering;
(3) ROM is an excellent physics-based data-driven surrogate model that can capture the mechanism of surrounding rock mass. GWO is an efficient metaheuristic method developed recently and is suitable for solving the black-box problem. However, ROM depends on the numerical fidelity model, and the parameters of the GWO algorithm influence the optimal performance. In a future study, the authors will further improve the developed framework by absorbing and combining the advantages and merits of various methods.

Author Contributions: Conceptualization, L.Z. and H.Z.; methodology, L.Z. and H.Z.; software, L.Z. and H.Z.; validation, X.L. and X.Z.; formal analysis, L.Z. and H.Z.; investigation, L.Z. and H.Z.; data curation, L.Z. and H.Z.; writing—original draft preparation, L.Z. and H.Z.; writing—review and editing, X.L. and X.Z.; visualization, H.Z. All authors have read and agreed to the published version of the manuscript.

Funding: This study is sponsored in part by the Shandong Provincial Natural Science Foundation, China under grant ZR2022ME198, to which the authors are very grateful.

Institutional Review Board Statement: Not applicable.

Informed Consent Statement: Not applicable.

Data Availability Statement: The data presented in this study are available on request from the corresponding author.

Conflicts of Interest: The authors declare no conflict of interest.

References

1. Zhao, Z. A practical and efficient reliability-based design optimization method for rock tunnel support. *Tunn. Undergr. Space Technol.* **2022**, *127*, 104587. [CrossRef]
2. Zhao, H.; Chen, B.; Li, S.; Li, Z.; Zhu, C. Updating models and the uncertainty of mechanical parameters for rock tunnels using Bayesian inference. *Geosci. Front.* **2021**, *12*, 101198. [CrossRef]
3. Zhao, H.; Chen, B.; Li, S. Determination of geomaterial mechanical parameters based on back analysis and reduced-order model. *Comput. Geotech.* **2021**, *132*, 104013. [CrossRef]
4. Jing, L.; Hudson, J.A. Numerical methods in rock mechanics. *Int. J. Rock Mech. Min. Sci.* **2002**, *39*, 409–427. [CrossRef]
5. Choi, Y.-H.; Lee, S.S. Reliability and efficiency of metamodel for numerical back analysis of tunnel excavation. *Appl. Sci.* **2022**, *12*, 6851. [CrossRef]
6. Zhao, H.; Chen, B. Inverse analysis for rock mechanics based on a high dimensional model representation. *Inverse Probl. Sci. Eng.* **2021**, *29*, 1565–1585. [CrossRef]
7. Gao, W. A comprehensive review on identification of the geomaterial constitutive model using the computational intelligence method. *Adv. Eng. Inform.* **2018**, *38*, 420–440. [CrossRef]
8. Sakurai, S.; Takeuchi, K. Back analysis of measured displacements of tunnels. *Rock Mech. Rock Eng.* **1983**, *16*, 173–180. [CrossRef]
9. Zhang, W.; Zhang, R.; Wu, C.; Goh, A.T.C.; Lacasse, S.; Liu, Z.; Liu, H. State-of-the-art review of soft computing applications in underground excavations. *Geosci. Front.* **2020**, *11*, 1095–1106. [CrossRef]
10. Zhang, W.; Wu, C.; Zhong, H.; Li, Y.; Wang, L. Prediction of undrained shear strength using extreme gradient boosting and random forest based on Bayesian optimization. *Geosci. Front.* **2021**, *12*, 469–477. [CrossRef]
11. Pan, S.; Li, T.; Shi, G.; Cui, Z.; Zhang, H.; Yuan, L. The Inversion Analysis and Material Parameter Optimization of a High Earth-Rockfill Dam during Construction Periods. *Appl. Sci.* **2022**, *12*, 4991. [CrossRef]
12. Deng, J.H.; Lee, C.F. Displacement back analysis for a steep slope at the Three Gorges Project site. *Int. J. Rock Mech. Min. Sci.* **2001**, *38*, 259–268. [CrossRef]
13. Shang, Y.J.; Cai, J.G.; Hao, W.D.; Wu, X.Y.; Li, S.H. Intelligent back analysis of displacements using precedent type analysis for tunneling. *Tunn. Undergr. Space Technol.* **2002**, *17*, 381–389. [CrossRef]
14. Yu, Y.Z.; Zhang, B.Y.; Yuan, H.N. An intelligent displacement back–analysis method for earth–rockfill dams. *Comput. Geotech.* **2007**, *34*, 423–434. [CrossRef]
15. Gao, W. Back analysis for mechanical parameters of surrounding rock for underground roadways based on new neural network. *Eng. Comput.* **2018**, *34*, 25–36. [CrossRef]
16. Feng, X.T.; Zhao, H.; Li, S.J. A new displacement back analysis to identify mechanical geo–material parameters based on hybrid intelligent methodology. *Int. J. Numer. Anal. Method Geomech.* **2004**, *28*, 1141–1165. [CrossRef]
17. Zhao, H.; Ru, Z.; Yin, S. A practical indirect back analysis approach for geomechanical parameters identification. *Mar. Georesources Geotechnol.* **2015**, *33*, 212–221. [CrossRef]
18. Zhao, H.; Yin, S. Inverse analysis of geomechanical parameters by artificial bee colony algorithm and multi-output support vector machine. *Inverse Probl. Sci. Eng.* **2016**, *24*, 1266–1281. [CrossRef]
19. Pichler, B.; Lackner, R.; Mang, H.A. Back analysis of model parameters in geotechnical engineering by means of soft computing. *Int. J. Numer. Method Eng.* **2003**, *57*, 1943–1978. [CrossRef]
20. Vardakos, S.; Gutierrez, M.; Xia, C.C. Parameter identification in numerical modeling of tunneling using the Differential Evolution Genetic Algorithm DEGA. *Tunn. Undergr. Space Technol.* **2012**, *28*, 109–123. [CrossRef]
21. Zhao, H.B.; Yin, S.D. Geomechanical parameters identification by particle swarm optimization and support vector machine. *Appl. Math. Model.* **2009**, *33*, 3997–4012. [CrossRef]
22. Yazdi, J.S.; Kalantary, F.; Yazdi, H.S. Calibration of Soil Model Parameters Using Particle Swarm Optimization. *Int. J. Geomech.* **2012**, *12*, 229–238. [CrossRef]
23. Mirjalili, S.; Mirjalili, S.M.; Lewis, A. Grey wolf optimizer. *Adv. Eng. Softw.* **2014**, *69*, 46–61. [CrossRef]
24. Audouze, C.; Vuyst, F.D.; Nair, P.B. Reduced-order modeling of parameterized PDEs using time-space-parameter principal component analysis. *Int. J. Numer. Methods Eng.* **2009**, *80*, 1025–1057. [CrossRef]

25. Sakurai, S. *Back Analysis in Rock Engineering*; CRC Press: Boca Raton, FL, USA, 2017.
26. Duncan, F.M.E. Numerical modeling of yield zones in weak rocks. In *Comprehensive Rock Engineering*; Hudson, J.A., Ed.; Pergamon: Oxford, UK, 1993; Volume 2, pp. 49–75.
27. Chen, B.R. Back Analysis of Rheological Parameters of Rock Mass Using Intelligent Method. Master's Thesis, Northeastern University, Shenyang, China, 2003.
28. Fahimifar, A.; Tehrani, F.M.; Hedayat, A.; Vakilzadeh, A. Analytical solution for the excavation of circular tunnels in a visco-elastic Burger's material under hydrostatic stress field. *Tunn. Undergr. Sp. Tech.* **2010**, *25*, 297–304. [CrossRef]
29. Goodman, R.E. *Introduction to Rock Mechanics*, 2nd ed.; Wiley: New York, NY, USA, 1989.

applied sciences

MDPI

Communication

SAR and Optical Image Registration Based on Uniform Feature Points Extraction and Consistency Gradient Calculation

Wannan Zhang and Yuqian Zhao *

School of Automation, Central South University, Changsha 410017, China
* Correspondence: zyq@csu.edu.cn

Abstract: Synthetic aperture radar (SAR) satellites have an active sensor on board, which emits electromagnetic signals and measures the strength and time delay of the returned signal backscattered from ground objects. Optical images have rich spectral information, but it is easily affected by atmospheric attenuation and weather conditions. Thus, the study of the registration between these two images is of great significance. We present a novel method for SAR and optical image registration. In the stage of feature points extraction, the method combines phase consistency intensity screening and scale space grid division to obtain stable and uniform feature points from the image. During the stage of feature description, the method employs the extended phase consistency method to calculate the gradient amplitude and direction of the image, and improves the correctness of the main direction calculation and descriptor construction. Experimental results demonstrate its superior matching performance with respect to the state-of-the-art methods.

Keywords: remote sensing; image registration; SAR; SIFT

Citation: Zhang, W.; Zhao, Y. SAR and Optical Image Registration Based on Uniform Feature Points Extraction and Consistency Gradient Calculation. *Appl. Sci.* **2023**, *13*, 1238. https://doi.org/10.3390/app13031238

Academic Editor: Wei Gao

Received: 4 November 2022
Revised: 3 January 2023
Accepted: 16 January 2023
Published: 17 January 2023

1. Introduction

Remote sensing technology is a critical means of earth observation from remote sensing devices mounted on artificial satellites, airplanes and other aviation or spacecraft [1], because it can collect electromagnetic information of ground targets on a large scale to help humans obtain observations that traditional technology cannot achieve. As a result, this technology is widely used in national economic and military aspects such as meteorological observation, map surveying and mapping and military investigation [2–7]. Remote sensing image registration is a method to establish a matching relationship between two or more remote sensing images captured by different angles or different sensors in the same scene at different times. It is a pioneer task for subsequent remote sensing image splicing, fusion or transformation detection.

With the rapid development of remote sensing technology, the acquisition methods of remote sensing images are gradually showing a diversified trend. In practical applications, the imaging methods of heterogeneous remote sensing images are different and the information contained is not the same. How to use heterogeneous remote sensing for effective fusion and complementation of information in images has attracted more and more attention in recent years. Synthetic Aperture Radar (SAR), as an active imaging radar, has the advantages of all-weather imaging due to its strong penetrating ability [8–10]. It can penetrate clouds and haze and other occlusions to break through the limitations of optical imaging, but is not easy to obtain the perceived characteristics of ground targets. In contrast, optical remote sensing images can obtain spectral information such as rich gray-scale textures of ground targets under good imaging conditions, good visual interpretation functions and have great advantages in ground target recognition and classification. Therefore, it has become an important research topic in the field of remote sensing image processing to realize the complementary advantages of multi-source images through effective heterogeneous remote sensing image registration and fusion technology [11]. At present, due to the great difference in imaging mechanism between SAR and optical images, the geometric and radiation characteristics of the images are different. In addition, the multiplicative speckle noise inherent in the SAR image brings many difficulties to the registration of SAR and optical images. Therefore, it is of great significance to study the registration of SAR and optical images.

The SIFT algorithm [12] is a classic local invariant feature extraction method. The algorithm mainly detects extreme points in the Gaussian difference scale space as stable feature points, and at the same time it calculates the gradient direction histogram in the neighborhood of the image to construct the feature vector. In terms of its similar algorithms, the extraction of robust feature points and the calculation of basic gradient features are of great significance to the performance of this method. As mentioned above, due to the different imaging mechanisms and acquisition methods of optical and SAR images, there is a large non-linear radiation difference between these two kinds of images [13]. At the same time, there is an inherent multiplicative noise in SAR images, which will cause different extractions. The source image has a small number of feature points with the same name and poor robustness [14]. In addition, the main direction of the calculated feature points and the constructed descriptor are unreliable due to gradient calculation errors, which leads to a lower correct matching rate in the feature point matching stage.

To solve the problems mentioned above, this paper conducts research from the two aspects of obtaining robust feature points with the same name and calculating the consistency gradient of heterogeneous images, and proposes a SIFT registration method based on uniform extraction of feature points and gradient consistency. In the stage of feature points extraction, the method combines phase consistency intensity screening and scale space grid division to obtain stable and uniform feature points from the image. In the feature description stage, the method employs the extended phase consistency method to calculate the gradient amplitude and direction of the image, and improves the correctness of the main direction calculation and descriptor construction compared with the original SIFT algorithm.

2. Proposed Method

Compared with the original SIFT algorithm, the method in this chapter first uses a combination of phase consistency intensity screening and uniformly distributed feature point detection to extract uniform and robust candidate points from the image as stable feature points. Secondly, the extended phase consistency method is used to calculate the gradient magnitude and direction of the image, and the main direction and descriptor of the feature points are calculated, accordingly. Finally, in order to improve the uniqueness of the feature point descriptor, a 136-dimensional GLOH-like descriptor is constructed by collecting the histogram of the neighborhood gradient around the feature point in polar coordinates. Finally, it is according to the dual-match and two-way matching and the eigenvector matching strategy combined with the RANSAC method, which obtains the deformed model parameters and realizes the registration between SAR and optical images. The workflow of the proposed algorithm is shown as Figure 1.

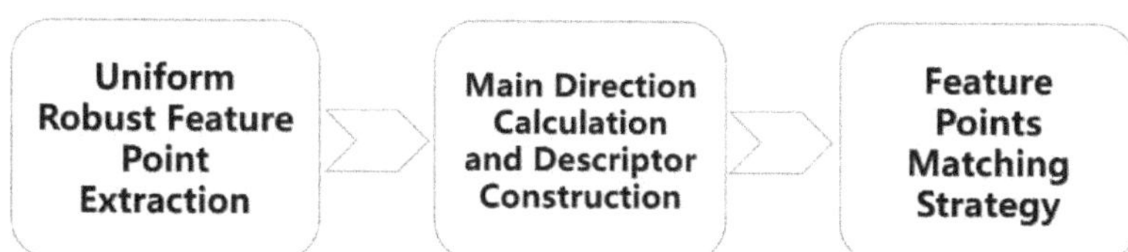

Figure 1. The workflow of the proposed algorithm.

2.1. Uniform Robust Feature Point Extraction

When processing the registration of SAR and optical images, the distribution of feature points has a great influence on the registration result, which obtains a suitable number of evenly distributed feature points and retains stable and highly repeatable feature points in heterogeneous images. It can accurately estimate the geometric transformation model between images and improve the registration accuracy. In order to solve the problem of uneven distribution of feature points in a single image, this section introduces the scale space scale factor and coordinate space blocking strategy, and uses the phase consistency intensity value to filter the feature points, and obtain as much as possible from heterogeneous images with robust feature points of the same name. The steps of the algorithm are as follows:

1. The upper limit of the total number of feature points is determined by the original image size N.
2. Calculate the phase consistency response intensity map of the original image I_{pc}.
3. Calculate the scale parameters of each group of each layer to construct a Gaussian pyramid.
4. Calculate the feature point set of each group of each layer in the scale space I_{ol}: (a) Determine the upper limit of the number of feature points in this layer; (b) detect extreme points as initial candidate points of interest; (c) divide the layer of the Gaussian image into regular grid cells and determine the features that each cell needs to retain the number of points in the n_cell_i; (d) position the coordinates of the candidate points of interest and analyze the principal curvature to eliminate unstable points on the edge; (e) find the I_{pc} according to the coordinates to obtain the PC intensity value of the reserved interest point, and according to the PC intensity, the degree value and the absolute value of the Gaussian difference response are sorted, and the top n_cell_i are retained as the feature points in the cell; (f) summarize the feature points in all cells to obtain the feature point set P_{ol} of the current layer I_{ol}.
5. Summarize the feature points of all scale layers in the scale space to obtain a set P.

According to the SIFT feature points obtained by the above feature point detection algorithm, because in the process of feature point detection, the number of feature points in the scale space distribution and coordinate space distribution is consciously restricted, and then the stability and saliency of the feature points can be characterized by the index screens of the feature points, and finally obtains a uniform and robust feature point set.

2.2. Main Direction Calculation and Descriptor Construction

In SIFT and its improved algorithm, specifying the main direction for the feature point can make it a rotation invariant, and the calculation of the main direction and the construction of the descriptor are all through the feature points in the differential pyramid space image to count the gradient histogram of the neighborhood. The image is obtained, so the gradient direction between the SAR and optical images should be consistent. However, as for the registration of multi-sensor images, due to the huge difference in radiation characteristics between each other, for example, the gradient direction of the matching area are often different. In order to avoid this problem, which may cause registration failure, in this paper, the gradient calculation method based on phase consistency is used to extract the consistent gradient features of SAR and optical images, and we use this basic feature to calculate the main direction of feature points by means of statistical histograms and build descriptors.

In this section, when calculating the main directions of feature points, we choose the method that approximates the original SIFT algorithm. After the precise coordinates of the feature points and the corresponding scale coordinates are obtained through the detection algorithm, the consistent gradient algorithm is used to calculate the neighboring neighbors of the feature points from the corresponding scale space. In the gradient direction histogram of the domain, the direction corresponding to the maximum peak value is selected as the main direction of the feature point, and the direction with the peak value in the histogram greater than 80% of the main direction peak value is used as the auxiliary direction to improve the stability of the algorithm.

When constructing the feature point descriptor, this paper makes some modifications to the selection and division of the neighborhood block of the SIFT feature point. The SIFT native structure description method divides the pixels in the radius neighborhood around the feature points into 4∗4 sub-blocks in space, and counts the gradient direction histograms of eight directions in each sub-block, and finally divides the directions of all sub-regions into 4∗4 sub-blocks. The gradient information combination is a 128-dimensional descriptor. According to the research of Krystian Mikolajczyk et al. this division method ignores the positional relationship between the pixels within the neighborhood block. Therefore, this article adopts the method that selects the image within a certain radius around the feature point as its neighborhood block. After rotating the neighborhood block to the main direction, the pixels in the neighborhood block are divided in the polar coordinate system according to their distance from the center point into different fan-shaped grids, so that 17 grid areas are obtained. For each fan-shaped sub-region and each pixel in the center circle, use the corresponding gradient algorithm to calculate the gradient magnitude and direction, and then assign the gradient value in the sub-block to eight directions, and finally after the gradient information of each direction in the 17 sub-regions is concatenated and normalized, 136 dimensions are obtained, so that the descriptors obtained have better robustness and uniqueness.

2.3. *Feature Points Matching Strategy*

After the feature point descriptors are constructed, the feature vector sets of SAR and optical images are obtained, respectively. It is necessary to establish a correct match based on the similarity measurement criterion between feature vector descriptors and the matching relationship. This chapter will first use the two-way matching algorithm to establish the initial matching relationship, and then use the random sampling consensus algorithm to remove the wrong matching point pairs, and calculate the parameters of the transformation model by retaining the correct matching point pairs. The main process is as follows:

2.3.1. Dual-Match

Assuming that the feature vector sets obtained in the previous feature extraction stage are V_s and V_o, respectively, the simplest way to calculate the similarity between feature vectors is to calculate the Euclidean distance. However, in the actual matching process, due to noise and scene occlusion problems, some feature points do not have correct matching points corresponding to them, and the points with the closest Euclidean distance found in the corresponding images may be mismatched points. Therefore, Lowe proposed the k-nearest neighbor ratio method to eliminate the false matches [10].

According to experiments, this method can achieve good results in most scenes. However, this one-way matching method is prone to "one-to-many" wrong matching phenomenon due to similar neighborhoods in different locations. A certain eigenvector in V_s has multiple eigenvectors that satisfy the k-nearest neighbor ratio method in the V_o set, which will cause a lot of trouble for the subsequent error matching elimination. Therefore, this paper adopts a two-way matching strategy [15] to improve the accuracy of matching point pairs.

The specific process is as follows: (a) For each feature vector L_a in V_s, calculate its Euclidean distance with each feature vector in V_o by traversing. Then, sort and find the closest Euclidean distance $Dist_{f1}$ and the next closest Euclidean distance $Dist_{f2}$ and the corresponding vectors L_b and L_b'. If $Dist_{f1}/Dist_{f2} \leq$ Threshold, then L_a and L_b match. Traverse all the feature vectors in V_s to obtain the forward matching set of $M_{forward}$. (b) For each feature vector L_b in V_o, calculate its Euclidean distance with each feature vector in V_s by traversing. Then, sort and find the closest Euclidean distance $Dist_{f1}$ and the next closest Euclidean distance $Dist_{f2}$ and the corresponding vectors L_a and L_a'. If $Dist_{b1}/Dist_{b2}$ $\leq$ Threshold, then L_a and L_b match. Traverse all the feature vectors in V_o to obtain the forward matching set $M_{backward}$. (c) Set up two loops, traversing $M_{forward}$ and $M_{backward}$, respectively. If the set of matching satisfies the relationship $\mathrm{Dist}_{f1} = \mathrm{Dist}_{b1}$, then the set of matching point pairs is retained to obtain a set of bidirectional matching point pairs.

2.3.2. Random Sampling Consistency Method Purification

In the rough matching point set after the previous two-way matching, there are still some wrong matching point pairs. This includes two types. The first type of mismatched points come from unmatched regional scenes, which are mismatched due to approximate descriptors; the second type of mismatched point pairs come from matched regional scenes, however, due to noise and local areas of the image deformation leads to a deviation in positioning. If such matching point pairs are not eliminated, the calculated transformation model will deviate from the actual situation, resulting in a decrease in registration accuracy. For the above two considerations, it is necessary to adopt a certain method to purify the set of rough matching points. Among the many methods, the random sampling consensus algorithm proposed by Fischler et al. [16] is the most commonly used. This method is a robust parameter estimation method. Its main idea is to fit the model parameters in an iterative manner in the search set, set a threshold to filter the interior points, continuously expand the set of support points, and find all the data pairs that meet the model parameters to obtain the optimal solution.

3. Experimental Results and Analysis

To evaluate the performance of the proposed method, three pairs of SAR and optical images are experimented. The test data consists of different characteristics including different resolutions, incidence angles, seasons etc. Experimental results are shown in Figures 2–4. To quantitatively evaluate the registration performances, we adopt the root-mean-square error (RMSE) [12] between the corresponding matching keypoints, and it can be expressed as:

$$\mathrm{RMSE} = \sqrt{\frac{1}{n}\sum_{i=1}^{n}(x_i - x_i')^2 + (y_i - y_i')^2} \tag{1}$$

where (x_i, y_i) and (x_i', y_i') are the coordinates of the ith matching keypoint pair; n means the total number of matching points. In addition, the correct matching ratio (CMR) is another effective measure, which is defined as:

$$CMR = \frac{correctMatches}{correspondences} \tag{2}$$

"*correspondences*" is the number of matches after using PROSAC, "*correctMatches*" is the number of correct matches after removing false ones. The results of quantitative evaluation for each method are listed in Table 1.

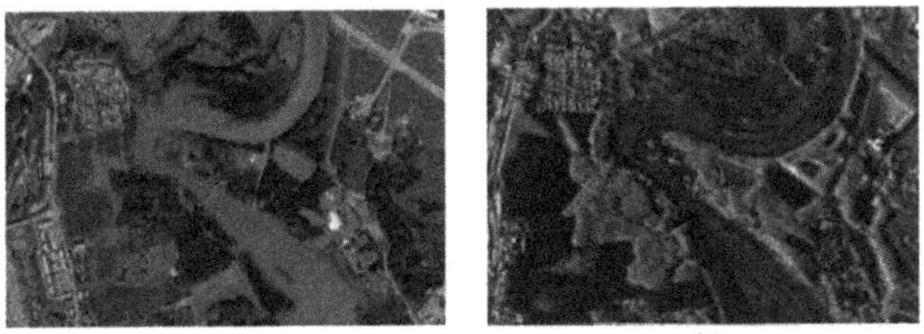

(a) (b)

Figure 2. *Cont.*

(c)

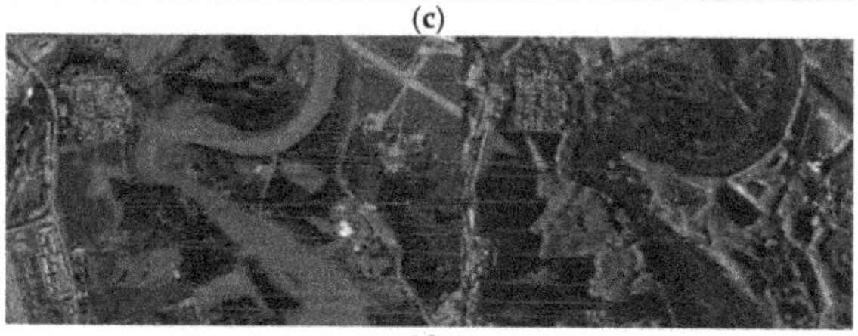

(d)

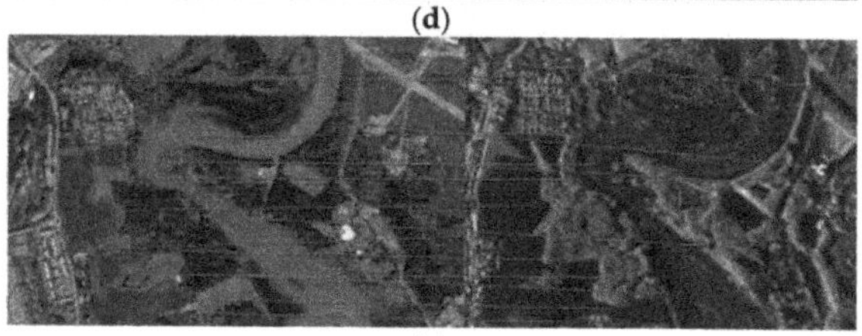

(e)

Figure 2. (**a**) Optical image; (**b**) SAR image; matches found in pair 1 using (**c**) SIFT, (**d**) SAR-SIFT, and (**e**) the proposed method. The reference image is shown on the left and the sensed image on the right.

(a)

(b)

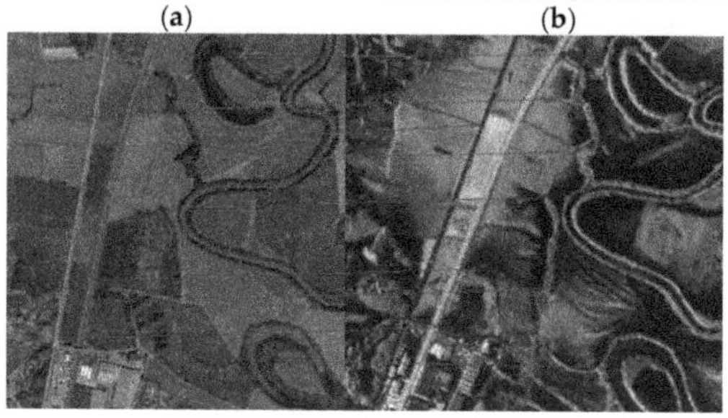

(c)

Figure 3. *Cont.*

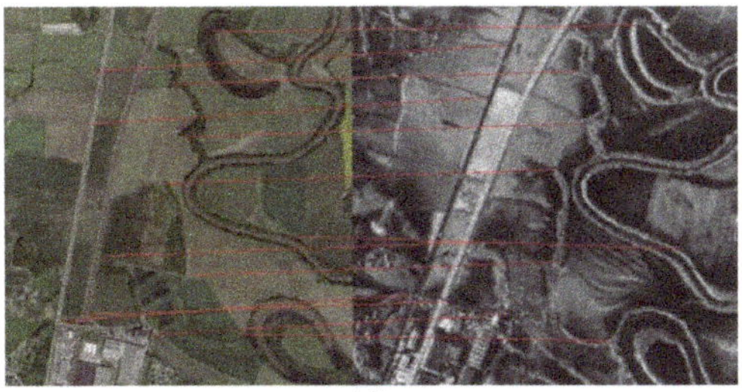

(**d**)

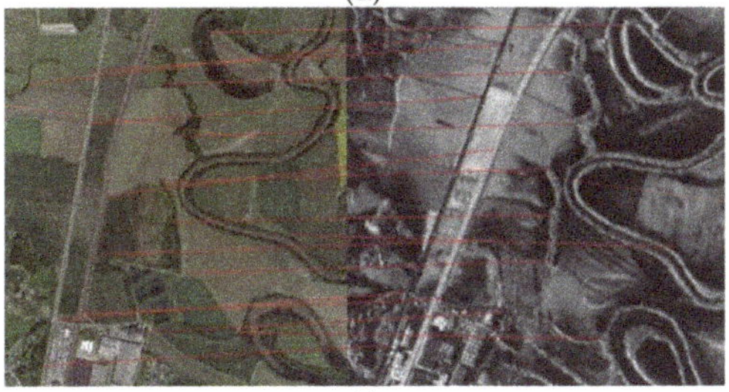

(**e**)

Figure 3. (**a**) Optical image; (**b**) SAR image; matches found in pair 2 using (**c**) SIFT, (**d**) SAR-SIFT, and (**e**) the proposed method. The reference image is shown on the left and the sensed image on the right.

(**a**)

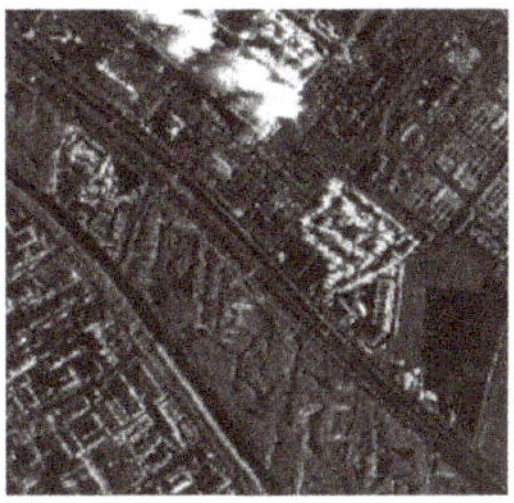

(**b**)

(**c**)

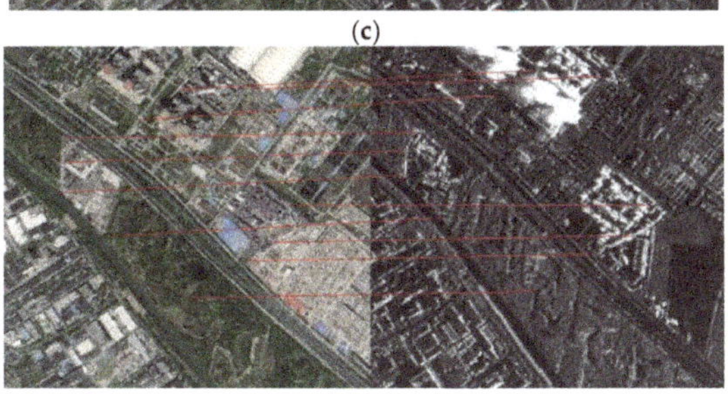

(**d**)

Figure 4. *Cont.*

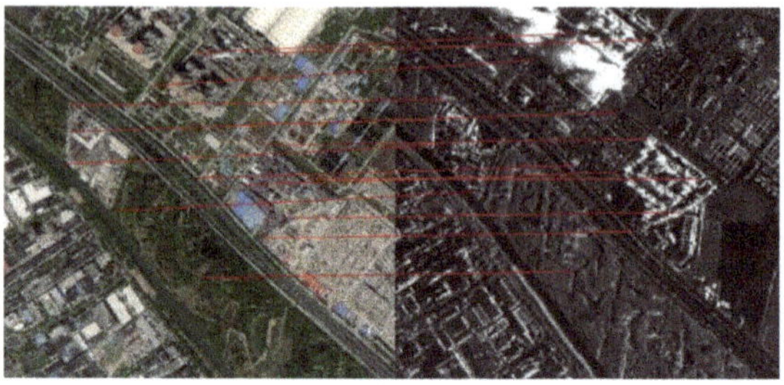

(e)

Figure 4. (**a**) Optical image; (**b**) SAR image; matches found in pair 3 using (**c**) SIFT, (**d**) SAR-SIFT, and (**e**) the proposed method. The reference image is shown on the left and the sensed image on the right.

Table 1. Quantitative comparison of the proposed method with other SIFT-based algorithms.

Image No.	Method	CMR/%	RMSE/Pixel
1	SIFT	64.79	1.1344
	SAR-SIFT	79.65	0.9276
	Proposed	88.21	0.5128
2	SIFT	67.58	1.3153
	SAR-SIFT	71.29	1.1452
	Proposed	81.89	0.8127
3	SIFT	62.46	0.9008
	SAR-SIFT	75.34	0.7097
	Proposed	89.73	0.4982

It can be observed from Table 1 that the correct matching rate obtained by the SIFT and SAR-SIFT algorithms is relatively low. The proposed algorithm can obtain more correct matching point pairs, which shows that this algorithm is better in suppressing the radiation difference between optical and SAR images. Compared with the other two algorithms, the CMR of this algorithm is improved by 88.21%, 81.89% and 89.73%, respectively in the three groups of images, and the RMSE is reduced by 0.5128, 0.8127 and 0.4982, respectively.

4. Discussion

The original SIFT method uses the gradient calculation method of the difference operator to process the SAR and optical images. Due to the influence of the non-linear radiation difference of the heterogeneous image, the gradient magnitude and direction calculated in the SAR image by this calculation method are quite different from the calculation result in the optical image. At the same time, due to the SAR due to the influence of image noise, the feature points detected by this method from heterogeneous images are often poor in stability, and the ratio of control points with the same name is low, which is easy to cause mismatches. These all cause the phenomenon that the correct rate of feature points matching by this method is not high.

In this chapter, the SIFT registration method based on uniform extraction of feature points and gradient consistency has successfully dealt with the shortcomings of the SIFT algorithm. Using the combination of phase consistency intensity screening and uniform distribution feature point detection, it improves the robustness of feature points, while taking into account the uniformity of the feature points in the coordinate space. In addition, the extended phase consistency method is used to calculate the gradient amplitude of the image; thus, the accuracy of the main direction calculation of the original SIFT method is improved. In the descriptor construction process, the descriptor constructed by dividing the sector area under polar coordinates has better reliability. In this way, the feature point matching method is optimized from the feature point extraction to the feature point description link, so the highest matching success rate and registration accuracy can be obtained.

5. Conclusions

In this paper, we present a novel method for the SAR and optical image registration. In the stage of the feature points extraction, the method combines the phase consistency intensity screening and scale space grid division to obtain stable and uniform feature points from the image. In the feature description stage, the method employs the extended phase consistency method to calculate the gradient amplitude and direction of the image, and improves the correctness of the original SIFT algorithm of the main direction calculation and descriptor construction. Experimental results demonstrate its superior matching performance with respect to the state-of-the-art methods.

Author Contributions: Writing—original draft preparation, W.Z.; Writing—review and editing, Y.Z. All authors have read and agreed to the published version of the manuscript.

Funding: This research received no external funding.

Institutional Review Board Statement: Not applicable.

Informed Consent Statement: Not applicable.

Data Availability Statement: The data presented in this study are available on request from the corresponding author.

Conflicts of Interest: The authors declare no conflict of interest.

References

1. Gao, G.; Liu, L.; Zhao, L.; Shi, G.; Kuang, G. An Adaptive and Fast CFAR Algorithm Based on Automatic Censoring for Target Detection in High-Resolution SAR Images. *IEEE Trans. Geosci. Remote Sens.* **2009**, *47*, 1685–1697. [CrossRef]
2. Zhao, Y.; Zhang, M.; Zhao, Y.W.; Geng, X.-P. A Bistatic SAR Image Intensity Model for the Composite Ship–Ocean Scene. *IEEE Trans. Geosci. Remote Sens.* **2015**, *53*, 4250–4258. [CrossRef]
3. Chini, M.; Pierdicca, N.; Emery, W.J. Exploiting SAR and VHR Optical Images to Quantify Damage Caused by the 2003 Bam Earthquake. *IEEE Trans. Geosci. Remote Sens.* **2008**, *47*, 145–152. [CrossRef]
4. Werninghaus, R.; Buckreuss, S. The TerraSAR-X Mission and System Design. *IEEE Trans. Geosci. Remote Sens.* **2010**, *48*, 606–614. [CrossRef]
5. Myronenko, A.; Song, X.B. Intensity-Based Image Registration by Minimizing Residual Complexity. *IEEE Trans. Med. Imaging* **2010**, *29*, 1882–1891. [CrossRef] [PubMed]
6. Wachinger, C.; Navab, N. Entropy and Laplacian images: Structural representations for multi-modal registration. *Med. Image Anal.* **2012**, *16*, 1–17. [CrossRef] [PubMed]
7. Pluim, J.P.W. Image registration by maximization of combined mutual information and gradient information. *IEEE Trans. Med. Imaging* **2000**, *19*, 809–814. [CrossRef]
8. Indra, R.; Rizkinia, M.; Arief, R.; Sudiana, D. Three-Dimensional Convolutional Neural Network on Multi-Temporal Synthetic Aperture Radar Images for Urban Flood Potential Mapping in Jakarta. *Appl. Sci.* **2022**, *12*, 1679. [CrossRef]
9. Eineder, M.; Minet, C.; Steigenberger, P.; Cong, X.; Fritz, T. Imaging Geodesy—Toward Centimeter-Level Ranging Accuracy with TerraSAR-X. *IEEE Trans. Geosci. Remote Sens.* **2011**, *49*, 661–671. [CrossRef]
10. Xiong, Z.; Zhang, Y. A Critical Review of Image Registration Methods. *Int. J. Image Data Fusion* **2010**, *1*, 137–158. [CrossRef]
11. Kern, J.P.; Pattichis, M.S. Robust Multispectral Image Registration Using Mutual-Information Models. *IEEE Trans. Geosci. Remote Sens.* **2007**, *45*, 1494–1505. [CrossRef]
12. Lowe, D.G. Distinctive Image Features from Scale-Invariant Keypoints. *Int. J. Comput. Vis.* **2004**, *60*, 91–110. [CrossRef]
13. Zhu, H.; Ma, W.; Hou, B. SAR image registration based on multifeature dection and arborescence network matching. *IEEE Geosci. Remote Sens.* **2016**, *13*, 706–710. [CrossRef]
14. Fan, J.; Wu, Y.; Wang, F.; Zhang, Q.; Liao, G.; Li, M. SAR image registration using phase congruency and nonlinear diffusion-based SIFT. *IEEE Geosci. Remote Sens. Lett.* **2015**, *12*, 562–566.
15. Wang, S. BFSIFT: A Novel Method to Find Feature Matches for SAR Image Registration. *IEEE Geosci. Remote Sens. Lett.* **2012**, *9*, 649–653. [CrossRef]
16. Fischler, M.A.; Bolles, R.C. Random Sample Consensus: A Paradigm for Model Fitting with Applications to Image Analysis and Automated Cartography. *Commun. ACM* **1981**, *24*, 381–395. [CrossRef]

applied sciences

MDPI

Article

Intelligent Feedback Analysis of Fluid–Solid Coupling of Surrounding Rock of Tunnel in Water-Rich Areas

Tao Zhan [1], Xinping Guo [2,*], Tengfei Jiang [2] and Annan Jiang [2]

1 Nanchang Rail Transit Group Co., Ltd., Nanchang 330013, China
2 Highway and Bridge Institute, Dalian Maritime University, Dalian 116026, China
* Correspondence: guoxinping113@163.com; Tel.: +86-15542519113

Abstract: To realize parameter feedback optimization of tunnel construction in water-rich areas, a feedback analysis method for tunnel parameters under fluid–solid coupling conditions was established based on an intelligent optimization algorithm. Firstly, the numerical calculation model was established and solved using the fluid–solid coupling model. In orthogonal design analysis, the displacement of surrounding rock and pore water pressure distribution in different rock mass parameter combinations were obtained, and the learning samples needed for machine learning were established. The input group was surrounding rock displacement and pore water pressure, and the output was rock mass parameters. Then, the Gaussian process algorithm was used to obtain the nonlinear mapping relationship contained in the learning samples. A differential evolution algorithm was used to optimize the critical parameters involved in this process. Furthermore, according to the established regression model and the measured displacement and pore water pressure in the research area, differential evolution was used again to optimize the rock mass parameters and obtain the parameter feedback analysis results. Finally, the inversion values were compared with the actual measured values, and the reliability of the surrounding rock parameters obtained from the feedback analysis was verified, providing an effective method for obtaining surrounding rock parameters for similar projects.

Keywords: tunnel engineering; fluid–solid coupling; intelligent feedback analysis; Gaussian process; difference evolution algorithm; water-rich tunnel

Citation: Zhan, T.; Guo, X.; Jiang, T.; Jiang, A. Intelligent Feedback Analysis of Fluid–Solid Coupling of Surrounding Rock of Tunnel in Water-Rich Areas. *Appl. Sci.* **2023**, *13*, 1479. https://doi.org/10.3390/app13031479

Academic Editor: Wei Gao

Received: 22 November 2022
Revised: 17 January 2023
Accepted: 19 January 2023
Published: 22 January 2023

1. Introduction

The interaction of water and rock in a tunnel constitutes a complex geological system. Due to the discontinuity and heterogeneity of rock and the singleness and randomness of laboratory tests, it is more difficult to obtain accurate surrounding rock parameters in tunnel design and numerical simulation. The displacement back analysis method based on an intelligent algorithm can better solve the above problems [1–4]. In 1971, Kavanagh et al. [5] proposed a method of back analysis of elastic modulus. Lu et al. [6] back-analyzed the surrounding rock elastic modulus, Poisson's ratio and other stratum parameters through actual deformation monitoring data. In the process of back analysis, the selection of a reasonable, intelligent algorithm is helpful in improving the inversion accuracy and efficiency of parameters.

Artificial neural network (ANN), genetic algorithm (GA) and particle swarm optimization (PSO) have been widely used in parameter back analysis [7–12]. Feng et al. [13] combined ANN and GA to form an evolutionary neural network method to identify surrounding rock parameters. Deng et al. [14] used the BP network and GA to back-analyze the elastic modulus of three different geologies by using slope displacement, which improved the calculation efficiency and overcame the defects of narrow application range and slow convergence speed of traditional optimization algorithms. Zhou [15] constructed a GA-BP intelligent feedback system to predict the parameters of the tunnel-surrounding rock. Wang et al. [16] realized the back analysis of fluid–solid coupling parameters through the hybrid intelligent algorithm of differential evolution algorithm (DE) and PSO. PSO and GA can effectively solve global optimization problems [17]. However, when using a stochastic global optimization algorithm, it is often necessary to evaluate the fitness of a large number of random solutions many times to determine the better solution. The Gaussian process (GP) algorithm is a machine learning regression method developed in recent years. It is mainly based on the statistical theory under the Bayesian framework, and has strong generalization ability and good adaptability in solving small nonlinear sample and high-dimensional regression problems [18,19]. Sun et al. [20] established a probabilistic back analysis method based on Bayesian theory. The research results provide a basis for the establishment of a probabilistic back analysis method of geotechnical engineering parameters. In order to realize the dynamic uncertainty inverse analysis of rock mass parameters with the construction process, Zhang et al. [21] introduced the multi-output support vector machine method and Bayesian theory into the dynamic uncertainty inverse analysis of rock mass parameters. Tao [22] established the probabilistic back analysis and deformation prediction method of rock and soil parameters based on Bayesian theory. Sun [23] studied the geotechnical engineering back analysis method based on multi-objective optimization and Bayesian theory. DE is a global search method based on population, which can evolve the population to the optimal solution through mutation crossover and selection. It has the advantages of fewer control parameters, fast convergence and strong robustness [24,25]. The DE algorithm has been comprehensively developed in recent decades, producing, for example, the DREAM algorithm. Luo et al. [26] studied the identification of the spatial variability of aquifer hydraulic conductivity based on the DREAM algorithm, providing a new idea for the study of spatial variability of aquifer parameters. Yang et al. [27] used the DREAM algorithm to analyze the factors affecting the uncertainty of groundwater numerical simulation. Zhang et al. [28] studied the probabilistic back analysis of soil parameters and displacement prediction of unsaturated slopes using Bayesian updating.

The problem of fluid–structure coupling is a key concern in engineering construction, and back analysis based on the fluid–structure coupling problem has been studied in recent years. Wu et al. [29] proposed a probabilistic back analysis method based on polynomial chaos expansion. Based on stochastic polynomial expansion, the probabilistic back analysis of fluid–structure coupling for an unsaturated soil slope was developed. Wang et al. [30] studied the inversion method of dams' seepage characteristics based on fluid–structure coupling. Based on the fluid–solid coupling theory and Bayesian theory, Zheng et al. [31] established a coupled probabilistic back analysis model for an unsaturated soil slope. A method of multi-objective probabilistic inverse analysis using time-varied data of displacement and pore water pressure was proposed based on Markov chain theory. Xu et al. [32] studied the coupled grouting reinforcement mechanism and displacement back analysis of mechanical parameters of surrounding rock.

In this paper, the GP algorithm and DE algorithm (GP-DE) are introduced in the parameter identification of fluid–solid coupling of the surrounding rock of a tunnel. Firstly, the numerical calculation model was established and solved using the fluid–solid coupling model. In orthogonal design analysis, the displacement of surrounding rock and pore water pressure distribution in different rock mass parameter combinations were obtained, and the learning samples needed for machine learning were established. Then, GP was used to obtain the nonlinear mapping relationship in the learning samples, and DE was used to optimize the critical parameters involved in this process. Furthermore, according to the established regression model and the measured displacement and pore water pressure in the research area, DE was used again to optimize the rock mass parameters and obtain the parameter feedback analysis results. In addition, in order to improve the mapping effect, the super-parameters of the GP model were optimized by the DE algorithm. Then, the trained GP model was integrated into the DE algorithm to identify the parameters of the tunnel-surrounding rock. Finally, the method was applied to the Chenjiadian tunnel in the city of Dalian, China. Through this method, the optimization of tunnel excavation footage is realized, and the construction efficiency is effectively improved on the premise of ensuring safety.

2. The Parameter Identification Method of Fluid–Solid Coupling of Surrounding Rock Based on GP-DE

2.1. The Problem of Parameter Inversion of Fluid–Solid Coupling of Surrounding Rock

The identification of parameters is essentially an optimization problem. The optimization process can be expressed as:

$$\begin{aligned} &\min E(x_1, x_2, \cdots, x_N) = \min\left(\frac{1}{m}\sum_{k=1}^{m}\left|Y_k^0 - Y_k\right|\right) \\ &x_k^a \le x_k \le x_k^b \qquad (k = 1, 2, \cdots, N) \end{aligned} \tag{1}$$

where E represents the mapping function between surrounding rock parameters and tunnel displacement, ${Y_k}^0$ is the field monitoring result of tunnel displacement and Y_k is the tunnel displacement calculated through numerical simulation. m is the number of observed values, x_k is the surrounding rock parameter, N is the number of parameters and ${x_k}^a$ and ${x_k}^b$ are the upper and lower limits.

2.2. The GP Algorithm

In the process of back analysis, the operation process of GP is as follows.

Assume $\mathbf{X} = [x_1, x_2, \ldots, x_n]$ is the $d \times n$ input matrix, and $\mathbf{y} = [y_1, y_2, \ldots, y_n]$ is the output vector, then the training dataset can be expressed as $\{\mathbf{X}, \mathbf{y}\}$; thus, the standard linear regression model with Gaussian white noise can be expressed as:

$$y_i = f(x_i) + \varepsilon \tag{2}$$

where ε denotes an independent random variable, and $\varepsilon \sim N(0, \sigma_n^2)$, while σ_n^2 represents the variance.

The prior distribution of the observed target value $\mathbf{y}$ can be expressed as:

$$\mathbf{y} \sim N\left(0, \mathbf{C} + \sigma_n^2 \boldsymbol{I}\right) \tag{3}$$

where $\mathbf{C} = C(\mathbf{X}, \mathbf{X})$ denotes a symmetric positive definite covariance matrix of the nth order.

For the test sample $(\mathbf{x}^*, \mathbf{y}^*)$, where $\mathbf{x}^* = ({x_1}^*, {x_2}^*, {x_3}^*, \ldots, {x_n}^*)$, $\mathbf{y}^* = ({y_1}^*, {y_2}^*, {y_3}^*, \ldots, {y_n}^*)$, the joint Gaussian prior distribution of $\mathbf{y}$ and $\mathbf{y}^*$ can be obtained and expressed as:

$$\begin{bmatrix} \mathbf{y} \\ \mathbf{y}^* \end{bmatrix} \sim N\left(0, \begin{bmatrix} C(\mathbf{X}, \mathbf{X}) + \sigma_n^2 \boldsymbol{I} & C(\mathbf{X}, \mathbf{x}^*) \\ C(\mathbf{x}^*, \mathbf{X}) & C(\mathbf{x}^*, \mathbf{x}^*) \end{bmatrix}\right) \tag{4}$$

where $C(\mathbf{X},\mathbf{X})$ denotes an $n \times n$ symmetric positive definite covariance matrix, and $\boldsymbol{I}$ represents the identity matrix. $C(\mathbf{X}, \mathbf{x}^*) = C(\mathbf{x}^*, \mathbf{X})^{\mathrm{T}}$ is an $n \times 1$ covariance matrix consisting of new input test points $\mathbf{x}^*$ and all input points; $\mathbf{C}(\mathbf{x}^*, \mathbf{x}^*)$ is the covariance matrix consisting of new input test points $\mathbf{x}^*$.

When the training set $\boldsymbol{D}$ and the input value $\mathbf{x}^*$ of a test sample are known, the GP can use the posterior probability formula to calculate the output value $\mathbf{y}^*$ of the test sample, which can be expressed as:

$$\mathbf{y}^* | \mathbf{x}^*, \boldsymbol{D} \sim N(u_{y^*}, \sigma_{y^*}^2) \tag{5}$$

$$u_{y^*} = C(\mathbf{x}^*, \mathbf{X})\left(C(\mathbf{X}, \mathbf{X}) + \sigma_n^2 \boldsymbol{I}\right)^{-1} \boldsymbol{y} \tag{6}$$

where u_{y^*} and $\sigma_{y^*}^2$ denote the expectation and variance of $\mathbf{y}^*$, respectively.

According to the Gaussian process, the covariance function is used to measure the degree of similarity between the learning sample and the prediction sample. In this case, the covariance function is similar to the kernel function of support vector machine, which plays an important role in Gaussian process machine learning methods. For the rest of the calculation, the constructed covariance function can meet the requirements of symmetry and positive qualitative. In the early trial calculation process, it is determined to choose the square index covariance function, and its prediction effect is better. It is expressed as:

$$k_{se}(x_p, x_q) = \sigma_f^2 \exp\left(-\frac{1}{2J^2}\|x_p - x_q\|^2\right) + \sigma_n^2 \delta_{pq} \tag{7}$$

where x_p and x_q can represent the learning samples, prediction samples or combinations of learning and prediction samples depending on a particular situation; J is the distance correlation between the two data points x_p and x_q; σ_f is the local correlation; σ_n is the standard deviation of the noise; and lastly, δ_{pq} is a sign function. When $p = q$, then $\delta_{pq} = 0$; otherwise, $\delta_{pq} = 1$.

The GP-based surface should be trained by representative data samples before it can map the complex nonlinear relation between the jointed parameters and displacements. The data samples can be obtained by model tests, field tests, numerical simulation and other methods. In this study, the data samples were collected using the orthogonal design, uniform design and numerical simulation. In the GP training process, hyper-parameters σ_f and σ_n affect the GP training effect and prediction accuracy, so this process can be described as an optimization problem, which is expressed as:

$$\min \mathrm{E}(\theta) = \min\left(\sum_{h=1}^{K} \frac{\mathrm{GP}_h(\theta) - Y_h}{Y_h}\right), h = 1, 2, \ldots K \tag{8}$$

where $GP_h(\theta)$ and Y_h denote the estimated output data of the tentative GP and the real output corresponding to the hth test sample. The test sample number is $h = 1, 2, \ldots, K$. $\theta = (\sigma_f, \sigma_n)$ represents the hyper-parametric vector.

2.3. The GP Optimized by DE

During the feedback analysis process, an intelligent optimization algorithm, DE, was used to optimize θ in this study. The basic operations of the algorithm include four steps:

(1) Generating initial population

Generate the initial search point; that is, generate the original population P_G:

$$P_G = \left\{\vec{x}_1, \cdots \vec{x}_i \cdots \vec{x}_{Np}\right\}, i = 1, \cdots, NP \tag{9}$$

where G is evolutionary algebra and NP is the population size, and its value does not change with evolution. Individual B is expressed as:

$$\vec{x}_i = (x_{i,1}, \cdots x_{i,j}, \cdots x_{i,n}), i = 1, \cdots, NP, j = 1, \cdots, n \tag{10}$$

The jth component $x_{i,j}$ of the individual $\vec{x}_i = (x_{i,1}, \cdots x_{i,j}, \cdots x_{i,n})$ in the initial population P_0 is randomly generated in the search space S, where S refers to the boundary constraint condition of the problem to be optimized:

$$\begin{gathered} x_{i,j} = Lbound_j + rand \times (Ubound_j - Lbound_j), \\ i = 1, \cdots, NP, j = 1, \cdots n \end{gathered} \tag{11}$$

where n represents the individual dimension, $Ubound_j$ and $Lbound_j$ represent the upper and lower limits of components, respectively, and rand represents the random number that follows uniform distribution within the range of [0, 1].

(2) Mutation operation

Perform the mutation operation. Two target individuals are taken as a group to generate variation vectors:

$$v_i = ax_1 + bx_2 \tag{12}$$

where a and b are randomly generated weight coefficients, and $a + b = 1$.

(3) Crossover operation

Crossover operation is performed on the variation vector obtained in the previous step and its corresponding target vector, and then the test vector is obtained:

$$u_{i,j} = \begin{cases} v_{i,j},\ for\ j = \langle l \rangle_n, \langle l+1 \rangle_n, \cdots, \langle l+L+1 \rangle_n \\ x_{i,j},\ otherwise \end{cases} \tag{13}$$

where $i = 1..., NP, j = 1,...,n$ is the modulo taking function with modulo n; l is an integer, which is randomly selected and generated in the interval $[1, n]$. L is the number of experimental vectors generated through crossover operations.

(4) Selection operation

The fitness of the test vectors generated by crossover was evaluated and compared with the original vector, and the vectors with better fitness were reserved for entering the new iteration process.

2.4. The Parameters Identification Flowchart

For Equation (1), the Yk can be calculated by the GP model, and then it is expressed as Equation (14). Adopting DE, the parameters of rock mass can be identified.

$$\begin{aligned} &\min E(x_1, x_2, \cdots, x_N) = \min\left(\frac{1}{m}\sum_{j=1}^{m}\left|\mathrm{GP}(x_1, x_2, \cdots, x_N)_j^0 - Y_j\right|\right) \\ &x_k^a \le x_k \le x_k^b \quad (k = 1, 2, \ldots, N) \end{aligned} \tag{14}$$

The process of back analysis of surrounding rock parameters is shown in Figure 1. The specific algorithm is as follows:

(1) Orthogonal samples are obtained by numerical simulation, and learning samples are established according to the samples.
(2) GP is used to learn the rules of learning samples.
(3) The DE method is used to generate the initial population.
(4) The mapping established in step 2 is called to calculate the output variables corresponding to the initial population in step 3.
(5) Compare the calculated results of the previous step with the field-measured results. Enter step 7 when it meets the fitness requirements; otherwise, enter step 6.
(6) Perform the DE optimization operation described above to generate a new initial population, and return to step 4.
(7) Obtain and record the population at this time, and this result is the target parameter of the required back analysis.

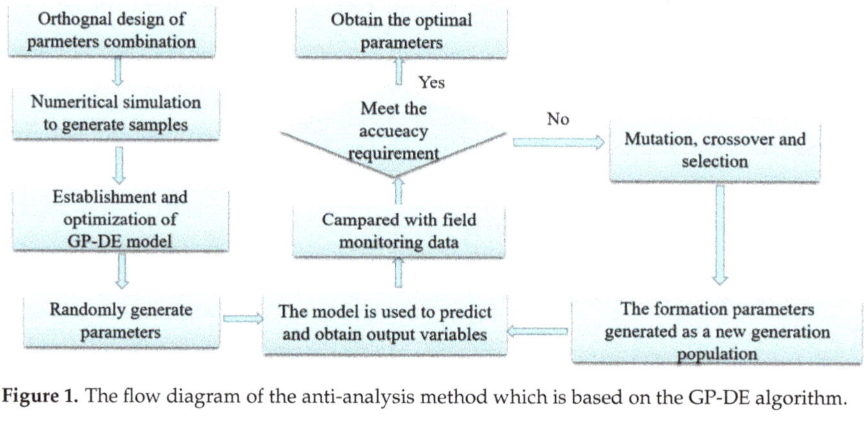

Figure 1. The flow diagram of the anti-analysis method which is based on the GP-DE algorithm.

3. Engineering Application

3.1. Engineering Overview

The Chenjiadian tunnel is 1500 m long, 10.5 m high and 12.7 m wide. From top to bottom, the tunnel geology contains a local surrounding rock fracture zone and abundant groundwater. The coupling action of water and rock and soil reduces the strength of the surrounding rock, which seriously affects the stability of the tunnel (Figure 2). Due to the complexity of geology and the limitations of exploration conditions, it is necessary to use an intelligent algorithm to determine the hydrogeological parameters.

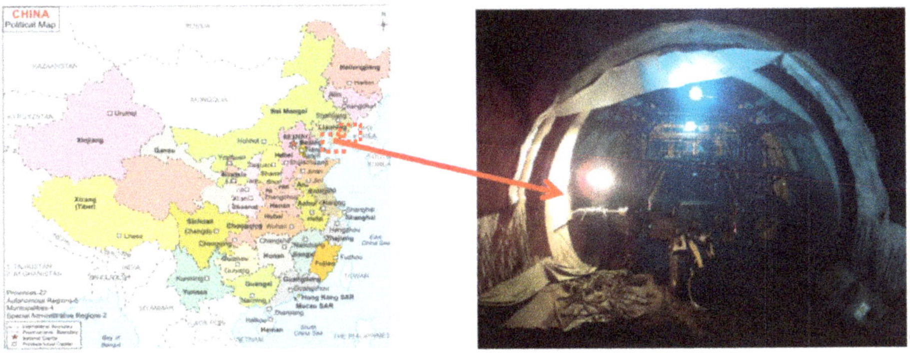

Figure 2. The location of the Chenjiadian tunnel.

3.2. The Principle of Fluid–Solid Coupling Modeling

FLAC3D software conducts fluid–structure coupling calculations based on the finite difference method. It defaults that the rock and soil mass are continuous media. The fluid seepage follows Darcy's law and satisfies Biot's equation, mainly including the following equations:

(1) Equilibrium equation

For small deformation, the fluid particle equilibrium equation is:

$$-q_{i,i} + q_v = \frac{\partial \varsigma}{\partial t} \tag{15}$$

where $q_{\mathrm{i,i}}$ is seepage velocity (m/s); q_v is the volume fluid source intensity (s^{-1}); ς is the change in fluid volume per unit volume of porous media.

$$\frac{\partial \varsigma}{\partial t} = \frac{1}{M}\frac{\partial p}{\partial t} + \alpha\frac{\partial \varepsilon}{\partial t} - \beta\frac{\partial T}{\partial t} \tag{16}$$

where M is the Biot modulus ($\mathrm{N/m^2}$); α is the Biot coefficient; β is the coefficient of thermal expansion ($^{\circ}\mathrm{C}^{-1}$), which considers liquid and solid particles. p is the pore water pressure (Pa); ε is the volume strain; T is the temperature.

The momentum balance equation is:

$$\sigma_{ij,j} + \rho g_i = \rho\frac{\mathrm{d}v_i}{\mathrm{d}t} \tag{17}$$

$$\rho = (1-n)\rho_{\mathrm{s}} + n\rho_{\mathrm{w}} \tag{18}$$

Among them, ρ is the bulk density ($\mathrm{kg/m^3}$); ρ_{s} and ρ_{w} are the density of solid and liquid, respectively; n is porosity; g_{i} is the component of gravity acceleration ($\mathrm{m/s^2}$); v_{i} is the velocity component of the medium (m/s).

$$q_i = -k\left[p - \rho_{\mathrm{f}} x_j g_j\right] \tag{19}$$

where k is the permeability coefficient of the medium ($m^2/(Pa{\cdot}s)$); ρ_f is the fluid density (kg/m^3); g_i is the component of gravity acceleration (m/s^2).

(2) Constitutive equation

The volume strain and the pore pressure of the fluid interact with each other. The change in the strain makes the pore pressure readjust, and the change in the pore pressure also affects the occurrence of the strain. The descriptive equation is:

$$\Delta\sigma_{ij} + \alpha\Delta p\delta_{ij} = H_{ij}(\sigma_{ij}, \Delta\xi_{ij}) \tag{20}$$

Among them, $\Delta\sigma_{ij}$ is the stress increment; Δp is the pore water pressure increment; δ_{ij} is the Kronecher factor. H_{ij} is the given function. $\Delta\xi_{ij}$ is the total strain increment.

(3) Compatibility equation

The relationship between strain rate and velocity gradient is:

$$\varepsilon_{ij} = \frac{1}{2}(v_{i,j} + v_{j,i}) \tag{21}$$

where v is the velocity of a point in the medium (m/s).

(4) Boundary condition

There are four types of boundary conditions in seepage calculation: (1) the given pore water pressure, (2) the given velocity vector outside the boundary normal direction, (3) the impervious boundary given by default in the program and (4) the pervious boundary. The form of pervious boundary is as follows:

$$q_n = h(p - p_e) \tag{22}$$

where q_n is the velocity component in the normal direction outside the boundary, h is the leakage coefficient ($m^3/(N{\cdot}s)$) and p_e is the pore water pressure at the seepage outlet.

(5) Time scale

The fluid and mechanics processes are involved in the fluid–structure coupling calculation, and the time scales in these two states need to be considered. The characteristic time can generally characterize the size of the time scale. The characteristic time of the mechanical process is expressed as follows:

$$t_c^m = \sqrt{\frac{\rho}{K_u + 4/3G}} L_c \tag{23}$$

where K_u is the undrained bulk modulus; G is the shear modulus; ρ is the density; L_c is the feature length (the average size of the model). The characteristic time of the fluid diffusion process is defined as:

$$t_c^f = \frac{L_c^2}{c} \tag{24}$$

where L_c is the characteristic length of seepage (the average size of seepage path in the model) and c is the diffusion rate, defined as the ratio of permeability coefficient to water storage coefficient:

$$c = \frac{k}{S} \tag{25}$$

3.3. Numerical Simulation Model

FLAC3D software was used for the numerical simulation of excavation and support. The model included the embankment of the He-Da Expressway with a slope of 1:1.5 (Figure 3a). At the bottom of the model were X, Y and Z constraints, and around them were normal constraints. The model consisted of 67,336 nodes and 62,784 elements, and the

Mohr–Coulomb yield criterion was adopted. Shell and cable structural elements simulated the primary support and bolt, respectively. Groundwater was located 6 m above the tunnel roof arch.

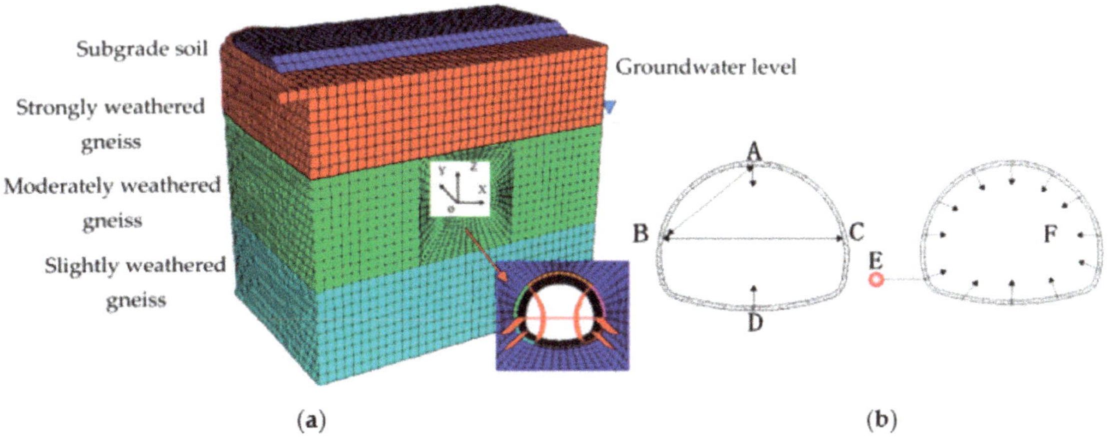

Figure 3. The numerical calculation model of the Chenjiadian tunnel. (**a**) Numerical model. (**b**) Distribution of monitoring points.

In the early stages of construction, the rock and soil masses within the construction range of the station are drilled and sampled. The geological parameters of the station are investigated and the stratigraphic parameters are obtained. However, there are some limitations in the geological surveys. The process of geological surveying is interfered with by many factors. The rock mass and overlying rock mass parameters of the station are essential references in the construction process. The main body of the tunnel in the study area is in moderately weathered gneiss, and the strongly weathered gneiss above the tunnel affects the stability of the tunnel to a certain extent. Therefore, the relevant parameters of moderately weathered gneiss and strongly weathered gneiss are mainly identified.

The elastic modulus (E1) and Poisson's (μ1) ratio of moderately weathered gneiss and the elastic modulus (E2) and Poisson's ratio (μ2) of strongly weathered gneiss were selected as the back analysis parameters. In addition, the permeability coefficients of moderately weathered gneiss (K1) and the permeability coefficients of strongly weathered gneiss (K2) that are difficult to measure were also added as the back analysis parameters. Other mechanical parameters are shown in Table 1. Input values were tunnel deformation value, pore water pressure and water inflow, and output values were surrounding rock parameters.

The arch crown settlement AZ, arch bottom uplift DZ, arch waist convergence BC and the relative displacement AB of measuring points A and B were taken as the displacement monitoring values (Figure 3b). The pore water pressure P at point E and the unit seepage volume F of the tunnel were taken as the seepage monitoring values. Among them, point E was 1 m away from the arch foot of the tunnel, the pore water pressure P was measured by the pore water pressure gauge and the unit seepage F of the tunnel was calculated by dividing the sum of the seepage of all outlet points in a particular mileage section of the tunnel by the length.

According to relevant specifications for tunnel engineering design and geological survey data, the value ranges of six parameters were as follows: E1 is 2.85 GPa~6.53 GPa; μ1 is 0.21~0.41; E2 is 1.81 GPa~3.49 GPa; μ2 is 0.25~0.45; K1 is 0.196 m/d~0.372 m/d; K2 is 0.4 m/d~0.78 m/d. The orthogonal design scheme and uniform design scheme were established through these six parameters for numerical calculation. The calculation results are shown in Table 2 (training samples) and Table 3 (test samples).

Table 1. The parameters of calculation.

	Elastic Modulus/GPa	Poisson's Ratio	Cohesion /kPa	Internal Friction Angle/°	Permeability Coefficient/(m/d)
Slightly weathered gneiss	4.19	0.26	27	38	0.025
Moderately weathered gneiss	——	——	21	42	——
Strongly weathered gneiss	——	——	15	46	——
Subgrade soil	0.15	0.35	23	19	0.843
Primary support	25	0.18	20,000	34	6.3×10^{-4}
Bolt	200	——	——	25	——
Middle wall	25	0.18	20,000	34	——

Table 2. Tunnel-surrounding rock parameters' orthogonal scheme and the calculation results.

	E1 (GPa)	μ1	E2 (GPa)	μ2	K1 (m/d)	K2 (m/d)	AZ (mm)	AB (mm)	BC (mm)	DZ (mm)	P (10^5) (Pa)	F(m^3/m × d)
1	2.85	0.21	1.81	0.25	0.196	0.4	5.961	3.64	0.355	1.313	1.337	9.07
2	3.52	0.26	1.81	0.3	0.24	0.495	5.445	3.481	0.291	1.186	1.264	8.866
3	4.19	0.31	1.81	0.35	0.284	0.59	5.002	3.342	0.763	1.097	1.16	5.34
4	5.86	0.36	1.81	0.4	0.328	0.685	4.513	3.224	1.425	0.982	1.026	3.1
5	6.53	0.41	1.81	0.45	0.372	0.78	4.12	3.175	2.149	1.015	0.84	3.04
6	4.19	0.21	2.23	0.3	0.328	0.78	5.193	3.471	0.165	1.496	1.348	4.31
7	5.86	0.26	2.23	0.35	0.372	0.4	4.717	3.288	0.002	0.921	1.311	3.74
8	6.53	0.31	2.23	0.4	0.196	0.495	4.431	3.202	0.753	0.883	1.162	3.52
9	2.85	0.36	2.23	0.45	0.24	0.59	4.362	3.234	1.035	1.656	1.045	3.33
10	3.52	0.41	2.23	0.25	0.284	0.685	4.773	3.167	1.696	0.706	0.721	6.34
11	6.53	0.21	2.65	0.35	0.24	0.685	4.598	3.306	0.018	0.9	1.358	3.23
12	2.85	0.26	2.65	0.4	0.284	0.78	4.819	3.367	0.012	1.494	1.231	3.38
13	3.52	0.31	2.65	0.45	0.328	0.4	4.283	3.209	0.511	1.414	1.217	3.45
14	4.19	0.36	2.65	0.25	0.372	0.495	4.663	3.135	0.941	0.66	0.882	6.4
15	5.86	0.41	2.65	0.3	0.196	0.59	4.205	3.049	1.57	0.613	0.72	3.79
16	3.52	0.21	3.07	0.4	0.372	0.59	4.639	3.343	0.233	1.318	1.355	3.37
17	4.19	0.26	3.07	0.45	0.196	0.685	4.24	3.232	0.151	1.287	1.283	3.18
18	5.86	0.31	3.07	0.25	0.24	0.78	4.462	3.143	0.48	0.644	0.977	3.35
19	6.53	0.36	3.07	0.3	0.284	0.4	4.184	3.039	0.927	0.572	0.895	3.89
20	2.85	0.41	3.07	0.35	0.328	0.495	4.469	3.097	1.261	0.959	0.749	7.38
21	5.86	0.21	3.49	0.45	0.284	0.495	4.123	3.2	0.03	1.028	1.398	4.5
22	6.53	0.26	3.49	0.25	0.328	0.59	4.38	3.156	0.151	0.642	1.114	3.36
23	2.85	0.31	3.49	0.3	0.372	0.685	4.857	3.226	0.227	0.89	0.95	6.56
24	3.52	0.36	3.49	0.35	0.196	0.78	4.39	3.111	0.692	0.88	0.829	3.86
25	4.19	0.41	3.49	0.4	0.24	0.4	3.959	3.004	1.214	0.908	0.761	6.17

Table 3. Uniform parameter test methods and the results of numerical calculation.

	E1 (GPa)	μ1	E2 (GPa)	μ2	K1 (m/d)	K2 (m/d)	AZ (mm)	AB (mm)	BC (mm)	DZ (mm)	P (10^5) (Pa)	F (m^3/m × d)
1	4.19	0.26	1.81	0.035	0.372	0.875	5.053	3.352	1.175	1.164	1.090	6.335
2	5.86	0.36	2.23	0.20	0.284	0.780	4.374	3.136	0.447	0.797	0.861	4.291
3	3.52	0.46	2.65	0.40	0.196	0.685	4.049	3.032	0.698	0.621	0.750	3.310
4	6.53	0.21	3.07	0.25	0.416	0.590	4.503	3.177	0.586	0.867	0.904	4.680
5	2.85	0.31	3.49	0.45	0.328	0.495	4.249	3.096	0.313	0.729	0.818	3.915

3.4. Parameters Identification Results

During the GP-DE feedback analysis process, the population size NP was 100, the variation factor F was 0.7, the cross factor CR was 0.9, the maximum evolutionary algebra was 200 and SE was selected as the kernel function. After tunnel excavation, the measured values of AZ, DZ, BC, AB, P and F were 8 mm, 3.17 mm, 3.02 mm, 6.31 mm, 0.122 MPa and 3.77 m3/m × d, respectively. The optimal parameters obtained by back analysis were E1 = 2.83 GPa, μ1 = 0.33, E2 = 1.24 GPa, μ2 = 0.36, K1 = 0.285 m/d and K2 = 0.658 m/d.

According to Table 4, compared with the numerical simulation results, the maximum relative error of the back analysis is 9.40%, which meets the requirements of engineering construction. This back analysis method can be used for surrounding rock parameter prediction.

Table 4. Parameters of surrounding rock of uniform testing scheme analysis results.

	Back Analysis Result						Relative Error					
	E1 (GPa)	μ1	E2 (GPa)	μ2	K1 (m/d)	K2 (m/d)	E1 (%)	μ1 (%)	E2 (%)	μ2 (%)	K1 (%)	K2 (%)
1	3.85	0.29	1.81	0.33	0.36	0.82	0.00	−9.40	8.74	6.81	4.49	7.24
2	6.96	0.35	2.39	0.22	0.26	0.78	−6.69	3.79	3.39	−8.19	9.15	0.00
3	3.37	0.47	2.49	0.37	0.20	0.73	6.43	−1.72	4.54	7.50	0.00	−6.16
4	6.19	0.23	3.26	0.27	0.45	0.58	−5.80	−8.41	5.56	−8.45	−7.01	2.02
5	3.07	0.34	3.49	0.41	0.33	0.52	0.00	−7.98	−7.14	9.22	−1.60	−4.81
6	6.43	0.39	3.74	0.28	0.25	0.38	4.55	6.33	−8.85	7.14	−4.57	5.26

Figure 4 shows the fitness values in the iterative process. It can be seen from the figure that with the increase in evolutionary algebra, the distribution of solution vector in space tends to converge. The iterative process of parameter acquisition is shown in Figure 5. In the initial stage of iteration, the fluctuation range of parameters is extensive. When the number of iterations reaches 45, the obtained parameters no longer fluctuate and the optimal solution is generated.

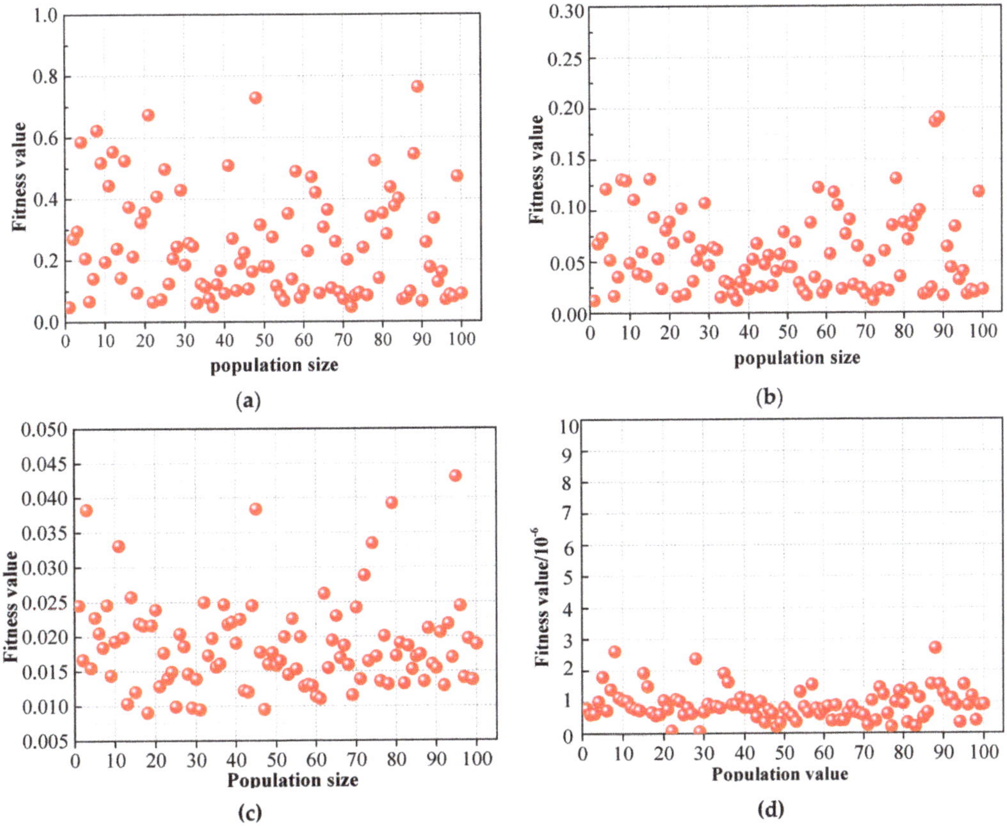

Figure 4. Fitness values with different evolutions. (**a**) 1st evolution. (**b**) 15th evolution. (**c**) 30th evolution. (**d**) 45th evolution.

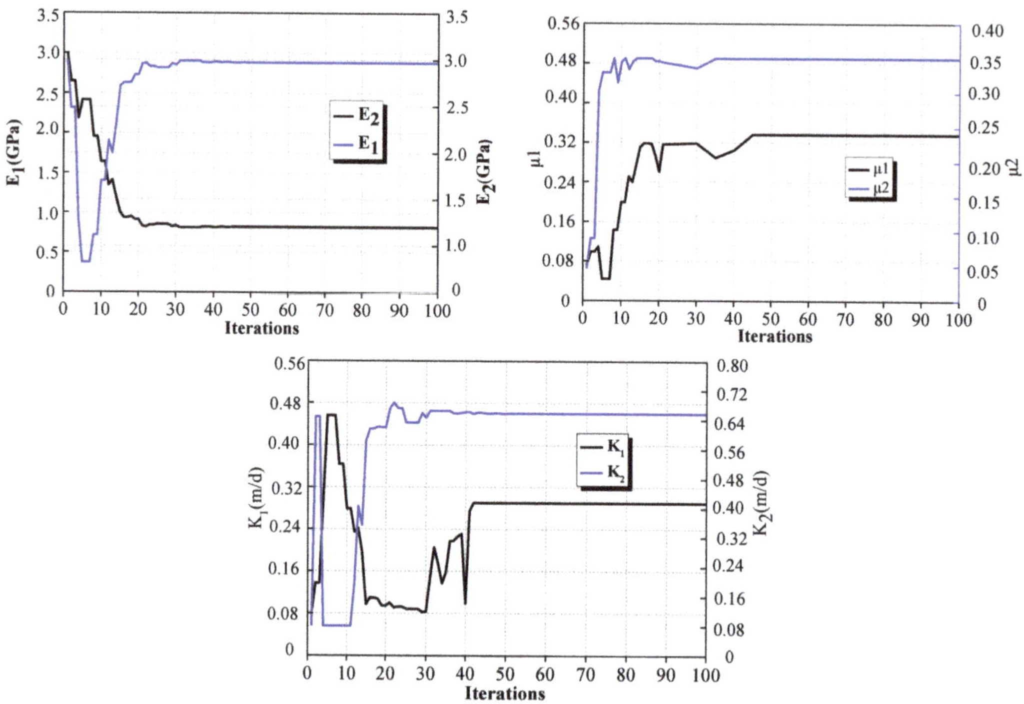

Figure 5. Variation in the recognized parameters.

3.5. Analysis of Tunnel Excavation Footage Based on Back Analysis Results

Based on the back analysis results, different amounts of excavation footage of the tunnel were selected to analyze their impact on the tunnel. The optimal amounts of excavation footage under four working conditions were selected by analyzing the distribution of the plastic zone. The distribution of the plastic zone under four cyclic excavation footage conditions is shown in Figure 6.

It can be seen from Figure 6 that the plastic zones in the four cyclic excavation footage conditions are distributed differently. Due to the reinforcement effect of the advanced small pipe, the plastic zone of the arch crown is reduced, and the plastic zone is mainly distributed on both sides of the arch foot and arch shoulder. With the increase in excavation footage, the pressure release from surrounding rock also increases, so the area of the plastic zone increases obviously. The area of the plastic zone is the largest under the condition of 2.5 m excavation footage. Considering the actual situation of the project, the excavation footage of 1.5 m should be selected for excavation. At the same time, grouting reinforcement should be strengthened on both sides of the arch foot and the arch shoulder of the tunnel to ensure construction safety.

According to the analysis results, the excavation footage of the construction site was determined to be 1.5 m. Figure 7 shows the site construction condition of the tunnel when it is constructed according to the excavation footage of 1.5 m. In the construction process, the surrounding rock of the tunnel is relatively stable and the construction environment is safe. The monitoring data of the arch settlement change obviously in the early stage of excavation, and gradually tend to be stable in the later stage. The arch settlement value is always within the monitoring control range in the monitoring process. In the construction process, the original 1.0 m excavation footage is adjusted to 1.5 m, which effectively improves the construction efficiency.

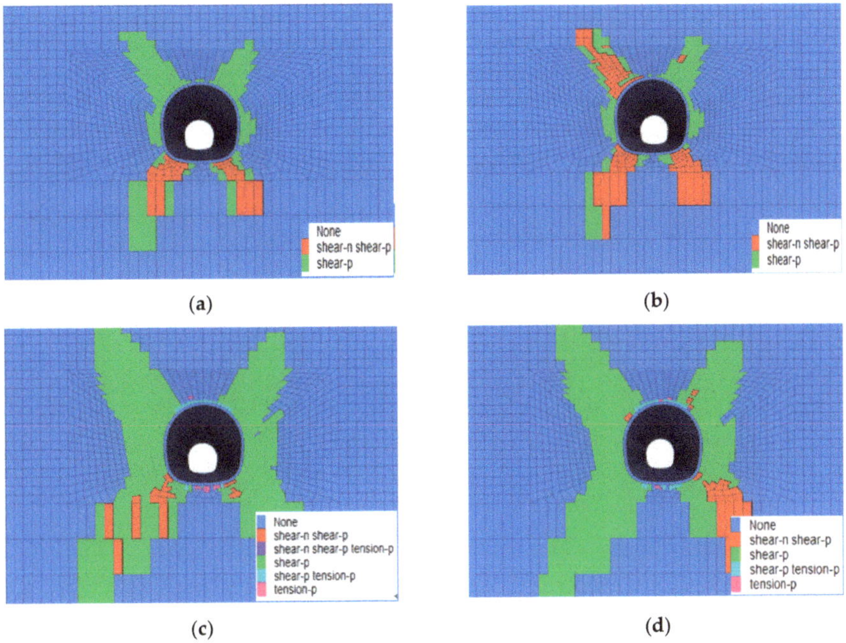

Figure 6. Plastic differentiation layout under various working conditions. (**a**) 1 m. (**b**) 1.5 m. (**c**) 2.0 m. (**d**) 2.5 m.

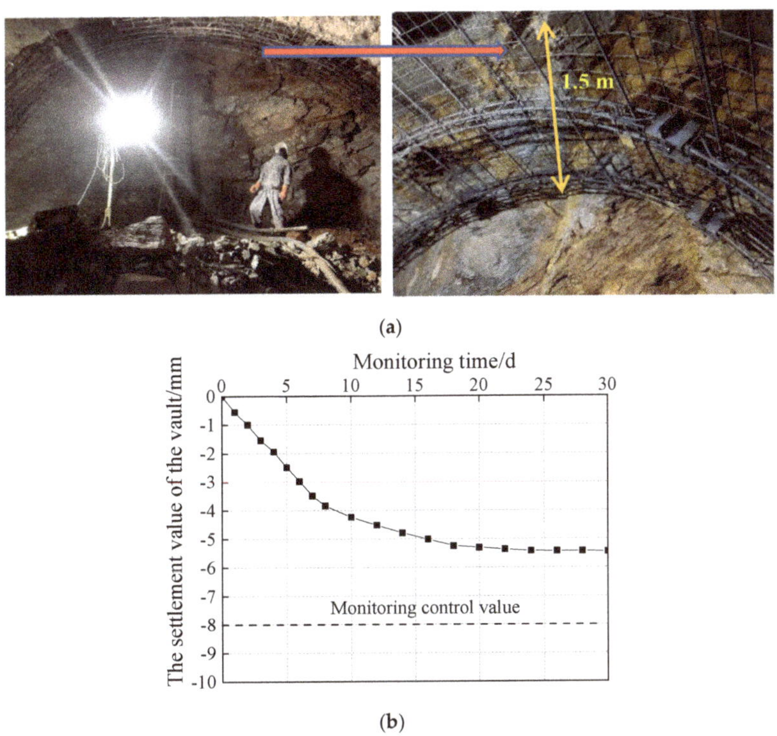

Figure 7. Optimization of tunneling excavation footage. (**a**) Optimized results of tunnel excavation at construction site. (**b**) Field monitoring data of vault settlement.

Figure 8 shows the results of vault settlement under different working conditions. In the simulation process, without considering the fluid–structure coupling, the arch settlement result caused by tunnel excavation is 4.26 mm, and when considering the fluid–structure coupling, the simulation result shows that the arch settlement is 5.32 mm. In the actual construction process, the maximum settlement of the vault is 5.45 mm. Therefore, when tunnel construction is carried out in water-rich areas, the numerical simulation results are closer to the actual on-site construction conditions when considering the fluid–structure coupling. Thus, the impact of groundwater on construction cannot be ignored. In the actual construction process, the corresponding water stop measures should be taken to ensure the safety of construction.

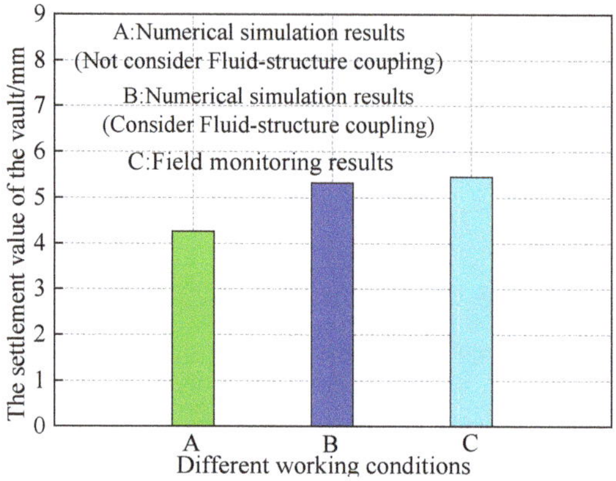

Figure 8. The vault settlement under different working conditions.

4. Discussion

4.1. The Influence of GP Parameters on the Results of Back Analysis

σ_f and σ_n are important super-parameters of the GP model. Figure 9 shows the prediction accuracy under different parameters. When $\ln\sigma_f$ = 3.56 and $\ln\sigma_n$ = 8.72, the relative error is 3.56%; thus, the accuracy of prediction is affected by the parameter selection. Therefore, in the back analysis of parameters, choosing the appropriate parameters for GP is important.

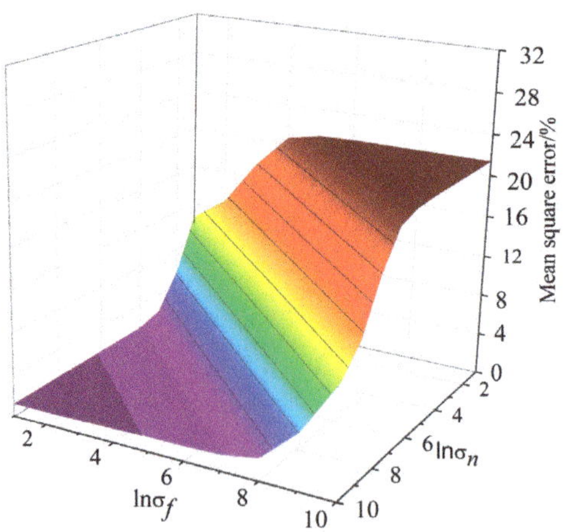

Figure 9. The influence of GP parameters on the prediction accuracy.

4.2. The Influence of DE Parameters

DE is more complex and involves many influencing factors in the GP-DE algorithm. F, CR, N and other difference strategies perhaps have an impact on the convergence speed.

The DE/Best/1 difference strategy and NP = 100 were selected, with F = 0.6 and CR between 0.5 and 0.9. There was a difference in convergence speed in the process of optimization. When CR = 0.9, the number of iterative steps required to achieve convergence is the lowest. Selecting CR as 0.9 and F as 0.5~0.9, the convergence rate is the fastest when F = 0.7. It is shown that the appropriate initial parameters can improve the convergence speed. CR = 0.9 and F = 0.7 were selected for this study (Figure 10).

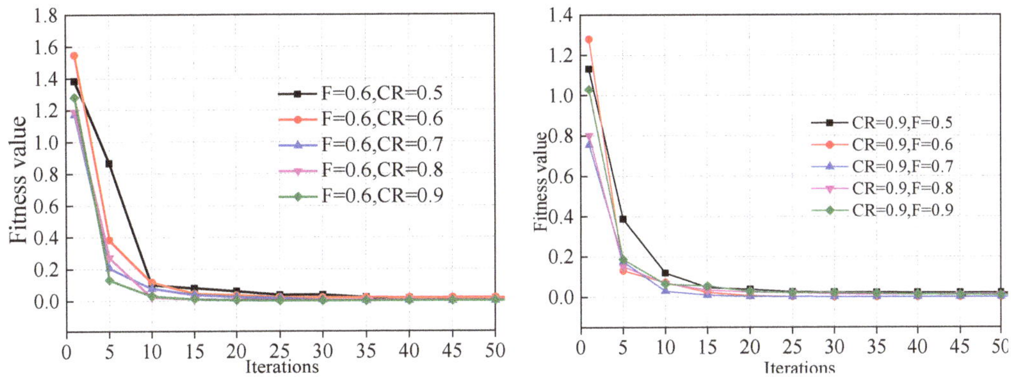

Figure 10. Iteration curve.

The DE/Best/1 difference strategy was selected, with CR = 0.9 and F = 0.7, and NP changed. As seen from Figure 11, when the population size reaches 100, the precision of parameter optimization no longer changes significantly with the increase in population.

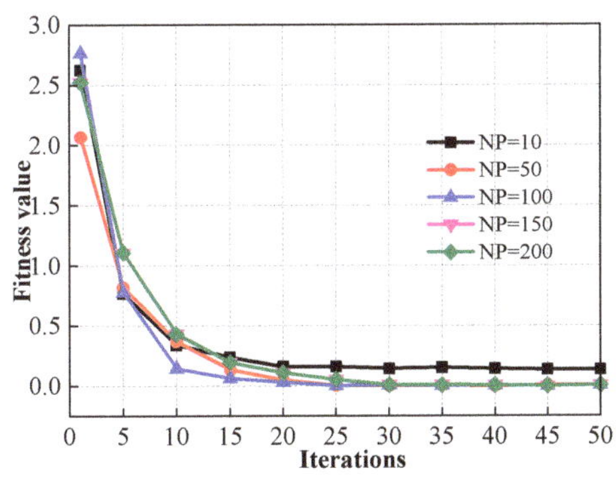

Figure 11. Iteration curve of different populations.

In the DE algorithm, there are various difference strategies, shown in Equation (26). Selecting F = 0.7, CR = 0.9 and NP = 100, the different strategies are compared. It can be seen from Figure 12, compared with other strategies, that DE/Best/1 is the best strategy for optimization.

$$\begin{cases} DE/rand/1: v_{i,g} = x_{r1,g} + F(x_{r2,g} - x_{r3,g}) \\ DE/best/1: v_{i,g} = x_{best,g} + F(x_{r2,g} - x_{r3,g}) \\ DE/rand/2: v_{i,g} = x_{r1,g} + F(x_{r2,g} - x_{r3,g} + x_{r4,g} - x_{r5,g}) \\ DE/best/2: v_{i,g} = x_{best,g} + F(x_{r1,g} - x_{r2,g} + x_{r3,g} - x_{r4,g}) \\ DE/rand-to-best/2: v_{i,g} = x_{r1,g} + F\left(x_{best,g} - x_{r2,g} + x_{r3,g} - x_{r4,g}\right) \end{cases} \quad (26)$$

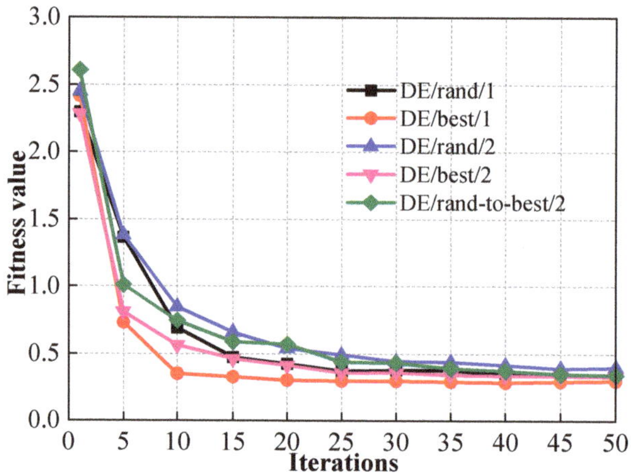

Figure 12. Various difference strategies.

5. Conclusions

Through back analysis of the surrounding rock parameters of the Chenjiadian tunnel and with the numerical calculation of the fluid–structure interaction, the following conclusions are obtained:

1. To realize parameter feedback optimization of tunnel construction in water-rich areas, a feedback analysis method of tunnel parameters under fluid–solid coupling conditions based on GP and DE was established based on an intelligent optimization algorithm.
2. Choosing the appropriate parameters of GP by DE is important to improve the accuracy of the back analysis results. The variation parameters of DE have an impact on the convergence speed. CR = 0.9, F = 0.7, N = 100 and the difference strategy DE/Best/1 were selected for this study.
3. The optimal hydrogeological parameters of the surrounding rock were obtained by a back analysis algorithm based on GP-DE. The optimal parameters from back analysis are E1 = 2.83 GPa, μ1 = 0.33, E2 = 1.24 GPa, μ2 = 0.36, K1 = 0.285 m/d and K2 = 0.658 m/d, providing an effective method for obtaining the surrounding rock parameters of similar projects.
4. Based on the back analysis results, different amounts of excavation footage of the tunnel were selected to analyze their impact on the tunnel. The optimal excavation footage under four working conditions was selected by analyzing the distribution of the plastic zone.

Author Contributions: Conceptualization, T.Z., T.J. and X.G.; methodology, T.Z., T.J. and A.J.; software, X.G., T.J. and A.J.; formal analysis, T.Z. and A.J.; resources, T.J. and A.J.; data curation, T.Z., T.J. and X.G.; writing—original draft preparation, T.Z., X.G. and A.J. All authors have read and agreed to the published version of the manuscript.

Funding: This work was supported by the National Natural Science Foundation of China (no. 52078093) and the Cultivation Program for the Excellent Doctoral Dissertation of Dalian Maritime University (no. 2022YBPY009).

Institutional Review Board Statement: Not applicable.

Informed Consent Statement: Informed consent was obtained from all subjects involved in the study.

Conflicts of Interest: The authors declare that they have no conflict of interest regarding the publication of this article.

References

1. Liang, Z.H.; Zhang, Y.B.; Tang, S.B. Size effect of rock messes and associated representative element properties. *Chin. J. Rock Mech. Eng.* **2013**, *32*, 1157–1166.
2. Xu, K.; Shen, C.H.; Hu, Y.T. The Optimization Research on the Foundation Treatment of Sand Drain on Vietnam Highway. *Highw. Eng.* **2017**, 240–243.
3. Shao, Y.; Macari, E. Information Feedback Analysis in Deep Excavations. *Int. J. Geomech.* **2008**, *8*, 91–103. [CrossRef]
4. Zeng, Y.Z.; Hu, L.; Huang, M. Research of Surrounding Rock Parametric Inversion on Mountain Tunnel. *Wuhan Ligong Daxue Xuebao (Jiaotong Kexue Yu Gongcheng Ban)/J. Wuhan Univ. Technol. (Transp. Sci. Eng.)* **2018**, 72–76.
5. Kavanagh, K.T.; Clough, R.W. Finite element applications in the characterization of elastic solids. *Int. J. Solids Struct.* **1971**, *7*, 11–23. [CrossRef]
6. Lu, F.; Yuan, B.Y.; Yuan, L. Rock parameters inversion for estimating the maximum heights of two failure zones in overburden strata of a coal seam. *Min. Sci. Technol. (China)* **2011**, *21*, 41–47.
7. Ge, H.W.; Liang, Y.C.; Liu, W. Applications of artificial neural networks and genetic algorithms to rock mechanics. *Chin. J. Rock Mech. Eng.* **2004**, 1542–1550.
8. Zhang, X.; Yang, F.; Ren, X.H.; Zhang, D.F. Back Analysis of Surrounding Rock Mechanical Parameters of Underground Caverns Based on Genetic Algorithm. *Mod. Tunn. Technol.* **2018**, 53–58.
9. Zhang, T.J.; Xu, L. Inversion of Mechanical Parameters of Slope Rock Based on Asynchronous Particle Swarm Optimization. *J. China Three Gorges Univ.* **2014**, 38–41.
10. Goh, Y.; Zhang, R.; Zhang, W.; Zhang, Y. Evaluating stability of underground entry-type excavations using multivariate adaptive regression splines and logistic regression. *Tunn. Undergr. Space Technol.* **2017**, *70*, 148–154. [CrossRef]
11. Lü, Q.; Chan, C.; Low, B.K. Probabilistic evaluation of ground-support interaction for deep rock excavation using artificial neural network and uniform design. *Tunn. Undergr. Space Technol.* **2012**, *32*, 1–18. [CrossRef]
12. Sarkar, K.; Tiwary, A.; Singh, T.N. Estimation of strength parameters of rock using artificial neural networks. *Bull. Eng. Geol. Environ.* **2010**, *69*, 599–606. [CrossRef]
13. Feng, X.T.; Zhang, Z.Q.; Yang, C.X. Study on genetic-neural net work method of displacement back analysis. *Chin. J. Rock Mech. Eng.* **1999**, *18*, 529–533.
14. Deng, J.; Lee, C.F.; Ge, X. Application of BP networkand genetic algorithm to displacement back analysis of rock slopes. *Chin. J. Rock Mech. Eng.* **2001**, *20*, 1–5.
15. Zhou, G.N.; Sun, Y.Y. Application of Genetic Algorithm Based BP Neural Network to Parameter Inversion of Surrounding Rock and Deformation Prediction. *Mod. Tunn. Technol.* **2018**, *55*, 107–113.
16. Wang, J.X.; Dong, J.H.; Chen, S.L. Multi-parameters Inversion of Stress-Seepage-Damage Coupling Model Based on DEPSO Intelligent Algorithm. *J. Basic Sci. Eng.* **2018**, 872–887.
17. Li, Y.G.; Jiao, P.P.; Qiao, W.D. Prediction of Steering Behaviors on Curves Based on BP Neural Network Optimized by Modified PSO. *J. Highw. Transp. Res. Dev.* **2019**, *36*, 128–136.
18. Rasmussen, C.E.; Williams, C.K.I. *Gaussian Processes for Machine Learning*; MIT Press: Cambridge, MA, USA, 2006.
19. Li, S.C.; Zhao, Y.; Xu, B.S.; Li, L.P.; Liu, Q.; Wang, Y.K. Study of determining permeability coefficient in water inrush numerical calculation of subsea tunnel. *Rock Soil Mech.* **2012**, *33*, 1497–1504.
20. Sun, Q.C.; Li, S.J.; Zhao, H.B.; Zheng, M.Z.; Yang, Z.Y. Probabilistic back analysis of rock mechanical parameters based on displacement and relaxation depth. *Chin. J. Rock Mech. Eng.* **2019**, *38*, 1884–1894.
21. Zhang, Y.X.; Hou, Z.J. Back analysis of rock mass parameters based on support vector machine and Bayesian method. *Yangtze River* **2022**, 186–192.
22. Zhang, Y.X. *Probabilistic Inverse Analysis of Geotechnical Parameters and Deformation Prediction based on Bayesian Theory*; Zhejiang University: Hangzhou, China, 2022.
23. Sun, Y. *Study on Back Analysis Method Based on Multi-Objective Optimization and Bayesian Theory for Geotechnical Engineering*; Wuhan University: Wuhan, China, 2019.
24. Storn, R.; Price, K. Differential Evolution—A Simple and Efficient Heuristic for global Optimization over Continuous Spaces. *J. Glob. Optim.* **1997**, *11*, 341–359. [CrossRef]
25. Wang, L.; Li, L. An effective differential evolution with level comparison for constrained engineering design. *Struct. Multidiscip. Optim.* **2009**, *41*, 947–996. [CrossRef]
26. Luo, Q.K.; Wu, J.F.; Yang, Y.; Wu, J.C.; Ma, S.F. Identification of the spatial variability of aquifer hydraulic conductivity. *J. Nanjing Univ. (Nat. Sci.)* **2016**, 448–455.
27. Yang, Y.; Wu, J.C.; Luo, Q.K.; Wu, J.F. Analysis of the Uncertainty of Groundwater Numerical Simulation Based on the DREAM Algorithm. *Geol. Rev.* **2016**, *62*, 353–361.
28. Zhang, W.G.; Gu, H.L.; Zhang, Q.; Wang, L.; Wang, L.Q. Probabilistic back analysis of soil parameters and displacement prediction of unsaturated slopes using Bayesian updating. *Rock Soil Mech.* **2022**, 1112–1122.
29. Wu, F.; Zhang, L.L.; Zheng, W.T.; Wei, X. Probabilistic back analysis method for unsaturated soil slopes with fluid-solid coupling process based on polynomial chaos expansion. *Chin. J. Geotech. Eng.* **2018**, 2215–2222.
30. Wang, F.M.; Miao, L.; Guo, X.M. Inverse analysis on seepage of dams based on fluid-solid coupling. *J. Hydroelectr. Eng.* **2008**, *27*, 60–64.

31. Zheng, Y.F.; Zhang, L.L.; Zhang, J.; Zhang, J.G.; Yu, Y.T. Multi-objective probabilistic inverse analysis of rainfall-induced landslide based on time-varied data. *Rock Soil Mech.* **2017**, *38*, 3371–3384.
32. Xu, C.Y.; Han, L.J.; Tian, M.L.; Wang, Y.J. Coupled Grouting Reinforcement Mechanism and Displacement Back Analysis of Mechanical Parameters of Surrounding Rock. *Saf. Coal Mines* **2020**, 155–160.

applied sciences

MDPI

Article

Assessment of the Rock Elasticity Modulus Using Four Hybrid RF Models: A Combination of Data-Driven and Soft Techniques

Chuanqi Li and Daniel Dias *

Laboratory 3SR, CNRS UMR 5521, Grenoble Alpes University, 38000 Grenoble, France
* Correspondence: daniel.dias@univ-grenoble-alpes.fr

Abstract: The determination of the rock elasticity modulus (EM) is an indispensable key step for the design of rock engineering problems. Traditional experimental analysis can accurately measure the rock EM, but it requires manpower and material resources, and it is time consuming. The EM estimation of new rocks using former published empirical formulas is also a possibility but can be attached of high uncertainties. In this paper, four types of metaheuristic optimization algorithms (MOA), named the backtracking search optimization algorithm (BSA), multi-verse optimizer (MVO), golden eagle optimizer (GEO) and poor and rich optimization algorithm (PRO), were utilized to optimize the random forest (RF) model for predicting the rock EM. A data-driven technology was used to generate an integrated database consisting of 120 rock samples from the literature. To verify the predictive performance of the proposed models, five common machine-learning models and one empirical formula were also developed to predict the rock EM. Four popular performance indices, including the root-mean-square error (RMSE), mean absolute error (MAE), the coefficient of determination (R^2) and Willmott's index (WI), were adopted to evaluate all models. The results showed that the PRO-RF model has obtained the most satisfactory prediction accuracy. The porosity (Pn) is the most important variable for predicting the rock EM based on the sensitive analysis. This paper compares the performance of the RF models optimized by using four MOA for the rock EM prediction. It provides a good example for the subsequent application of soft techniques on the EM and other important rock parameter estimations.

Keywords: elasticity modulus; rock materials; data-driven; soft techniques; poor and rich optimization algorithm

Citation: Li, C.; Dias, D. Assessment of the Rock Elasticity Modulus Using Four Hybrid RF Models: A Combination of Data-Driven and Soft Techniques. *Appl. Sci.* **2023**, *13*, 2373. https://doi.org/10.3390/app13042373

Academic Editor: Arcady Dyskin

Received: 7 January 2023
Revised: 10 February 2023
Accepted: 10 February 2023
Published: 12 February 2023

1. Introduction

In rock engineering, the rock elasticity modulus (EM) plays an important role for structure designs [1–4]. The EM is an important index for quantifying the rock behavior. It is also closely related to the rock's durability, which can determine rock applications to a large extent [5]. Numerous experiments were developed according to the international society of rock mechanics (ISRM) to calculate the rock EM [6]. Nevertheless, the expensive sample costs and time-consuming laboratory operations have forced engineers to develop other methods for estimating the rock EM.

The empirical formula based on statistics is a popular method to estimate the rock EM in preliminary design phases [7–12]. The aim of the empirical formulas (simple-regression (SR) or multiple-regression (MR) formulas) is to establish a relationship between one or more rock properties and the EM. Numerous researchers proposed various empirical formulas to estimate the rock EM [3,4,13–16]. Beiki et al. [17] used the porosity (Pn) to predict the rock EM with a low prediction accuracy of the coefficient of determination (R^2). Yasar and Erdogan [18] provided a linear SR formular using P-wave velocity (Vp) to estimate EM. Dincer et al. [19] established an SR formula between the Schmidt hammer rebound number (SHRN) and EM. The prediction accuracy of R^2 was equal to 0.85. Behzadafshar et al. [20]

only utilized the point load index (PLI) to forecast the granite EM. The prediction accuracy was not satisfactory, resulting in a low R^2 of 0.58. Dehghan et al. [21] developed an MR formula consisting of the above four variables to predict the travertine EM. By reviewing the developed empirical formula, the Pn, Vp, SHRN and PLI are usually considered into these empirical equations, which can be obtained directly from tests [13]. However, the original data used to develop each empirical formula are fixed and different, and the formula performance decreases once it is used to predict the EM of a new rock [22]. From the prediction accuracy perspective, the intrinsic relationship of the SR and MR formulas cannot well describe the complex and nonlinear correlation between the rock properties and the EM [23].

In recent years, soft techniques have been widely used to solve prediction problems in rock engineering [24–30], especially machine-learning (ML) methods. For the rock EM prediction, Ocak and Seker [31] proposed an artificial neural network (ANN) model to predict the EM of intact rock. The results showed that the ANN model obtained a higher accuracy for the root-mean-square error (RMSE is equal to 0.191) than those of previous approaches. Pappalardo and Mineo [32] utilized ANN models to estimate the EM statis of rock samples. The results indicated that this artificial intelligence model is of more practical value in estimating the rock EM. Singh et al. [33] used an adaptive neurofuzzy inference system (ANFIS) to predict the EM values of 85 rock samples. The performance evaluation results illustrated that the ANFIS model has a better performance than the initial ANN model and fuzzy inference system (FIS) model by means of the lower RMSE value of 6.799. Umrao et al. [34] developed an ANFIS to estimate the EM of 45 heterogenous sedimentary rocks. This model has obtained a satisfactory prediction accuracy of R^2 = 0.935. Acar and Kaya [35] adopted a least-square support vector machine (LS-SVM) model to find the EM of weak rocks by considering the Vp, unit weight (r), PLI and tensile strength (Ts). The prediction results showed that the LS-SVM models can be a good substitute for experiments to measure the weak rock EM. Al-Anazi and Gates [36] used a support vector regression (SVR) model and a backpropagation neural network (BPNN) model to forecast the EM of reservoir rocks. The prediction results indicated that the former has a better performance than the latter. Matin et al. [37] used an integrated ML model, named the random forest (RF) model, to predict the EM and the uniaxial compressive strength (UCS) of various rocks. Based on the prediction results, the RF model not only achieves a satisfactory accuracy (R^2 = 0.91) for the EM prediction but can also accurately estimate the UCS (R^2 = 0.93). Other similar studies on the rock EM prediction by ML models can be referred to in the literature [38–42]. To improve the ML model's performance, the metaheuristic optimization algorithms (MOA) are used to select the model hyperparameters. Tian et al. [43] used the imperialism competitive algorithm (ICA) and the particle-swarm optimization (PSO) to optimize an ANN model for predicting the EM of rock materials. The optimization results illustrated that the ICA-ANN has the best prediction accuracy for both the training and testing phases (R^2: 0.952 and 0.955). Mokhtari and Behnia [44] combined the cuckoo optimization algorithm (COA) and ANN model to estimate the EM of limestone rocks. The results showed that the COA can obviously improve the prediction performance of the ANN model. Other optimized ML models for the EM prediction can be found in the literature [45–50].

Among common ML models, the RF model has unique advantages in resisting the overfitting phenomenon, and its combination with MOA can effectively solve the hyperparameter selection problem [51]. In general, MOA can be divided into four groups, i.e., the based evolutionary, based swarm intelligence, based human behavior and based physicochemical groups. Therefore, the aim of this paper is to generate four optimized RF models using four different MOA strategies for predicting the rock EM. An integrated rock database was established using the data-driven technology to train and test the proposed models. In addition, five widely used ML models and one empirical equation were also developed to compare the predictive performance with four hybrid RF models for predicting the rock EM. Finally, a sensitive analysis is carried out to calculate the variable importance.

2. Data Preparation

Reviewing the published studies on the rock EM prediction, the used rock samples are generally different, resulting in a loss of accuracy in predicting other rock sample properties from original empirical formulas or other models. To overpass this limitation, an integrated rock database consisting of 120 rock samples (e.g., granite and travertine) was established using the data-driven technology. The EM values of these rock samples was investigated by Armaghani et al. [3], Dehghan et al. [21] and Tuğrul and Zarif [52]. The reason for choosing these data is that the EM value of each rock is determined jointly by the porosity (Pn), P-wave velocity (Vp), Schmidt hammer rebound number (SHRN) and point load index (PLI). Before generating the prediction models, a correlation analysis needs to be conducted by outputting the correlation coefficient (CC) to determine the final variables used in the EM prediction. As illustrated in Figure 1, the scatter plot at the lower right shows the distribution of the four input variables and of one output variable. The histograms in the diagonal showed the data range of all variables, and the number at the upper left represents the CC between the two corresponding variables. If the number value is negative, the correlation between the two variables is negative. Otherwise, the correlation is positive. The absolute number value is used to evaluate the correlation between any of two variables. If the CC between two input parameters (or between each input and output parameter) is very high (or very low), one of them needs to be removed to increase the prediction efficiency. The results showed that the CC between the four variables is not high; Pn and EM have the highest CC value (−0.651). Especially, the low correlation between PLI and EM is caused by the diverse sources of databases used to predict EM in this paper, while the PLI is beneficial to accurately predict the rock EM [3]. Therefore, the Pn, Vp, SHRN and PLI are used as input variables to predict the EM (output variable). Their detailed information is shown in Table 1.

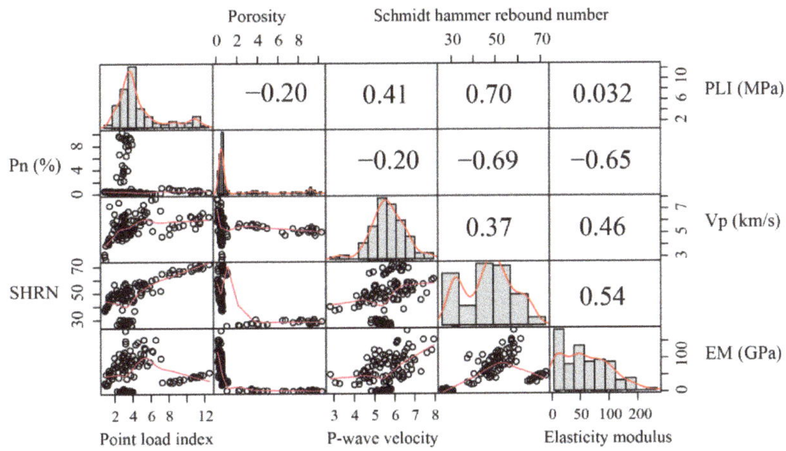

Figure 1. Correlation analysis results of four input variables and EM.

Table 1. Detailed description of the input and output variables.

Variables	Statistical Information					
	Sign	Unit	Min	Max	Mean	St. D
Point load index	PLI	MPa	0.890	12.530	4.365	2.839
Porosity	Pn	%	0.100	10.270	1.957	3.047
P-wave velocity	V_P	km/s	2.823	7.943	5.575	0.892
Schmidt hammer rebound number	SHRN	/	25.630	72.000	47.093	13.795
Elasticity modulus	EM	GPa	3.050	183.300	60.139	44.832

Note: Min—minimum value; Max—maximum value; St. D—standard deviation value.

3. Development of Hybrid RF Models for Predicting the Rock EM

3.1. Metaheuristic Optimization Algorithms

3.1.1. Backtracking Search Optimization Algorithm (BSA)

Civicioglu [53] proposed a MOA-based evolutionary method to solve optimization problems, namely, the backtracking search optimization algorithm (BSA). The algorithm is still inspired by an individual evolution, especially the traditional mutation, and crossover operators are uniformly replaced by breeding operators. The optimization process of the BSA can be described using five strategies: initialization, selection, mutation, crossover and selection. These strategies are defined by using Equations (1)–(5).

$$P_i = Low + rand(N) \cdot (Up - Low) \tag{1}$$

$$P_{oi} = \begin{cases} P[a, b \sim P_i), a < b \\ permuting(P_{oi}), a > b \end{cases} \tag{2}$$

$$Mutant = P_i + F(P_{oi} - P_i) \tag{3}$$

$$T_i = P_i + map \cdot F(P_{oi} - P_i) \tag{4}$$

$$P_{ui} = T_i, fitness(T_i) < fitness(P_i) \tag{5}$$

where P_i and P_{oi} represent the initial and historical position of the i-th individual, respectively. *Low* and *Up* indicate the lower and upper bounds of the dimension space. a and b are random numbers within the range of [0, 1]. F and *map* represent the control parameter of the mutation operator and a binary matrix consisting of crossover probability parameters, respectively. T_i and P_{ui} are the current position of the i-th tested individual and the updated position of the i-th individual, respectively.

3.1.2. Multi-Verse Optimizer (MVO)

The multi-verse optimizer (MVO) is a MOA-based physics algorithm proposed by Mirjalili et al. [54], which is inspired by the idea that the universe moves from white holes to black holes through wormholes to achieve a stable situation. The white holes are believed to be an important part of the original universe, the black holes have an irresistible pull on everything including light beams, and the wormhole is a bridge or passage connecting different universes. In the MVO algorithm, the birth of the universe is always related to the objects transfer. The universe with a low inflation rate is more likely to take in more objects. The optimization process of MVO can be described as follows:

(1) Population—the initial population of the universes in the searching space is defined using the Equation (6).

$$U = \begin{bmatrix} u_1^1 & u_1^2 & \cdots & u_1^d \\ u_2^1 & u_2^2 & \cdots & u_2^d \\ \vdots & \vdots & \vdots & \vdots \\ u_n^1 & u_n^2 & \cdots & u_n^d \end{bmatrix} \tag{6}$$

where u_n^d indicates the parameter of the n-th universe in the d-dimension searching space.

(2) Exploration and exploitation—the function of wormholes is to help objects move from one universe to another (see Figure 2). Thus, this mechanism by which objects are exchanged between universes through wormholes can be described as:

$$u_j = \begin{cases} u_j + TDR \cdot ((ub - lb) \cdot r_3 + lb), & r_2 < 0.5 \text{ and } r_1 < WEP \\ u_j + TDR \cdot ((ub - lb) \cdot r_3 + lb), & r_2 > 0.5 \text{ and } r_1 < WEP \\ u_j, & r_1 > WEP \end{cases} \tag{7}$$

where u_j represents the j-th parameter of the best universe. ub and lb are the lower and upper bounds of the multi-universes space, respectively. r_1, r_2 and r_3 indicate three random numbers within the range of [0, 1]. TDR and WEP represent two coefficients, the former is the wormhole existence probability, and the latter is named the travelling-distance rate. These coefficients can be calculated using the Equations (8) and (9).

$$WEP = \min + t \cdot \left(\frac{\max - \min}{T}\right) \tag{8}$$

$$TDR = 1 - \frac{t^{1/e}}{T^{1/e}} \tag{9}$$

where min and max represent the minimum maximum values, respectively. t and T indicate the current iteration and the maximum iteration, respectively. e is the exploitation accuracy.

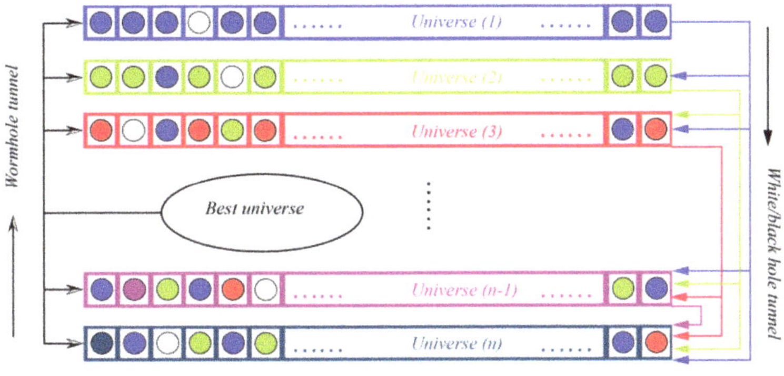

Figure 2. The conceptual model of wormholes in the MVO algorithm.

3.1.3. Golden Eagle Optimizer (GEO)

Mohammadi-Balani et al. [55] developed a novel MOA-based swarm intelligence named the golden eagle optimizer (GEO) to provide an effective scheme for solving optimization problems. This algorithm is inspired by the hunting behavior of golden eagles, who can adjust their speed to hunt. The hunting behavior can be divided into three parts: (a) selecting the prey; (b) attacking the prey; and (c) cruising. It is worth noting that each hunting behavior of a golden eagle is carefully considered. It allows striking a balance between attack and cruise. Once an attack is launched, the golden eagle is unable to obtain food or replenish enough energy.

(I) Selecting the prey—the selection can occur in a basic way, with each golden eagle randomly select a prey from the memory of any other group member to better explore the landscape. It is important to note that the chosen prey is not necessarily the nearest or furthest prey. Figure 3 shows how prey selection works.

(II) Exploration and exploitation—after determining the prey, each golden eagle carries out the attacking and cruising behaviors. The attacking behavior can be expressed by the following mathematical formula:

$$Y_i^G = Y_l^P - A_i \tag{10}$$

where Y_i^G and Y_l^P represent the position of the i-th golden eagle and the prey determined by the l-th golden eagle, respectively. A_i indicates the attacking distance between the prey and the i-th golden eagle. The cruising behavior is related to the attacking behavior, which can be expressed using the Equations (11) and (12).

$$h_1 y_1 + h_2 y_2 + \ldots + h_m y_m = d \Rightarrow \sum_{z=1}^{m} h_z R_z \tag{11}$$

$$\sum_{z=1}^{m} \alpha_z y_z = \sum_{z=1}^{m} \alpha_z^s y_z^* \quad (12)$$

where $[h_1, h_2, \ldots h_m]$, $[y_1, y_2, \ldots y_m]$ and $[R_1, R_2, \ldots R_m]$ represent the normal coefficients, variables and random points, respectively. s is the current iteration. $[y_1^*, y_2^*, \ldots y_m^*]$ indicates the position of the selected prey. $[\alpha_1, \alpha_2, \ldots \alpha_m]$ belongs to the A_i.

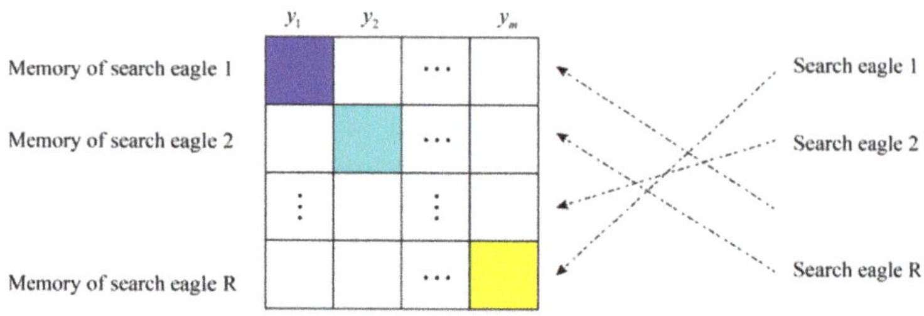

Figure 3. Prey selection in the GEO algorithm.

3.1.4. Poor and Rich Optimization Algorithm (PRO)

The poor and rich optimization algorithm (PRO) was proposed by Moosavi and Bardsiri [56]. It is inspired by people's desires and attitudes towards money. The wealth accumulated by individuals can preliminarily divide society into two classes, namely, rich and poor. For the rich (i.e., wealth level is obviously higher than the average one), observing the behavior of the poor (i.e., wealth level is obviously lower than the average one) can help them to increase wealth and consolidate their class position. The poor tend to narrow the gap by learning from rich ideas about wealth and approaches to making money. In the PRO algorithm, the population distribution of rich and poor can be expressed using Equation (13).

$$POP_{main} = POP_{poor} + POP_{rich} \quad (13)$$

where POP_{main} represents the main population, which is related to the POP_{poor} (poor population) and POP_{rich} (rich population). It should be noted that the position of the rich is better than the poor position. Their positions are calculated using the Equations (14) and (15), respectively.

$$P_{rich}^* = P_{rich} + c \cdot \left(P_{rich} - P_{poor}^{best}\right) \quad (14)$$

$$P_{poor}^* = P_{poor} + \left[c(\text{Pattern}) - P_{poor}\right] \quad (15)$$

where P_{rich}^* and P_{rich} represent the updated and old positions of the rich, respectively. P_{poor}^* and P_{poor} indicate the updated and old positions of the poor, respectively. P_{poor}^{best} is the current position of the best people in the poor population. c is a random number within the range of 0 to 1. The Pattern value is calculated by using the Equation (16).

$$\text{Pattern} = \frac{P_{rich}^{best} + P_{rich}^{mean} + P_{rich}^{worst}}{3} \quad (16)$$

where P_{rich}^{best} and P_{rich}^{worst} represent the current positions of the best and worst people in the rich population, respectively. P_{rich}^{mean} indicates the average position of the people in the rich population. The position distribution of rich and poor is shown in Figure 4.

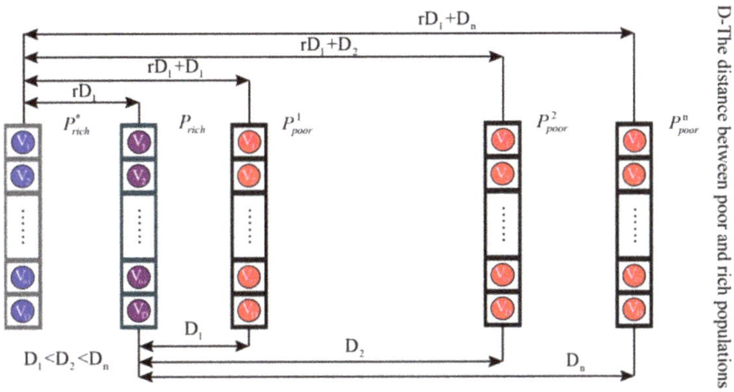

Figure 4. Position distribution for the rich and poor populations.

However, some sudden changes can occur for the rich and poor situations, such as the stock prices falling and rising, shortages of petroleum products and inflation [56]. Since the occurrence of the above situation is not predictable, P^*_{rich} and P^*_{poor} after mutation can be expressed as follows:

$$P^*_{rich} = P^*_{rich} + randn, if\ rand < Pmut \tag{17}$$

$$P^*_{poor} = P^*_{poor} + randn, if\ rand < Pmut \tag{18}$$

where *rand* is a random number within the range of 0 to 1. *randn* represents a value considering a normal distribution. *Pmut* represents the mutation probability.

3.2. Hybrid RF Models

In this paper, the BSA, MVO, GEO and PRO are used to optimize the RF model for predicting the rock EM. The definition of the RF model is described in the literature [57–60]. The hyperparameter combination selection (i.e., number of trees (Nt) and random features (Maxdepth)) is a key step to tap into the RF model prediction potential. Therefore, four hybrid RF models were generated to find the optimal hyperparameter combination according to the following process:

(i) Data preprocessing
A total of 120 rock samples with four input variables were used to predict the EM in this paper. All variables need to be extracted and normalized to [−1, 1]. The purpose of this step is to prevent a failure for establishing the accurate prediction relationships due to the parameter variability. After that, the train and test sets are separated from the initial database. The ratio of the train set to the test set is set equal to 4 to 1. It should be noted that the same train or test set is used to generate each hybrid RF model for predicting the rock EM and comparing their performance.

(ii) Parameter settings
Although the Nt increase will not cause an overfitting of the RF model, a large parameter selection range can greatly increase the computation time. Therefore, the ranges of Nt and Maxdepth are set equal to [1, 100] and [1, 10], respectively. For the four MOA algorithms, the number of initial solutions (i.e., individuals of BSA, candidates of MVO, population of GEO and human of PRO) and the iteration time are the core factors that affect the optimization performance of these algorithms. To better activate the optimization performance, the solutions are set equal to 30, 60, 90, 120 and 150 during the 200 iterations.

(iii) Optimization evaluation
The fitness function is utilized to evaluate the performance of each hybrid RF model with different solutions during the 200 iterations. The RMSE is adopted to represent the fitness values of all models in this paper. They do not need an absolute value to

evaluate the model performance [51]. In other words, the best-optimized RF model has the lowest RMSE value among all hybrid models based on the same MOA. The flowchart for developing four hybrid RF models for predicting the rock EM is shown in Figure 5.

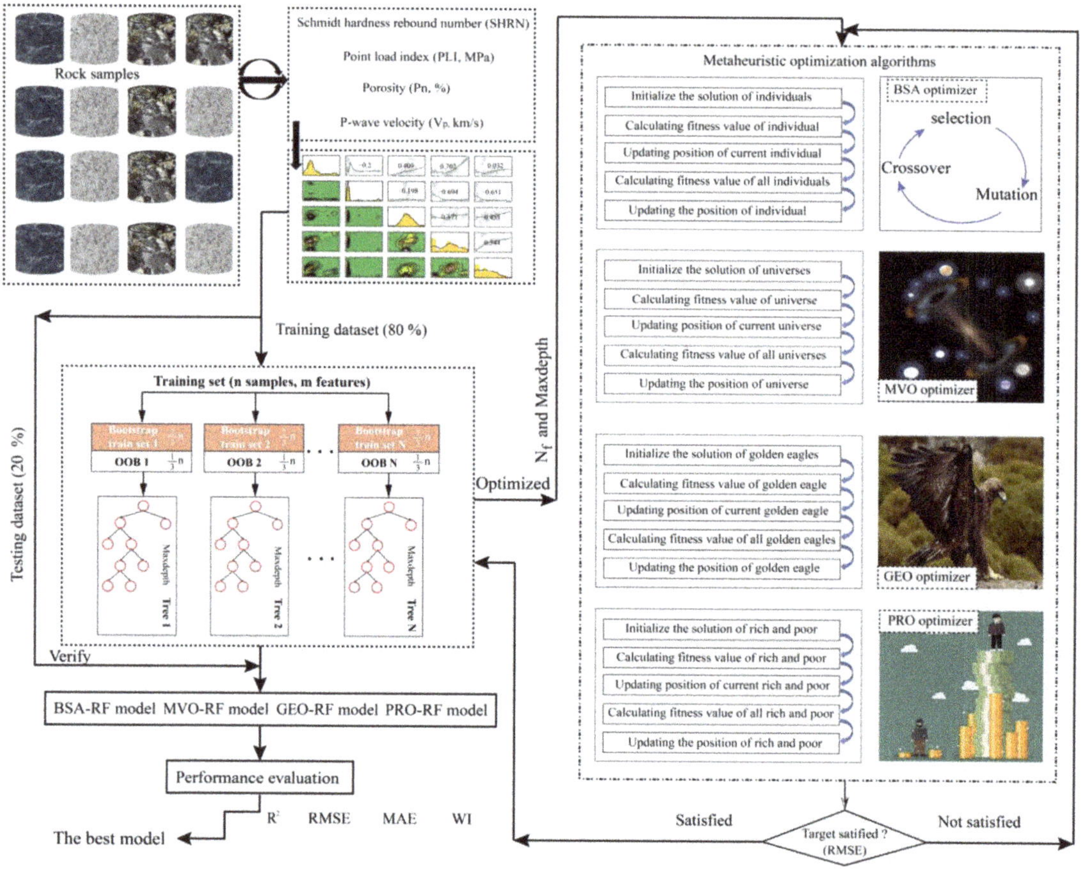

Figure 5. Flowchart of predicting the rock EM based on the four hybrid RF models.

4. Performance Evaluation

The statistical indices widely utilized to evaluate the performance of the prediction models include the root-mean-square error (RMSE), the mean absolute error (MAE), the coefficient of determination (R^2) and the Willmott's index (WI). The RMSE and MAE are able to reflect the error between the predicted and the measured values. On the other hand, the R^2 and WI describe the fitting performance of the prediction models. In terms of values, the lowest values of RMSE and MAE and the highest values of R^2 and WI represent the best prediction model.

$$\mathrm{RMSE} = \sqrt{\frac{1}{n}\sum_{i=1}^{n}(E_i - e_i)^2} \tag{19}$$

$$\mathrm{R}^2 = 1 - \frac{\left[\sum\limits_{i=1}^{n}(E_i - e_i)\right]^2}{\left[\sum\limits_{i=1}^{n}(E_i - \overline{E})\right]^2} \tag{20}$$

$$\mathrm{MAE} = \frac{1}{n}\sum_{i=1}^{n}|E_i - e_i| \tag{21}$$

$$\mathrm{WI} = 1 - \left[\frac{\sum_{i=1}^{n}(E_i - e_i)^2}{\sum_{i=1}^{n}\left(|e_i - \overline{E}| + |E_i - \overline{E}|\right)^2}\right] \tag{22}$$

where n is the number of the samples. E_i and e_i represent the measured and the predicted values of the rock EM, respectively. $\overline{E}$ is the average value of the measured rock EM.

5. Results and Discussion

5.1. Results of the Proposed Four Hybrid Models

To determine the optimal solution and corresponding hyperparameter combination (i.e., Nf and Maxdepth) of the RF model, all hybrid models were performed during 200 iterations. The iteration curves of each hybrid RF model with five solutions are shown in Figure 6. As it can be seen in Figure 6a, it is obvious that the BSA-RF with 60 solutions has a lower fitness value than the other four BSA-RF models during the 200 iterations. The solution of 90 is the most suitable for generating the MVO-RF model by means of the lowest fitness value (see Figure 6b). As illustrated in Figure 6c,d, the optimal solutions of the GEO-RF and PRO-RF models are equal to 120 and 90, respectively. Table 2 lists the results and the best hyperparameter combination of the four optimized RF models. The PRO-RF model has the lowest value of RMSE (0.1861) among all models. This optimized RF model shows that the optimal Nf and Maxdepth values are 17 and 2, respectively.

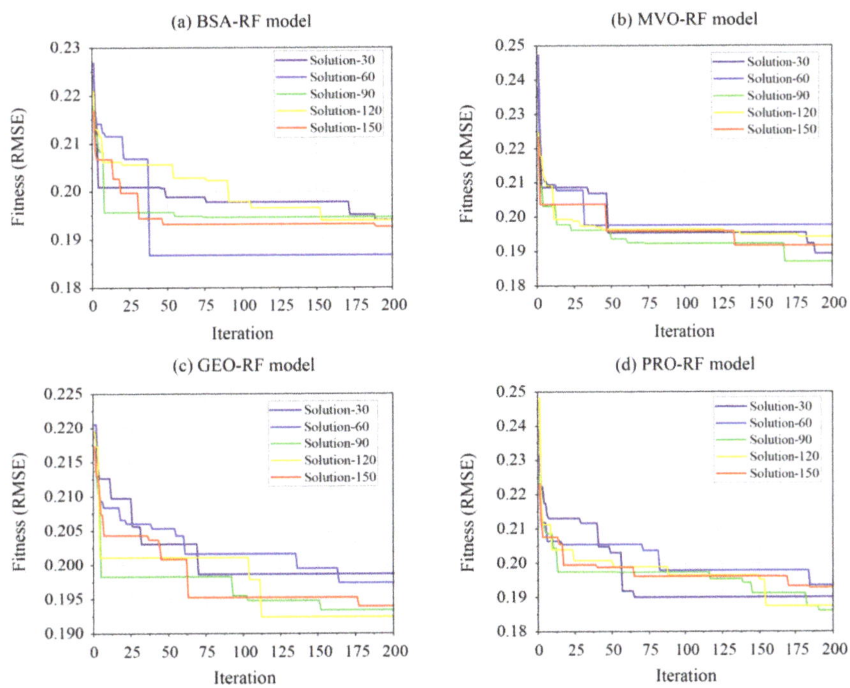

Figure 6. Iteration curves of four hybrid RF models for predicting the rock EM.

Table 2. Development results of all hybrid RF models.

Solutions	Fitness (RMSE)			
	BSA-RF	MVO-RF	GEO-RF	PRO-RF
30	0.1941	0.1893	0.1987	0.1901
60	0.1868	0.1977	0.1974	0.1935
90	0.1947	0.1870	0.1934	0.1861
120	0.1940	0.1942	0.1925	0.1875
150	0.1927	0.1917	0.1940	0.1928
Optimal hyperparameter combination				
Nf	19	21	20	17
MaxDepth	2	2	2	2

Four hybrid RF models with the optimal hyperparameter combinations were used to predict the rock EM in the training phase. The performance indices of each hybrid RF model are listed in Table 3. As illustrated in this table, four models achieved a good prediction accuracy with high values of R^2 and WI and low values of RMSE and MAE. Compared with the other three models, the PRO-RF is the best prediction model for predicting the EM by means of the best performance indices, i.e., R^2 is equal to 0.9423, RMSE to 10.7420, MAE to 7.6514 and WI is equal to 0.9843. The ranking score results of all models indicated that the GEO-RF model with the lowest total score of 4 is the worst model among all hybrid models for predicting EM in the training phase. The score of the MVO-RF model is the second model in the rank just after the PRO-RF model.

Table 3. Performance and ranking results of four hybrid RF models in the training phase.

Models	Performance Indices and Ranking Scores								Total
	R^2	Score	RMSE	Score	MAE	Score	WI	Score	
BSA-RF	0.9359	2	11.3203	2	7.8165	2	0.9824	2	8
MVO-RF	0.9407	3	10.8867	3	7.7471	3	0.9837	3	12
GEO-RF	0.9317	1	11.6807	1	8.0452	1	0.9809	1	4
PRO-RF	0.9423	4	10.7420	4	7.6514	4	0.9843	4	16

As illustrated in Figure 7, the regression distribution of the predicted and the measured EM values represents the performance of the prediction models. The position of each rock EM data in the two-dimension regression diagram is determined by the predicted EM values (horizontal axis, x) and the measured EM values (vertical axis, y). If y = x, it means that the predicted EM value is equal to the measured EM value. The rock EM data are then located on the diagonal line. To this end, the PRO-RF model has obtained the most rock EM data on the diagonal or close to the line among the four hybrid RF models (Figure 7d). After the PRO-RF model, the MVO-RF model and the BSA-RF model have similar regression distributions for both the small and big rock EM data. The MVO-RF model has obtained more data close to the diagonal line than the BSA-RF model in the interval between 20 and 70.

To further determine the performance of all hybrid RF models, the performance indices were calculated again using the test set, as shown in Table 4. The best model is still the PRO-RF, which has the lowest values of RMSE and MAE (10.1548 and 6.0423) and the highest values of R^2 and WI (0.9410 and 0.9840). The predictive performance of the BSA-RF model is better than the MVO-RF model in the testing phase, the former not only has the better performance indices but also has a higher score (12) than the latter (8). In addition, the GEO-RF model still does not achieve better predictive performances than the other models using the test set.

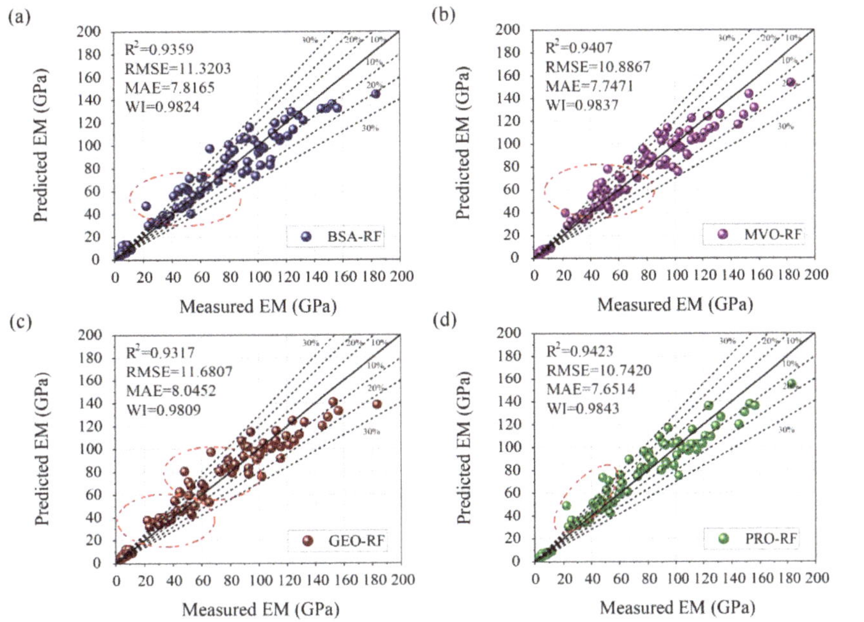

Figure 7. Regression diagrams of the four hybrid RF models using the train set: (**a**) BSA-RF model; (**b**) MVO-RF model; (**c**) GEO-RF model; (**d**) PRO-RF model.

Table 4. Performance and ranking results of four hybrid RF models in the testing phase.

Models	Performance Indices and Ranking Scores								Total
	R^2	Score	RMSE	Score	MAE	Score	WI	Score	
BSA-RF	0.9322	3	10.8902	3	7.2155	3	0.9812	3	12
MVO-RF	0.9236	2	11.5567	2	7.5280	2	0.9785	2	8
GEO-RF	0.9123	1	12.3826	1	7.9739	1	0.9746	1	4
PRO-RF	0.9410	4	10.1548	4	6.0423	4	0.9840	4	16

Figure 8 illustrates the regression results of the four hybrid RF models in the testing phase. As it can be seen in Figure 8a–c, large errors between the predicted EM values by the BSA-RF, MVO-RF and GEO-RF models and the measured EM values indicate that the data point move away from the diagonal, especially when the EM values are in the range of [50, 80] and [0, 20]. The PRO-RF model has obtained the most EM data points close to the diagonal based on the best predictive performance in the testing phase. Therefore, the PRO-RF model is the best hybrid RF model for predicting the rock EM in this paper.

5.2. *Performance Comparison between the Proposed and Other Models*

To compare the predictive performance with the proposed hybrid models, five common ML models, named the ANN, SVR, extreme learning machine (ELM), kernel-extreme learning machine (KELM) and generalized regression neural network (GRNN), and one empirical formula, were also developed to predict the rock EM. The definition and the hyperparameter settings of the five ML models can be found in [61–65]. The multivariate-quadratic equation (MQE) of MR was used to generate an empirical formula as expressed using Equation (23).

$$\begin{aligned} EM = \; & -136.609 + 7.864PLI - 1.056PLI^2 - 17.038P_n + 1.378{P_n}^2 \\ & + 4.265V_p + 1.084{V_p}^2 + 5.297SHRN - 0.042SHRN^2 \end{aligned} \quad (23)$$

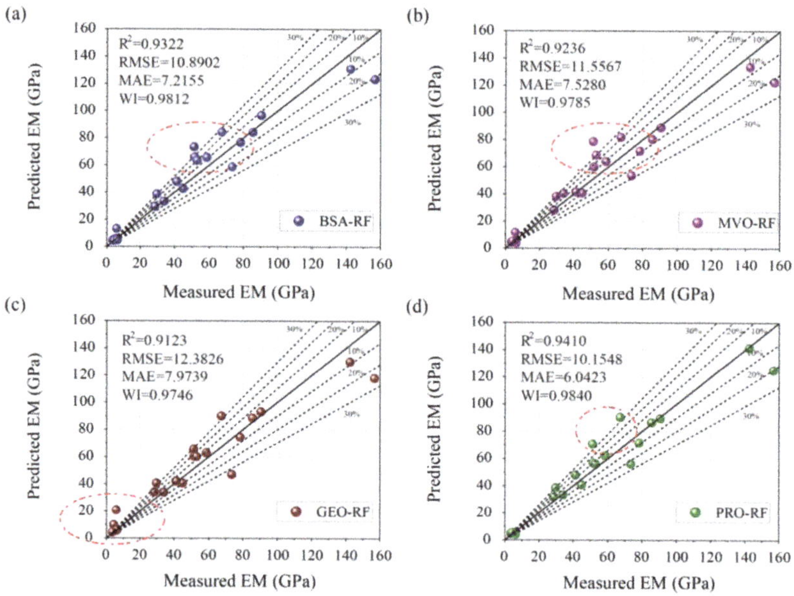

Figure 8. Regression diagrams of the four hybrid RF models using the test set: (**a**) BSA-RF model; (**b**) MVO-RF model; (**c**) GEO-RF model; (**d**) PRO-RF model.

The optimal hyperparameter combination of each model and the corresponding performance results are shown in Table 5. The GRNN obtained better performance indices (R^2: 0.9010; RMSE: 13.1593; MAE: 8.2674 and WI: 0.9717) than the other models using the test set. After this model, the prediction accuracy of the KELM and the ELM models is superior to the ANN and the SVR models. The worst prediction model is the MQE; its index values of R^2, RMSE, MAE and WI are, respectively, equal to 0.8318, 17.1476, 13.5849 and 0.9497. Figure 9 shows the rank scores of all comparison models. It can be obviously observed that the performance ranking of the six models is GRNN (24), KELM (20), ELM (16), ANN (12), SVR (8) and MQE (4).

Table 5. Performance results of five common ML models using test set.

Models	Performance Indices				Hyperparameters
	R^2	RMSE	MAE	WI	
ANN	0.8683	15.1724	10.8323	0.9619	$N_h = 2; N_e = 4{,}4$
SVR	0.8592	15.6918	11.7625	0.9591	$C = 128; R_k = 0.25$
ELM	0.8795	14.5124	10.2086	0.9665	$N_{es} = 65$
KELM	0.8987	13.3074	8.4755	0.9716	$R_c = 128; R_k = 1.0$
GRNN	0.9010	13.1593	8.2674	0.9717	$S_f = 0.3$
MQE	0.8318	17.1476	13.5849	0.9497	Equation (7)

Note: N_h—the hidden layers number; N_e—the number of neurons in the corresponding hidden layer; C—penalty parameter; R_k—RBF kernel parameter; N_{es}—the number of neurons in a single hidden layer; R_c—regularization coefficient; S_f—smoothing factor.

The regression diagrams of the six comparison models in the testing phase are presented in Figure 10. As it can be seen in these diagrams, all models have a finite number of rock EM data points near the diagonal line. Compared with the other five models, the GRNN model obtained a good EM regression distribution in the range of 70 to 90 (Figure 10e). For the MQE model, most of the rock EM data points are away from the diagonal, which also indicates a large error between the EM value predicted by this model and the measured EM value. As a result, the GRNN model is the best prediction model among the six other models by means of the best performance indices. It obtains the highest rank score value and the best regression distribution.

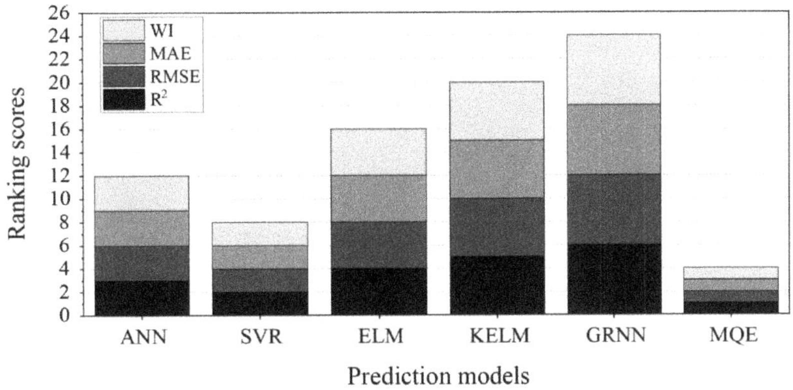

Figure 9. Ranking scores of the six comparison models using the test set.

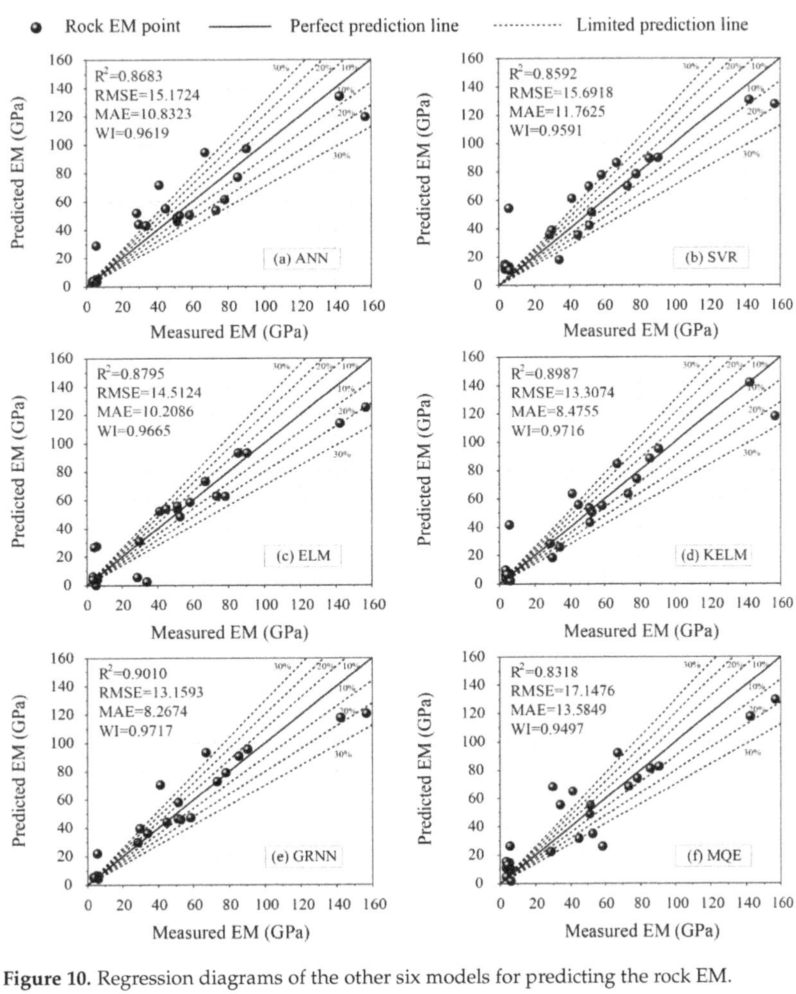

Figure 10. Regression diagrams of the other six models for predicting the rock EM.

Thus, the GRNN model is used to compare the performance with the PRO-RF model for predicting the rock EM. The bar chart tool is utilized to make an intuitive comparison between the measured EM and the EM predicted by the GRNN model and PRO-RF model, as shown in Figure 11. In Figure 11a, the predicted EM values of the No. 1, No. 7, No. 11, No. 13 and No. 15 samples by the PRO-RF model showed deviations. On the other hand, the GRNN model also does not accurately predict the EM values of the No. 2, No. 3, No. 4 and No. 19 samples, in addition to the above samples. The error analysis results of the PRO-RF model and the GRNN model for predicting the rock EM are shown in Table 6. It can be intuitively observed that the error in the statistical indices based on the PRO-RF model is better than the GRNN model ones, such as the error sum of the PRO-RF model (145.014) being lower than the GRNN model (198.417).

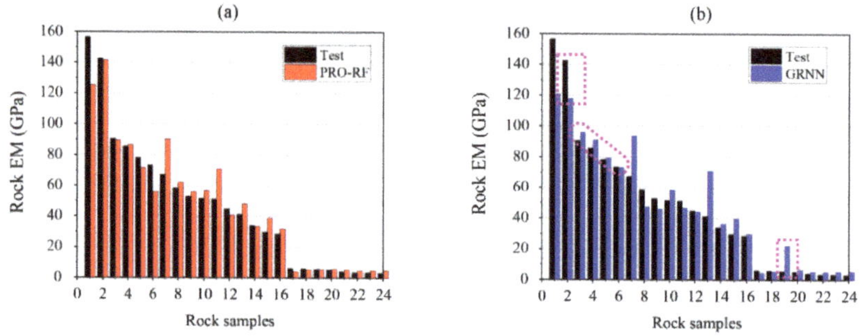

Figure 11. Intuitive comparison diagram of the measured EM and predicted EM by the proposed models: (**a**) PRO-RF model; (**b**) GRNN model.

Table 6. Error analysis comparison between the PRO-RF model and the GRNN model.

Models	Statistical Indices						
	Min	**Max**	**Median**	**Mean**	**St. E**	**St. D**	**Sum**
PRO-RF	0.134	31.524	2.678	6.042	1.702	8.337	145.014
GRNN	0.328	35.992	3.477	8.267	2.135	10.458	198.417

Note: St. E—standard error value. St. D—standard deviation value.

The relative deviation (RD) distribution is also an effective tool to evaluate the model performances [64]. The definition of RD is the ratio of the error between the predicted value and the measured value to the measured value. This also means that the models with a better performance have lower RD values. As illustrated in Figure 12a, the maximum RD of the PRO-RF model is 44.77%, and most of the RD values are lower than 40%. Especially, there are 11 RD values lower than 10%. It should be noted that the maximum RD of the GRNN model is higher than 70%, and only 9 RD values are lower than 10%. Therefore, the PRO-RF model has a better predictive performance than the GRNN model for calculating the rock EM.

5.3. Sensitive Analysis

After determining the best prediction model, the sensitive analysis was carried out to calculate the parameter importance of the rock EM prediction. The average impact on model output magnitude (mean) is one of indices in the Shapley additive explanations (SHAP), which is used to represent the parameter importance, as shown in Figure 13. As can be seen in this graph, the porosity (Pn) has the larger importance value (30.52) than the other input variables in the EM prediction based on the PRO-RF model. After the Pn, the importance ranking of other three variables are the SHRN (3.8), Vp (4.09) and PLI (2.08). In addition, another SHAP index is the impact on the model output (SHAP value), which is used to describe the correlation between the input and output variables. As illustrated in

Figure 14, only the Pn has the greatest negative correlation with EM. The SHRN, Vp and PLI are positively correlated with the EM, and the correlation decreased successively.

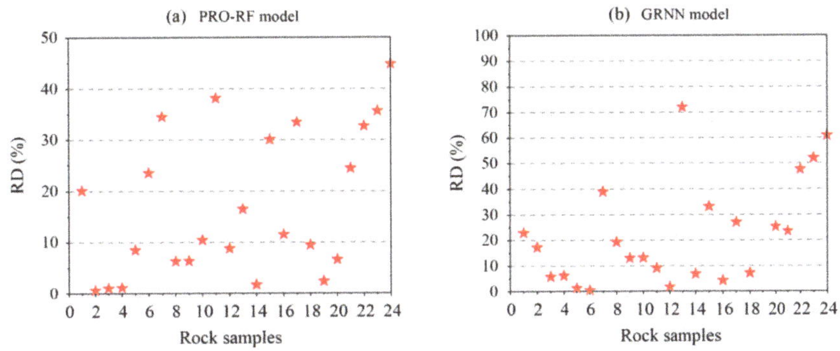

Figure 12. The RD distribution diagrams of the PRO-RF model and the GRNN model.

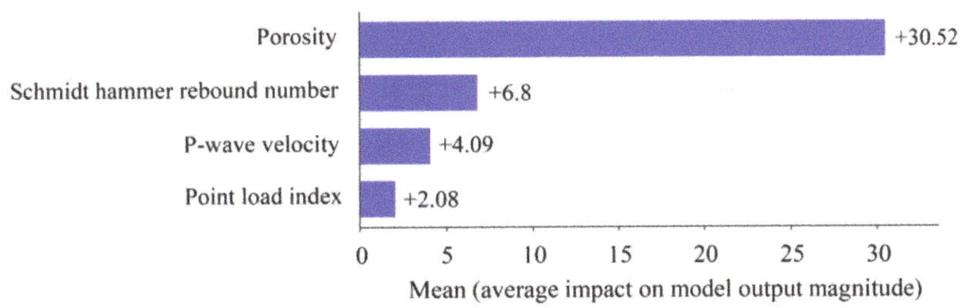

Figure 13. Parameter importance results based on the PRO-RF model.

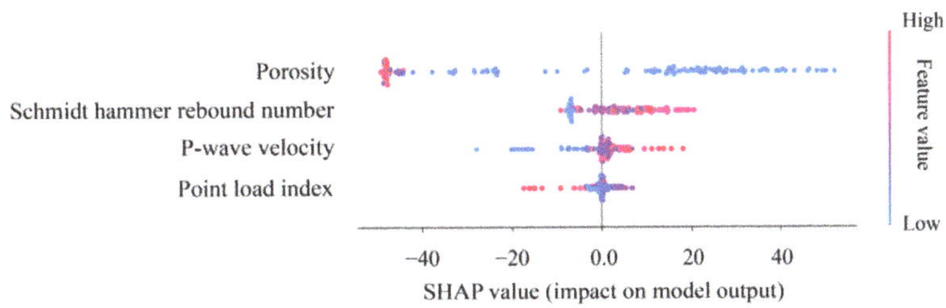

Figure 14. The SHAP values results of all input parameters based on the PRO-RF model.

6. Conclusions

In this study, four hybrid RF models, named BSA-RF, MVO-RF, GEO-RF and PRO-RF were developed to predict the elasticity modulus (EM) of 120 rock samples. Four rock properties, named the porosity (Pn), P-wave velocity (Vp), Schmidt hammer rebound number (SHRN) and point load index (PLI), were considered as the main factors for the EM prediction. In addition, five ML models (i.e., ANN, SVR, ELM, KELM and GRNN) and one empirical formula were also developed to predict the rock EM and compare the predictive performance with the proposed hybrid RF models. The main conclusion of this paper can be listed as follows:

i. Four hybrid RF models have obtained a good prediction accuracy by means of four performance indices. In particular, the PRO-RF model is the best model among them.
ii. The GRNN model has a better predictive performance than the other ML models and the empirical formula. It results in the higher values of R^2 (0.9010) and WI (0.9717)

and the lower values of RMSE (13.1593) and MAE (8.2674). However, these four optimized RF models are superior to the GRNN model.

iii. The porosity (Pn) is the most important variable by means of the highest average impact value of 30.52 for predicting the rock EM. Meanwhile, the Pn is also the only variable negatively correlated with EM.

This paper proposes four effective hybrid RF models to predict the rock EM. It shows a successful application of soft techniques for a rock parameter prediction. Nevertheless, more various rocks should be collected into the integrated database to increase the prediction model's accuracy. Furthermore, other rock properties such as the density, water content and UCS could be considered in the rock EM prediction.

Author Contributions: Conceptualization, C.L. and D.D.; methodology, C.L.; investigation, D.D.; writing—original draft preparation, C.L.; writing—review and editing, C.L. and D.D.; visualization, C.L.; funding acquisition, C.L. All authors have read and agreed to the published version of the manuscript.

Funding: The first author was funded by the China Scholarship Council (Grant No. 202106370038).

Institutional Review Board Statement: Not applicable.

Informed Consent Statement: Not applicable.

Data Availability Statement: The data used in this study are from published research: Armaghani et al. [3] (https://doi.org/10.1007/s12517-015-2057-3), Dehghan et al. [21] (https://doi.org/10.1016/S1674-5264(09)60158-7) and Tuğrul and Zarif [52] (https://doi.org/10.1016/S0013-7952(98)00071-4).

Conflicts of Interest: The authors declare that they have no known competing financial interest or personal relationships that could have appeared to influence the work reported in this paper.

References

1. Ersoy, H.; Kanik, D. Multicriteria decision-making analysis based methodology for predicting carbonate rocks' uniaxial compressive strength. *Earth Sci. Res. J.* **2012**, *16*, 65–74.
2. Armaghani, D.J.; Tonnizam Mohamad, E.; Momeni, E.; Narayanasamy, M.S. An adaptive neuro-fuzzy inference system for predicting unconfined compressive strength and Young's modulus: A study on Main Range granite. *Bull. Eng. Geol. Environ.* **2015**, *74*, 1301–1319. [CrossRef]
3. Armaghani, D.J.; Tonnizam Mohamad, E.; Momeni, E.; Monjezi, M.; Sundaram Narayanasamy, M. Prediction of the strength and elasticity modulus of granite through an expert artificial neural network. *Arab. J. Geosci.* **2016**, *9*, 48. [CrossRef]
4. Madhubabu, N.; Singh, P.K.; Kainthola, A.; Mahanta, B.; Tripathy, A.; Singh, T.N. Prediction of compressive strength and elastic modulus of carbonate rocks. *Measurement* **2016**, *88*, 202–213. [CrossRef]
5. Nasiri, H.; Homafar, A.; Chelgani, S.C. Prediction of uniaxial compressive strength and modulus of elasticity for Travertine samples using an explainable artificial intelligence. *Results Geophys. Sci.* **2021**, *8*, 100034. [CrossRef]
6. Jamshidi, A.; Nikudel, M.R.; Khamehchiyan, M.; Sahamieh, R.Z. The effect of specimen diameter size on uniaxial compressive strength, P-wave velocity and the correlation between them. *Geomech. Geoengin.* **2016**, *11*, 13–19. [CrossRef]
7. Sonmez, H.; Gokceoglu, C.; Nefeslioglu, H.A.; Kayabasi, A. Estimation of rock modulus: For intact rocks with an artificial neural network and for rock masses with a new empirical equation. *Int. J. Rock Mech. Min. Sci.* **2006**, *43*, 224–235. [CrossRef]
8. Palchik, V. On the ratios between elastic modulus and uniaxial compressive strength of heterogeneous carbonate rocks. *Rock Mech. Rock Eng.* **2011**, *44*, 121–128. [CrossRef]
9. Najibi, A.R.; Ghafoori, M.; Lashkaripour, G.R.; Asef, M.R. Empirical relations between strength and static and dynamic elastic properties of Asmari and Sarvak limestones, two main oil reservoirs in Iran. *J. Pet. Sci. Eng.* **2015**, *126*, 78–82. [CrossRef]
10. Hoek, E.; Diederichs, M.S. Empirical estimation of rock mass modulus. *Int. J. Rock Mech. Min. Sci.* **2006**, *43*, 203–215. [CrossRef]
11. Alemdag, S.; Gurocak, Z.; Gokceoglu, C. A simple regression based approach to estimate deformation modulus of rock masses. *J. Afr. Earth Sci.* **2015**, *110*, 75–80. [CrossRef]
12. Kayabasi, A.; Gokceoglu, C. Deformation modulus of rock masses: An assessment of the existing empirical equations. *Geotech. Geol. Eng.* **2018**, *36*, 2683–2699. [CrossRef]
13. Yilmaz, I.; Yuksek, G. Prediction of the strength and elasticity modulus of gypsum using multiple regression, ANN, and ANFIS models. *Int. J. Rock Mech. Min. Sci.* **2009**, *46*, 803–810. [CrossRef]
14. Moradian, Z.A.; Behnia, M. Predicting the uniaxial compressive strength and static Young's modulus of intact sedimentary rocks using the ultrasonic test. *Int. J. Geomech.* **2009**, *9*, 14. [CrossRef]
15. Yılmaz, I.; Sendır, H. Correlation of Schmidt hardness with unconfined compressive strength and Young's modulus in gypsum from Sivas (Turkey). *Eng. Geol.* **2002**, *66*, 211–219. [CrossRef]

16. Saedi, B.; Mohammadi, S.D.; Shahbazi, H. Application of fuzzy inference system to predict uniaxial compressive strength and elastic modulus of migmatites. *Environ. Earth Sci.* **2019**, *78*, 208. [CrossRef]
17. Beiki, M.; Majdi, A.; Givshad, A.D. Application of genetic programming to predict the uniaxial compressive strength and elastic modulus of carbonate rocks. *Int. J. Rock Mech. Min. Sci.* **2013**, *63*, 159–169. [CrossRef]
18. Yasar, E.; Erdogan, Y. Correlating sound velocity with the density, compressive strength and Young's modulus of carbonate rocks. *Int. J. Rock Mech. Min. Sci.* **2004**, *41*, 871–875. [CrossRef]
19. Dinçer, I.; Acar, A.; Çobanoğlu, I.; Uras, Y. Correlation between Schmidt hardness, uniaxial compressive strength and Young's modulus for andesites, basalts and tuffs. *Bull. Eng. Geol. Environ.* **2004**, *63*, 141–148. [CrossRef]
20. Behzadafshar, K.; Sarafraz, M.E.; Hasanipanah, M.; Mojtahedi, S.F.F.; Tahir, M.M. Proposing a new model to approximate the elasticity modulus of granite rock samples based on laboratory tests results. *Bull. Eng. Geol. Environ.* **2019**, *78*, 1527–1536. [CrossRef]
21. Dehghan, S.; Sattari, G.H.; Chelgani, S.C.; Aliabadi, M.A. Prediction of uniaxial compressive strength and modulus of elasticity for Travertine samples using regression and artificial neural networks. *Min. Sci. Technol.* **2010**, *20*, 41–46. [CrossRef]
22. Rezaei, M.; Majdi, A.; Monjezi, M. An intelligent approach to predict unconfined compressive strength of rock surrounding access tunnels in longwall coal mining. *Neural Comput. Appl.* **2014**, *24*, 233–241. [CrossRef]
23. Jin, X.; Zhao, R.; Ma, Y. Application of a Hybrid Machine Learning Model for the Prediction of Compressive Strength and Elastic Modulus of Rocks. *Minerals* **2022**, *12*, 1506. [CrossRef]
24. Ceryan, N.; Ozkat, E.C.; Korkmaz Can, N.; Ceryan, S. Machine learning models to estimate the elastic modulus of weathered magmatic rocks. *Environ. Earth Sci.* **2021**, *80*, 448. [CrossRef]
25. Shahani, N.M.; Zheng, X.; Liu, C.; Hassan, F.U.; Li, P. Developing an XGBoost regression model for predicting young's modulus of intact sedimentary rocks for the stability of surface and subsurface structures. *Front. Earth Sci.* **2021**, *9*, 761990. [CrossRef]
26. Li, C.; Zhou, J.; Dias, D.; Gui, Y. A Kernel Extreme Learning Machine-Grey Wolf Optimizer (KELM-GWO) Model to Predict Uniaxial Compressive Strength of Rock. *Appl. Sci.* **2022**, *12*, 8468. [CrossRef]
27. Elkatatny, S.; Tariq, Z.; Mahmoud, M.; Abdulraheem, A.; Mohamed, I. An integrated approach for estimating static Young's modulus using artificial intelligence tools. *Neural Comput. Appl.* **2019**, *31*, 4123–4135. [CrossRef]
28. Aboutaleb, S.; Behnia, M.; Bagherpour, R.; Bluekian, B. Using non-destructive tests for estimating uniaxial compressive strength and static Young's modulus of carbonate rocks via some modeling techniques. *Bull. Eng. Geol. Environ.* **2018**, *77*, 1717–1728. [CrossRef]
29. Siddig, O.; Elkatatny, S. Workflow to build a continuous static elastic moduli profile from the drilling data using artificial intelligence techniques. *J. Pet. Explor. Prod. Technol.* **2021**, *11*, 3713–3722. [CrossRef]
30. Mei, X.; Li, C.; Sheng, Q.; Cui, Z.; Zhou, J.; Dias, D. Development of a hybrid artificial intelligence model to predict the uniaxial compressive strength of a new aseismic layer made of rubber-sand concrete. *Mech. Adv. Mater. Struct.* **2022**, 1–18. [CrossRef]
31. Ocak, I.; Seker, S.E. Estimation of elastic modulus of intact rocks by artificial neural network. *Rock Mech. Rock Eng.* **2012**, *45*, 1047–1054. [CrossRef]
32. Pappalardo, G.; Mineo, S. Static elastic modulus of rocks predicted through regression models and Artificial Neural Network. *Eng. Geol.* **2022**, *308*, 106829. [CrossRef]
33. Singh, R.; Kainthola, A.; Singh, T.N. Estimation of elastic constant of rocks using an ANFIS approach. *Appl. Soft Comput.* **2012**, *12*, 40–45. [CrossRef]
34. Umrao, R.K.; Sharma, L.K.; Singh, R.; Singh, T.N. Determination of strength and modulus of elasticity of heterogenous sedimentary rocks: An ANFIS predictive technique. *Measurement* **2018**, *126*, 194–201. [CrossRef]
35. Acar, M.C.; Kaya, B. Models to estimate the elastic modulus of weak rocks based on least square support vector machine. *Arab. J. Geosci.* **2020**, *13*, 590. [CrossRef]
36. Al-Anazi, A.F.; Gates, I.D. On support vector regression to predict Poisson's ratio and Young's Modulus of reservoir rock. In *Artificial Intelligent Approaches in Petroleum Geosciences*; Springer: Cham, Switzerland, 2015; pp. 167–189.
37. Matin, S.S.; Farahzadi, L.; Makaremi, S.; Chelgani, S.C.; Sattari, G.H. Variable selection and prediction of uniaxial compressive strength and modulus of elasticity by random forest. *Appl. Soft Comput.* **2018**, *70*, 980–987. [CrossRef]
38. Khan, N.M.; Cao, K.; Yuan, Q.; Bin Mohd Hashim, M.H.; Rehman, H.; Hussain, S.; Khan, S. Application of Machine Learning and Multivariate Statistics to Predict Uniaxial Compressive Strength and Static Young's Modulus Using Physical Properties under Different Thermal Conditions. *Sustainability* **2022**, *14*, 9901. [CrossRef]
39. Mahmoud, A.A.; Elkatatny, S.; Al Shehri, D. Application of machine learning in evaluation of the static young's modulus for sandstone formations. *Sustainability* **2020**, *12*, 1880. [CrossRef]
40. Shahani, N.M.; Zheng, X.; Guo, X.; Wei, X. Machine Learning-Based Intelligent Prediction of Elastic Modulus of Rocks at Thar Coalfield. *Sustainability* **2022**, *14*, 3689. [CrossRef]
41. Tsang, L.; He, B.; Rashid, A.S.A.; Jalil, A.T.; Sabri, M.M.S. Predicting the Young's Modulus of Rock Material Based on Petrographic and Rock Index Tests Using Boosting and Bagging Intelligence Techniques. *Appl. Sci.* **2022**, *12*, 10258. [CrossRef]
42. Ghasemi, E.; Kalhori, H.; Bagherpour, R.; Yagiz, S. Model tree approach for predicting uniaxial compressive strength and Young's modulus of carbonate rocks. *Bull. Eng. Geol. Environ.* **2018**, *77*, 331–343. [CrossRef]
43. Tian, H.; Shu, J.; Han, L. The effect of ICA and PSO on ANN results in approximating elasticity modulus of rock material. *Eng. Comput.* **2019**, *35*, 305–314. [CrossRef]

44. Mokhtari, M.; Behnia, M. Comparison of LLNF, ANN, and COA-ANN techniques in modeling the uniaxial compressive strength and static Young's modulus of limestone of the Dalan formation. *Nat. Resour. Res.* **2019**, *28*, 223–239. [CrossRef]
45. Majdi, A.; Beiki, M. Evolving neural network using a genetic algorithm for predicting the deformation modulus of rock masses. *Int. J. Rock Mech. Min. Sci.* **2010**, *47*, 246–253. [CrossRef]
46. Gowida, A.; Moussa, T.; Elkatatny, S.; Ali, A. A hybrid artificial intelligence model to predict the elastic behavior of sandstone rocks. *Sustainability* **2019**, *11*, 5283. [CrossRef]
47. Cao, J.; Gao, J.; Nikafshan Rad, H.; Mohammed, A.S.; Hasanipanah, M.; Zhou, J. A novel systematic and evolved approach based on XGBoost-firefly algorithm to predict Young's modulus and unconfined compressive strength of rock. *Eng. Comput.* **2022**, *38*, 3829–3845. [CrossRef]
48. Fattahi, H. Application of improved support vector regression model for prediction of deformation modulus of a rock mass. *Eng. Comput.* **2016**, *32*, 567–580. [CrossRef]
49. Feng, X.T.; Chen, B.R.; Yang, C.; Zhou, H.; Ding, X. Identification of visco-elastic models for rocks using genetic programming coupled with the modified particle swarm optimization algorithm. *Int. J. Rock Mech. Min. Sci.* **2006**, *43*, 789–801. [CrossRef]
50. Shahani, N.M.; Zheng, X.; Liu, C.; Li, P.; Hassan, F.U. Application of soft computing methods to estimate uniaxial compressive strength and elastic modulus of soft sedimentary rocks. *Arab. J. Geosci.* **2022**, *15*, 384. [CrossRef]
51. Li, J.; Li, C.; Zhang, S. Application of Six Metaheuristic Optimization Algorithms and Random Forest in the uniaxial compressive strength of rock prediction. *Appl. Soft Comput.* **2022**, *131*, 109729. [CrossRef]
52. Tuğrul, A.T.İ.Y.E.; Zarif, I.H. Correlation of mineralogical and textural characteristics with engineering properties of selected granitic rocks from Turkey. *Eng. Geol.* **1999**, *51*, 303–317. [CrossRef]
53. Civicioglu, P. Backtracking search optimization algorithm for numerical optimization problems. *Appl. Math. Comput.* **2013**, *219*, 8121–8144. [CrossRef]
54. Mirjalili, S.; Mirjalili, S.M.; Hatamlou, A. Multi-verse optimizer: A nature-inspired algorithm for global optimization. *Neural Comput. Appl.* **2016**, *27*, 495–513. [CrossRef]
55. Mohammadi-Balani, A.; Nayeri, M.D.; Azar, A.; Taghizadeh-Yazdi, M. Golden eagle optimizer: A nature-inspired metaheuristic algorithm. *Comput. Ind. Eng.* **2021**, *152*, 107050. [CrossRef]
56. Moosavi, S.H.S.; Bardsiri, V.K. Poor and rich optimization algorithm: A new human-based and multi populations algorithm. *Eng. Appl. Artif. Intell.* **2019**, *86*, 165–181. [CrossRef]
57. Zhou, J.; Dai, Y.; Du, K.; Khandelwal, M.; Li, C.; Qiu, Y. COSMA-RF: New intelligent model based on chaos optimized slime mould algorithm and random forest for estimating the peak cutting force of conical picks. *Transp. Geotech.* **2022**, *36*, 100806. [CrossRef]
58. Dai, Y.; Khandelwal, M.; Qiu, Y.; Zhou, J.; Monjezi, M.; Yang, P. A hybrid metaheuristic approach using random forest and particle swarm optimization to study and evaluate backbreak in open-pit blasting. *Neural Comput. Appl.* **2022**, *34*, 6273–6288. [CrossRef]
59. Zhou, J.; Huang, S.; Zhou, T.; Armaghani, D.J.; Qiu, Y. Employing a genetic algorithm and grey wolf optimizer for optimizing RF models to evaluate soil liquefaction potential. *Artif. Intell. Rev.* **2022**, *55*, 5673–5705. [CrossRef]
60. Zhou, J.; Dai, Y.; Khandelwal, M.; Monjezi, M.; Yu, Z.; Qiu, Y. Performance of hybrid SCA-RF and HHO-RF models for predicting backbreak in open-pit mine blasting operations. *Nat. Resour. Res.* **2021**, *30*, 4753–4771. [CrossRef]
61. Koopialipoor, M.; Ghaleini, E.N.; Tootoonchi, H.; Jahed Armaghani, D.; Haghighi, M.; Hedayat, A. Developing a new intelligent technique to predict overbreak in tunnels using an artificial bee colony-based ANN. *Environ. Earth Sci.* **2019**, *78*, 165. [CrossRef]
62. Fathipour-Azar, H. Hybrid machine learning-based triaxial jointed rock mass strength. *Environ. Earth Sci.* **2022**, *81*, 118. [CrossRef]
63. Yu, C.; Koopialipoor, M.; Murlidhar, B.R.; Mohammed, A.S.; Armaghani, D.J.; Mohamad, E.T.; Wang, Z. Optimal ELM–Harris Hawks optimization and ELM–Grasshopper optimization models to forecast peak particle velocity resulting from mine blasting. *Nat. Resour. Res.* **2021**, *30*, 2647–2662. [CrossRef]
64. Jamei, M.; Hasanipanah, M.; Karbasi, M.; Ahmadianfar, I.; Taherifar, S. Prediction of flyrock induced by mine blasting using a novel kernel-based extreme learning machine. *J. Rock Mech. Geotech. Eng.* **2021**, *13*, 1438–1451. [CrossRef]
65. Ceryan, N. Prediction of Young's modulus of weathered igneous rocks using GRNN, RVM, and MPMR models with a new index. *J. Mt. Sci.* **2021**, *18*, 233–251. [CrossRef]

applied sciences

MDPI

Article

CatBoost–Bayesian Hybrid Model Adaptively Coupled with Modified Theoretical Equations for Estimating the Undrained Shear Strength of Clay

Huajian Yang, Zhikui Liu *, Yuantao Li, Haixia Wei and Nengsheng Huang

Department of Civil Engineering and Architecture, Guilin University of Technology, Guilin 541004, China
* Correspondence: 1998009@glut.edu.cn

Abstract: The undrained shear strength of clay is an important index for the calculation of the bearing capacity of the foundation soil, the calculation of the soil pressure of the foundation pit, and the analysis of the slope stability. Therefore, the purpose of this paper is to conduct a comprehensive study of the combined use of machine learning with clay theoretical equations to estimate it. Under the Bayesian framework, the CatBoost algorithm (CatBoost–Bayesian) based on Bayesian optimization algorithm was developed to obtain the feature importance level of soil parameters affecting the undrained shear strength of clay, so as to adaptively couple the theoretical equation of undrained shear strength of K_0 consolidated clay, which was derived from the modified Cambridge model. Then, the theoretical equation of undrained shear strength of the isotropically consolidated clay was established from the critical state of the clay parameters. Finally, it was illustrated and verified using the experimental samples of Finnish clay. The results indicate that the theoretical equation established by the overconsolidation ratio and effective overburden pressure parameters can well estimate the undrained shear strength of isotropically consolidated clays, and the parameter uncertainty can be considered explicitly and rigorously.

Keywords: clay; undrained shear strength; CatBoost–Bayesian; overconsolidation ratio; effective overburden pressure

Citation: Yang, H.; Liu, Z.; Li, Y.; Wei, H.; Huang, N. CatBoost–Bayesian Hybrid Model Adaptively Coupled with Modified Theoretical Equations for Estimating the Undrained Shear Strength of Clay. *Appl. Sci.* **2023**, *13*, 5418. https://doi.org/10.3390/app13095418

Academic Editor: Daniel Dias

Received: 12 April 2023
Revised: 24 April 2023
Accepted: 24 April 2023
Published: 26 April 2023

1. Introduction

The composition and physical and mechanical properties of clayey soil are important indexes in geotechnical engineering, among which the undrained shear strength index of clay is the most important index for calculation of the bearing capacity of foundation soil, calculation of the soil pressure of the foundation pit, and analysis of slope stability. Cross plate shear tests in the field, direct shear tests in the laboratory, triaxial compression tests, and other such direct measurement methods are time-consuming and expensive [1]. To evaluate through indirect methods, theoretical derivation and empirical methods are effective methods. Based on many experiments, Mesri and Ladd et al., respectively, proposed empirical equations for the undrained shear strength of clay [2,3]. Jiang Shuihua et al. simulated the uncertainty of soil parameters and proposed an effective non-stationary random field model for undrained shear strength parameters [4]. Indirect methods also include measurements obtained from dilatometer tests (DMT) based on empirical equations [5,6].

The application of machine learning in the field of geotechnical engineering has gradually become a research hotspot, including slope stability testing, TBM performance evaluation, rockburst vibration estimation, and pile foundation evaluation, etc., and has been proposed, expanded upon and applied by many researchers [7–10]. The emergence of artificial intelligence techniques holds great potential for solving soil parameter estimation problems involving complex soil–structure interactions [11–13]. However, compared with other directions in the field of geotechnical engineering, research devoted to soil parameter estimation is still very limited. The CatBoost algorithm can effectively solve gradient offset and prediction offset problems, improving the accuracy and generalization ability of the calculation [14]. Therefore, it has been widely used in crop evapotranspiration estimation in hydrology, data-driven seismic performance evaluation of corroded RC columns in civil engineering, and groundwater salinity prediction in multi-layer coastal aquifers in geotechnical engineering [14–18].

Moreover, the combination of Bayesian optimization with machine learning algorithms is still relatively rare, and Zhang et al. use extreme gradient boosting and random forest based on Bayesian optimization in geotechnical engineering to effectively capture the relationship between undrained shear strength and various fundamental soil parameters [19]. Ho et al. used the good performance of Bayesian, functional, and meta-ensemble machine learning models to generate land subsidence susceptibility (LSS) maps [20]. The Cambridge model and modified Cambridge model are two elastic–plastic constitutive models commonly used in geotechnical engineering to describe the shear properties of soils [21]. They are typical of the constitutive relationships of soils under undrained conditions. The Cambridge model describes the shear properties of soils under circular stress paths, while the modified Cambridge model modifies the stress paths in the plastic potential function to better describe the shear properties of soils under non-circular stress paths [22].

Under the Bayesian framework, the CatBoost algorithm (CatBoost–Bayesian) based on a Bayesian optimization algorithm was developed to obtain the feature importance level of soil parameters affecting the undrained shear strength of clay by using the experimental samples of Finnish clay, so as to adaptively couple the theoretical equation of undrained shear strength of K_0 consolidated clay, which was derived from the modified Cambridge model. Then, the theoretical equation of undrained shear strength of isotropically consolidated clay was established from the critical state of clay parameters. Finally, this paper analyzed the uncertainty of the parameters of the equation and verified the rationality of the calculation results. The rest of this study was arranged as follows. In Section 2, the CatBoost algorithm, the principle of Bayesian optimization algorithm based on random forest and K-fold cross-validation, as well as the theoretical equation derived from the revised Cambridge model, and the selection of quantitative evaluation indicators are introduced, respectively. The Finnish clay database used for training and validating the CatBoost–Bayesian hybrid model and theoretical equations is presented in Section 3, and a sensitivity analysis of different types of clay was performed to show the results of the importance of clay parameters. Finally, various verifications of the calculation results of the undrained shear strength theoretical equation of isotropically consolidated clay were carried out, and the results of the similar mixed model of CatBoost–Bayesian were compared and analyzed. Section 4 discusses the conclusions of this study, followed by references.

2. Materials and Methods

2.1. CatBoost Algorithm

The CatBoost algorithm was developed by researchers and engineers at Yandex, the largest Russian search engine company, in 2017, and opened source code in April [14,23]. It is the latest algorithm in the open-source field in the world today, and surpasses the XGBoost algorithm and the LightGBM algorithm in terms of performance. The name comes from "Category" and "Boosting", and belongs to the family of Boosting algorithms. The CatBoost algorithm is an improvement in the framework of the GBDT algorithm. It effectively solves the problem of gradient bias and prediction shift, avoids the occurrence of overfitting, and improves calculation accuracy and generalization ability (Figure 1). The details are as follows.

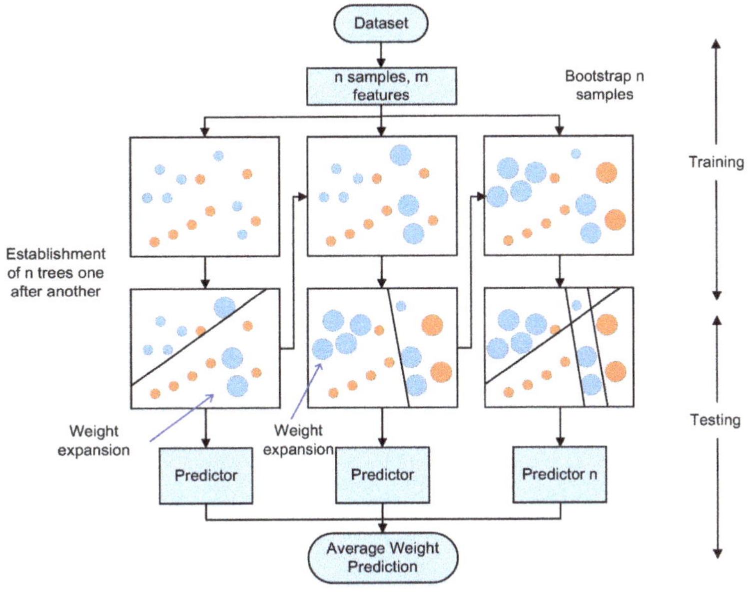

Figure 1. The structure of the CatBoost algorithm.

1. The CatBoost algorithm can handle categorical features in GBDT features better, and the simplest way is to use the average value of the corresponding labels to replace them. In the decision tree, the label average value will be used as the criterion for node splitting. This method is known as greedy target-based statistics, or greedy TS. However, this method has obvious drawbacks, so greedy TS is improved by adding prior distribution terms to reduce the effect of noise and low frequency categorical data on the data distribution [14,23].

$$\hat{x}_k^i = \frac{\sum_{j=1}^{p-1}\left[x_{\sigma_{j,k}} = x_{\sigma_{p,k}}\right] \cdot Y_{\sigma_j} + a \cdot p}{\sum_{j=1}^{p-1}\left[x_{\sigma_{j,k}} = x_{\sigma_{p,k}}\right] + a} \quad (1)$$

where p is the added prior term and a is usually a weighting factor greater than 0. For regression problems, the prior term can be taken as the average value of the dataset in general.

2. The prediction shift is caused by the gradient bias. To overcome this problem, CatBoost proposes a new algorithm called ordered boosting (Algorithm 1).

Algorithm 1: Ordered boosting pseudo-code algorithm

input: $\{(X_k, Y_k)\}_{k=1}^{n}, I$;
$\sigma \leftarrow$ random permutation of $[1, n]$;
$M_i \leftarrow 0$ for $i = 1 \ldots n$;
for $t \leftarrow 1$ to I do
    for $i \leftarrow 1$ to n do
$r_i \leftarrow y_i - M_{\sigma(i)-1}(X_i)$;
    for $i \leftarrow 1$ to n do
$\Delta M \leftarrow$ learn model $\left(\left(X_j, r_j\right) : \sigma(j) \leq i\right)$;
$M_i \leftarrow M_i + \Delta M$;

From the Algorithm 1, X_i denotes a sample, M_i denotes a separate model, and model M_i is obtained by training with a training set that does not contain sample X_i.

2.2. *Bayesian Optimization Algorithm (SMAC) and k-Fold Cross-Validation*

Hyperparameter optimization is a combinatorial optimization problem that cannot be optimized by gradient descent as general parameters. Evaluating a set of hyperparameter problems often involves issues such as resource allocation and efficiency. The simpler hyperparameter optimization methods include manual search, grid search, and random search. In this study, we used the currently popular adaptive hyperparameter search method of Bayesian optimization [24–26]. Based on the combinations of hyperparameters that were already tested, this method can predict the next combination that is more likely to bring the greatest benefit. This study is implemented using the optuna automatic hyperparameter optimization framework.

SMBO stands for sequential model-based optimization. The so-called serialization refers to the optimization by iterative method one trial at a time. SMBO is a specific implementation form of Bayesian optimization.

SMAC, proposed by Hutter et al. [27], stands for sequential model-based optimization for general algorithm configuration; the model originates from the random forest model. The random forest (RF) algorithm is a combination of the Bootstrap Aggregating algorithm proposed by Breiman in 1996 and the stochastic subspace algorithm proposed by Ho [28] in 1998. SMAC was initially designed to remove some of the limitations of SMBO to make it applicable to general algorithm configuration problems with many classification parameters and benchmark sets of instances, and to solve the case in which the parameter type cannot be discrete in the Gaussian regression process.

During the process of model training, the problem of data overfitting often occurs. The model can match the training data well, but cannot predict the data outside the training set well. This affects the accuracy of the final evaluation results.

K-fold cross-validation (CV) is widely used as a criterion for model selection [29], and was originally created to reduce the computational cost of leave-one-out cross-validation (LOOCV) [30]. It divides a dataset into K equal samples, of which K-1 samples are used to build the model for training, and the remaining 1 sample is used for validation. During this K-time iteration, each part of K is successively assigned as validation data.

Jung [31] proposed a new method of choosing K-fold to ensure K = ln(n) and guarantee $n/\mathrm{K} > 3d$ simultaneously. After verifying the feasibility of it, where n denotes a data point and d denotes the number of parameters (in this study, n = 202, d = 12, K = 5.3), and combining it with Pham [32] method, K was setted to 5. Therefore, this study used a five-fold cross-validation method for validation (Figure 2).

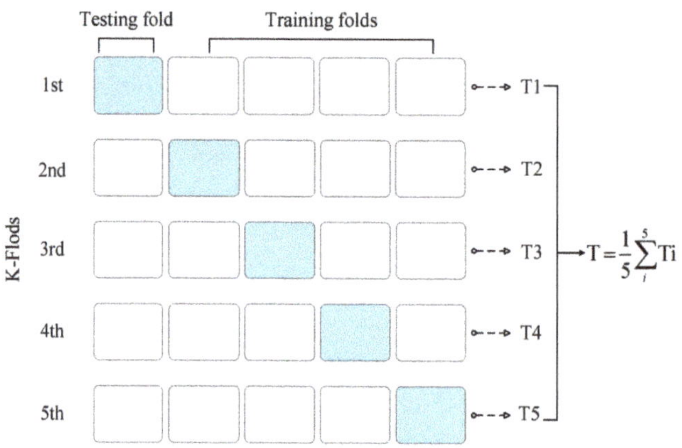

Figure 2. Five-fold cross-validation.

To reduce the reliance on empirical rules and inefficient brute force search, Bayesian optimization methods are applied to determine the appropriate model hyperparameters for the CatBoost algorithm (CatBoost–Bayesian). With appropriate model hyperparameters, the CatBoost–Bayesian hybrid model is able to provide a data-driven ranking of clay feature importance and properly uncover the intrinsic linkage of soil parameter essences; however, it cannot give full interpretability. To address this issue, the clay feature importance is adaptively coupled with the theoretical equations derived from the modified Cambridge model to estimate the undrained shear strength of isotropically consolidated clay under a Bayesian probabilistic framework, taking into account the parameter and model uncertainties, which are explained in detail in Sections 2.3 and 2.4.

2.3. Theoretical Equation

Deriving the undrained shear strength ratio S_u/σ'_{vi} from the soil intrinsic relationship is a common method for solving soil strength. Based on the Cambridge model, Ohta et al. [33] proposed the undrained shear strength derivation equation under the condition of triaxial K_0 (K_0 is the coefficient of earth pressure at rest) consolidation.

$$\left(\frac{S_u}{\sigma'_{vi}}\right)_{\text{OCA}} = OCR^{\overline{\Lambda}} \cdot \left(\frac{S_u}{\sigma'_{v0}}\right)_{\text{NCA}} \tag{2}$$

where OCA is the overconsolidated state of anisotropic clay, NCA is the normally consolidated state of anisotropic clay, the overconsolidation ratio is $OCR = \sigma'_{v0}/\sigma'_{vi}$, $\overline{\Lambda} = 1 - \overline{\kappa}/\lambda$, $\overline{\kappa}$ is the slope of the swelling lines described by the effective overburden pressure, and λ is the slope of the anisotropical normally consolidated soil compression curve:

$$\overline{\kappa} = \frac{(e_i - e_0)}{(ln\sigma'_{v0} - ln\sigma'_{vi})} \tag{3}$$

$$\kappa = \frac{(e_i - e_0)}{(lnp'_0 - lnp'_i)} \tag{4}$$

where κ is the slope of the swelling lines described by the effective average stress, p'_0 is the effective average vertical pressure in the anisotropical normally consolidated state, and p'_i is the effective average overburden pressure. e_i is the void ratio, and e_0 is the initial void ratio.

$$\frac{p'_0}{p'_i} = OCR^{(\overline{\kappa}/\kappa)} \tag{5}$$

Substitute $p'_0 = \frac{1}{3}(1 + 2K_{0nc}) \cdot \sigma'_{v0}$ and $p'_i = \frac{1}{3}(1 + 2K_0) \cdot \sigma'_{vi}$ into Equation (5), to obtain the following:

$$\overline{\kappa} = \kappa \cdot \frac{ln\left(\frac{1+2K_{0nc}}{1+2K_0}\right) + ln(OCR)}{ln(OCR)} \tag{6}$$

where K_0 is the coefficient of earth pressure at rest in the soil in the overconsolidated state, K_{0nc} is the coefficient of earth pressure at rest in the soil in the normally consolidated state, and the undrained shear strength ratio of the normally consolidated soil is

$$\left(\frac{S_u}{\sigma'_{v0}}\right)_{\text{NCA}} = \frac{1 + 2K_0}{3}\frac{M}{2}exp\left(-\Lambda \pm \frac{\Lambda}{M}\eta_0\right) \tag{7}$$

where $\Lambda = 1 - \kappa/\lambda$, $\eta_0 = 3 \cdot (1 - K_0)/(1 + 2 \cdot K_0)$, M is the critical state stress ratio, M= q/p, q is generalized shear stress, $q = (\sigma_1 - \sigma_3)$, p is the average principal stress, and $p = (\sigma_1 + \sigma_2 + \sigma_3)/3$. When $\eta_0 = 0$ and $K_0 = 1$, the ratio of undrained shear strength of the isotropical normally consolidated state and isotropically overconsolidated state clay is obtained, respectively.

$$\left(\frac{S_u}{\sigma'_{v0}}\right)_{\mathrm{NCI}} = \frac{M}{2^{\Lambda+1}} \tag{8}$$

$$\left(\frac{S_u}{\sigma'_{vi}}\right)_{\mathrm{OCI}} = OCR^{\Lambda}\cdot\left(\frac{S_u}{\sigma'_{v0}}\right)_{\mathrm{NCI}} = OCR^{\Lambda}\frac{M}{2^{\Lambda+1}} \tag{9}$$

where OCI is the overconsolidated state of isotropic clay, and NCI is the normally consolidated state of isotropic clay. Based on the experimental results of isotropic consolidation and swelling tests, Karube [34] proposed an empirical equation $M = 1.75\cdot\Lambda$.

From Equations (8) and (9), the equations for calculating the undrained shear strength of the isotropical normally consolidated and isotropically overconsolidated clays can be obtained.

$$(S_u)_{\mathrm{NCI}} = \sigma'_{v0}\cdot\frac{1.75\cdot\Lambda}{2^{\Lambda+1}} = \sigma'_{v0}\cdot\frac{1.75\cdot\left(1-\frac{\kappa}{\lambda}\right)}{2^{\left(2-\frac{\kappa}{\lambda}\right)}} \tag{10}$$

$$(S_u)_{\mathrm{OCI}} = \sigma'_{vi}\cdot OCR^{\Lambda}\cdot\frac{1.75\cdot\Lambda}{2^{\Lambda+1}} = \sigma'_{vi}\cdot OCR^{\left(1-\frac{\kappa}{\lambda}\right)}\cdot\frac{1.75\cdot\left(1-\frac{\kappa}{\lambda}\right)}{2^{\left(2-\frac{\kappa}{\lambda}\right)}} \tag{11}$$

In the Cambridge model, in isotropically consolidated clays, $\lambda = 0.434\cdot Cc$, $\kappa = 0.434\cdot Cs$ [35].The initial void ratio can be solved by equation $e_0 = 2.5\cdot C_c + 0.25$ [36]. Ladd [37] proposed an empirical equation for computing the undrained shear strength of clay based on the results of indoor and outdoor tests and examples of foundation failure.

$$\frac{S_u}{\sigma'_{vi}} = S\cdot(OCR)^m \tag{12}$$

where S is under normal consolidation conditions, m is the strength growth index, and m is affected by $1-\kappa/\lambda$ or $1-C_s/C_c$ according to the critical state theory [21].

2.4. Bayesian Perspective of Unified Undrained Shear Strength Equation

Unlike classical statistics, which is based on the frequentist approach, the CatBoost–Bayesian hybrid model is based on Bayesian statistics and machine learning. The CatBoost–Bayesian hybrid model uses Bayesian theory to combine engineering judgment and empirical prior knowledge with data to derive posterior distributions. In geotechnical analysis and prediction, engineers usually consider data from multiple sources [38]. This study updates the intrinsic model parameters (posterior distribution) with information from laboratory experimental data, field test data, and engineering experience [39]. Considering the differences in soil area and experimental methods, according to Equations (7)–(12), the effective overlying pressure σ'_{v_i} is used together, and σ'_{vi} and σ'_{v0} are converted by the correction coefficient β to complete the unification of the undrained shear strength equations of isotropic normal consolidation and isotropic overconsolidation, derived from the constitutive model.

$$S_u = f(\xi) + \varepsilon \tag{13}$$

where $\xi = \left[\alpha, \beta, m, \sigma'_{v_i}, OCR\right]$ represents the constitutive model parameter, $f(\xi) = \alpha\left(\sigma'_{v_i}\right)^{\beta}(OCR)^m$, $\alpha = 1.75m/2^{(1+m)}$, $n = 1-k/\lambda = 1-c_s/c_c$, and ε is a Gaussian random variable with mean u_ε=0 and standard deviation σ_ε = 0.215 [40]. Conditional probability theory is used to update the S_u posterior probability density function (PDF) based on prior information and laboratory test data of random variables. The calculation is as follows:

$$p(\xi|S_u) = KL(S_u|\xi)p(\xi) \tag{14}$$

where K is the normalization constant; $p(\xi)$ is the prior distribution of the key constitutive parameter ξ, and $L(S_u|\xi)$ is the likelihood function reflecting the probability relationship between the S_u of the laboratory test (site test) and constitutive model parameters.

2.4.1. Prior Distribution

Information about model parameters is usually limited given laboratory/field tests. Therefore, non-informative prior knowledge is used in this study. Assuming that a single random variable is independent, the prior distribution is equal to the product of the prior distributions of all constitutive model parameters. In order that there is no preference for any value in the possible range of parameters, uniform prior distribution is usually used. Uniform prior distribution is used to reflect the knowledge state of the model parameters in the Bayesian framework [41], expressed as

$$p(\xi) = \prod_{i=1}^{n} p(\xi_i) \tag{15}$$

$$p(\xi) = \begin{cases} \frac{1}{\mu_{imax}-\mu_{imin}} \cdot \frac{1}{\sigma_{imax}-\sigma_{imin}} & \mu_i \epsilon [\mu_{imin}, \mu_{imax}], \sigma_i \epsilon [\sigma_{imin}, \sigma_{imax}] \\ 0 & others \end{cases}. \tag{16}$$

where n is the number of constitutive model parameters, and μ_{imin}, μ_{imax}, σ_{imin} and σ_{imax} are the lower and upper limits of μ_i and σ_i respectively.

2.4.2. Likelihood Function

The likelihood function reflects the degree of consistency between constitutive model parameter A and the statistical characteristics of laboratory test/field test b, which can be expressed by multivariate normal distribution function [42]:

$$L(S_u|\xi) = \left(\frac{1}{\sqrt{2\pi}\sigma_{S_u}}\right)^N \prod_{i=1}^{N} \exp\left[\frac{(f_i(\xi) - \mu_{S_u})^2}{2\sigma_{S_u}^2}\right] \tag{17}$$

where N is the number of laboratory tests/field tests. μS_u and σS_u are the mean and standard deviation of S_u respectively, and $f_i(\xi)$ is the value estimated by the parameters of the constitutive model. With the continuous development of the field of geotechnical engineering driven by data, it is less difficult to obtain prior information than in previous engineering practices. In order to obtain updated posterior ξ, Markov chain Monte Carlo simulation (MCMC) is used to obtain discrete samples based on prior distribution and likelihood function simulation [43,44]. The basic idea of MCMC simulation is to repeatedly extract samples from the target distribution by converging with the Markov chain of the target distribution.

2.4.3. Uncertainty Analysis of the Unified Constitutive Model

The uncertainty of the undrained shear strength equations of isotropic normal consolidation and isotropic overconsolidation after updating the constitutive model parameters is studied by using a first-order approximation method. The mean value and standard deviation of a can be approximated as [45,46]

$$\mu_{S_u} = \alpha\left(\mu_{\sigma'_{vi}}\right)^{\beta} (\mu_{OCR})^m \tag{18}$$

$$\begin{aligned} \sigma_{S_u}^2 &= \left(\frac{\partial S_u}{\partial \sigma'_{vi}}\right)^2 \sigma_{\sigma'_{vi}}^2 + \left(\frac{\partial S_u}{\partial OCR}\right)^2 \sigma_{OCR}^2 + \sigma_\varepsilon^2 \\ &\approx \left[\frac{S_u\left(\mu_{\sigma'_{vi}} + \sigma_{\sigma'_{vi}}\right) - S_u\left(\mu_{\sigma'_{vi}} - \sigma_{\sigma'_{vi}}\right)}{2}\right]^2 + \left[\frac{S_u(\mu_{OCR}+\sigma_{OCR}) - S_u(\mu_{OCR}-\sigma_{OCR})}{2}\right]^2 + \sigma_\varepsilon^2 \end{aligned} \tag{19}$$

where S_u is a function of σ'_{vi} and OCR with means and standard deviations, $\mu_{\sigma'_{vi}}$ and $\sigma_{\sigma'_{vi}}$, μ_{OCR} and σ_{OCR}, respectively, and where ε is the model factor that takes into account the uncertainty of the intrinsic structure model and the effect of sample differences.

2.5. Quantitative Evaluation Indicators

R^2, Evar, RMSE, and MAE, the evaluation indicators used in this study, are commonly used to evaluate the prediction performance of the regression model, where R^2 and Evar explain the variance score of the regression model, and their values take the range of [0,1]; values closer to 1 indicate that the independent variable is more able to explain the variance change in the dependent variable, and smaller values indicate the opposite [47–49].

$$R^2 = 1 - \frac{\sum_{i=1}^{n}(y_i - \hat{y}_i)^2}{\sum_{i=1}^{n}(y_i - \overline{y})^2} \tag{20}$$

$$\text{Evar} = \text{Explain variance} = 1 - \frac{\text{Var}\{y_i - \hat{y}_i\}}{\text{Var}\{y_i\}} \tag{21}$$

RMSE indicates the average value of the square root of the error of the fitted data and the original data corresponding to the sample points; MAE assesses the degree of closeness between the prediction results and the real dataset. The smaller the value of these two, the better the fitting effect. The statistical indicators are computed as follows:

$$\text{RMSE} = \sqrt{\frac{\sum_{i=1}^{n}(y_i - \hat{y}_i)^2}{n}} \tag{22}$$

$$\text{MAE} = \frac{\sum_{i=1}^{n}|y_i - \hat{y}_i|}{n} \tag{23}$$

where y_i denotes the true value of the undrained shear strength of the clay sample, $\hat{y}_i$ denotes the predicted value of the undrained shear strength of the sample clay, $\overline{y}$ denotes the value of the undrained shear strength of the clay sample, and n denotes the number of the clay sample. The coefficient of variation (COV) is commonly used in engineering to perform an analysis of variance, showing the variability of a sample population relative to the sample mean.

$$\text{COV} = \frac{\sigma}{\mu} \tag{24}$$

where σ represents the standard deviation of the undrained shear strength of the sample clay, and μ represents the mean undrained shear strength of the sample clay.

3. Results and Discussion

A comprehensive study of the use of the CatBoost–Bayesian hybrid model with clay-corrected theoretical equations to estimate clay undrained shear strength values is shown in Figure 3.

3.1. Properties of Clays and the Database

The data used in this study were derived from the FI-CLAY/14/856 dataset in the TC304 database, made up of laboratory test data of clay parameters in 33 regions in Finland [50,51]. In this study, 11 parameters of 202 Finnish clay samples were selected for research, including organic content (Org), clay content (CI), void ratio (e), natural water content (W), liquid limit (LL), plastic limit (PL), effective overburden pressure (σ'_{vi}), preconsolidation pressure (σ'_{v0}), overconsolidation ratio (OCR), compression index (C_c), sensitivity (S_t), and undrained shear strength (S_u); the statistical results of this data set are as follows (Table 1 [51], Figure 4).

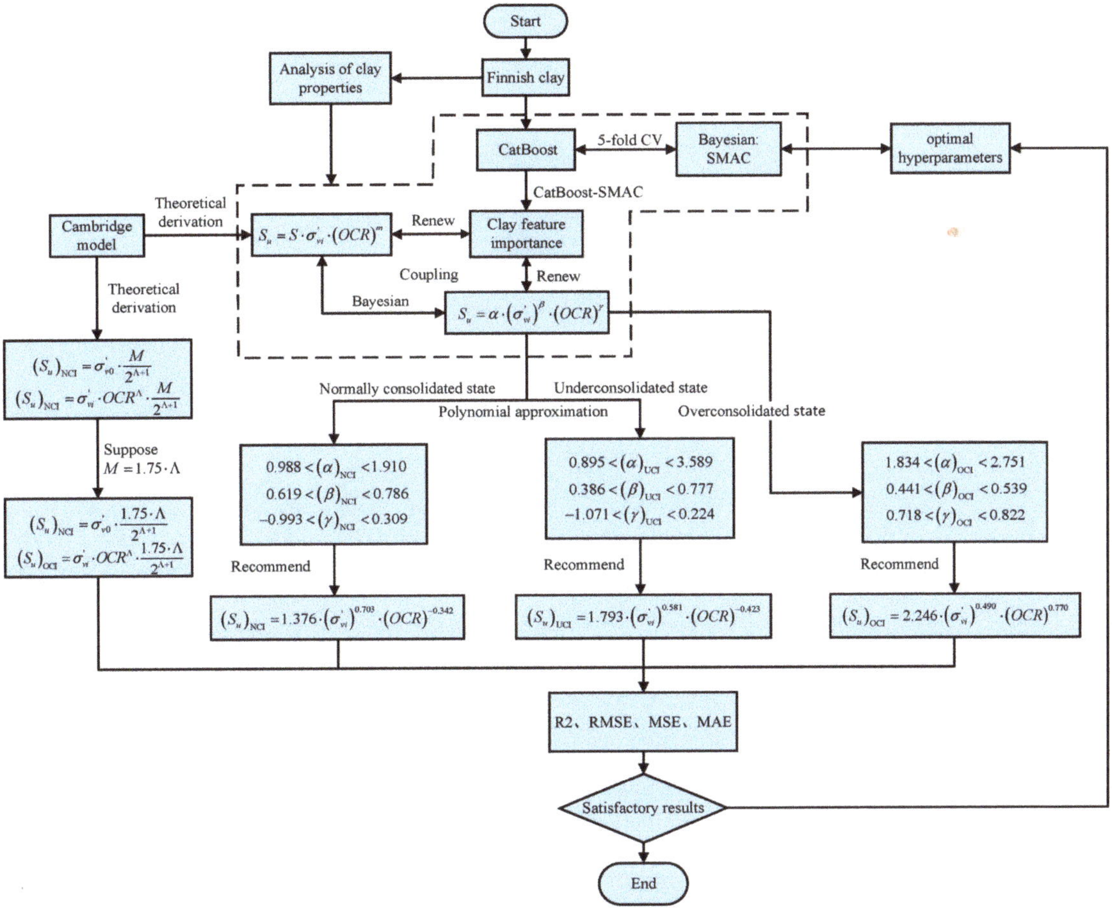

Figure 3. Flow chart of CatBoost–Bayesian adaptively coupled modified theoretical equations.

Table 1. Statistical analysis of dataset.

Parameters	Symbol	Min	Max	Std	Mean	COV	Unit
Organic content	Org	0.00	7.10	1.48	1.23	1.20	%
Clay content	CI	12.70	95.00	20.61	58.99	0.35	%
Void ratio	e	0.81	3.88	0.69	2.13	0.32	-
Natural water content	W	28.00	155.00	25.52	77.45	0.33	%
Liquid limit	LL	24.40	166.00	24.73	76.78	0.32	%
Plastic limit	PL	17.70	42.00	4.90	27.68	0.18	%
Effective in situ stress	σ'_{vi}	4.00	130.00	27.31	41.68	0.66	kPa
Preconsolidation pressure	σ'_{v0}	13.00	198.00	37.89	69.35	0.55	kPa
Overconsolidation ratio	OCR	0.46	20.00	1.81	2.12	0.85	-
Compression index	C_c	0.10	4.22	0.86	1.29	0.67	-
Sensitivity	S_t	1.69	163	19.97	24.26	0.84	-
Undrained shear strength	S_u	5.21	240.00	31.32	28.95	1.08	kPa

Std: Standard deviation.

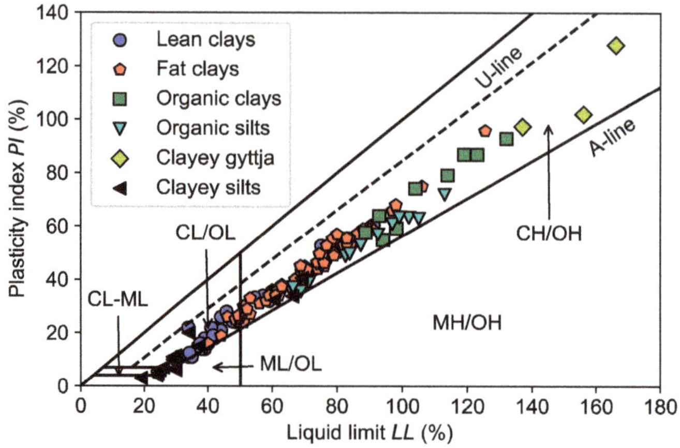

Figure 4. Plasticity chart and different soil types.

Figure 5 shows the normal distribution and fitting effect of organic content (Org), clay content (CI), void ratio (e), natural water content (W), liquid limit (LL), plastic limit (PL), effective overburden pressure (σ'_{vi}), preconsolidation pressure (σ'_{v0}), overconsolidation ratio (OCR), compression index (C_c), and undrained shear strength (S_u), and an analysis of the abnormal value of clay undrained shear strength.

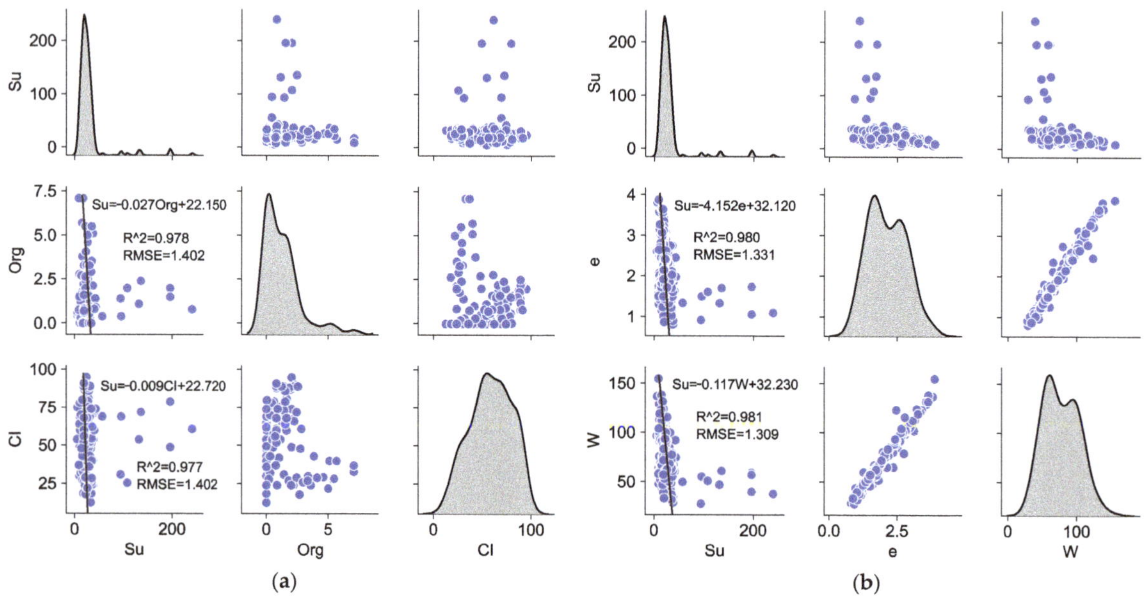

Figure 5. *Cont.*

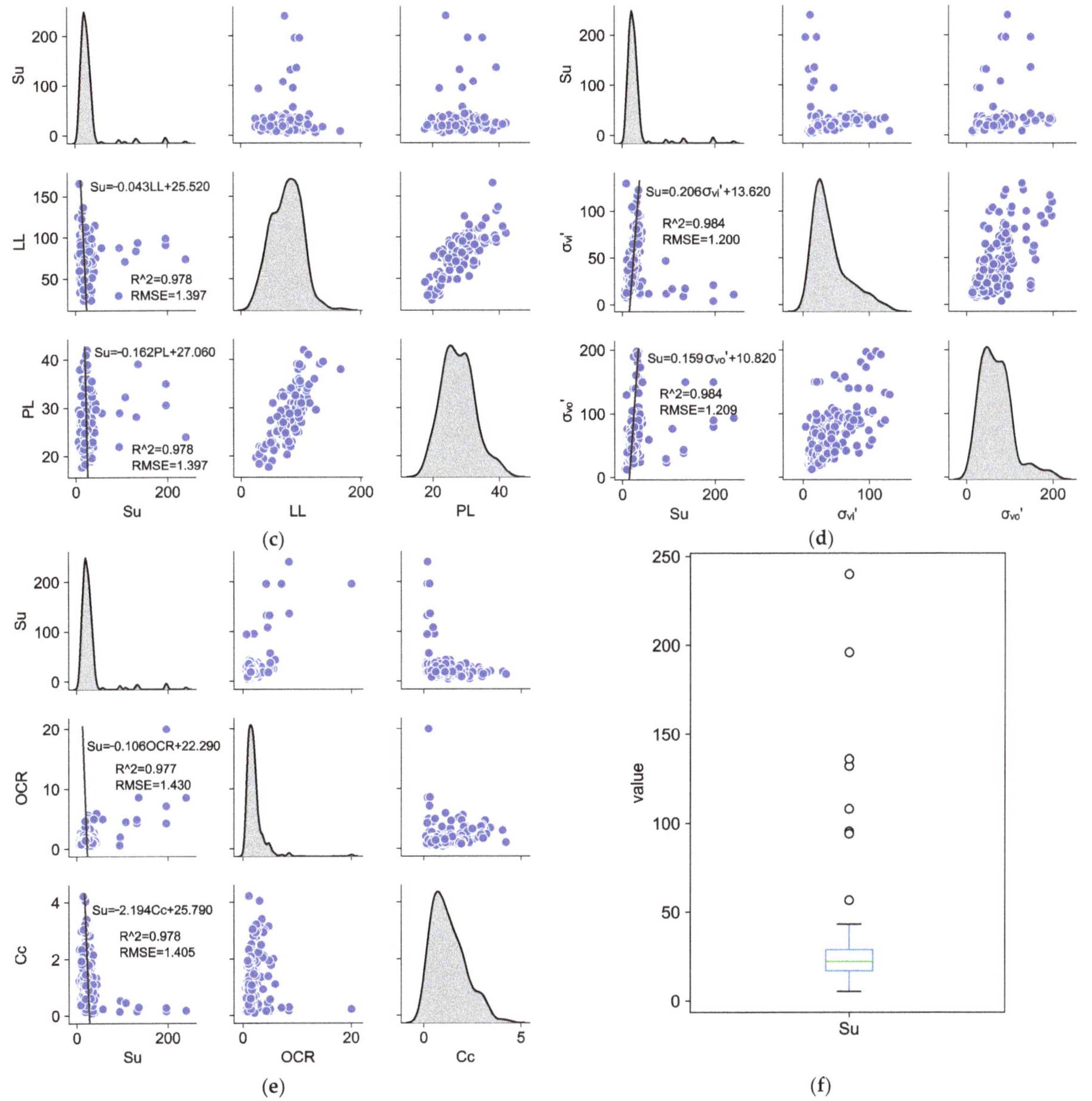

Figure 5. Visualization of normal distribution of 11 clay parameters. (**a**) Fitted relationship of S_u, Org, CI. (**b**) Fitted relationship of S_u, e, W. (**c**) Fitted relationship of S_u, LL, PL. (**d**) Fitted relationship of S_u, σ'_{vi}, σ'_{v0}. (**e**) Fitted relationship of S_u, OCR, C_c. (**f**) S_u abnormal value analysis.

Considering the abnormal value of clay undrained shear strength to reduce its influence on the fitting, a simple linear fitting can be performed to obtain the empirical equation of clay undrained shear strength parameters (Figure 5). It could be observed that between the single parameters of clay, the linear relationship is obvious, but the fitting error is large. At the same time, the empirical equation could not be verified. There is a certain particularity, and the constitutive relationship of the clay could not be obtained. Therefore, further research on the relationship between the multi-parameters of the clay is needed.

3.2. Feature Importance of CatBoost–Bayesian Hybrid Model

The CatBoost algorithm (CatBoost–Bayesian hybrid model) was optimized using the Bayesian optimization algorithm under the five-fold cross-validation, and the optimal hyperparameters of the CatBoost algorithm were obtained (Table 2); this study was carried out under the optimal parameters to avoid the excessive model fit, and was beneficial to the CatBoost–Bayesian hybrid model's adaptively coupled modified theoretical equations for stable estimation of the undrained shear strength of clay. Through the CatBoost–Bayesian hybrid model, an importance ranking of soil parameters that affect the undrained shear strength of clay was obtained.

Table 2. CatBoost optimal hyperparameters.

Optuna_Parameters of CatBoost	Description
loss function	MAE
n_estimators	1000
learning rate	0.153
random state	2019
l2_leaf_reg	0.030
colsample_bylevel	0.098
depth	1
boosting type	Plain
bootstrap type	MVS
min_data_in_leaf	4
one_hot_max_size	3
early_stopping_rounds	100

When the CatBoost–Bayesian hybrid model was used to estimate the undrained shear strength of clay, the feature importance of the model input parameters to the hybrid model under five-fold cross-validation was obtained, and it was explained whether the input parameters contribute positively (positive correlation) or negatively (negative correlation). Each point in the graph represents a data point from the training set. The color represents the value of the feature parameter; red represents the sample with a higher value of the feature parameter, and blue represents the sample with a lower value of the feature parameter. The length of the horizontal line represents the importance of the estimation of the undrained shear strength of the clay.

It can be seen from the whole that the overconsolidation ratio (OCR) is the most important parameter for estimating the undrained shear strength of clay, followed by preconsolidation pressure (σ'_{v0}), effective overburden pressure (σ'_{vi}), etc. The effective overburden pressure (σ'_{vi}) increases and the undrained shear strength depends on the consolidation stress before shearing, that is, the research status of the preconsolidation pressure (σ'_{v0}) [52–54]. The void ratio (e) and clay content (CI) features are the least important. At the same time, the characteristic variables significantly show that the compression index (C_c) makes a negative contribution to the computed value of the drainage shear strength, while the preconsolidation pressure (σ'_{v0}) and the overconsolidation ratio (OCR) make a positive contribution to the computed value of the drainage shear strength (Figure 6). The model's interpretability method is expected to help geotechnical engineers in the selection of soil parameters in practical engineering. However, the CatBoost–Bayesian hybrid model is still a black-box model, which finds it difficult to explain the internal mechanism of the model and the feature importance; therefore, the combination of the feature importance results with the theoretical equations derived from the constitutive model can provide a reference for geotechnical analysis.

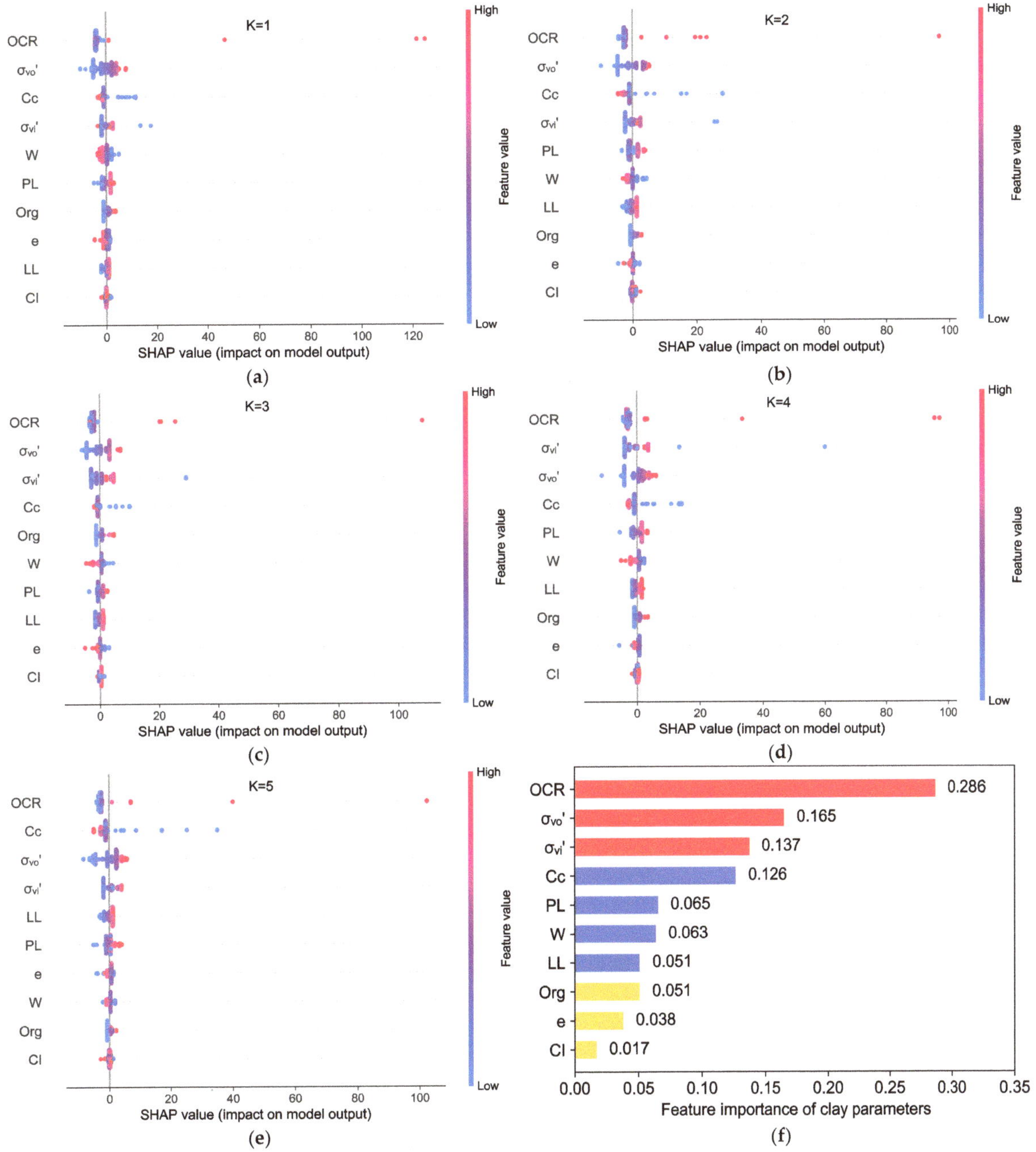

Figure 6. Clay parameter importance ranking of CatBoost–Bayesian. (**a**) Feature importance analysis of K = 1. (**b**) Feature importance analysis of K = 2. (**c**) Feature importance analysis of K = 3. (**d**) Feature importance analysis of K = 4. (**e**) Feature importance analysis of K = 5. (**f**) Feature importance ranking.

3.3. Estimation of Clay Undrained Shear Strength

3.3.1. Uncertainty Analysis of Equation Parameters

Under the Bayesian framework, the CatBoost algorithm (CatBoost–Bayesian) based on a Bayesian optimization algorithm was developed to obtain the feature importance level of soil parameters affecting the undrained shear strength of clay, so as to adaptively couple the theoretical equation of undrained shear strength of K_0 consolidated clay, which was derived from the modified Cambridge model; then, the theoretical equation of undrained shear strength of isotropically consolidated clay was established from the critical state of the overconsolidation ratio (OCR) and overburden effective pressure (σ'_{vi}).

Considering the constitutive relation of clay, the calculation results of the established undrained shear strength theoretical equation of isotropically overconsolidated clay were used for inversion calculation. The equation parameter γ of the clay in the overconsolidated state is affected by $(1 - C_s/C_c)$, $\gamma = m = 0.88 \cdot (1 - C_s/C_c) \pm 0.06 \cdot SD$, and the equation constant parameter is affected by the plasticity index, $\alpha = 2 + 0.5 \cdot I_p$. For the measured parameters of clay that are difficult to obtain, it is recommended that when the clay depth is 1.9 m $< D <$ 17 m, the calculation parameters of undrained shear strength should be computed according to the following recommended values $(\alpha)_{\text{OCI}} = 2.246$, $(\beta)_{\text{OCI}} = 0.490$; the value range of the correction coefficient of σ'_{vi} is $0.441 < (\beta)_{\text{OCI}} < 0.539$, and the average value is $(\beta)_{\text{OCI}} = 0.490$. The theoretical equation is $(S_u)_{\text{OCI}} = 2.246 \cdot (\sigma'_{vi})^{0.490} \cdot (OCR)^{0.770}$. This results in a stable estimate of the undrained shear strength of isotropically consolidated clays (Figure 7a,d).

$$(S_u)_{\text{OCI}} = 2.246 \cdot (\sigma'_{vi})^{0.490} \cdot (OCR)^{0.770} \tag{25}$$

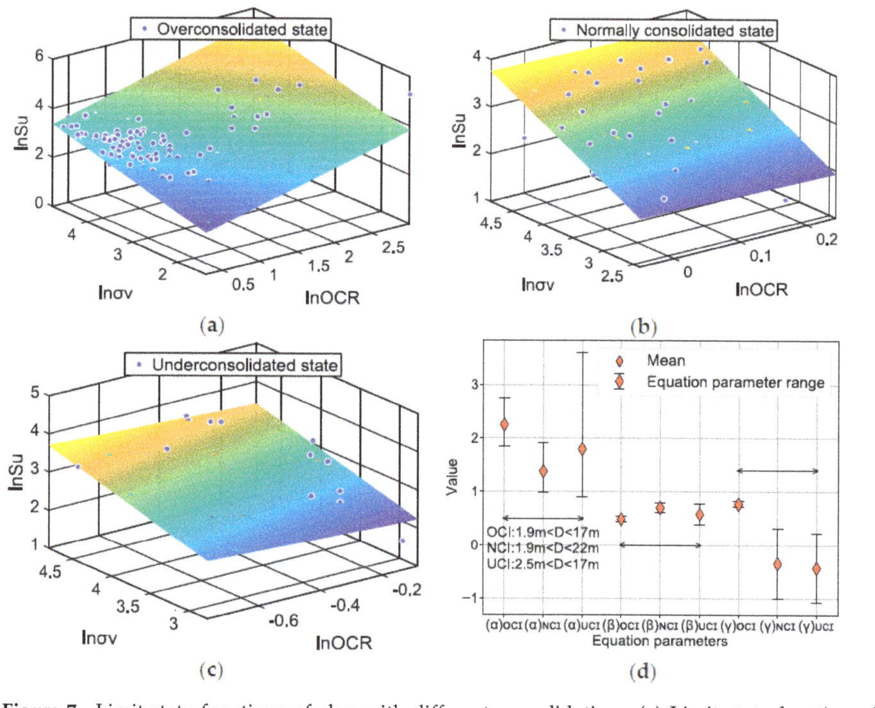

Figure 7. Limit state functions of clay with different consolidation. (**a**) Limit state function of overconsolidated clay. (**b**) Limit state function of normally consolidated clay. (**c**) Limit state function of underconsolidated clay. (**d**) Parameter analysis of equation of state for different consolidations.

In the normally consolidated state, the fitting parameter ranges were $0.988 < (\alpha)_{\text{NCI}} < 1.910$, $0.619 < (\beta)_{\text{NCI}} < 0.786$, and $-0.993 < (\gamma)_{\text{NCI}} < 0.309$ in this study (Figure 7d). There is a relatively stable estimation result when the clay depth is $1.9m < D < 22m$. Therefore, for the calculation of the parameters of undrained shear strength, it is recommended to take the means $(\alpha)_{\text{NCI}} = 1.376$, $(\beta)_{\text{NCI}} = 0.703$ and $(\gamma)_{\text{NCI}} = -0.342$ (Figure 6b).

$$(S_u)_{\text{NCI}} = 1.376 \cdot (\sigma'_{vi})^{0.703} \cdot (OCR)^{-0.342} \tag{26}$$

In the underconsolidated state, the parameter ranges $0.895 < (\alpha)_{\text{UCI}} < 3.589$, $0.386 < (\beta)_{\text{UCI}} < 0.777$, and $0.386 < (\beta)_{\text{UCI}} < 0.777$ were fitted in this study (Figure 7d). When the clay depth was 2.5 m $< D <$ 17 m , the undrained shear strength has a relatively stable estimation result. It is recommended to take the means $(\alpha)_{\text{UCI}} = 1.793$, $(\beta)_{\text{NCI}} = 0.581$ and $(\gamma)_{\text{NCI}} = -0.423$ for calculation (Figure 7c). The underconsolidated state of clay is the result of the interaction between the strength characteristics of the soil itself and excess pore water pressure, and its undrained shear resistance exhibits structural properties [55]. Therefore, it cannot be verified by the previous results, and it needs to be verified by the relationship between the measured value and the estimated value.

$$(S_u)_{\text{UCI}} = 1.793 \cdot (\sigma'_{vi})^{0.581} \cdot (OCR)^{-0.423} \tag{27}$$

where UCI is the underconsolidated state of isotropic clay.

It can be seen from the whole that the estimated undrained shear strength of clay is affected by the clay depth (D) in different consolidation states, which is in line with the actual situation to a certain extent. Asaoka, Guo, and Jiang et al. verified that the undrained shear strength of clay exhibits an obvious linear trend along the depth, through cross-plate shear tests [56,57].

3.3.2. Verification of the Feasibility of the Theoretical Equation

The Bayesian framework concept is an effective means of correcting the original judgments using the new information collected. Based on the new information of the S_u and OCR ofthe Finnish clay database, as well as the σ'_{vi} test index, priori equations are corrected so that the generated posterior equations are more realistic and have fewer errors. The mean and standard deviation of S_u are also calculated to verify the distribution form of the posterior equation.

The form of the probability distribution of S_u for the posterior equation was performed to verify the rationality of the Bayesian framework. By testing, the posterior equation estimated the undrained shear strength with small standard deviation and low variability (Figure 8).

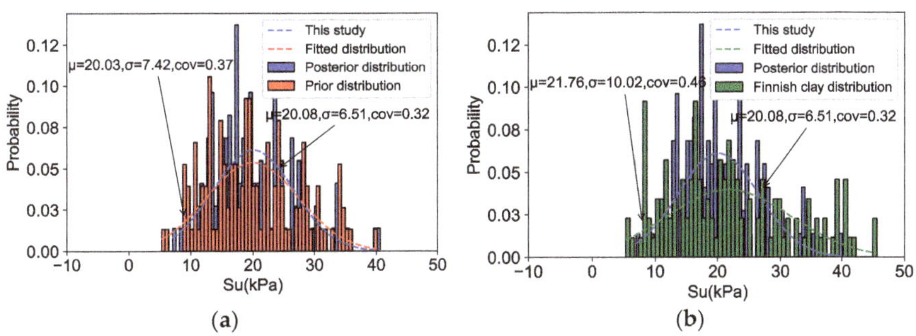

Figure 8. Test of probability distribution of undrained shear strength. (**a**) Test of prior and posterior probability distribution. (**b**) Test of Finnish clay and posterior probability distribution.

The results visualized in Figure 9 were computed by the theoretical equation mentioned above. The measured and theoretically computed values of the undrained shear strength of clay were compared, in which the abscissa is the measured value, and the ordinate is the theoretical computed value. For the underconsolidated state of clay (Figure 9a), since there was no previous result to verify, Equation (10) of Ohta and Wang et al. was selected to better estimate the undrained shear strength of clay. The theoretically computed value is in good agreement with the measured value. For the normally consolidated state of clay (Figure 9b), the theoretically computed results in this study are slightly larger or smaller than the measured values; meanwhile, for the overconsolidated state of clay (Figure 9c), the theoretically computed results in this study are in good agreement with the measured values, and the computed values of Ohta and Wang et al. are obviously larger. It is found that the R^2 values in the underconsolidated state, the normally consolidated state and the overconsolidated state are 0.88, 0.91 and 0.97, respectively.

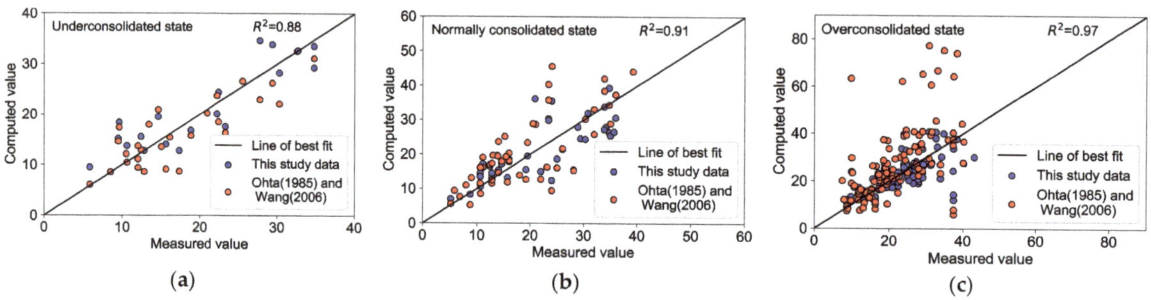

Figure 9. Comparison between the theoretically computed and measured values of S_u. (**a**) 2.5 m < D < 17 m. (**b**) 1.9 m < D < 22 m. (**c**) 1.9 m < D < 17 m.

Under the double logarithmic coordinate, the undrained strength ratio $S_u/(\sigma'_{vi})^{\beta}$, corrected by β, and the overconsolidation ratio OCR both show an obvious linear relationship, which verifies the theory proposed by our predecessors. $S_u/(\sigma'_{vi})^{\beta}$ increases linearly with the increase in OCR in the overconsolidated state (Figure 10a), and $S_u/(\sigma'_{vi})^{\beta}$ decreases linearly with the increase in OCR in the underconsolidated state and normally consolidated state (Figure 10b). The reliability of the theoretical equation proposed in this study is further verified. It is not difficult to find that the gap between the theoretical computed value and the experimental point in this study is significantly smaller than the gap between the theoretically computed value and the measured value studied by Ohta and Wang et al. (Figure 10a).

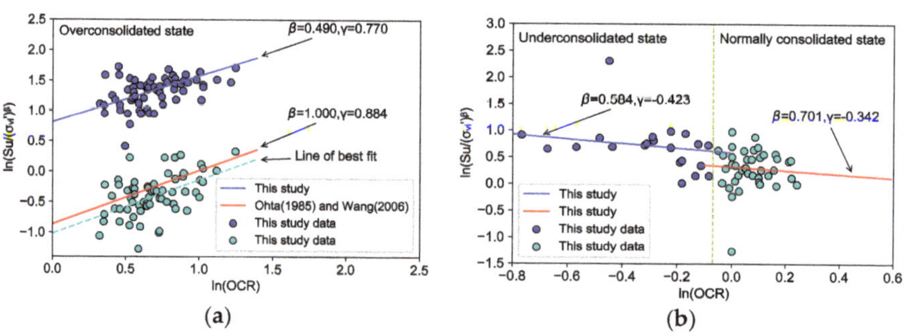

Figure 10. The relationship between $S_u/(\sigma'_{vi})^{\beta}$ and OCR. (**a**) Overconsolidated state. (**b**) Underconsolidated and normally consolidated state.

This study found that $(S_u)_{NCI}$ has a stable estimation result when the clay depth is 1.9 m < D < 22 m, and $(S_u)_{OCI}$ has stable estimation result when the clay depth is 1.9 m < D < 17 m. When the clay depth is 0 m < D < 1.9 m, the estimated value of $(S_u)_{OCI}$ fluctuates greatly. The smaller the compressibility index (C_c) value, the lower the compressibility of the soil, and the C_c value of the low compressibility soil is generally less than 0.2. When the depth is less than 1.9 m, the clay in this study is in a state of low compressibility; at the same time, the natural water content is low and the degree of looseness is high, so it is not conducive to estimating the undrained shear strength of the clay. It may be seen from the feature importance ranking of the CatBoost–Bayesian hybrid model that the natural water content (W) and the compression index (C_c) are important parameters of clay, which can explain the large error in the estimation of $(S_u)_{\text{OCI}}$ by the theoretical equation (Figure 11). Combined with the analysis of the geological tectonic environment, it is because the deposition time of the upper part of the clay depth is relatively short, which is affected by long-term evaporation and water loss. The lower clay has a long deposition time and is affected by the rise and fall in the groundwater level for a long time, which is equivalent to continuous loading and unloading, and finally shows the overconsolidation characteristics and the abrupt change of the clay properties. Therefore, the computed value of the theoretical equation should be much smaller than the measured value.

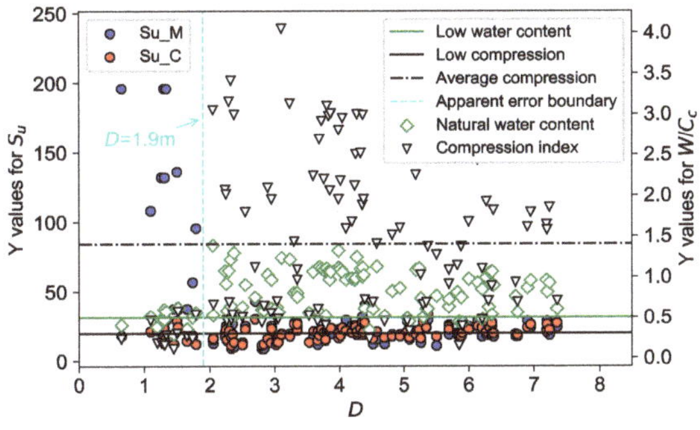

Figure 11. Error analysis of undrained shear strength of clay.

3.3.3. Comparative Analysis of Estimation Results

Considering the effect of clay depth (D), the combination of the feature importance of the CatBoost–Bayesian hybrid model and the calculation of the theoretical equation derived from the modified Cambridge model can well estimate the undrained shear strength of clay, and the theoretical computed value is in good agreement with the measured value (Figure 12a,b). The computed average properties of the clay at different consolidation states were $\text{R}^2 = 0.92$, $\text{Evar} = 0.92$, $\text{RMSE} = 0.19$, $\text{MAE} = 0.03$. This indicated that the uncertainty of the theoretical equation is significantly reduced when the overconsolidation ratio was combined with the preconsolidation pressure or the effective overburden pressure.

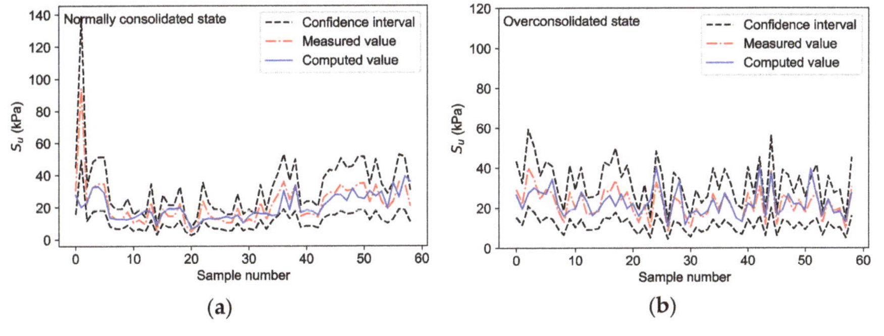

Figure 12. Comparison of S_u computed results. (**a**) Recommended equation: $R^2 = 0.91$. (**b**) Recommended equation: $R^2 = 0.97$.

The computed results of the test set of the CatBoost–Bayesian hybrid model under five-fold cross-validation are within the 95% confidence interval, and the computed average performance is training set $R^2 = 0.91$, Evar $= 0.91$, RMSE $= 0.30$, MAE $= 0.12$; test set $R^2 = 0.86$, Evar $= 0.86$, RMSE $= 0.37$, MAE $= 0.20$. Only a very small number of undrained shear strength estimates at K3 in the test set exceed the 95% confidence interval, proving the reliability of the computed results [58]. In the five-fold cross-validation, there is a gap between the computed value of a small amount of undrained shear strength and the real value, which leads to the large value of RMSE and MAE, and the fitting effect is not significant enough (Figure 13a–d).

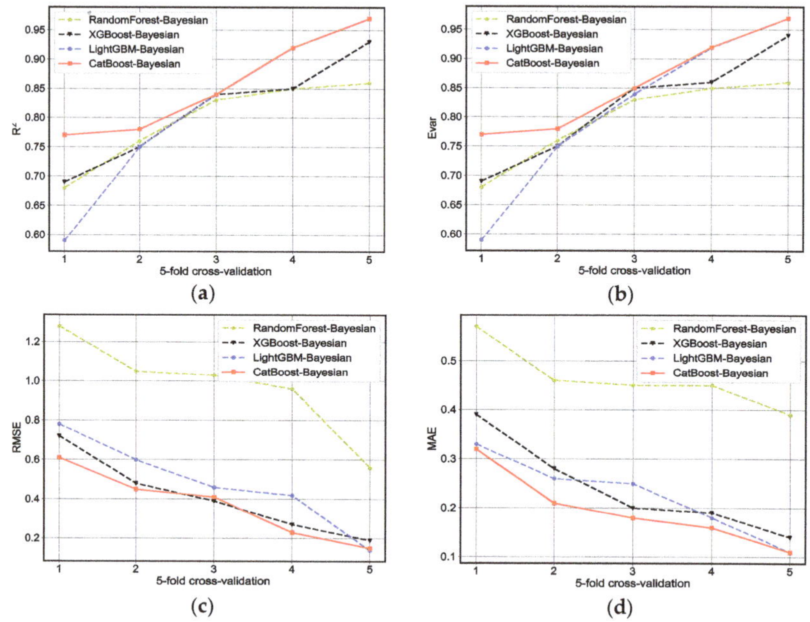

Figure 13. Comparison of quantitative evaluation indicators. (**a**) R^2. (**b**) Evar. (**c**) RMSE. (**d**) MAE.

As shown in Figure 14, UCI_S_u, NCI_S_u, and OCI_S_u demonstrate the performance of CatBoost–Bayesian hybrid model for estimating the undrained shear strength of clay in different consolidation states of clay, as well as the overall CatBoost–Bayesian hybrid model performance, which is subsequently compared with LightGBM-Bayesian, XGBoost-Bayesian hybrid model for comparison. The training and test set performance evaluations are shown in Table 3.

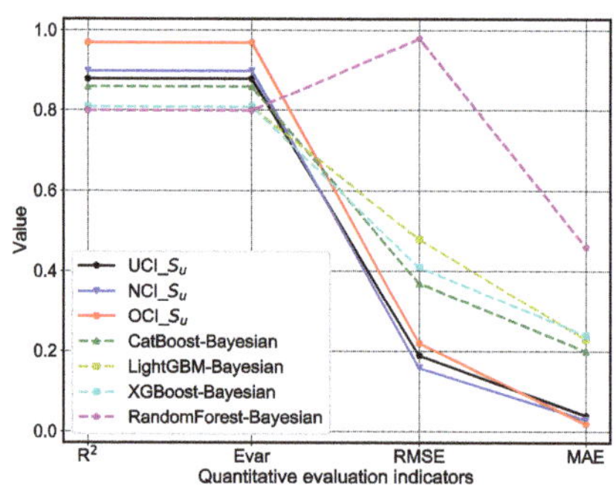

Figure 14. Estimation performance visualization.

Table 3. Estimation performance comparison of four hybrid models.

	R^2_Mean		Evar_Mean		RMSE_Mean		MAE_Mean	
	Train	Test	Train	Test	Train	Test	Train	Test
UCI_*Su*	-	0.88	-	0.88	-	0.19	-	0.04
NCI_*Su*	-	0.90	-	0.90	-	0.16	-	0.03
OCI_*Su*	-	0.97	-	0.97	-	0.22	-	0.02
CatBoost–Bayesian	0.91	0.86	0.91	0.86	0.30	0.37	0.12	0.20
LightGBM-Bayesian	0.99	0.81	0.99	0.81	0.14	0.48	0.08	0.23
XGBoost-Bayesian	0.94	0.81	0.94	0.82	0.24	0.41	0.16	0.24
RandomForest -Bayesian	0.95	0.80	0.95	0.80	0.99	0.98	0.46	0.46

On the R^2 and Evar curves explaining the variance score of the model, the CatBoost–Bayesian hybrid model has the strongest ability to explain the variance in the undrained shear strength under five-fold cross-validation, which could explain almost 86% of the 202 undrained shear strengths of clay. The sample variability shows that the hybrid model has a better effect; RMSE and MAE evaluate the closeness of the predicted clay undrained shear strength to its experimental value, and the smaller the value, the better the model fitting effect. It could be observed that the estimated results for clays in different isotropically consolidated states could explain almost 92% of the 202 sample variability in the undrained shear strength of clays. The RMSE and MAE of this study are all lower than other intelligent mixed models, and the results show that the combination of CatBoost–Bayesian feature importance and theoretical formula has the best fitting effect (Figure 14).

The estimated performance results in Table 3 show that the theoretical equations of UCI_*Su*, NCI_*Su* and OCI_*Su* derived by combining the CatBoost–Bayesian importance parameters are better than other models in different consolidation states of clay. Among the models of the same type, the estimation performance of the CatBoost–Bayesian hybrid model in the training set and test set is better than other models.

4. Conclusions

Unlike the extensive traditional study of transformation models based on empirical evidence, data-driven ensemble learning methods combined with traditional empirical models have received limited research attention in geotechnical engineering. Moreover, the comprehensive use of Bayesian theory for probabilistic characterization of soil parameter uncertainties and algorithmic optimization problems has not been fully leveraged in a systematic and coherent manner. To address the above challenges, the following study has been conducted. Under the Bayesian framework, the CatBoost algorithm (CatBoost–Bayesian) based on Bayesian optimization algorithm was developed to obtain the feature importance level of soil parameters affecting the undrained shear strength of clay, so as to adaptively couple the theoretical equation of undrained shear strength of K_0 consolidated clay, which was derived from the modified Cambridge model, and then the theoretical equation of undrained shear strength of isotropically consolidated clay was established from the critical state of clay parameters, and the calculation results were verified later.

1. From the feature importance ranking of the CatBoost–Bayesian hybrid model, parameters with high importance and ease of measurement were selected; the overconsolidation ratio (OCR) and the effective overburden pressure (σ'_{vi}) could reasonably explain the model and indirectly estimate the undrained shear strength of the clay.
2. The equation parameter γ of the clay in the overconsolidated state was affected by $(1 - C_s/C_c), \gamma = m = 0.88 \cdot (1 - C_s/C_c) \pm 0.06 \cdot SD$, and the equation parameter α was affected by the plasticity index, $\alpha = 2 + 0.5 \cdot I_p$. For the measured parameters of clay that were difficult to obtain, it was recommended that when the clay depth is $1.9m < D < 17m$, the calculation parameters of undrained shear strength should be computed according to the following recommended values $(\alpha)_{\text{OCI}} = 2.246$, $(\beta)_{\text{OCI}} = 0.490$ and $(\gamma)_{\text{OCI}} = 0.770$. The theoretical equation was $(S_u)_{\text{OCI}} = 2.246 \cdot (\sigma'_{vi})^{0.490} \cdot (OCR)^{0.770}$.

3. When the undrained shear strength of clay in the normally consolidated state was estimated at a depth of $1.9m < D < 22m$, the recommended theoretical equation was $(S_u)_{\text{NCI}} = 1.376 \cdot (\sigma'_{vi})^{0.703} \cdot (OCR)^{-0.342}$; when the undrained shear strength of clay in the underconsolidated state was estimated at a depth of $1.9m < D < 22m$, the recommended theoretical equation was $(S_u)_{\text{UCI}} = 1.793 \cdot (\sigma'_{vi})^{0.581} \cdot (OCR)^{-0.423}$.
4. Compared with the calculation results of Ohta and Wang et al., it was found that the theoretical equation in this study can well estimate the undrained shear strength of isotropically consolidated clay. When the clay depth is $0\ m < D < 1.9\ m$, the huge fluctuation of the estimated value of $(S_u)_{\text{OCI}}$ is mainly due to the long-term influence of evaporative water loss in the upper part of the clay.
5. The CatBoost–Bayesian hybrid model could excavate the intrinsic relationship of the soil parameters, but it could not give a comprehensive interpretability. The undrained shear strength of isotropic clays was estimated and is to a certain extent interpretable by the CatBoost–Bayesian hybrid model feature importance, adaptively coupled to the theoretical equation derived from the modified Cambridge model. Comparing the results of the CatBoost–Bayesian hybrid model and its similar hybrid models, this study ensured that the average R^2 reaches 0.92, the average RMSE and MAE were 0.19 and 0.03, respectively, and the overall performance was good.

Author Contributions: Conceptualization, H.Y. and Z.L.; methodology, H.Y. and Z.L.; writing—original draft preparation, H.Y. and Y.L.; writing—review and editing, Y.L., H.W. and N.H.; funding acquisition, Z.L. All authors have read and agreed to the published version of the manuscript.

Funding: This study was supported by the National Natural Science Foundation of China (41867039), Guangxi Key Laboratory of Geomechanics and Geotechnical Engineering (20-Y-XT-03), and the Foundation Project of South China Mine Geological Environment Technology Innovation Center (CXZX2020002).

Institutional Review Board Statement: Not applicable.

Informed Consent Statement: Not applicable.

Data Availability Statement: The datasets generated during and/or analyzed during the current study are available from the corresponding author on reasonable request.

Acknowledgments: The authors acknowledge the members of ISSMGE-TC304 for developing the database 304 dB. Thanks to Monica Löfman and Leena Korkiala-Tanttu for contributing this data to make this study possible.

Conflicts of Interest: The authors declare no conflict of interest.

References

1. Motaghedi, H.; Eslami, A. Analytical Approach for Determination of Soil Shear Strength Parameters from CPT and CPTu Data. *Arabian J. Sci. Eng.* **2014**, *39*, 4363–4376. [CrossRef]
2. Ladd, C.C.; Foott, R. New Design Procedure for Stability of Soft Clays. *J. Geotech. Eng. Div.* **1974**, *100*, 763–786. [CrossRef]
3. Mesri, G. Discussion of "New design procedure for stability of soft clays". *J. Geotech. Eng. Div.* **1975**, *101*, 409–412. [CrossRef]
4. Jiang, S.H.; Zeng, S.H.; Yang, J.H.; Yao, C.; Huang, J.S.; Zhou, C.B. Slope Reliability Analysis by Simulation of Non-Stationary Random Field of Undrained Shear Strength. *Yantu Lixue* **2018**, *39*, 1071–1081. (In Chinese). Available online: http://ytlx.whrsm.ac.cn/EN/10.16285/j.rsm.2016.0609 (accessed on 20 May 2022).
5. Marchetti, S.; Monaco, P.; Totani, G.; Calabrese, M. *The Flat Dilatometer Test (DMT) in Soil Investigation*; ISSMGE TC 16 Report; ISSMGE: London, UK, 2001; pp. 1–26.
6. Robertson, P.K. Soil Behavior Type Using the DMT. In Proceedings of the 3rd International Flat Dilatometer Conference, Roma, Italy, 14–16 June 2015; pp. 14–16. Available online: https://www.cpt-robertson.com/PublicationsPDF/Robertson%20DMT15%202015.pdf (accessed on 20 May 2022).
7. Nguyen, H.; Bui, X.-N.; Tran, Q.-H.; Mai, N.-L. A New Soft Computing Model for Estimating and Controlling Blast-Produced Ground Vibration Based on Hierarchical K-Means Clustering and Cubist Algorithms. *Appl. Soft Comput.* **2019**, *77*, 376–386. [CrossRef]
8. Xu, H.; Zhou, J.; Asteris, P.G.; Armaghani, D.J.; Tahir, M.M. Supervised Machine Learning Techniques to the Prediction of Tunnel Boring Machine Penetration Rate. *Appl. Sci.* **2019**, *9*, 3715. [CrossRef]

9. Zhou, J.; Li, E.; Yang, S.; Wang, M.; Shi, X.; Yao, S.; Mitri, H.S. Slope Stability Prediction for Circular Mode Failure Using Gradient Boosting Machine Approach Based on an Updated Database of Case Histories. *Saf. Sci.* **2019**, *118*, 505–518. [CrossRef]
10. Jiao, P.; Alavi, A.H. Artificial Intelligence in Seismology: Advent, Performance and Future Trends. *Geosci. Front.* **2020**, *11*, 739–744. [CrossRef]
11. Cui, K.; Jing, X. Research on Prediction Model of Geotechnical Parameters Based on BP Neural Network. *Neural. Comput. Appl.* **2019**, *31*, 8205–8215. [CrossRef]
12. Tran, Q.A.; Ho, L.S.; Le, H.V.; Prakash, I.; Pham, B.T. Estimation of the Undrained Shear Strength of Sensitive Clays Using Optimized Inference Intelligence System. *Neural. Comput. Appl.* **2022**, *34*, 7835–7849. [CrossRef]
13. Jong, S.; Ong, D.; Oh, E. State-of-the-Art Review of Geotechnical-Driven Artificial Intelligence Techniques in Underground Soil-Structure Interaction. *Tunn. Undergr. Space Technol.* **2021**, *113*, 103946. [CrossRef]
14. Dorogush, A.V.; Ershov, V.; Gulin, A. CatBoost: Gradient Boosting with Categorical Features Support. *arXiv* **2018**, arXiv:1810.11363.
15. Tran, D.A.; Tsujimura, M.; Ha, N.T.; Van Binh, D.; Dang, T.D.; Doan, Q.-V.; Bui, D.T.; Ngoc, T.A.; Thuc, P.T.B.; Pham, T.D.; et al. Evaluating the Predictive Power of Different Machine Learning Algorithms for Groundwater Salinity Prediction of Multi-Layer Coastal Aquifers in the Mekong Delta, Vietnam. *Ecol. Indic.* **2021**, *127*, 107790. [CrossRef]
16. Xu, J.-G.; Hong, W.; Zhang, J.; Hou, S.-T.; Wu, G. Seismic Performance Assessment of Corroded RC Columns Based on Data-Driven Machine-Learning Approach. *Eng. Struct.* **2022**, *255*, 113936. [CrossRef]
17. Huang, G.; Wu, L.; Ma, X.; Zhang, W.; Fan, J.; Yu, X.; Zeng, W.; Zhou, H. Evaluation of CatBoost Method for Prediction of Reference Evapotranspiration in Humid Regions. *J. Hydrol.* **2019**, *574*, 1029–1041. [CrossRef]
18. Zhang, Y.X.; Zhao, Z.G.; Zheng, J.H. CatBoost: A New Approach for Estimating Daily Reference Crop Evapotranspiration in Arid and Semi-Arid Regions of Northern China. *J. Hydrol.* **2020**, *588*, 125087. [CrossRef]
19. Zhang, W.; Wu, C.; Zhong, H.; Li, Y.; Wang, L. Prediction of Undrained Shear Strength Using Extreme Gradient Boosting and Random Forest Based on Bayesian Optimization. *Geosci. Front* **2021**, *12*, 469–477. [CrossRef]
20. Oh, H.-J.; Syifa, M.; Lee, C.-W.; Lee, S. Land Subsidence Susceptibility Mapping Using Bayesian, Functional, and Meta-Ensemble Machine Learning Models. *Appl. Sci.* **2019**, *9*, 1248. [CrossRef]
21. Roscoe, K.H.; Burland, J.B. On the Generalised Stress-Strain Behaviour of "wet" Clay. *Eng. Plast.* **1968**, 535–609. Available online: https://trid.trb.org/view/124868 (accessed on 20 May 2022).
22. Wang, L.; Ye, S.; Shen, K.; Hu, Y. Undrained Shear Strength of K0 Consolidated Soft Clays. *Chin. J. Geotech. Eng.* **2006**, *28*, 971–977. [CrossRef]
23. Prokhorenkova, L.; Gusev, G.; Vorobev, A.; Dorogush, A.V.; Gulin, A. CatBoost: Unbiased Boosting with Categorical Features. In *Advances in Neural Information Processing Systems*; Curran Associates, Inc.: Dutchess County, NY, USA, 2018; Volume 31.
24. Mockus, J. The Application of Bayesian Methods for Seeking the Extremum. *J. Glob. Optim.* **1998**, *2*, 117. Available online: https://cir.nii.ac.jp/crid/1370576118710356l1 (accessed on 20 May 2022).
25. Katakami, S.; Sakamoto, H.; Okada, M. Bayesian Hyperparameter Estimation Using Gaussian Process and Bayesian Optimization. *J. Phys. Soc. Jpn.* **2019**, *88*, 074001. [CrossRef]
26. Lindauer, M.; Eggensperger, K.; Feurer, M.; Biedenkapp, A.; Deng, D.; Benjamins, C.; Ruhkopf, T.; Sass, R.; Hutter, F. SMAC3: A Versatile Bayesian Optimization Package for Hyperparameter Optimization. *J. Mach. Learn. Res.* **2022**, *23*, 1–9. [CrossRef]
27. Hutter, F.; Hoos, H.H.; Leyton-Brown, K. Sequential Model-Based Optimization for General Algorithm Configuration. In Proceedings of the Learning and Intelligent Optimization: 5th International Conference, LION 5, Rome, Italy, 17–21 January 2011. [CrossRef]
28. Breiman, L. Random Forests. *Mach Learn* **2001**, *45*, 5–32. [CrossRef]
29. Geisser, S. The Predictive Sample Reuse Method with Applications. *J. Am. Stat. Assoc.* **1975**, *70*, 320–328. [CrossRef]
30. Stone, M. Cross-Validatory Choice and Assessment of Statistical Predictions (with Discussion). *J. R. Stat. Soc. B* **1976**, *38*, 102. [CrossRef]
31. Jung, Y. Multiple Predicting K-Fold Cross-Validation for Model Selection. *J. Nonparametr. Stat.* **2018**, *30*, 197–215. [CrossRef]
32. Pham, B.T.; Qi, C.; Ho, L.S.; Nguyen-Thoi, T.; Al-Ansari, N.; Nguyen, M.D.; Nguyen, H.D.; Ly, H.-B.; Le, H.V.; Prakash, I. A Novel Hybrid Soft Computing Model Using Random Forest and Particle Swarm Optimization for Estimation of Undrained Shear Strength of Soil. *Sustainability* **2020**, *12*, 2218. [CrossRef]
33. Ohta, H.; Nishihara, A. Anisotropy of Undrained Shear Strength of Clays under Axi-Symmetric Loading Conditions. *Soils Found.* **1985**, *25*, 73–86. [CrossRef]
34. Karube, D. Nonstandard Triaxial Testing Method and Its Problems. Proceedings of the 20th Symposium of the International Society for Rock Mechanics, JSSMFE. 1975, pp. 45–60. Available online: https://cir.nii.ac.jp/crid/1572261549455735296 (accessed on 20 May 2022).
35. He, P.; Wang, W.; Xu, Z. Empirical Correlations of Compression Index and Swelling Index for Shanghai Clay. *Yantu Lixue* **2018**, *39*, 1–10. (In Chinese) [CrossRef]
36. Azzouz, A.S.; Krizek, R.J.; Corotis, R.B. Regression Analysis of Soil Compressibility. *Soils Found.* **1976**, *16*, 19–29. [CrossRef]
37. Ladd, C.C. Stability Evaluation during Staged Construction. *J. Geotech. Eng.* **1991**, *117*, 540–615. [CrossRef]
38. Zhang, J. *Bayesian Method: A Natural Tool for Processing Geotechnical Information*; TC205/TC304 Discussion Groups; ISSMGE: London, UK, 2016.

39. Fu, Y.; Ma, C.; Bian, Y.; Lv, G.; Hu, Y.; Wang, C. Stochastic Mechanics-Based Bayesian Method Calibrating the Constitutive Parameters of the Unified Model for Clay and Sand with CPTU Data. *Acta Geotech.* **2022**, *17*, 4577–4598. [CrossRef]
40. Cao, Z.; Wang, Y. Bayesian Model Comparison and Characterization of Undrained Shear Strength. *J. Geotech. Geoenviron. Eng.* **2014**, *140*, 04014018. [CrossRef]
41. Zhao, Z.; Duan, W.; Cai, G.; Wu, M.; Liu, S. CPT-Based Fully Probabilistic Seismic Liquefaction Potential Assessment to Reduce Uncertainty: Integrating XGBoost Algorithm with Bayesian Theorem. *Comput. Geotech.* **2022**, *149*, 104868. [CrossRef]
42. Juang, C.H.; Zhang, J. *Bayesian Methods for Geotechnical Applications—A Practical Guide*; ASCE: Reston, VA, USA, 2017; pp. 215–246. [CrossRef]
43. Guan, Z.; Wang, Y. SPT-Based Probabilistic Evaluation of Soil Liquefaction Potential Considering Design Life of Civil Infrastructures. *Comput. Geotech.* **2022**, *148*, 104807. [CrossRef]
44. Guan, Z.; Wang, Y. CPT-Based Probabilistic Liquefaction Assessment Considering Soil Spatial Variability, Interpolation Uncertainty and Model Uncertainty. *Comput. Geotech.* **2022**, *141*, 104504. [CrossRef]
45. Juang, C.H.; Ching, J.; Ku, C.-S.; Hsieh, Y.-H. Unified CPTu-Based Probabilistic Model for Assessing Probability of Liquefaction of Sand and Clay. *Geotechnique* **2012**, *62*, 877–892. [CrossRef]
46. Ku, C.-S.; Juang, C.H.; Chang, C.-W.; Ching, J. Probabilistic Version of the Robertson and Wride Method for Liquefaction Evaluation: Development and Application. *Can. Geotech. J.* **2012**, *49*, 27–44. [CrossRef]
47. Draper, N.R.; Smith, H. *Applied Regression Analysis*; John Wiley & Sons: Hoboken, NJ, USA, 1998; Volume 326. [CrossRef]
48. Huang, J.-C.; Ko, K.-M.; Shu, M.-H.; Hsu, B.-M. Application and Comparison of Several Machine Learning Algorithms and Their Integration Models in Regression Problems. *Neural. Comput. Appl.* **2020**, *32*, 5461–5469. [CrossRef]
49. Zhang, R.; Li, Y.; Goh, A.T.; Zhang, W.; Chen, Z. Analysis of Ground Surface Settlement in Anisotropic Clays Using Extreme Gradient Boosting and Random Forest Regression Models. *J. Rock Mech. Geotech. Eng.* **2021**, *13*, 1478–1484. [CrossRef]
50. FI-CLAY/14/856 Finland Clays. Available online: http://140.112.12.21/issmge/tc304.htm (accessed on 15 June 2022).
51. Löfman, M.S.; Korkiala-Tanttu, L.K. Transformation Models for the Compressibility Properties of Finnish Clays Using a Multivariate Database. *Georisk* **2022**, *16*, 330–346. [CrossRef]
52. Rutledge, P.C. Cooperative Triaxial Shear Research Program of the Corps of Engineers. 1947. Available online: https://trid.trb.org/view/119101 (accessed on 20 May 2022).
53. Jamiolkowski, M.; Ladd, C.C.; Germaine, J.T.; Lancellotta, R. New developments in field and laboratory testing of soils. In Proceedings of the XI the International Conference on Soil Mechanics & Foundation Engineering, San Francisco, CA, USA, 12–16 August 1985.
54. Yuchun, C. A Comparison of Simplified Calculation Methods of Undrained Shear Strength of Soft Clays after Consolidation. *China. Civil. Eng.* **2014**, *47*, 107–116. (In Chinese) [CrossRef]
55. Qiao, Y.F.; Lu, X.B.; Huang, J.; Ding, W.Q. Simplified calculation method for lateral pressure at rest in the under-consolidation stratum. *Yantu Lixue* **2020**, *41*, 3722–3729. (In Chinese). Available online: http://ytlx.whrsm.ac.cn/CN/10.16285/j.rsm.2020.0124 (accessed on 20 May 2022).
56. Asaoka, A.; A-Grivas, D. Spatial Variability of the Undrained Strength of Clays. *J. Geotech. Eng. Div.* **1982**, *108*, 743–756. [CrossRef]
57. Xiao-qing, G.; Bin, Z.; Jin-chao, L. Others Experimental Study of Undrained Shear Strength and Cyclic Degradation Behaviors of Marine Clay in Pearl River Estuary. *Yantu Lixue* **2016**, *37*, 1005–1012. (In Chinese). Available online: http://ytlx.whrsm.ac.cn/CN/10.16285/j.rsm.2016.04.013 (accessed on 20 May 2022).
58. Ching, J.; Arroyo, M.; Chen, J.; Jorge, C.; Lansivaara, T.; Li, D.; Mayne, P.; Phoon, K.; Prakoso, W.; Uzielli, M. Transformation Models and Multivariate Soil Databases. In *Final Report of Joint TC205/TC304 Working Group on "Discussion of Statistical/Reliability Methods for Eurocodes"*; International Society for Soil Mechanics and Geotechnical Engineering (ISSMGE): London, UK, 2017; p. 372.

applied sciences

MDPI

Article

Prediction Modeling of Ground Subsidence Risk Based on Machine Learning Using the Attribute Information of Underground Utilities in Urban Areas in Korea

Sungyeol Lee *, Jaemo Kang and Jinyoung Kim

Department of Geotechnical Engineering Research, Korea Institute of Civil Engineering and Building Technology, Goyang-si 10223, Republic of Korea; jmkang@kict.re.kr (J.K.); goldcamp@kict.re.kr (J.K.)
* Correspondence: leesy@kict.re.kr; Tel.: +82-31-910-0645

Abstract: As ground subsidence accidents in urban areas that occur due to damage to underground utilities can cause great damage, it is necessary to predict and prepare for such accidents in order to minimize such damage. It has been reported that the main cause of ground subsidence in urban areas is cavities in the ground formed by damage to underground utilities. Thus, in this study, attribute information and historical ground subsidence information of six types of underground utility lines (water supply, sewage, power, gas, heating, and communication) were collected to develop a ground subsidence risk prediction model based on machine learning. To predict the risk of ground subsidence in the target area, it was divided into a grid with a square size of 500 m × 500 m, and attribute information of underground utility lines and historical information of ground subsidence included in the grid were extracted. Six types of underground utility lines were merged into single-type attribute information, and the risk of ground subsidence was categorized into three levels using the number of ground subsidence occurrences to develop a dataset. In addition, 12 datasets, which were developed based on the conditions of certain divided ranges of attribute information and risk levels, and 12 additional datasets, which were developed using the Synthetic Minority Oversampling Technique to resolve the imbalance of data, were built. Then, factors that represented significant correlations between input and output data were singled out and were then applied to the RandomForest, XGBoost, and LightGBM algorithms to select a model that produced the best performance. By classifying the ground subsidence risk levels through the selected model, it was found that density was the most important influencing factor used in the model. A risk map of ground subsidence in the target area was made through the model; the map showed the trend of well-predicted risk levels in the area where ground subsidence was concentrated.

Keywords: ground subsidence; machine learning; ground subsidence risk prediction model; risk map

Citation: Lee, S.; Kang, J.; Kim, J. Prediction Modeling of Ground Subsidence Risk Based on Machine Learning Using the Attribute Information of Underground Utilities in Urban Areas in Korea. *Appl. Sci.* **2023**, *13*, 5566. https://doi.org/10.3390/app13095566

Academic Editor: Wei Gao

Received: 10 April 2023
Revised: 24 April 2023
Accepted: 26 April 2023
Published: 30 April 2023

1. Introduction

Damage to underground utility lines is known to be one of the main causes of ground subsidence. Since underground utility lines are concentrated in urban areas with highly dense populations, accidents due to ground subsidence can cause significant social chaos [1]. As such, it is necessary to prevent accidents related to ground subsidence by analyzing their fundamental causes and mechanisms.

An investigation of the causes and the number of ground subsidence occurrences from 2010 to July 2014 in Seoul showed that the number of accidents has steadily increased, and their main cause was found to be damage to water supply and sewage lines [2]. A mechanism by which ground subsidence occurs is often when pipes are damaged by external impacts and deterioration due to aging, causing water channels to form around the damaged location. Soil particles in the ground can then move along the channels, creating and expanding cavities around the pipes [3]. Thus, ground subsidence is likely to increase as excavation construction work is repeatedly performed over time.

Extensive research has been performed on ways to prevent accidents related to ground subsidence. In Japan, a study using indoor model experiments simulating ground subsidence using standard sand was published to identify the mechanism of how cavities, a precursor to ground subsidence, were generated inside the ground, while a study on the identification of a cavity generation mechanism inside the ground by simulating a crack in the sewage pipeline under the soil box and visualization of the cavity generation through equipment such as X-rays and computed tomography has also been published [4,5].

Research that aims to identify the mechanism of ground subsidence occurrence using numerical analysis has also been active. Using the finite element method to simulate the ground cavity and relaxation zone, several published studies have shown that the location of the underground utility damage, the relative density of the ground, and the ground layer conditions have a significant effect on the ground subsidence [6–8].

In addition, studies on performing a decision tree, which is one of the machine learning algorithms, and the analytic hierarchy process, were published to derive factors influencing ground subsidence and calculate the weights of influencing factors [9,10].

Studies aiming to predict the risk level of ground subsidence have also been steadily conducted. One study on the evaluation of ground subsidence risk level that was announced uses surveyed CCTV data based on sewage pipelines, which is the main cause of ground subsidence, as well as cavity exploration data by underground exploration radar. In addition, a study was conducted to propose a regression equation of the ground subsidence risk level in urban areas in Korea through logistic regression analysis [11]. Moreover, studies have been conducted to select a model for predicting the ground subsidence risk level in urban areas in Korea through machine learning, after selecting influencing factors such as the number of years used and pipeline diameter among attribute values of underground utilities, and then to suggest a risk map [1].

Researchers have used various ways to predict risk levels in order to prevent accidents related to ground subsidence. However, they have had difficulty deriving highly accurate and reliable results, as ground subsidence occurs in complex ways and is caused by various factors in a wide range of areas. Thus, this study aims to propose a machine learning-based ground subsidence risk prediction model by selecting the following as influence factors among the attribute information of underground utility lines in representative urban areas in Korea: the number of years used, pipeline diameter and length, and the density of pipelines, which are likely to have a close correlation with ground subsidence. We compared the results of machine learning models by applying datasets with a range of conditions and selected the model that exhibited optimal performance. Furthermore, we aimed to present the importance of each influencing factor, which was used when classifying the ground subsidence risk levels by the machine learning model through the selected model.

2. Method and Data Characteristics

2.1. Subsection Flow of the Study

In this study, a representative urban area in Korea was selected as the target region. To develop a ground subsidence risk level prediction model based on machine learning, the historical information of ground subsidence, and attribute information of underground utility lines in the target region were used to build a dataset and then applied to the machine learning algorithm. The target region was divided into a grid with a total of 2391 squares of 500 m × 500 m in size, using the ArcGIS program to predict risk level. Six types of underground utility lines included in each grid square were merged into a single type to extract the attribute information and density. The dataset was built using a method that calculated a risk level based on the number of ground subsidence occurrences in the grid using the historical ground subsidence information.

The developed dataset was divided into training and test datasets at an 80:20 ratio to prevent overfitting of the model and to test the model. To mitigate the data imbalance, the Synthetic Minority Oversampling Technique (SMOTE) was applied to the training data. This developed training dataset was applied to machine learning algorithms: RandomForest (RF), XGBoost (XGB), and LightGBM (LGBM) to check the model results by adjusting the hyperparameters that exhibited the optimal performance. Using the 20% test data, the model's performance was validated through the test indices of accuracy, F1-score, and area under the curve (AUC). Based on the test results, dataset types and machine learning models that exhibited optimal performance were selected, and the importance of the influencing factors was derived. Figure 1 shows the flow chart of the study.

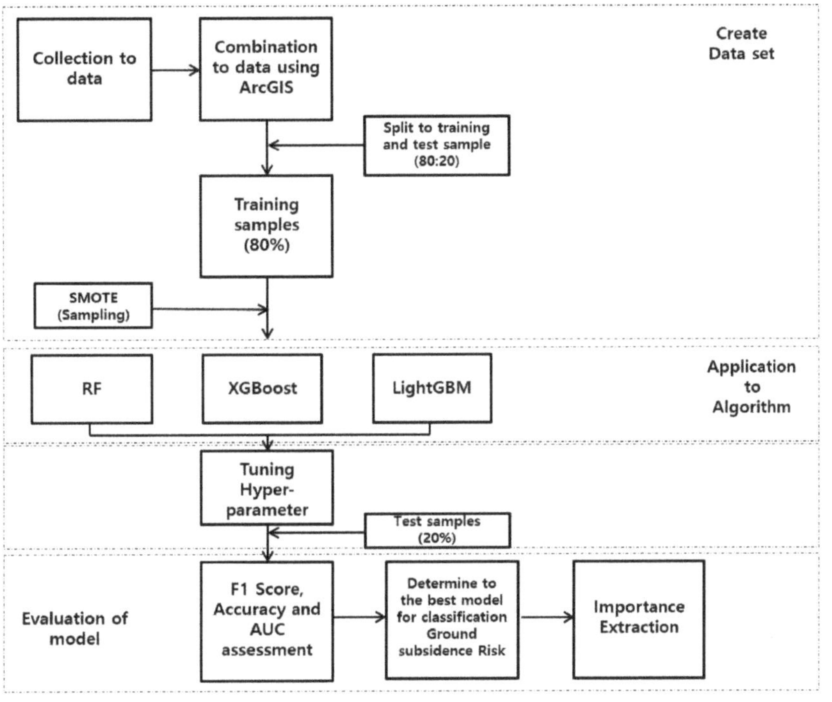

Figure 1. Flow chart.

2.2. Characteristics of the Data

A representative area in Korea was selected as the target region for the prediction of ground subsidence risk level. The target region was divided into a grid with a total of 2391 squares of 500 m × 500 m in size for the risk level evaluation. For each grid square, the attribute data of six types of underground utility lines (water supply, sewage, power, gas, heating, and communication cables) and the historical data of ground subsidence were compiled. As described above, six types of underground utility lines were merged into a single type to extract attribute data. In the attribute information of underground utility lines to build a dataset, the number of years used, pipeline type, diameter, length, burial depth, slope, etc., were included, but there were many missing and erroneous values as well. Thus, as the data that could be usable, the number of years used, pipeline diameter, and length were selected. Then, the density of all pipelines was calculated to be used as a factor influencing the occurrence of ground subsidence. To improve the model's performance, raw data were not directly used, but they were preprocessed to divide the attribute information of underground utility lines by a certain range. The years used were divided into 5- and 10-year units, and the pipeline diameter was divided into 50 mm and 100 mm units. The basic unit of data that belongs to the corresponding range was set to the pipeline's length. For example, an underground pipeline that was used for three years in the grid was assigned to a class corresponding to an age of 1 to 4 years, and the length of the pipeline was reflected.

For the output data, the risk level of ground subsidence was calculated by summing the number of ground subsidence occurrences in the grid using the historical information of ground subsidence occurrences. It is difficult to provide a quantifiable measure of ground subsidence risk. Thus, multiple datasets of ground subsidence risk levels, categorized by the number of ground subsidence occurrences, were developed. The developed datasets were applied to the machine learning algorithms to select a condition of the risk level of ground subsidence that exhibited good performance. The ground subsidence risk was categorized into three levels. Risk Level 1 means an area where the number of ground subsidence occurrences in the grid is "0". The conditions of Risk Levels 2 and 3 were adjusted depending on the number of ground subsidence occurrences in the grid. If the number of ground subsidence occurrences in the grid of Risk Level 2 is one, the number of ground subsidence occurrences of Risk Level 3 was set to two or more. If the number of ground subsidence occurrences in the grid of Risk Level 2 is set to a range of one to two, the number of ground subsidence occurrences of Risk Level 3 was set to three or more. If the number of ground subsidence occurrences in the grid of Risk Level 2 is set to a range of one to three, the number of ground subsidence occurrences of Risk Level 3 was set to four or more. Risk Level 1 of ground subsidence means a relatively safer area from ground subsidence. The boundary between Levels 2 and 3 varies depending on the conditions, but Level 2 means an area that needs attention, and Level 3 is an area that is at the highest risk. Table 1 presents the categories of factors in the datasets. Table 2 presents the dataset which is set according to the data category condition. A total of 24 datasets were built according to whether or not SMOTE was applied to each dataset.

Table 1. Category of factors.

Factors	Unit	Category
Year (year)	5	1~5, 6~10, 11~15, 16~20, 21~25, 26~30, 31~35, 36~40, 41~45, 46~50
	10	1~10, 11~20, 21~30, 31~40, 41~50
Diameter (mm)	50	1~50, 51~100, 101~150, 151~200, 201~250, 251~300, 301~350, 351~400, 401~450, 451~500, 501~550, 551~600
	100	1~100, 101~200, 201~300, 301~400, 401~500, 501~600
Risk level (Sum of occurrences of ground subsidence in grid)	1	0
	2	1, 1~2, 1~3
	3	2~, 3~, 4~

Table 2. Category of Factors.

No.	Grid	Year (Year)	Diameter (mm)	Risk Level (Level 2′s Range)
1	500 m × 500 m	5	50	3 (1)
2				3 (1–2)
3				3 (1–3)
4			100	3 (1)
5				3 (1–2)
6				3 (1–3)
7		10	50	3 (1)
8				3 (1–2)
9				3 (1–3)
10			100	3 (1)
11				3 (1–2)
12				3 (1–3)

2.3. Density

A previous study proved that the density of pipelines was significantly correlated with ground subsidence [9] Accordingly, we used the density of the pipeline as the influencing factor of the model to predict the ground subsidence risk level. The density was calculated using a linear density analysis on the pipelines in the grid using ArcGIS. This method calculated the length of the pipeline that corresponded to the unit area.

2.4. Risk Level of Ground Subsidence

The risk levels used as the output data in this study were divided into three levels according to the number of times ground subsidence occurred in the grid. Since there are no quantifiable measures to categorize the risk level, we build datasets by selecting different numbers of data belonging to risk level 2 according to the number of occurrences of ground subsidence in the grid. Thus, the number of data varies according to the category based on the number of occurrences of ground subsidence of each dataset, which is presented in Table 3. As presented in Table 3, the ratio of Risk Level 1 data was the highest (57%), and the ratios of Risk Level 2 and 3 data varied according to the conditions. As such, the composition of the data shows unbalanced features, and we applied SMOTE, an over-sampling technique, to the 12 datasets to balance the data [12–14].

Table 3. The ratio of data according to the risk level of ground subsidence.

Range of Risk Level 2 \ Risk Level	1	2	3
1	1374 (57%)	348 (15%)	669 (28%)
1–2	1374 (57%)	635 (27%)	382 (16%)
1–3	1374 (57%)	706 (30%)	311 (13%)

2.5. Data Correlation Analysis

A Pearson correlation analysis was conducted to verify the correlation between the input and output data of the dataset which was developed according to the data category conditions. The results are presented in Table 4.

Table 4. Results of correlation analysis of the influencing factors.

Model No.	1		2		3	
Factor	Corr	*p*-Value	Corr	*p*-Value	Corr	*p*-Value
5Y_5	−0.138	0.000	−0.149	0.000	−0.150	0.000
5Y_10	−0.108	0.000	−0.098	0.000	−0.096	0.000
5Y_15	−0.004	0.858	−0.007	0.724	−0.049	0.017
5Y_20	−0.141	0.000	−0.178	0.000	−0.171	0.000
5Y_25	−0.108	0.000	−0.161	0.000	−0.154	0.000
5Y_30	−0.065	0.002	−0.099	0.000	−0.106	0.000
5Y_35	−0.150	0.000	−0.165	0.000	−0.173	0.000
5Y_40	−0.150	0.000	−0.169	0.000	−0.168	0.000
5Y_45	−0.167	0.000	−0.187	0.000	−0.167	0.000
5Y_50	−0.135	0.000	−0.134	0.000	−0.144	0.000
50DTR_50	0.057	0.005	0.061	0.003	0.064	0.002
50DTR_100	0.146	0.000	0.147	0.000	0.148	0.000
50DTR_150	0.159	0.000	0.163	0.000	0.169	0.000
50DTR_200	0.117	0.000	0.116	0.000	0.112	0.000
50DTR_250	0.015	0.478	0.008	0.698	0.013	0.522
50DTR_300	0.153	0.000	0.155	0.000	0.158	0.000
50DTR_350	0.038	0.067	0.026	0.198	0.022	0.274
50DTR_400	0.059	0.004	0.062	0.002	0.059	0.004
50DTR_450	0.099	0.000	0.109	0.000	0.107	0.000
50DTR_500	0.043	0.035	0.044	0.032	0.052	0.011
50DTR_550	0.014	0.494	−0.006	0.783	−0.003	0.895
50DTR_600	0.082	0.000	0.090	0.000	0.089	0.000
Density	0.544	0.000	0.534	0.000	0.526	0.000

Table 4. *Cont.*

Model No.	4		5		6	
Factor	Corr	*p*-Value	Corr	*p*-Value	Corr	*p*-Value
5Y_5	−0.138	0.000	−0.149	0.000	−0.150	0.000
5Y_10	−0.108	0.000	−0.098	0.000	−0.096	0.000
5Y_15	−0.004	0.858	−0.007	0.724	−0.049	0.017
5Y_20	−0.141	0.000	−0.178	0.000	−0.171	0.000
5Y_25	−0.108	0.000	−0.161	0.000	−0.154	0.000
5Y_30	−0.065	0.002	−0.099	0.000	−0.106	0.000
5Y_35	−0.150	0.000	−0.165	0.000	−0.173	0.000
5Y_40	−0.150	0.000	−0.169	0.000	−0.168	0.000
5Y_45	−0.167	0.000	−0.187	0.000	−0.167	0.000
5Y_50	−0.135	0.000	−0.134	0.000	−0.144	0.000
100DTR_100	0.131	0.000	0.134	0.000	0.136	0.000
100DTR_200	0.152	0.000	0.154	0.000	0.155	0.000
100DTR_300	0.128	0.000	0.127	0.000	0.132	0.000
100DTR_400	0.067	0.001	0.064	0.002	0.060	0.004
100DTR_500	0.103	0.000	0.111	0.000	0.113	0.000
100DTR_600	0.083	0.000	0.089	0.000	0.088	0.000
Density	0.544	0.000	0.534	0.000	0.526	0.000
Model No.	7		8		9	
Factor	Corr	*p*-Value	Corr	*p*-Value	Corr	*p*-Value
10Y_10	0.085	0.000	0.077	0.000	0.071	0.000
10Y_20	0.123	0.000	0.126	0.000	0.131	0.000
10Y_30	0.150	0.000	0.156	0.000	0.159	0.000
10Y_40	0.108	0.000	0.113	0.000	0.116	0.000
10Y_50	0.107	0.000	0.117	0.000	0.118	0.000
50DTR_50	0.057	0.005	0.061	0.003	0.064	0.002
50DTR_100	0.146	0.000	0.147	0.000	0.148	0.000
50DTR_150	0.159	0.000	0.163	0.000	0.169	0.000
50DTR_200	0.117	0.000	0.116	0.000	0.112	0.000
50DTR_250	0.015	0.478	0.008	0.698	0.013	0.522
50DTR_300	0.153	0.000	0.155	0.000	0.158	0.000
50DTR_350	0.038	0.067	0.026	0.198	0.022	0.274
50DTR_400	0.059	0.004	0.062	0.002	0.059	0.004
50DTR_450	0.099	0.000	0.109	0.000	0.107	0.000
50DTR_500	0.043	0.035	0.044	0.032	0.052	0.011
50DTR_550	0.014	0.494	−0.006	0.783	−0.003	0.895
50DTR_600	0.082	0.000	0.090	0.000	0.089	0.000
Density	0.544	0.000	0.534	0.000	0.526	0.000
Model No.	10		11		12	
Factor	Corr	*p*-Value	Corr	*p*-Value	Corr	*p*-Value
10Y_10	0.085	0.000	0.077	0.000	0.071	0.000
10Y_20	0.123	0.000	0.126	0.000	0.131	0.000
10Y_30	0.150	0.000	0.156	0.000	0.159	0.000
10Y_40	0.108	0.000	0.113	0.000	0.116	0.000
10Y_50	0.107	0.000	0.117	0.000	0.118	0.000
100DTR_100	0.131	0.000	0.134	0.000	0.136	0.000
100DTR_200	0.152	0.000	0.154	0.000	0.155	0.000
100DTR_300	0.128	0.000	0.127	0.000	0.132	0.000
100DTR_400	0.067	0.001	0.064	0.002	0.060	0.004
100DTR_500	0.103	0.000	0.111	0.000	0.113	0.000
100DTR_600	0.083	0.000	0.089	0.000	0.088	0.000
Density	0.544	0.000	0.534	0.000	0.526	0.000

In Table 4, Y refers to the number of years used, and 5Y and 10Y mean the five-year and 10-year units, respectively (5–50 refers to the data range). In addition, DTR refers to the pipeline diameter; 50 and 100 refer to the pipeline diameter of 50 mm and 100 mm, and 50–600 refers to the range of the pipeline diameters. In this study, the presence of data correlation was verified by *p*-value in the correlation analysis. If the *p*-value was less than 0.05, it was interpreted as showing a significant correlation, so it was used as input data. Conversely, if the *p*-value is more than 0.05, it was interpreted as not showing a significant correlation, so it was excluded from the input data.

3. Results of Analysis of Ground Subsidence Risk Levels Using Machine Learning

In this study, a machine learning algorithm was used to develop a model to predict the risk level of ground subsidence, focusing on urban areas in South Korea. The machine learning algorithms used were RF, XGB, and LGBM, which have produced good results in previous studies [1,15].

3.1. Random Forest

The Random Forest (RF) algorithm is a tree-based ensemble model that is developed to solve regression and classification problems in machine learning [16]. An ensemble model derives better results than a model that trains a single model once, as it trains multiple algorithms iteratively. It includes techniques such as voting and bagging.

RF presents the best result among the results derived from the trees after creating multiple tree-based algorithms as the representative result. RF is based on a tree algorithm, has a low overfitting risk, and can be easily applied to various data. It is widely used in problem-solving through machine learning to derive a good result [17–20].

RF predicts the outcome as a binary value of 0 or 1, as presented in (1), after extracting an arbitrary number of input data from a number of single-algorithm predictors and performing a final decision by majority vote on the results derived from each predictor, where $y_i = f_i(\mathrm{X})$, and w_i refers to the weight. If the calculated value is larger than the threshold value, the predicted value is 1, otherwise it is 0 [21].

$$\mathrm{F(X)} = \sum w_i y_i \quad (1)$$

3.2. XGBoost (eXtreme Gradient Boosting)

XGBoost (XGB) is a typical algorithm of a boosting technique where a result is derived by learning a single model sequentially, and the result of the previous model affects the next result. XGB is a tree-based algorithm used in solving regression and classification problems. It is effective in preventing overfitting due to its different regularization penalties. In addition, it has the advantage of being able to process big data in a short period of time, so it has been actively used in various fields [22,23].

The calculation equation for the decision-making of XGBoost is presented in (2), where $\hat{y}_i$ refers to the i-th sample's prediction value and f_k refers to the prediction value where the k-th tree's sigmoid function is applied. The output is derived by summing all prediction values. The prediction value can be calculated using (3).

$$\hat{y}_i = \sum_{K=1}^{K} f_i(x_i) \quad (2)$$

$$\hat{y}_i = \frac{1}{1 + e^{-f(x_i)}} \quad (3)$$

The error is calculated using the difference between the prediction and real values in the tree, and the weight is calculated to reduce the error as presented in (4). $\hat{y}_i^{(t-1)}$ refers to the prediction value of the previous model, $h_t(x_i)$ refers to the tree trained by the current model, and η refers to the learning rate, which is the percentage of reflections from the prior model. The model's error is reduced by iterating this method [24,25].

$$\hat{y}_i^{(t)} = \hat{y}_i^{(t-1)} + \eta h_t(x_i) \quad (4)$$

3.3. LightGBM (Light Gradient Boosting Machine)

LightGBM (LGBM) is an algorithm in which a tree-based boosting technique is applied in the same manner as XGB. It has been used in solving regression and classification problems and in selecting the priority of importance of influencing factors. LGBM is advantageous for its fast operation speed because it derives a result using a method that reduces data characteristics by employing partial data only. Thus, LGBM processes big data quickly and with a high level of accuracy, and can derive the importance among the influencing factors used, advantages which have made it a popular choice [26].

LightGBM calculates the loss function using cross-entropy. The equation for calculating the cross-entropy is presented in (5), where N is the number of samples, K is the number of classes, $y_{i,j}$ refers to the binary variable indicating whether the i-th sample belongs to the j-th class, and $p_{i,j}$ refers to the probability that the i-th sample belongs to the j-th class. LightGBM derives its results by learning to update the model while minimizing the CE received from the previous model [27].

$$\mathrm{CE} = \frac{1}{N}\sum_{i=1}^{N}\sum_{j=1}^{N} y_{i,j}\log\left(p_{i,j}\right) \tag{5}$$

3.4. Evaluation Indexes of Machine Learning Algorithms

For evaluation indexes of machine learning models to solve a classification problem, accuracy, F1-score, and AUC are generally used. The results of these evaluation indexes can be calculated using Equations (6)–(10) via the confusion matrix.

$$\mathrm{Accuracy} = \frac{\mathrm{TP}+\mathrm{TN}}{\mathrm{TP}+\mathrm{TN}+\mathrm{FP}+\mathrm{FN}} \tag{6}$$

$$\mathrm{Recall}(\mathrm{Sensitivity}) = \frac{\mathrm{TP}}{\mathrm{TP}+\mathrm{FN}} \tag{7}$$

$$\mathrm{Precision} = \frac{\mathrm{TP}}{\mathrm{TP}+\mathrm{FP}} \tag{8}$$

$$\mathrm{F1Score} = 2 \times \frac{\mathrm{Precision} \times \mathrm{Recall}}{\mathrm{Precision} + \mathrm{Recall}} \tag{9}$$

$$\mathrm{Specificity} = \frac{\mathrm{TN}}{\mathrm{TN}+\mathrm{FP}} \tag{10}$$

Intuitively, it is highly convenient if the model's performance is evaluated through the model's accuracy, but it is also difficult to identify the objective model performance for imbalanced data. Thus, a model using imbalanced data is evaluated by employing the F1-score, which uses a harmonic mean of the data. The model confidence is evaluated using the AUC that uses the receiver operation characteristic (ROC) [28–34].

3.5. Results of Applying Machine Learning

To build a machine learning-based model for the prediction of ground subsidence risk levels in urban areas, we selected a model that exhibited the best performance by applying 24 datasets, which were developed using the attribute information of underground utility lines and the historical information of ground subsidence, to RF, XGB, and LGBM classifiers. To implement machine learning, Python 3.8 was used, and the Scikit-learn library was employed.

The model's evaluation indexes, accuracy, F1-score, and AUC were selected. The accuracy was used to determine the presence of overfitting by comparing the results of the training set with those of the test set. If the difference between the training and test scores is equal to or less than 0.1, it was determined that overfitting was avoided. In addition, the model's performance was identified using the F1-score and AUC indices to select the optimal model.

The results of an evaluation of the machine learning models derived in this study are presented in Tables 5 and 6. Table 5 shows the model results where SMOTE was not applied, and Table 6 presents the model results where SMOTE was applied.

Table 5. Results of machine learning model (SMOTE not applied).

Model	RF				XGB				LGBM			
No.	Train Score	Test Score	F1-Score (Macro)	AUC (Macro)	Train Score	Test Score	F1-Score (Macro)	AUC (Macro)	Train Score	Test Score	F1-Score (Macro)	AUC (Macro)
1	0.742	0.670	0.450	0.780	0.725	0.668	0.480	0.770	0.765	0.676	0.480	0.800
2	0.766	0.645	0.500	0.800	0.719	0.628	0.490	0.790	0.714	0.666	0.550	0.800
3	0.745	0.649	0.490	0.810	0.768	0.660	0.560	0.800	0.759	0.643	0.560	0.810
4	0.791	0.676	0.470	0.780	0.724	0.674	0.490	0.770	0.763	0.670	0.480	0.790
5	0.764	0.664	0.530	0.800	0.719	0.628	0.500	0.790	0.758	0.660	0.550	0.810
6	0.751	0.664	0.520	0.810	0.732	0.658	0.550	0.810	0.768	0.666	0.570	0.820
7	0.736	0.639	0.420	0.750	0.696	0.641	0.410	0.750	0.714	0.645	0.440	0.750
8	0.681	0.591	0.310	0.750	0.694	0.601	0.390	0.750	0.655	0.591	0.360	0.760
9	0.732	0.635	0.390	0.770	0.680	0.620	0.410	0.770	0.715	0.616	0.410	0.770
10	0.729	0.643	0.430	0.740	0.697	0.647	0.420	0.740	0.715	0.635	0.420	0.750
11	0.651	0.597	0.330	0.750	0.686	0.599	0.380	0.740	0.681	0.603	0.360	0.750
12	0.729	0.635	0.400	0.770	0.683	0.599	0.350	0.760	0.706	0.610	0.350	0.760

Table 6. Results of machine learning model (SMOTE applied).

Model	RF				XGB				LGBM			
No.	Train Score	Test Score	F1-Score (Macro)	AUC (Macro)	Train Score	Test Score	F1-Score (Macro)	AUC (Macro)	Train Score	Test Score	F1-Score (Macro)	AUC (Macro)
1	0.676	0.626	0.560	0.790	0.648	0.608	0.560	0.770	0.716	0.628	0.560	0.790
2	0.683	0.593	0.550	0.790	0.705	0.608	0.580	0.800	0.656	0.593	0.560	0.800
3	0.718	0.608	0.550	0.790	0.744	0.644	0.590	0.800	0.706	0.620	0.570	0.810
4	0.662	0.624	0.560	0.790	0.679	0.585	0.510	0.770	0.656	0.582	0.520	0.790
5	0.687	0.595	0.560	0.800	0.666	0.587	0.550	0.790	0.690	0.597	0.550	0.800
6	0.699	0.585	0.540	0.800	0.712	0.612	0.560	0.800	0.729	0.624	0.570	0.820
7	0.671	0.603	0.530	0.760	0.655	0.580	0.520	0.770	0.668	0.580	0.520	0.770
8	0.645	0.543	0.460	0.750	0.615	0.553	0.500	0.740	0.655	0.545	0.490	0.750
9	0.632	0.501	0.380	0.740	0.628	0.553	0.490	0.770	0.627	0.551	0.490	0.760
10	0.682	0.597	0.520	0.770	0.660	0.578	0.520	0.760	0.669	0.578	0.510	0.770
11	0.608	0.541	0.450	0.750	0.647	0.555	0.490	0.740	0.651	0.570	0.500	0.750
12	0.636	0.511	0.390	0.750	0.615	0.532	0.460	0.760	0.676	0.568	0.490	0.750

Based on the evaluation results, the optimal model for the prediction of ground subsidence risk levels in the target region was determined. It was SMOTE-applied XGB (No. 3 model) when the number of years used was a five-year unit, the pipeline diameter was 50mm, and the number of ground subsidence occurrences in the grid of risk level 2 was set to 1 to 3. In this model, the F1-score (0.590) and AUC (0.800) were the best, and the difference between the training (0.744) and test (0.644) scores was equal to or less than 0.1, which meant overfitting was avoided. Thus, this model was selected as the fittest classifier for the prediction model of ground subsidence risk level in the target region.

The model results, according to whether or not SMOTE was applied, revealed that when SMOTE was not applied there was an F1-score of 0.310 to 0.570, and when SMOTE was applied there was an F1-score of 0.380 to 0.590. This meant that the imbalance in the number of ground subsidence occurrences, which was the output data, was resolved through SMOTE, thereby obtaining an efficient classification of the model. F1-Score and AUC of the XGB classifier in this study were 0.590 and 0.8 (Figure 2). Thus, XGB was found to not be a very good model from a computer science perspective. This result is due to the deepening of the data imbalance caused by the wide range of target areas and the limited use of influencing factors (underground utility attribute information). As ground subsidence is a phenomenon caused by various causes (underground structures, ground conditions, ground layer, etc.) in addition to the damage to underground utility lines, it is expected that the performance of the model will be improved in the future by obtaining more data on the underground space.

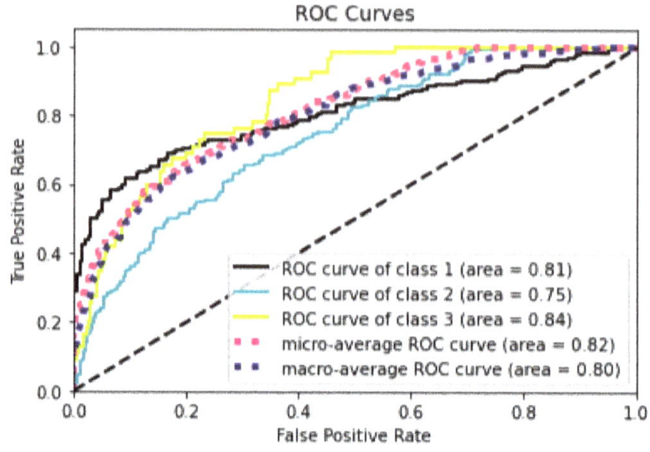

Figure 2. ROC Curves of XGB Model.

In addition, the tuning of the hyperparameters of each classifier was set to the hyperparameter that produces the optimal result using a trial-and-error method. Table 7 summarizes the main hyperparameters of the selected XGB model.

Table 7. Summary of hyperparameters in the model.

Model	Hyper Parameter
XGB	Estimators (300), learning rate (0.002), max depth (4)

The XGB model included a function to derive the importance of the input data employed in the process of solving the classification problem. Using this function, we selected the main influencing factors used to classify the ground subsidence risk levels. Figure 3 shows a graph that exhibits the importance of the factors used in the model, in which Y refers to the number of years used and DTR refers to the pipeline diameter. Density was the most importantly used factor in the classification of ground subsidence risk levels in the XGB model. The number of years used was found to be more important than the diameter of the pipeline. In pipelines used between 20 and 40 years, it was found to be relatively more important.

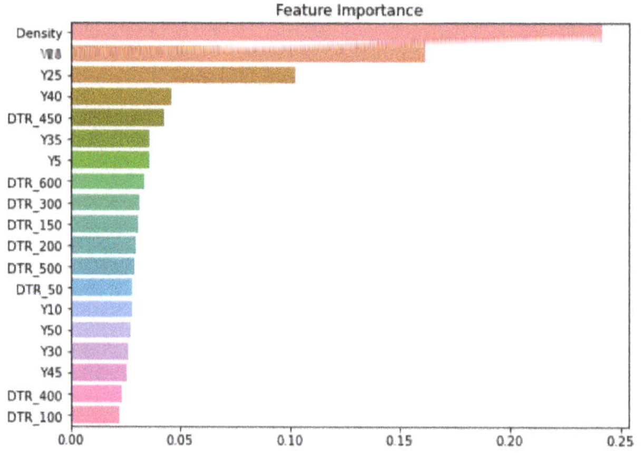

Figure 3. Importance of influencing factors in the XGB model.

3.6. Map of Ground Subsidence Risk

Figure 4 shows a prediction map of the ground subsidence risk level in the target region using the selected prediction model of ground subsidence risk level, as well as a map of ground subsidence risk level based on the historical data of past ground subsidence. In the Figure, the red, yellow, and green colors refer to Level 3, Level 2, and Level 1 ground subsidence risk, respectively. The points on the map indicate the regions where ground subsidence occurred.

(a) (b)

Figure 4. Map of ground subsidence risk. (**a**) Prediction map of ground subsidence risk level. (**b**) Map of ground subsidence risk using real data.

When comparing the prediction map using the model and the map drawn based on the past ground subsidence data, the prediction map had relatively higher risk levels. The prediction model classified the region in which ground subsidence was concentrated in the past as the high-risk region. The prediction map of ground subsidence risk levels in the region will be used as a basis for the management entity to prioritize the areas to be inspected when investigating cavities inside the ground for the prevention of ground subsidence.

4. Conclusions

To develop a model that predicts the risk level of ground subsidence and create a risk level map targeting the urban area in South Korea, a dataset was built using the pipeline length, the number of years used, and the diameter and density of pipelines in the target area. The developed datasets were applied to machine learning algorithms RF, XGB, and LGBM, to comparatively analyze the evaluation indexes. Through this process, the best performance was found in the model with the following dataset conditions applied to the XGB classifier: the number of years used was five years, the pipeline diameter was 50 mm, and the number of ground subsidence occurrences in the grid with risk Level 2 of ground subsidence was set to 1 to 3, when using SMOTE applied data (F1 Score = 0.590, AUC = 0.8). Previously, a machine learning-based ground subsidence risk prediction model has been developed for a small subset of urban areas (two districts) in South Korea [15]. However, since the model was trained using data from a very small area, it is not reliable enough to be applied to a wide range of target areas. Thus, in this study, we collected a large number of data for the entire city and proposed a model for predicting the risk of ground subsidence. As a result, it is now possible to create a reliable ground subsidence risk map for urban areas in Korea through the ground subsidence risk prediction model presented in this study.

The ground subsidence risk prediction model presented in this study derives the importance of influencing factors used when classifying the risk level of ground subsidence. Our study results verified that the density had the highest importance, and the number of years used was more important than the pipeline diameter. This result is similar to that of a previous study which found the density and the number of ground subsidence occurrences were highly correlated [9], as well as another study where the aging of pipelines had an impact on the ground subsidence occurrence as the number of years used increased [3]. Thus, excavation work to bury underground utility lines should be minimized, and aged pipelines should be managed to cope with ground subsidence.

The risk level map of ground subsidence in the target area was created using the ground subsidence risk prediction model. This map predicted a number of spots with higher risk levels than that in the risk map based on the historical data of past ground subsidence. The ground subsidence risk prediction classifier presented in this study predicted the risk level of the area in which ground subsidence was concentrated in the past relatively well.

It is expected that the results presented in this study can be used as foundational data for a proactive response to the occurrence of ground subsidence in urban areas. In future research, we will add underground structures (subway tunnels, etc.) and high-rise building information in the target region to develop a more reliable prediction model of ground subsidence risk level in urban areas.

Author Contributions: Conceptualization, J.K. (Jaemo Kang) and J.K. (Jinyoung Kim); developed the models and carried out the model simulations, S.L.; writing—original draft preparation, S.L.; writing—review and editing, J.K. (Jaemo Kang) and J.K. (Jinyoung Kim). All authors have read and agreed to the published version of the manuscript.

Funding: This research was supported by a grant from the project "Underground Utilities Diagnosis and Assessment Technology (4/4)," which was funded by the Korea Institute of Civil Engineering and Building Technology (KICT).

Institutional Review Board Statement: Not applicable.

Informed Consent Statement: Not applicable.

Data Availability Statement: Not applicable.

Conflicts of Interest: The authors declare no conflict of interest.

References

1. Lee, S.Y.; Kang, J.M.; Kim, J.Y. Development of Machine Learning Model to predict the ground subsidence risk grade according to the Characteristics of underground facility. *J. Korean Geo-Environ. Soc.* **2022**, *23*, 5–10.
2. Seoul City, Cause Analysis of Cavity at Seokchon Underground Roadway and Road Cavity, Seokchon-dong Cavity Cause Investigation Committee, 2014.
3. Kim, J.Y.; Kang, J.M.; Choi, C.H.; Park, D.H. Correlation Analysis of Sewer Integrity and Ground Subsidence. *J. Korean Geo-Environ. Soc.* **2017**, *18*, 31–37.
4. Kuwano, R.; Horii, T.; Kohashi, H.; Yamauchi, K. Defects of sewer pipes causing cave-in's in the road. In Proceedings of the 5th International Symposium on New Technologies for Urban Safety of Mega Cities in Asia, Phuket, Thailand, 16–17 November 2006; pp. 347–353.
5. Mukunoki, T.; Kuwano, N.; Otani, J.; Kuwano, R. Visualization of three dimensional failure in sand due to water inflow and soil drainage from defected underground pipe using X-ray CT. *Soils Found.* **2009**, *49*, 959–968. [CrossRef]
6. Masud, M.; Bairagi, A.K.; Nahid, A.A.; Sikder, N.; Rubaiee, S.; Ahmed, A.; Anand, D. A Pneumonia Diagnosis Scheme Based on Hybrid Features Extracted from Chest Radiographs Using an Ensemble Learning Algorithm. *J. Healthc. Eng.* **2021**, *2021*, 11. [CrossRef] [PubMed]
7. Takeuchi, D.; Fukatani, W.; Miyamoto, T.; Yokota, T. Using decision tree analysis to extract factors affecting road subsidence. *J. Jpn. Sew. Work. Assoc.* **2007**, *54*, 124–133.
8. Jin, Y.S. The Analysis on Correlation of Precipitation and Risk Factors to the Soil Subsidence. Ph.D. Dissertation, Chonnam National University, Gwangju, Republic of Korea, 2018; pp. 104–105.
9. Kim, K.Y. Susceptibility Model for Sinkholes Caused by Damaged Sewer Pipes Based on Logistic Regression. Master's Thesis, Seoul National University, Seoul, Republic of Korea, 2018.

10. Han, M.S. A Risk Assessment of Ground Subsidence by GPR and CCTV Investigation. Master's Thesis, Seoul National University of Science and Technology, Seoul, Republic of Korea, 2017.
11. Kim, J.Y.; Kang, J.M.; Choi, C.H. Correlation Analysis of the Occurrence of Ground Subsidence According to the Density of Underground Pipelines. *J. Korean Geo-Environ. Soc.* **2021**, *22*, 23–29.
12. Muhammad, F.I.; Ganjar, A.; Muhammad, S.; Rhee, J. Hybrid Prediction Model for Type 2 Diabetes and Hypertension Using DBSCAN-Based Outlier Detection, Synthetic Minority Over Sampling Technique (SMOTE), and Random Forest. *Appl. Sci.* **2018**, *8*, 1325. [CrossRef]
13. Mimi, M.; Matloob, K. SMOTE-ENC: A Novel SMOTE-Based Method to Generate Synthetic Data for Nominal and Continuous Features. *Appl. Syst. Innov.* **2021**, *4*, 18. [CrossRef]
14. Georgios, D.; Fernado, B.; Joao, F.; Manvel, K. Imbalanced Learning in Land Cover Classification: Improving Minority Classes' Prediction Accuracy Using the Geometric SMOTE Algorithm. *Remote Sens.* **2019**, *11*, 3040. [CrossRef]
15. Lee, S.Y.; Kang, J.M.; Kim, J.Y. Ground Subsidence Risk Grade Prediction Model Based on Machine Learning According to the Underground Facility Properties and Density. *J. Korean Geo-Environ. Soc.* **2023**, *24*, 23–29.
16. Breiman, L.; Friedman, J.; Stone, C.; Olshen, R. *Classification and Regression Trees*; Taylor & Francis: Abingdon, UK, 1984.
17. Pal, M. Random forest classifier for remote sensing classification. *Int. J. Remote Sens.* **2005**, *26*, 217–222. [CrossRef]
18. Park, E.J.; Park, J.H.; Kim, H.H. Mapping Species-Specific Optimal Plantation Sites Using Random Forest in Gyeongsangnam-do Province, South Korea. *J. Agric. Life Sci.* **2019**, *53*, 65–74. [CrossRef]
19. Hastie, T.; Tibshirani, R.; Friedman, J. *The Elements of Statistical Learning: Data Mining, Inference, and Prediction*, 2nd ed.; Springer: Berlin/Heidelberg, Germany, 2009; p. 745.
20. Lee, S.H.; Yoon, Y.A.; Jung, J.H.; Sim, H.S.; Chang, T.W.; Kim, Y.S. A Machine Learning Model for Predicting Silica Concentrations through Time Series Analysis of Mining Data. *J. Korean Soc. Qual. Manag.* **2020**, *48*, 511–520.
21. Louppe, G. *Understanding Random Forests*; University of Liege: Liege, Belgium, 2014; p. 211.
22. Chen, T.; Guestrin, C. XGBoost: A Scalable Tree Boosting System, KDD'16. In Proceedings of the 22nd ACM SIGKDD International Conference on Knowledge Discovery and Data Mining, San Francisco, CA, USA, 13–17 August 2016; pp. 785–794.
23. Zhang, Y.; Haghani, A. A gradient boosting method to improve travel time prediction. Transportation Research Part C. *Emerg. Technol.* **2015**, *58*, 308–324. [CrossRef]
24. Zhang, D.; Chen, H.D.; Zulfiqar, H.; Yuan, S.S.; Huang, Q.L.; Zhang, Z.Y.; Deng, K.J. iBLP: An XGBoost-Based Predictor for Identifying Bioluminescent Proteins. *Comput. Math. Methods Med.* **2021**, *2021*, 15. [CrossRef]
25. Le NQ, K.; Do, D.T.; Le, Q.A. A sequence-based prediction of Kruppel-like factors proteins using XGBoost and optimized features. *Gene* **2021**, *787*, 145643. [CrossRef]
26. Ke, G.; Meng, Q.; Finley, T.; Wang, T.; Chen, W.; Ma, W.; Ye, Q.; Liu, T. LightGBM: A Highly Efficient Gradient Boosting Decision Tree, Part of Advances in Neural Information Processing Systems. *Adv. Neural Inf. Process. Syst.* **2017**, *30*, 1.
27. Lv, J.; Wang, C.; Gao, W.; Zhao, Q. An Economic Forecasting Method Based on the LightGBM-Optimized LSTM and Time-Series Model. *Comput. Intell. Neurosci.* **2021**, *2021*, 10. [CrossRef]
28. Sokolova, M.; Japkowicz, N.; Szpakowicz, S. Beyond Accuracy, F-Score and ROC: A Family of Discriminant Measures for Performance Evaluation. In Proceedings of the Advances in Artificial Intelligence (AI 2006) Lecture Notes in Computer Science; Springer: Heidelberg, Germany, 2006; Volume 4304, pp. 1015–1021.
29. Wang, L.; Chu, F.; Xie, W. Accurate cancer classification using expressions of very few genes. *IEEE/ACM Trans. Comput. Biol. Bioinf.* **2007**, *4*, 40–53. [CrossRef]
30. Gu, Q.; Zhu, L.; Cai, Z. Evaluation measures of the classification performance of imbalanced data sets. In Proceedings of the ISICA 2009—The 4th International Symposium on Computational Intelligence and Intelligent Systems, Communications in Computer and Information Science, Huangshi, China, 23–25 October 2009; Springer: Heidelberg, Germany, 2009; Volume 51, pp. 461–471.
31. Bekkar, M.; Djemaa, H.K.; Alitouche, T.A. Evaluation measures for models assessment over imbalanced data sets. *J. Inf. Eng. Appl.* **2013**, *3*, 27–38.
32. Akosa, J.S. Predictive accuracy: A misleading performance measure for highly imbalanced data. In Proceedings of the SAS Global Forum 2017 Conference, Orlando, FL, USA, 2–5 April 2017; SAS Institute Inc.: Cary, NC, USA, 2017; pp. 942–2017.
33. Davide, C.; Giuseppe, J. The advantages of the Matthews correlation coefficient (MCC) over F1 score and accuracy in binary classification evaluation. *BMC Genom.* **2020**, *21*, 6.
34. Nguyen, Q.K.L.; Nguyen, T.T.D.; Ou, Y.Y. Identifying the molecular functions of electron transport proteins using radial basis function networks and biochemical properties. *J. Mol. Graph. Model.* **2017**, *73*, 166–178.

applied sciences

MDPI

Article

Assessing the Performance of Machine Learning Algorithms for Soil Classification Using Cone Penetration Test Data

Ayele Tesema Chala * and Richard Ray

Structural and Geotechnical Engineering Department, Faculty of Architecture, Civil and Transport Sciences, Szechenyi Istvan University, Egyetem ter 1, H-9026 Gyor, Hungary
* Correspondence: chala.ayele.tesema@hallgato.sze.hu

Abstract: Conventional soil classification methods are expensive and demand extensive field and laboratory work. This research evaluates the efficiency of various machine learning (ML) algorithms in classifying soils based on Robertson's soil behavioral types. This study employs 4 ML algorithms, including artificial neural network (ANN), random forest (RF), support vector machine (SVM), and decision trees (DT), to classify soils from 232 cone penetration test (CPT) datasets. The datasets were randomly split into training and testing datasets to train and test the ML models. Metrics such as overall accuracy, sensitivity, precision, F1_score, and confusion matrices provided quantitative evaluations of each model. Our analysis showed that all the ML models accurately classified most soils. The SVM model achieved the highest accuracy of 99.84%, while the ANN model achieved an overall accuracy of 98.82%. The RF and DT models achieved overall accuracy scores of 99.23% and 95.67%, respectively. Additionally, most of the evaluation metrics indicated high scores, demonstrating that the ML models performed well. The SVM and RF models exhibited outstanding performance on both majority and minority soil classes, while the ANN model achieved lower sensitivity and F1_score for minority soil class. Based on these results, we conclude that the SVM and RF algorithms can be integrated into software programs for rapid and accurate soil classification.

Keywords: cone penetration test; soil classification; machine learning; artificial neural network; support vector machine; random forest; soil behavioral type

Citation: Chala, A.T.; Ray, R. Assessing the Performance of Machine Learning Algorithms for Soil Classification Using Cone Penetration Test Data. *Appl. Sci.* **2023**, *13*, 5758. https://doi.org/10.3390/app13095758

Academic Editor: Wei Gao

Received: 29 March 2023
Revised: 2 May 2023
Accepted: 4 May 2023
Published: 6 May 2023

1. Introduction

For many years, the cone penetration test (CPT) has been the predominant method for conducting field exploration in geotechnical engineering [1–5]. This test requires a cone-shaped instrument to be inserted into the soil at a consistent penetration rate, while measuring the cone tip resistance (qc) and sleeve friction (fs). The CPT continuously provides precise, repeatable results for its entire profile depth. Moreover, the CPT is a relatively quick and inexpensive means of acquiring field data for estimating parameters for many applications, such as soil classification, environmental studies, hydrological analysis, and seismic site response assessments.

Soil classification is essential in geotechnical engineering, especially when evaluating site response to seismic events. Accurate soil classification helps to understand the dynamic properties of soil and the effects of earthquakes on the soil's behavior. The traditional soil classification based on the CPT data involves analyzing 2D charts. Early research was aimed to predict the distribution of soil particles by using the CPT measurements, as outlined in the pioneering work of Begemann [6]. However, later work by Douglas and Olsen [7] suggested that a more useful soil classification approach in practical engineering projects would involve considering soil behavior, rather than solely relying on soil particle distribution. As a result, Robertson developed soil classification charts based on a soil behavior type index using CPT measurements [4,8]. Additionally, there are alternative methods for soil classification. The Unified Soil Classification System (USCS), for example, relies on extensive field and laboratory tests to classify soil [9].

Soil classification and parameter estimation using traditional methods can be costly and time-consuming. Field and laboratory testing is required, and soil samples need to be transported to a laboratory where particle size distribution and Atterberg limits are conducted. These tests take time to complete, and the results may not be immediately available. Additionally, soil properties can significantly change with variations in the temperature and moisture content. However, in recent years, machine learning (ML) techniques have shown great promise in soil classification. Many studies have demonstrated the potential of ML techniques in soil classification based on CPT measurements [10–15].

The study conducted by [15] explored the feasibility of utilizing a general regression neural network (GRNN) to predict soil composition and overall soil type employing CPT data. The research demonstrated that the GRNN model successfully categorized soils as either coarse-grained or fine-grained. Similarly, studies have demonstrated the effectiveness of artificial neural network (ANN) models in predicting complex soil profiles [16–18]. In addition, various machine learning techniques such as random forests (RFs) [19], support vector machines (SVMs) [20,21], decision trees (DTs) [22], gradient boosting machine (GBM) [23–26], and logistic regression (LR) [27] have been utilized for a variety of geotechnical engineering applications including classification and liquefaction.

ML techniques have been widely used in various fields, including image [28–30] and speech recognition, natural language processing, and data analysis, to extract insights from large datasets. For instance, ML algorithms are used to identify objects, faces, and patterns in images, which is crucial in facial recognition, autonomous driving, and object detection in security systems. ML algorithms are also used to transcribe speech into text, enabling the creation of voice assistants, language translation software, and speech-to-text dictation tools. In addition, ML algorithms are used to analyze large datasets, uncovering patterns and insights that would be difficult or impossible to identify manually. This technique has numerous applications in finance, healthcare, and scientific research.

Although machine learning (ML) techniques have been widely applied in various fields, there has been limited research on their use in geotechnical engineering. However, researchers have started exploring the potential of ML techniques for soil classification and estimation of soil parameters using CPT data. Some geotechnical researchers applied ML techniques to predict various geotechnical properties such as landslide [31], slope stability [32–34], soil type [12], and shear wave velocity [13] utilizing CPT data.

In our study, we aim to evaluate the performance of four commonly used ML algorithms, including artificial neural network (ANN), random forest (RF), support vector machine (SVM), and decision tree (DT), for soil classification using CPT data. This study has the potential to address the gap in the existing literature and offer valuable insights into the efficacy of ML algorithms for soil classification through CPT data. Furthermore, the findings of this study could help improve the efficiency and accuracy of soil classification in geotechnical engineering, which could have significant implications for geotechnical engineering.

The selection of the ML model for a classification task is based on several factors, including desired accuracy, dataset size, generalization ability, interpretability, and robustness. For our specific soil classification problem, we chose to evaluate the performance of the ANN, DT, SVM and RF algorithms, each with its own strength and weakness. ANNs are known for their ability to capture complex non-linear relationships in the data [35], while RF is the ML algorithm that utilizes multiple decision trees to enhance the accuracy and robustness of the model [36,37]. SVM can handle high-dimensional data and nonlinear decision boundaries [38,39], and DT is easy to interpret and visualize, and can handle both categorical and numerical data. By selecting these four algorithms, we aimed to strike a balance between complexity and interpretability and compare the performance of the models. Our choice of algorithms provides a diverse set of models that can handle various aspects of the classification task, including complex relationships, high-dimensional data, and interpretability. By evaluating their performances, we hope to gain insights into which algorithm is the most suitable for our specific soil classification problem.

The performance of the ML models can be compromised by various factors if not properly addressed. One of the critical factors that can affect the models' performance is the selection of hyperparameters. By carefully selecting and tuning the hyperparameters, we can improve the models' robustness and ensure that they can perform optimally in real-world applications [40]. Grid search (technique in which sets of predefined hyperparameter values are defined) is one of the most commonly used methods for hyperparameter tuning [41]. Bayesian optimization is another approach that uses a probabilistic model to estimate the performance of different hyperparameter configurations [41,42].

ML models rely on input features to make predictions, and the quality and relevance of those features can have a significant impact on the performance of the models [43]. Performing feature importance such as permutation feature importance [44] and eliminating irrelevant features from the dataset can significantly enhance the performance of ML models. Feature importance is the process of determining the most important features in a dataset for a given model.

In ML, outliers are one of the factors that contribute to the poor performance of ML models. According to the literature [45,46], outliers are data points that deviate significantly from the surrounding data points. Abnormal data readings during CPT operations can primarily occur due to human or procedural errors, such as the addition of a rod [47]. These outliers are not representative of the actual CPT measurements and should be detected and removed during the data preprocessing stage.

The structure of this paper comprises six sections. The first section provides a detailed discussion on the background of soil classification and ML models. In Section 2, the cone penetration test is explained, while Section 3 outlines the dataset preprocessing and methodology utilized. The ML models employed in this study are briefly summarized in Section 4. In Section 5, a detailed discussion is presented on the results obtained from the ML models. Finally, Section 6 provides a summary of the main points of the study and concludes the paper by proposing recommendations for future research.

2. Cone Penetration Tests

The CPT is a widely used in situ geotechnical testing method that involves inserting a cone-shaped penetrometer into the soil and recording the soil's resistance (i.e., qc and fs) to penetration. Figure 1 visually represents a graph that plots the recorded qc, fs, and friction ratio used in this study.

The CPT and its variations, such as the CPT with pore pressure measurement (CPTu) and the seismic cone penetration test with pore pressure measurements (SCPTu), are valuable tools for various engineering applications. These tests can estimate geotechnical parameters and classify soils over a broad range of soil types, from very soft soil to weak rock. Over the past few decades, various soil behavior charts have evolved for soil classification based on CPT-measured data [1–4,48].

One such chart was developed by Robertson [8] and can be used to classify soils into different categories, such as sand, clay, silt mixture, organic soil, and more. An example of such a chart is shown in Figure 2, which illustrates the classification of soil types ranging from sensitive clays to very stiff over-consolidated (OC) clays. The chart categorizes soils into various classes or zones based on their soil behavioral type index (Ic) determined by Equation (1) [48]. Table 1 lists boundaries for classification based on Ic values. In this study, the zone numbers (see Figure 2) are directly used as ML labels as they represent the soil types in a straightforward and intuitive way.

$$I_c = \sqrt{\left(3.47 - log(q_t/p_a)^2 + (logR_f + 1.22)^2\right)} \tag{1}$$

where q_t is corrected cone resistance or CPT cone resistance q_c, p_a is atmospheric pressure in the same unit as q_c, R_f is friction ratio, and f_s is CPT sleeve friction.

$$R_f = (f_s/q_c) \times 100\% \tag{2}$$

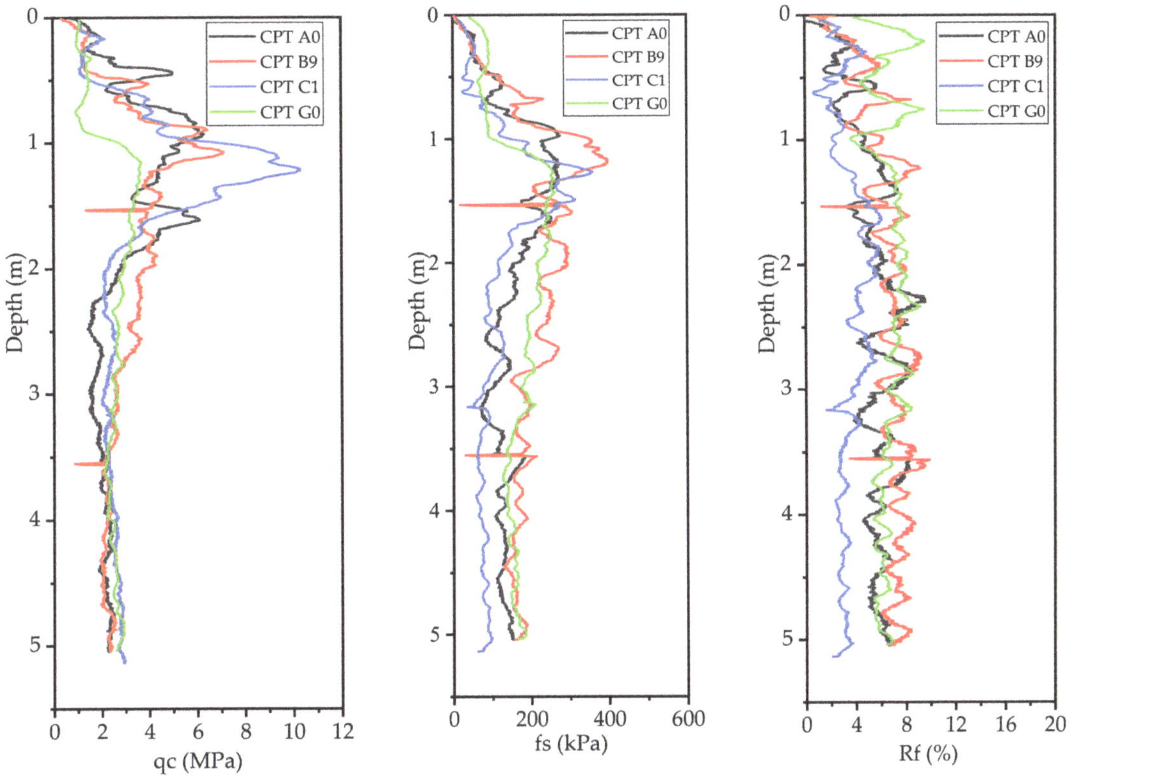

Figure 1. Sample cone tip resistance (qc), sleeve friction (fs), and friction ratio (Rf) vs. depth employed in this study.

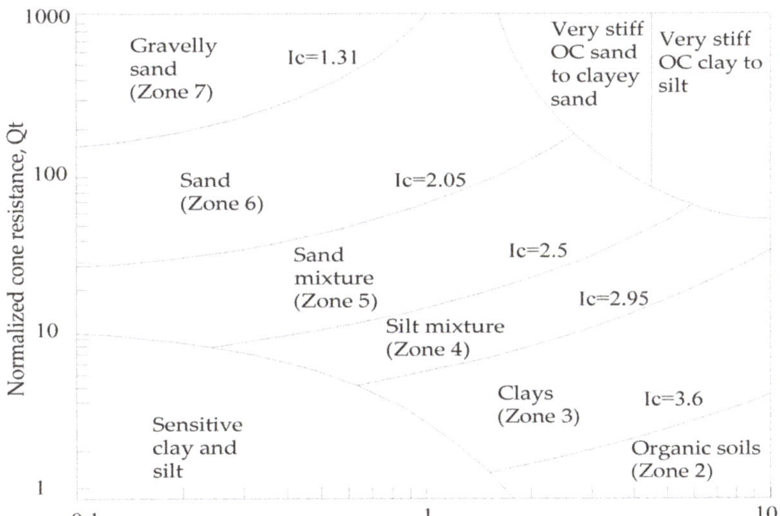

Figure 2. Robertson soil classification chart based on soil behavioral type index, Ic (adapted from [48]).

Table 1. Soil behavior type classification based on Ic boundaries (adapted from [8,48]).

SBT Classification	Ic Boundaries	Soil Type ID/ML Labels
Organic soils: peats	Ic > 3.6	2
Clays: clay to silty clay	2.95 < Ic < 3.6	3
Silt mixtures: clayey silt and silty clay	2.6 < Ic < 2.95	4
Sand mixtures: silty sand to sandy silt	2.05 < Ic < 2.6	5
Sands: clean sand to silty sand	1.31 < Ic < 2.05	6
Gravelly sand to dense sand	Ic < 1.31	7

Although existing empirical correlations work well with the CPT data, their applicability is limited to primarily fine-grained soils. Additionally, CPT and core drilling techniques work together to provide more detailed information about subsurface soil properties [12].

3. Datasets

For our study, we used publicly available CPT datasets contributed by [47], which were accessible in the International Society for Soil Mechanics and Geotechnical Engineering (ISSMGE) database. The CPTs were collected from an area measuring 50 by 50 m. Each CPT was performed to a depth of 5 m below the ground surface, and the measurement spacing of qc and fs was 5 mm. Further information about the specifics of the CPTs can be found in [47,49,50].

We preprocessed the datasets using MS Excel (see Figure 3) to categorize the soil behavior types based on Robertson's classification [48]. In order to reduce bias and ensure that the training and testing datasets are representative of the overall dataset, we shuffled the dataset and divided it into training and testing datasets using (80, 20) ratio.

The steps followed to preprocess data are as follows:

(1) Combine the individual CPT soundings into the appropriate columns (e.g. depth, qc, and fs) using Power Query in MS Excel and remove the missing values. (2) Calculate the inter quartile range (IQR) values for the qc and fs columns using Excel's built-in functions such as QUARTILE. (3) Determine the upper threshold values for outlier detection by multiplying the IQR by three and adding the third quartile. (4) Identify the outlier values in the qc and fs columns using conditional formatting. (5) Remove the outlier from the dataset and replace the values with the threshold value. (6) Estimate the R_f, total vertical stress (σ_v), effective vertical stresses (σ_v'), and Ic.

Table 2 presents a statistical summary of the datasets organized into 222,100 rows and 7 columns. The frequency distribution of each soil type in the dataset is shown in Figure 4. The distribution analysis demonstrated that soil type 5 has the highest frequency and represents over 50% of the total dataset. Soil type 4 has the second highest frequency and represents over 30% of the dataset. Soil types 2, 3, 6, and 7 (minority class) have much lower frequencies and represent less than 20% of the dataset combined, indicating an imbalanced dataset. Balancing this highly imbalanced dataset using oversampling or under sampling techniques may be possible, but it can also affect the natural variability of the soil, potentially leading to biased predictions and incorrect soil classification. To avoid this, we opted to train the ML models on the imbalanced datasets and evaluate their performances using appropriate evaluation metrics such as sensitivity, precision, and F1_score, instead of artificially generating or discarding soil samples that could impact the true variability of the soil.

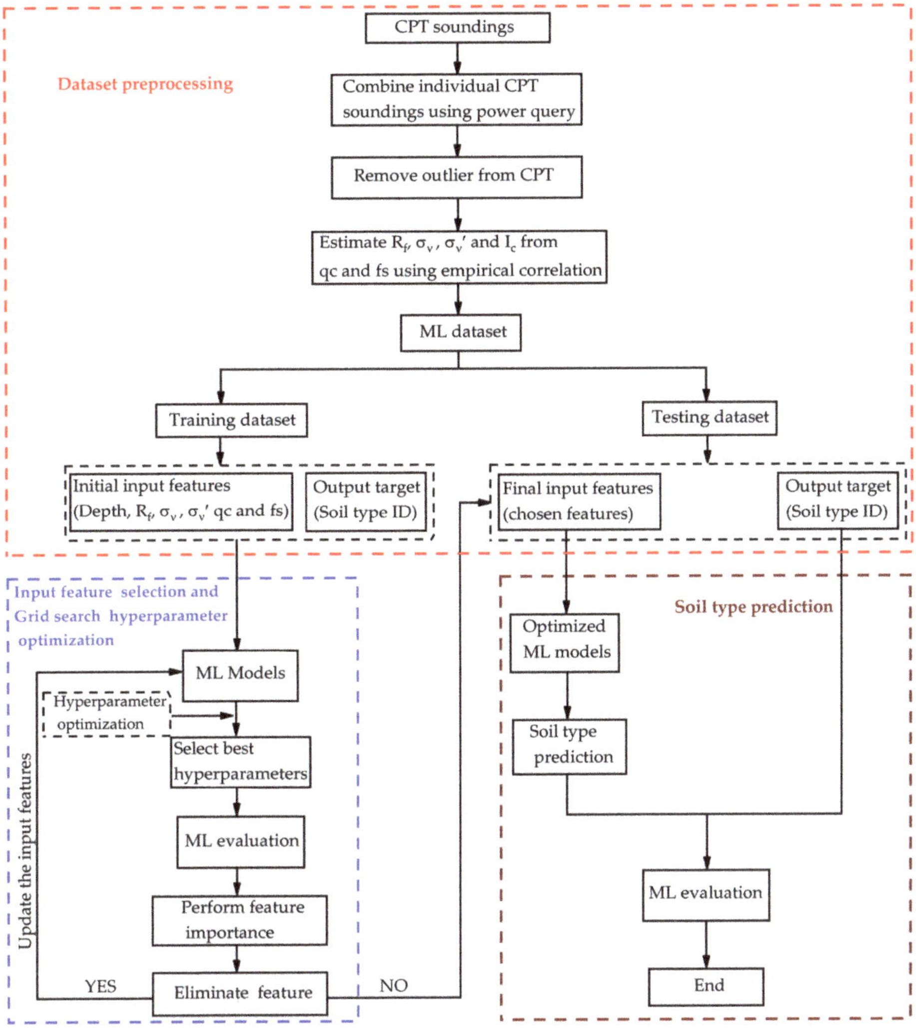

Figure 3. Flow diagram illustrating dataset preprocessing and machine learning model architecture.

Table 2. Summary statistics of dataset.

	Depth (m)	fs (kPa)	qc (kPa)	Rf (%)	σ_v (kPa)	σ'_v (kPa)	Soil Type ID
Mean	2.57	158.91	2535	9.53	49.75	24.00	4.68
Median	2.57	144.70	2360	6.14	49.87	23.89	5
Standard Deviation	1.43	84.31	1181	16.59	27.41	13.22	0.82
Kurtosis	−1.16	1.72	1.79	87.94	−1.11	−1.06	1.23
Skewness	0.00	1.12	0.99	5.86	−0.01	0.01	−0.53
Minimum	0.01	0.30	10.0	0.01	0.07	0.02	2
Maximum	5.63	438.90	6830	899	109.94	54.92	7

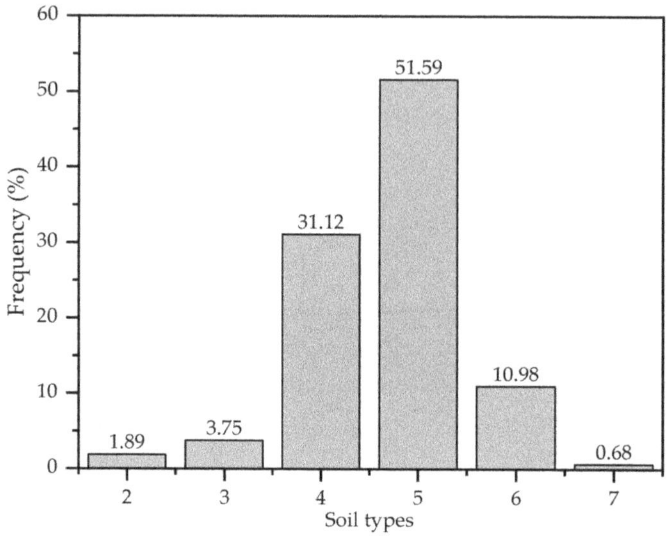

Figure 4. Frequency distribution of soil types considered for ML models.

The input features, which include depth, qc, and fs, are raw data directly obtained from the test. In contrast, the friction ratio R_f, total vertical stress (σ_v), and effective vertical stresses (σ_v') are results from empirical correlations (Equations (2), (3) and (5), respectively).

$$\sigma_v = \gamma \times h \tag{3}$$

where σ_v is total vertical stress, γ is unit weight of soil and h is depth of soil.

The unit weight of the soil is estimated using the following expression [51]:

$$\gamma = \gamma_w \left[0.27logR_f + 0.36log(q_c/p_a) + 1.236\right] \tag{4}$$

where γ is unit weight of soil, γ_w is unit weight of water in the same unit as γ, q_c is cone tip resistance, and p_a is atmospheric pressure in the same unit as q_c.

$$\sigma_v' = \sigma_v - \gamma_w h \tag{5}$$

where σ_v' is effective vertical stress, σ_v is total vertical stress, γ_w is unit weight of water in the same unit as σ_v, and h is depth of soil.

4. Machine Learning Models

ML is a subfield of artificial intelligence (AI) that aims to develop algorithms and statistical models to help computer systems improve their performance on specific tasks by learning from the data [52]. The types of learning include supervised, unsupervised, and reinforcement [53]. While the supervised and reinforcement learning algorithms can involve human supervision, the unsupervised learning algorithms do not rely on labeled data or human guidance.

Our study utilized the supervised ML algorithms to classify soils using the CPT datasets. We trained four different ML algorithms, ANN, RF, SVM, and DT, using training CPT datasets and tested their performance on test datasets via R programming language [54]. In the following section, we discuss each of the ML algorithms to gain insight into their strengths and limitations.

4.1. Artificial Neural Network Model

ANNs are ML models that draw inspiration from the human brain's structure and functions [35]. They comprise interconnected neurons that use weighted connections to process and transmit information. ANNs can learn data patterns and relationships by modifying the connection strength based on the input and output. ANN models typically contain three layers, including input, hidden, and output layers. Figure 5 presents an example of an ANN model that includes an input layer with 6 neurons, 2 hidden layers with 16 and 8 neurons, and an output layer. Deep learning is commonly used to describe neural networks with many hidden layers.

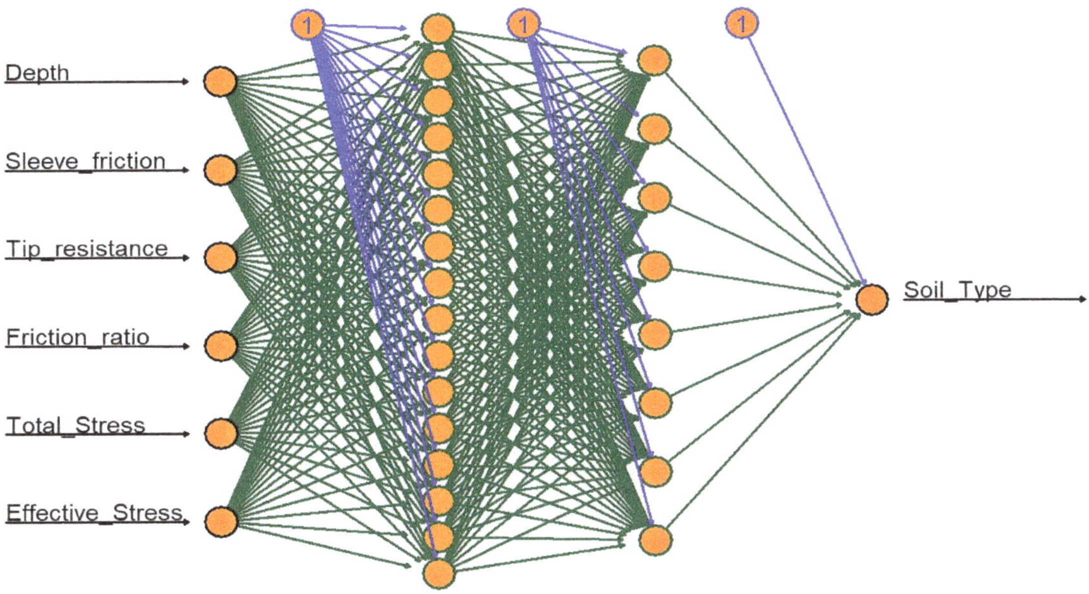

Figure 5. Example of visualized neural network plot with 2 hidden layers, 16 and 6 neurons, and 6 features.

In the ANN models, weights (the connection strength between neurons) and activations (output of a neuron in the network) are fundamental elements that enable the network to learn patterns and relationships in the data. The weights in an ANN are adjusted during the learning process to optimize the model's performance. At the same time, activation applies a mathematical operation to the input and transmits an output to the other neurons in the network.

Choosing the proper activation function is essential when dealing with an ANN model. There are several types of activation functions, namely, Sigmoid function (commonly used for binary classifications), ReLU (rectified linear unit) function, Tanh (hyperbolic tangent) function, and Softmax function (commonly used in the output layer).

Our study considers an ANN model with 2 hidden layers containing 128 and 32 neurons and an output layer. We implemented our models using the Keras package [55], which provides an easy-to-use interface for building and training neural networks. A multi-layer perceptron (MLP) model provided the soil classification with the ReLU activation function in both hidden layers, and the Softmax activation function in the output layer. The Keras library in R aided the model development, which was compiled using the categorical cross-entropy loss function, the Adam optimizer, and accuracy as the evaluation metric. The model was trained on the training data for 200 epochs, using a batch size of 32 and a validation split of 0.2. The model would learn from the data and adjust its weights and biases to minimize the loss, which measures the difference between the predicted and actual values.

The categorical cross-entropy loss function is used to measure the difference between the predicted and actual values in a classification task. To minimize this loss, the Adam optimizer adjusts the weights and biases of the model during training. Accuracy, on the other hand, is a metric that evaluates how well the model generalizes to new, unseen data by measuring the percentage of correct predictions. The performance of the model was improved through the Bayesian optimization fine-tune of its hyperparameters including dense units 1, dense units 2, dropout 1, dropout 2, and batch size.

4.2. Random Forest Model

Random forest is a widely used ensemble learning algorithm for both classification and regression tasks. The algorithm employs multiple decision trees to improve the model's accuracy and robustness. Unlike individual decision trees, random forest is less prone to overfitting as it combines multiple trees with varying biases and variances. Additionally, it can efficiently handle high-dimensional data with many features by randomly selecting a subset of features for each tree. As a result, the algorithm is capable of handling large and complex datasets [36,37,56].

In our research, we utilized the random forest algorithm to train a model using the random forest package [57] in the R programming language. We fine-tuned the model's hyperparameters, including the number of variables randomly sampled at each split of a decision tree (mtry), the minimum number of internal node size (min.node.size), and the number of decision trees (ntree), using a model-based Bayesian optimization technique. The performance of the model was evaluated using cross-validation, and we selected the optimal values of the hyperparameters based on its best performance.

4.3. Decision Tree Model

Decision tree (DT) is a widely used machine learning algorithm that can be applied to both classification and regression problems. It is a non-parametric algorithm that can handle large and complex datasets without imposing a rigid parametric structure, making it a versatile tool for various applications [57]. The DT algorithm builds a tree-like model where the internal nodes of the tree represent decisions based on input features, while each leaf node represents class labels or target values. DT models are particularly suitable for multi-class classification problems due to their ability to capture non-linear relationships between input features and target variables [58,59].

For our soil classification problem, we utilized the rpart package [60] in the R programming language to implement a decision tree model. We fine-tuned the model's hyperparameters, including the complexity parameter (cp), the maximum depth of trees, the minimum split, and the maximum number of competitor splits, using Bayesian optimization. We used cross-validation to prevent overfitting and improve the model's ability to generalize to new data.

4.4. Support Vector Machine Model

SVM is a well-known supervised ML algorithm frequently utilized for multi-class classification and regression problems [38,39]. The SVM algorithm operates by locating the optimal hyperplane that segregates the input data points into distinct classes. The hyperplane locates itself by maximizing the margin, which is the gap between the hyperplane and the nearest data points of each class. For our study, we employed the e1071 R package [61], which offers an SVM implementation in R. This allowed us to train a model using the training data and assess its effectiveness on the test data. To ensure a well-tuned and generalized model, we used cross-validation to optimize the hyper-parameters (cost and gamma).

5. Results and Discussion

In the following subsections, the results of the ML models are presented and discussed using confusion matrix and various performance metrics such as overall accuracy, sensitivity (ability to detect positive instances), specificity (ability to detect negative instances), negative predicted value (NPV), positive predicted value (PPV), and balanced accuracy (the average of sensitivity and specificity). Due to the imbalanced dataset used for the training and testing purposes, additional informative performance metrics such as precision, recall, and F1_score are utilized to assess the efficacy of the ML models.

$$Overall\ Accuracy = (TP + TN)\ /\ (TP + TN + FP + FN) \tag{6}$$

where TP = True Positive (number of samples correctly predicted as positive), TN = True Negative (number of samples correctly predicted as negative), FP = False Positive (number of samples incorrectly predicted as positive), and FN = False Negative (number of samples incorrectly predicted as negative).

$$Precision = TP\ /\ (TP + FP) \tag{7}$$

$$Sensitivity = TP\ /\ (TP + FN) \tag{8}$$

$$F1\_score = \frac{2 \times precision \times Sensitivity}{precision + Sensitivity} \tag{9}$$

5.1. Artificial Neural Network Model Results

The results of the ML model implemented utilizing ANN to classify different soil types are presented here. Figure 6 displays the accuracy and loss of the ANN model for 200 epochs on both the training and validation data. At the beginning of the training, the model has a low accuracy of 0.79 and a high loss of 0.63 values, indicating that it cannot make good predictions. However, as training progresses, the accuracy improves, and the loss decreases, indicating that it gradually improves its ability to make more accurate predictions. When the validation accuracy and loss metrics improve, it suggests that the model is generalizing well to new data, which is a desirable outcome.

Table 3 displays the confusion matrix of the ANN model, which provides insight into the model's performance on the test data. The rows correspond to the predicted values, while the columns correspond to the actual values. The diagonal elements of the confusion matrix represent the number of instances that the model correctly classified, while the off-diagonal elements correspond to the misclassifications made by the model.

Table 3. Artificial neural network confusion matrix.

Prediction	Actual					
	2	3	4	5	6	7
2	808	8	0	0	0	0
3	31	1594	23	1	0	0
4	0	63	13,708	69	0	0
5	0	0	91	22,781	55	0
6	0	0	0	62	4801	99
7	0	0	0	1	20	202

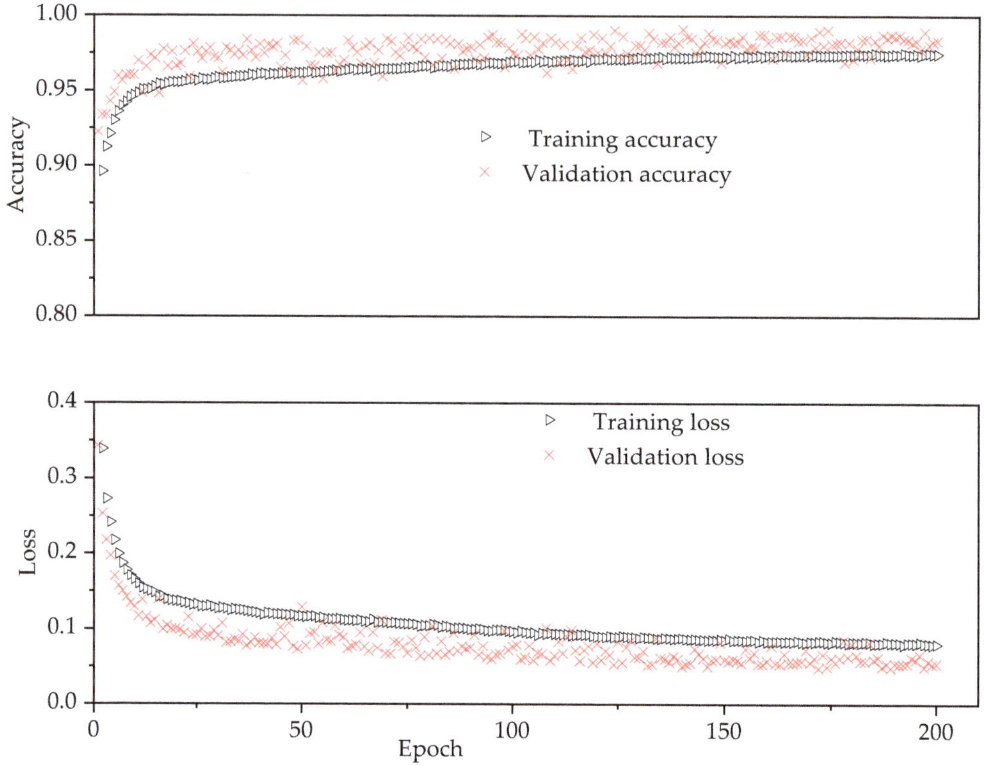

Figure 6. Change in loss and accuracy over epochs for a neural network model.

Statistics by class (Table 4) show that the model has a high sensitivity for all soil types except type 7, with a low sensitivity value of 0.67. Additionally, the model has a high specificity for all classes, with values ranging from 0.99 to 1.0.

Table 4. Class distribution summary for ANN model.

Performance Metrics	2	3	4	5	6	7
Sensitivity	0.96	0.96	0.99	0.99	0.98	0.67
Specificity	1.00	1.00	1.00	0.99	1.00	1.00
PPV	0.99	0.97	0.99	0.99	0.97	0.91
NPV	1.00	1.00	1.00	0.99	1.00	1.00
Prevalence	0.02	0.04	0.31	0.52	0.11	0.01
Detection Rate	0.02	0.04	0.31	0.51	0.11	0.00
Detection Prevalence	0.02	0.04	0.31	0.52	0.11	0.01
Balanced Accuracy	0.98	0.98	0.99	0.99	0.99	0.84

The positive predicted value (PPV) and negative predicted value (NPV) are important performance metrics in evaluating the effectiveness of a classifier. A high PPV indicates that it is likely correct when the model predicts a sample to belong to a particular class. On the other hand, a high NPV indicates that when the model predicts a sample to not belong to a particular class, it is likely to be correct. The ANN model results show a high PPV for all soil types, with values ranging from 0.91 to 0.99. Similarly, the NPV is high for all soil types, with values ranging from 0.99 to 1.

In summary, the ANN model has an overall accuracy of 98.82%, showing that the model performs well in classification tasks. However, the model struggles to predict class 7 (minority class), given a low sensitivity value.

5.2. Random Forest Model Results

Table 5 displays a confusion matrix that compares the soil types predicted by the RF model with the actual soil types. The confusion matrix shows that the model made some correct and incorrect predictions for each class. For example, the model correctly predicted 835 samples as class 2. The model has high diagonal values, signifying a high number of correct predictions, and low off-diagonal values, implying a low number of misclassifications. The model's overall accuracy is very high, indicating the model's performance. It achieved a 99.23% accuracy, indicating that it effectively predicts soil types.

Table 5. Random forest confusion matrix.

Prediction	Actual					
	2	**3**	**4**	**5**	**6**	**7**
2	835	2	0	0	0	0
3	4	1655	8	0	0	0
4	0	8	13,719	88	0	0
5	0	0	95	22,763	59	0
6	0	0	0	63	4806	5
7	0	0	0	0	11	296

The statistics by class (Table 6) show that the RF model has a high sensitivity and specificity values for all soil types. Overall, it performed well in the classification task, achieving high scores for multiple performance metrics such as PPV, NPV, and balanced accuracy.

Table 6. Class distribution summary for RF model.

Performance Metrics	2	3	4	5	6	7
Sensitivity	1.00	0.99	0.99	0.99	0.99	0.98
Specificity	1.00	1.00	1.00	0.99	1.00	1.00
PPV	1.00	0.99	0.99	0.99	0.99	0.96
NPV	1.00	1.00	1.00	0.99	1.00	1.00
Prevalence	0.02	0.04	0.31	0.52	0.11	0.01
Detection Rate	0.02	0.04	0.31	0.51	0.11	0.01
Detection Prevalence	0.02	0.04	0.31	0.52	0.11	0.01
Balanced Accuracy	1.00	1.00	0.99	0.99	0.99	0.99

5.3. Decision Tree Model Results

Table 7 presents the confusion matrix for the DT model utilized for the soil classification task. The table evaluates the performance of a predictive model in classifying different soil types based on input features. The number of observations that were accurately predicted by the model (diagonal entries) and the number of misclassifications (off-diagonal entries) appear in the table. Based on the model's confusion matrix, the model performed well in the classification task, as the number of correctly predicted values are significantly higher than the number of misclassifications. The overall accuracy of the model in predicting soil types on the test dataset was 95.67%.

Table 7. Confusion matrix for DT model.

Prediction	Actual					
	2	**3**	**4**	**5**	**6**	**7**
2	819	40	0	0	0	0
3	20	1554	45	0	0	0
4	0	71	13,240	591	0	0
5	0	0	537	22,049	269	0
6	0	0	0	274	4548	19
7	0	0	0	0	59	282

The statistics by class (Table 8) show that the DT model has a high sensitivity and specificity for all soil types. Moreover, the model exhibits a high balanced accuracy with values ranging from 0.96 to 0.99. Overall, the model performed well in the classification task, with high scores for multiple performance metrics across each soil type.

Table 8. Class distribution summary for DT model.

Performance Metrics	2	3	4	5	6	7
Sensitivity	0.98	0.93	0.96	0.96	0.93	0.94
Specificity	1.00	1.00	0.98	0.96	0.99	1.00
PPV	0.95	0.96	0.95	0.96	0.94	0.83
NPV	1.00	1.00	0.98	0.96	0.99	1.00
Prevalence	0.02	0.04	0.31	0.52	0.11	0.01
Detection Rate	0.02	0.03	0.30	0.50	0.10	0.01
Detection Prevalence	0.02	0.04	0.31	0.51	0.11	0.01
Balanced Accuracy	0.99	0.97	0.97	0.96	0.96	0.97

5.4. Support Vector Machine Model Results

The results of the ML model implemented utilizing the SVM to classify different soil types are presented here. The confusion matrix computed with the SVM model to evaluate its effectiveness is presented in Table 9. The confusion matrix shows that the model predicted almost all instances correctly, with a few misclassifications in each soil type. The overall accuracy of the model is very high (almost 100%), indicating that it is a high-performing model.

Table 9. Support vector machine confusion matrix.

Prediction	Actual					
	2	3	4	5	6	7
2	837	0	0	0	0	0
3	2	1662	0	0	0	0
4	0	3	13,812	6	0	0
5	0	0	10	22,893	7	0
6	0	0	0	15	4848	9
7	0	0	0	0	21	292

Table 10 shows the class distribution summary for the SVM model. The model's sensitivity is high for each soil type, indicating that the model is good at correctly identifying the positive cases for each soil type. The model's specificity is also high for all soil types, indicating that the model is good at correctly identifying the negative cases for all soil types. Moreover, the model's balanced accuracy (the average of sensitivity and specificity) is remarkably high (almost 1) for all soil types. This shows that the model can accurately identify both positive and negative cases, making it a reliable classifier for the soil classification task.

Table 10. Class distribution summary for SVM model.

Performance Metrics	2	3	4	5	6	7
Sensitivity	1	1	1	1	1	0.97
Specificity	1	1	1	1	1	1
PPV	1	1	1	1	1	0.93
NPV	1	1	1	1	1	1
Prevalence	0.02	0.04	0.31	0.52	0.11	0.01
Detection Rate	0.02	0.04	0.31	0.52	0.11	0.01
Detection Prevalence	0.02	0.04	0.31	0.52	0.11	0.01
Balanced Accuracy	1	1	1	1	1	0.98

The model exhibits high PPVs for all soil types, indicating its strong ability to predict the samples of specific soil types accurately. Similarly, the model shows high NPVs for all soil types, indicating its reliability in predicting the samples that do not belong to a particular soil class. Overall, the model performs exceptionally well on the dataset in terms of multiple performance metrics.

5.5. Comparison of ML Models' Performance

To compare the efficiency of the ML models, different performance metrics such as overall accuracy, sensitivity, precision, and F1_score are utilized. The results of this evaluation are summarized in Tables 11 and 12. Table 11 shows that the SVM model achieved the highest overall accuracy of 99.84%.

Table 11. Performance comparisons of ML models on test datasets.

	ML Models			
	ANN	RF	DT	SVM
Overall Accuracy (%)	98.82	99.23	95.67	99.84

Table 12. Performance metrics of ML models for each soil type.

ML Models	Soil Type	Sensitivity	Precision	F1_Score
ANN	2	0.96	0.99	0.98
	3	0.96	0.97	0.96
	4	0.99	0.99	0.99
	5	0.99	0.99	0.99
	6	0.98	0.97	0.98
	7	0.67	0.91	0.77
RF	2	1	0.99	0.99
	3	1	0.99	0.99
	4	1	0.99	0.99
	5	1	0.99	0.99
	6	1	0.99	0.99
	7	0.98	0.96	0.97
SVM	2	1	1	1
	3	1	1	1
	4	1	1	1
	5	1	1	1
	6	0.99	1	0.99
	7	0.97	0.93	0.95
DT	2	0.98	0.95	0.96
	3	0.93	0.96	0.95
	4	0.96	0.95	0.96
	5	0.96	0.96	0.96
	6	0.93	0.94	0.94
	7	0.94	0.83	0.88

The ANN, RF, and DT models also performed well, achieving overall accuracies of 98.82%, 99.23%, and 95.67%, respectively. It is important to note that the datasets were imbalanced, and therefore, it is necessary to consider both the overall accuracy and other performance metrics for each soil type to accurately assess the ML models' performance.

Table 12 presents the performance metrics of the ML models for each soil type. The table shows the sensitivity, precision, and F1_score values of each model and soil type. These metrics indicate the models' efficiency in correctly identifying the soil type. Across all models, the sensitivity, precision, and F1_score values for each soil type are very high, indicating that the models successfully identified instances of all classes. However, the efficiency of the ANN model on minority class 7 was low compared to the other models. It scored lower sensitivity and F1_score values of 0.67 and 0.77, respectively, compared to the

SVM and RF models with almost perfect scores for all metrics. This indicates that the ANN model needs additional data to better identify minority classes.

The SVM and RF models outperformed the ANN and DT models in terms of sensitivity, precision, and F1_score values for all soil types. These two models achieved almost perfect scores for all performance metrics for all soil types, indicating their high accuracy in the classification task. Overall, the performance of the ML models in classifying soils based on the CPT dataset is consistent with previous similar research carried out on ML techniques (e.g., see [12,21]).

6. Conclusions

In this paper, various ML algorithms, namely, ANN, RF, SVM, and DT are used to classify soils based on Robertson's soil behavioral types. To optimize the performance of these models, hyperparameter tuning was performed using Bayesian optimization. Additionally, cross-validation was employed to ensure the models' generalization ability and optimal performance. The study used 232 CPTs from ISSMGE's database and randomly divided the dataset into training and testing datasets. The performance of each ML model was evaluated using performance metrics such as sensitivity, precision, F1_score, and overall accuracy.

Based on the findings of this study, the following conclusions can be made:

- The ANN model achieved an overall accuracy of 98.39%. It achieved high scores in multiple performance metrics for majority soil classes. However, the model achieved lower sensitivity and F1_score values of 0.67 and 0.77, respectively, for minority soil class 7.
- The DT model achieved an overall accuracy of 95.67%, indicating high performance in classifying soils. Additionally, the model demonstrated excellent sensitivity, precision, and F1_score across all soil types, with scores ranging from 0.88–0.98.
- The SVM model outperformed the other models with the highest overall accuracy of 99.84%. It achieved almost perfect scores for all performance metrics across all soil types.
- The RF model achieved an overall accuracy of 99.23% and demonstrated high performance across all soil types. Similar to the SVM model, the RF model also achieved almost perfect scores for all performance metrics across all soil types.
- In general, the SVM and RF models achieved a high level of overall accuracy (almost 100%) in classifying soils, even when trained with imbalanced CPT datasets. These models exhibited outstanding performance on both majority and minority soil classes, indicating their potential as valuable tools in geotechnical engineering. Integrating these ML models into software programs for rapid and accurate soil classification in real-world projects can aid in making informed decisions.
- Future research could focus on improving the performance of the ANN and DT models in classifying soils based on CPT data. This could involve exploring other approaches such as training the models using balanced datasets.

Author Contributions: Conceptualization, A.T.C.; methodology, A.T.C.; software, A.T.C.; formal analysis, A.T.C.; writing—original draft preparation, A.T.C.; writing—review and editing, A.T.C.; supervision, R.R. All authors have read and agreed to the published version of the manuscript.

Funding: This research received no external funding.

Institutional Review Board Statement: Not applicable.

Informed Consent Statement: Not applicable.

Data Availability Statement: The datasets can be downloaded at the following link: http://140.112.12.21/issmge/tc304.htm?=6 (Accessed on 20 January 2023).

Acknowledgments: This publication was financially supported by Széchenyi István University.

Conflicts of Interest: The authors declare no conflict of interest.

References

1. Robertson, P.K. Interpretation of in-situ tests. In Proceedings of the J.K. Mitchell Lecture-Proceedings of ISC'4, Refice, Brazil, 17–21 September 2012; pp. 1–22.
2. Robertson, P.K. Soil Behaviour Type from the CPT: An Update. In Proceedings of the 2nd International Symposium on Cone Penetration Testing, Huntington Beach, CA, USA, 9–12 May 2010.
3. Robertson, P. Cone penetration test (CPT)-based soil behaviour type (SBT) classification system—An update. *Can. Geotech. J.* **2016**, *53*, 1910–1927. [CrossRef]
4. Robertson, P.K.; Campanella, R.G.; Gillespie, D.; Greig, J. Use of Piezometer Cone Data. In *Use of In Situ Tests in Geotechnical Engineering*; ASCE: Reston, VA, USA, 1986; pp. 1263–1280.
5. Laufer, I. Statistical analysis of CPT tip resistances. *Period. Polytech. Civ. Eng.* **2013**, *57*, 45–61. [CrossRef]
6. Begemann, H.K.S.P. The Friction Jacket Cone as an Aid in Determining the Soil Profile. In Proceedings of the 6th International Conference on Soil Mechanics and Foundation Engineering, Montreal, QC, Canada, 8–15 September 1965; pp. 17–20. Available online: https://cir.nii.ac.jp/crid/1573950399307239936 (accessed on 15 April 2023).
7. Douglas, B.J.; Olsen, R.S. Soil Classification Using Electric Cone Penetrometer. In Proceedings of the Symposium on Cone Penetration Testing and Experience, St. Louis, MO, USA, 26–30 October 1981; pp. 209–227.
8. Robertson, P.K. Soil classification using the cone penetration test. *Can. Geotech. J.* **1990**, *27*, 151–158. [CrossRef]
9. Rock, A.C.D. *Standard Practice for Classification of Soils for Engineering Purposes (Unified Soil Classification System) 1*; ASTM International: West Conshohocken, PA, USA, 2017.
10. Wang, H.; Wang, X.; Wellmann, F.; Liang, R.Y. A Bayesian unsupervised learning approach for identifying soil stratification using cone penetration data. *Can. Geotech. J.* **2019**, *56*, 1184–1205. [CrossRef]
11. Reale, C.; Gavin, K.; Librić, L.; Jurić-Kaćunić, D. Automatic classification of fine-grained soils using CPT measurements and Artificial Neural Networks. *Adv. Eng. Inform.* **2017**, *36*, 207–215. [CrossRef]
12. Rauter, S.; Tschuchnigg, F. CPT Data Interpretation Employing Different Machine Learning Techniques. *Geosciences* **2021**, *11*, 265. [CrossRef]
13. Tsiaousi, D.; Travasarou, T.; Drosos, V.; Ugalde, J.; Chacko, J. Machine Learning Applications for Site Characterization Based on CPT Data. In *Geotechnical Earthquake Engineering and Soil Dynamics V*; American Society of Civil Engineers: Reston, VA, USA, 2018; pp. 461–472.
14. Rogiers, B.; Mallants, D.; Batelaan, O.; Gedeon, M.; Huysmans, M.; Dassargues, A. Model-based classification of CPT data and automated lithostratigraphic mapping for high-resolution characterization of a heterogeneous sedimentary aquifer. *PLoS ONE* **2017**, *12*, e0176656. [CrossRef]
15. Kurup, P.U.; Griffin, E.P. Prediction of Soil Composition from CPT Data Using General Regression Neural Network. *J. Comput. Civ. Eng.* **2006**, *20*, 281–289. [CrossRef]
16. Bhattacharya, B.; Solomatine, D. Machine learning in soil classification. *Neural Netw.* **2006**, *19*, 186–195. [CrossRef]
17. Arel, E. Predicting the spatial distribution of soil profile in Adapazari/Turkey by artificial neural networks using CPT data. *Comput. Geosci.* **2012**, *43*, 90–100. [CrossRef]
18. Carvalho, L.; Ribeiro, D. Application of kernel k-means and kernel x-means clustering to obtain soil classes from cone penetration test data. *Soils Rocks* **2020**, *43*, 607–618. [CrossRef]
19. Kohestani, V.R.; Hassanlourad, M.; Ardakani, A. Evaluation of liquefaction potential based on CPT data using random forest. *Nat. Hazards* **2015**, *79*, 1079–1089. [CrossRef]
20. Goh, A.T.; Goh, S. Support vector machines: Their use in geotechnical engineering as illustrated using seismic liquefaction data. *Comput. Geotech.* **2007**, *34*, 410–421. [CrossRef]
21. Carvalho, L.O.; Ribeiro, D.B. A multiple model machine learning approach for soil classification from cone penetration test data. *Soils Rocks* **2021**, *44*, e2021072121. [CrossRef]
22. Livingston, G.; Piantedosi, M.; Kurup, P.; Sitharam, T.G. Using Decision-Tree Learning to Assess Liquefaction Potential from CPT and V_S. In *Geotechnical Earthquake Engineering and Soil Dynamics IV*; ASCE: Reston, VA, USA, 2008; pp. 1–10. [CrossRef]
23. Nhat-Duc, H.; Van-Duc, T. Comparison of histogram-based gradient boosting classification machine, random Forest, and deep convolutional neural network for pavement raveling severity classification. *Autom. Constr.* **2023**, *148*, 104767. [CrossRef]
24. Aydın, Y.; Işıkdağ, Ü.; Bekdaş, G.; Nigdeli, S.M.; Geem, Z.W. Use of Machine Learning Techniques in Soil Classification. *Sustainability* **2023**, *15*, 2374. [CrossRef]
25. Kang, T.-H.; Choi, S.-W.; Lee, C.; Chang, S.-H. Soil Classification by Machine Learning Using a Tunnel Boring Machine's Operating Parameters. *Appl. Sci.* **2022**, *12*, 11480. [CrossRef]
26. Hikouei, I.S.; Kim, S.S.; Mishra, D.R. Machine-Learning Classification of Soil Bulk Density in Salt Marsh Environments. *Sensors* **2021**, *21*, 4408. [CrossRef]
27. Eyo, E.; Abbey, S. Multiclass stand-alone and ensemble machine learning algorithms utilised to classify soils based on their physico-chemical characteristics. *J. Rock Mech. Geotech. Eng.* **2021**, *14*, 603–615. [CrossRef]
28. Huang, H.-W.; Li, Q.-T.; Zhang, D.-M. Deep learning based image recognition for crack and leakage defects of metro shield tunnel. *Tunn. Undergr. Space Technol.* **2018**, *77*, 166–176. [CrossRef]
29. Cheng, G.; Guo, W. Rock images classification by using deep convolution neural network. *J. Phys. Conf. Ser.* **2017**, *887*, 12089. [CrossRef]

30. Ran, X.; Xue, L.; Zhang, Y.; Liu, Z.; Sang, X.; He, J. Rock Classification from Field Image Patches Analyzed Using a Deep Convolutional Neural Network. *Mathematics* **2019**, *7*, 755. [CrossRef]
31. Xiao, L.; Zhang, Y.; Peng, G. Landslide Susceptibility Assessment Using Integrated Deep Learning Algorithm along the China-Nepal Highway. *Sensors* **2018**, *18*, 4436. [CrossRef] [PubMed]
32. Bui, D.T.; Tsangaratos, P.; Nguyen, V.-T.; Van Liem, N.; Trinh, P.T. Comparing the prediction performance of a Deep Learning Neural Network model with conventional machine learning models in landslide susceptibility assessment. *Catena* **2020**, *188*, 104426. [CrossRef]
33. Chakraborty, A.; Goswami, D. Prediction of slope stability using multiple linear regression (MLR) and artificial neural network (ANN). *Arab. J. Geosci.* **2017**, *10*, 385. [CrossRef]
34. Qi, C.; Tang, X. Slope stability prediction using integrated metaheuristic and machine learning approaches: A comparative study. *Comput. Ind. Eng.* **2018**, *118*, 112–122. [CrossRef]
35. Stock, D.J. *An Introduction to Neural Networks*; CRC Press: Boca Raton, FL, USA, 1992; Volume 23. [CrossRef]
36. Liu, Y.; Wang, Y.; Zhang, J. New machine learning algorithm: Random forest. In Proceedings of the Information Computing and Applications: Third International Conference, ICICA 2012, Chengde, China, 14–16 September 2012; pp. 246–252.
37. Friedman, J.H. Stochastic gradient boosting. *Comput. Stat. Data Anal.* **2002**, *38*, 367–378. [CrossRef]
38. Petropoulos, G.P.; Kalaitzidis, C.; Vadrevu, K.P. Support vector machines and object-based classification for obtaining land-use/cover cartography from Hyperion hyperspectral imagery. *Comput. Geosci.* **2012**, *41*, 99–107. [CrossRef]
39. Huo, L.-Z.; Tang, P. Spectral and spatial classification of hyperspectral data using SVMs and Gabor textures. *Int. Geosci. Remote Sens. Symp.* **2011**, *46*, 1708–1711. [CrossRef]
40. Meinshausen, N.; Ridgeway, G. Quantile regression forests. *J. Mach. Learn. Res.* **2006**, *7*, 983–999.
41. Sameen, M.I.; Pradhan, B.; Lee, S. Self-Learning Random Forests Model for Mapping Groundwater Yield in Data-Scarce Areas. *Nat. Resour. Res.* **2019**, *28*, 757–775. [CrossRef]
42. Zhang, Y.-M.; Wang, H.; Mao, J.-X.; Xu, Z.-D.; Zhang, Y.-F. Probabilistic Framework with Bayesian Optimization for Predicting Typhoon-Induced Dynamic Responses of a Long-Span Bridge. *J. Struct. Eng.* **2021**, *147*, 04020297. [CrossRef]
43. Stoppiglia, H.; Rémi Dubois, E.; Oussar Yacineoussar, Y. Ranking a Random Feature for Variable and Feature Selection Hervé Stoppiglia Gérard Dreyfus. *J. Mach. Learn. Res.* **2003**, *3*, 1399–1414.
44. Dai, B.; Gu, C.; Zhao, E.; Qin, X. Statistical model optimized random forest regression model for concrete dam deformation monitoring. *Struct. Control Health Monit.* **2018**, *25*, e2170. [CrossRef]
45. Kwak, S.K.; Kim, J.H. Statistical data preparation: Management of missing values and outliers. *Korean J. Anesthesiol.* **2017**, *70*, 407–411. [CrossRef]
46. Barnett, V.; Lewis, T. *Outliers in Statistical Data*; Wiley: New York, NY, USA, 1994; Volume 3.
47. Jaksa, M.B. The Influence of Spatial Variability on the Geotechnical Design Properties of a Stiff, Overconsolidated Clay. Available online: https://digital.library.adelaide.edu.au/dspace/handle/2440/37800 (accessed on 25 January 2023).
48. Robertson, P.K.; Wride, C. Evaluating cyclic liquefaction potential using the cone penetration test. *Can. Geotech. J.* **1998**, *35*, 442–459. [CrossRef]
49. Liu, J.; Liu, J.; Li, Z.; Hou, X.; Dai, G. Estimating CPT Parameters at Unsampled Locations Based on Kriging Interpolation Method. *Appl. Sci.* **2021**, *11*, 11264. [CrossRef]
50. Chala, A.; Ray, R. Generation and Evaluation of CPT Data Using Kriging Interpolation Technique. *Period. Polytech. Civ. Eng.* **2023**, *67*, 545–551. [CrossRef]
51. Robertson, P.K.; Cabal, K.L. Estimating soil unit weight from CPT. In Proceedings of the 2nd International Symposium on Cone Penetration Testing, Huntington Beach, CA, USA, 9–12 May 2010; p. 8. Available online: https://www.mendeley.com/catalogue/4c2ffa47-74a9-3ea8-b17c-5a8843514cd6/?utm_source=desktop&utm_medium=1.19.8&utm_campaign=open_catalog&userDocumentId=%7B2cb2fdcc-bb36-49ee-0cf3-a99ebf60b478%7D (accessed on 14 January 2023).
52. Géron, A.; Courville, A. *Hands-On Machine Learning with Scikit-Learn, Keras, and TensorFlow*, 2nd ed.; O'Reilly Media, Inc.: Sebastopol, CA, USA, 2011; Volume 44.
53. Vemuri, V.K. The Hundred-Page Machine Learning Book. *J. Inf. Technol. Case Appl. Res.* **2020**, *22*, 136–138. [CrossRef]
54. R Core Team. *R: A Language and Environment for Statistical Computing*; R Foundation for Statistical Computing: Vienna, Austria, 2022.
55. Venables, W.N.; Ripley, B.D. *Modern Applied Statistics with S*; Springer: New York, NY, USA, 2002.
56. Ren, Q.; Cheng, H.; Han, H. Research on machine learning framework based on random forest algorithm. *AIP Conf. Proc.* **2017**, *1820*, 80020. [CrossRef]
57. Liaw, A.; Wiener, M. Classification and regression by randomForest. *R News* **2002**, *2*, 18–22.
58. Quinlan, J.R. Induction of decision trees. *Mach. Learn.* **1986**, *1*, 81–106. [CrossRef]
59. Hastie, T.; Tibshirani, R.; Friedman, J. *The Elements of Statistical Learning: Data Mining, Inference, and Prediction*; Springer: New York, NY, USA, 2009; Volume 2.

60. Therneau, T.; Atkinson, B.; Ripley, B. *rpart: Recursive Partitioning and Regression Trees;* R Package Version; R Foundation for Statistical Computing: Vienna, Austria, 2015; Volume 4, pp. 1–9.
61. Meyer, D.; Dimitriadou, E.; Hornik, K.; Weingessel, A.; Leisch, F. e1071: Misc Functions of the Department of Statistics, Probability Theory Group (Formerly: E1071), TU Wien_R Package Version 1.7-13. 2023. Available online: https://cran.r-project.org/package=e1071 (accessed on 23 March 2023).

applied sciences

MDPI

Article

Application of an Artificial Neural Network (ANN) Model to Determine the Value of the Damping Ratio (D) of Clay Soils

Marzena Lendo-Siwicka, Karina Zabłocka, Emil Soból, Anna Markiewicz and Grzegorz Wrzesiński *

Institute of Civil Engineering, Warsaw University of Life Sciences, Nowoursynowska 159 St., 02-776 Warsaw, Poland; marzena_lendo_siwicka@sggw.edu.pl (M.L.-S.)

* Correspondence: grzegorz_wrzesinski@sggw.edu.pl; Tel.: +48-22-59-35-210

Abstract: The properties and behavior of soils depend on many factors. The interaction of individual factors is difficult to determine by traditional statistical methods due to their interdependence. The paper presents a procedure of creating an artificial neural network (ANN) model to determine the value of the damping ratio (D) of clay soils. The main purpose of this paper is to compare the appropriateness of ANN model application with empirical formulas described in the literature. The ANN model was developed using a series of laboratory tests of the damping ratio performed in the Resonance Column. Predicted values of the damping ratio of clay soils obtained from the ANN model are characterized by high convergence (coefficient of determination $R^2 = 0.976$). In comparison with other published empirical formulas, the ANN model showed an improvement in the prediction accuracy. What is more, ANN models proved to be more flexible compared to formulas and relationships with a predetermined structure, and they were well suited to modeling the complex behavior of most geotechnical engineering materials, which, by their very nature, exhibit extreme variability. In conclusion, ANNs have the potential to predict the damping ratio (D) of clay soils and can do much better than traditional statistical techniques.

Keywords: artificial neural networks (ANNs); damping ratio; cohesive soil; resonant column

Citation: Lendo-Siwicka, M.; Zabłocka, K.; Soból, E.; Markiewicz, A.; Wrzesiński, G. Application of an Artificial Neural Network (ANN) Model to Determine the Value of the Damping Ratio (D) of Clay Soils. *Appl. Sci.* **2023**, *13*, 6224. https://doi.org/10.3390/app13106224

Academic Editors: Daniel Dias and Wei Gao

Received: 30 March 2023
Revised: 12 May 2023
Accepted: 18 May 2023
Published: 19 May 2023

1. Introduction

Properties and behavior of soils are determined by the following factors: fabric, mineralogy, and pore water. Due to the interdependence of individual factors, it is difficult to determine their interaction using only traditional statistical methods [1]. Classical constitutive modeling based on the elasticity and plasticity theories is unable to properly simulate the behavior of geomaterials. This is due to the formulation complexity, the idealization of the behavior of the material, and excessive empirical parameters [2]. Therefore, in many papers, artificial neural networks (ANNs) have been proposed as a reliable and practical alternative to modeling the constitutive monotonic and hysteretic behavior of geomaterials [3–28].

This article concerns the use of artificial neural networks (ANNs) to determine the value of the damping ratio (D) of clay soils. Parameter D is one of the basic soil parameters. It characterizes the behavior of soil subjected to dynamic loads, such as the foundations of machines, the traffic of city trams and subways, and the driving of precast pile or sheet pile. These geotechnical problems are associated with significant damping, so its impact has to be taken into account in the dynamic analysis [29]. Unfortunately, the damping ratio, although important, is not easy to determine either on the basis of empirical formulas or in laboratory tests. The authors present a model of an artificial neural network (ANN) and compare the appropriateness of its application with empirical formulas described in the literature. The ANN model was developed on the basis of a series of laboratory tests of the damping ratio performed in the Resonance Column manufactured by GDS Instruments Ltd. Neural networks were to increase the accuracy of prediction. They provide an attractive solution for determining the behavior of a complex system and they are widely used for nonlinear pattern recognition and regression.

The paper also presents the procedure of creating an artificial neural network (ANN) model to determine the value of the damping ratio (D) of clay soils. The main purpose of this paper is to compare the appropriateness of ANN model application with empirical formulas described in the literature, and it also presents a method of selecting the structure of the ANN with the best predictive quality. To determine the architecture of the neural network, the following were determined: the number of input variables, the number of output variables, the number of hidden layers, the number of neurons in hidden layers, the type of activation function in the neurons of hidden layers and the output layer, and the number of learning epochs [30]. The Python programming language and the following libraries: TensorFlow, Keras, NumPy, Pandas, and Scikit-Learn, were used to create the ANN model. This article can be used as a guideline for creating other ANN models in the field of geotechnics.

2. Materials and Methods

Soils for the tests were collected from 3 test sites in Warsaw (Poland) from different depths (from 1.5 to 9.5 m). The soil cores were extracted in their undisturbed state using Shelby tubes.

Before performing the damping tests in the Resonance Column, the particle size distribution of the soil and physical properties were studied. The obtained results are presented in Table 1.

Table 1. Soil characteristics and mean effective stress range for tested soils.

Test Site	Test No.	Depth (m)	Soil Type	w_c (%)	LL (%)	PL (%)	IP (%)	p' (kPa)	e_o (-)	OCR (-)
Bartycka	S1	2.5	sandy clay	11.75	22.1	11.5	10.6	50–200	0.41	2.80–1.00
	S2	1.5	sandy silty clay	18.85	32.8	16.0	16.8	30–390	0.56	1.00
	S3	1.7	clayey sand	14.43	21.2	12.2	8.9	35–210	0.44	1.00
	S4	6.0	clayey sand	11.06	17.7	11.1	6.6	60–240	0.32	3.33–1.00
Pelczynskiego	S5	4.5	sandy silty clay	17.41	36.5	14.1	22.4	90	0.48	1.00
	S6	7.5	sandy silty clay	10.76	24.5	12.5	12.0	75–415	0.30	2.13–1.00
	S7	6.0	silty clay	17.53	37.3	17.1	20.1	120-410-120	0.47	1.00–3.42
Pory	K8	8.5	siCl	19.84	44.60	19.	25.11	85–310	0.5	2.00–1.00
	S9	2.2	sandy clay	12.23	37.0	11.5	25.6	45–315	0.37	1.00
Pelczynskiego	S10	2.2	clayey sand	15.57	41.7	14.3	27.4	90–315	0.40	1.00
	S11	2.2	clayey sand	10.50	18.2	9.1	9.1	45–315	0.43	1.00
	S12	7.2	silty clay	21.98	51.3	23.7	27.6	145–290	0.60	1.38–1.00
Pory	S13	8.0	silty clay	22.95	63.5	26.8	36.7	160–320	0.63	1.25–1.00
	S14	9.5	clay	26.04	71.0	33.1	37.8	95–285	0.75	6.32–2.11
Pelczynskiego	S15	2.7	sandy silty clay	12.68	27.1	12.3	14.8	55-165-55	0.39	1.00–3.00

Notes: w_c—initial water content, LL—liquid limit, PL—plasticity limit, IP—plasticity index, p'—mean effective stress, e_o—initial void ratio, OCR—overconsolidation ratio.

After the examination of the properties of the soil, the core sample was pushed out from the Shelby tube by a hydraulic press. Subsequently, from the undisturbed core, a cylindrical specimen at 140 mm high and 70 mm diameter was cut out. The prepared sample was placed on the resonant column pedestal. After that, the proper test was started. To prepare the specimen for testing, the resonant column chamber was closed, and the soil sample was saturated to a Skempton parameter B of at least 0.95. The consolidation process was then initiated, with dynamic tests conducted at every stage of the consolidation. First, the resonant frequency at torsional mode excitation was determined, followed by the application of a sinusoidal wave at the resonant frequency on top of the soil sample using the drive system. After seconds of excitation, the coils were switched off, and the specimen was left to vibrate freely. The damping of the soil sample movement was recorded by an accelerometer and displayed on the screen as a free vibration decay curve. The scientist then carefully selected the appropriate number of damping cycles at which specimen motions were completely damped based on the free vibration decay curve. A detailed description of the performed research is presented in [31].

To generate a damping ratio curve, the amplitude of the applied voltage on the drive system was increased, and the same processes as described above were repeated. The damping tests were performed following the procedures outlined in [32]. Furthermore, every damping test was repeated 10 times at each shear strain. Later, from ten results of the parameter D, the average value was evaluated and used in the performed analysis. More details about materials and methods are presented in [29].

3. Artificial Neural Network

Artificial neural networks (ANNs) are computational methods based on the principle of operation of human neurons or nerve cells. When certain information reaches them, they process it and remember it for future use. Artificial systems work in this way as well. Data are inserted into the system, processed, stored, and corrected.

In geotechnics, the use of neural networks allows to achieve progress in the interpretation of research results, especially in relation to such specific materials as soil. When analyzing problems related to geotechnics, we most often deal with regression issues, which means a problem in which we want to predict a numerical value. Regression ANNs predict an output variable as a function of the inputs. It is required that numeric variables be dependent in regression ANNs. The use of ANNs often allows to develop better predictive models than statistical ones and to improve the models as new data become available.

According to Geron [33], in order to create any machine learning project, the following series of steps should be performed:

- Analyze the whole task;
- Get the data;
- Discover and visualize the data to recognize patterns and additional information—Exploratory data analysis;
- Prepare data in terms of machine learning algorithms—Data preparation;
- Select and teach a model;
- Tune the model—Model adjustment;
- Present the solution—Results.

In this article, we will go through all these steps to create the ANN model which is used to predict the value of the damping ratio determined based on the basic geotechnical parameters of cohesive soils.

3.1. Analyze the Whole Task

The first step is to define the purpose for which we create the project. In our case, the goal is to determine the value of the damping ratio based on the available soil parameters. This will allow us to know the value of the coefficient of a given soil without the need to perform tedious, expensive, and not always available laboratory tests. The next step is to define the problem we are dealing with. This is essential to start designing the system. We faced a classic problem of supervised learning. The task was a multiple regression task because our system would use many features to forecast the result (shear strain, effective stress, soil type, etc.). We also used a batch learning mechanism (offline learning).

The root mean square error (RMSE) was chosen as the performance metric, which, in regression problems, is a classic measure of performance. The mean absolute error (MAE) and coefficient of determination (R^2) were also determined, but the RMSE metric was more sensitive to outliers than the MAE. Equation (1) shows the mathematical formula used to compute the RMSE error:

$$RMSE(\mathbf{X},h) = \sqrt{\frac{1}{m}\sum_{i=1}^{m}\left(h(\boldsymbol{x}^{(i)}) - y^{(i)}\right)^2} \quad (1)$$

where:

- m—the number of elements of the training set;
- $x^{(i)}$—the value vector of all the features of the i-th example;
- $y^{(i)}$—the label of i-th example;
- X—the matrix containing the value of all features;
- h—the predictive function.

3.2. Get the Data

The data that we used to create the ANN model comes from the laboratory tests described in Section 2 herein; "Materials and Methods". The test results were saved in the .xlsx format in Excel.

The data were loaded using the Jupyter notebook and the Python Pandas module. The first five lines of "raw" data are presented below (Table 2):

Table 2. First 5 lines of "raw" data from the damping ratio laboratory test.

Test No.	Cl (%)	Si (%)	Sa + Gr (%)	LL (%)	PL (%)	IP (%)	IL (%)	p′ (%)	eo (-)	OCR (-)	γ (%)	G (MPa)	G/Gmax (-)	D (%)
S1	14	26	60	22.1	11.5	10.6	0.024	50	0.40	2.8	0.0006	53.35	1.00	4.99
S1	14	26	60	22.1	11.5	10.6	0.024	50	0.40	2.8	0.0009	54.63	1.00	4.99
S1	14	26	60	22.1	11.5	10.6	0.024	50	0.40	2.8	0.0012	54.20	1.00	5.01
S1	14	26	60	22.1	11.5	10.6	0.024	50	0.40	2.8	0.0018	53.63	0.98	5.06
S1	14	26	60	22.1	11.5	10.6	0.024	50	0.40	2.8	0.0024	53.35	0.98	5.11

Notes: Cl—clay content, Si—silt content, Sa + Gr—sand + gravel content, LL—liquid limit, PL—plasticity limit, IP—plasticity index, IL—liquidity index, p′—mean effective stress, e_o—initial void ratio, OCR—overconsolidation ratio, γ—shear strain, G—shear modulus, G/G_{max}—normalized shear modulus.

Our dataset consists of 1227 examples obtained from testing 15 soil samples. All attributes (except "Test No.") have numeric values. A summary of all numerical attributes is provided in Table 3.

Table 3. Summary of all numeric attributes.

	Cl	Si	Sa + Gr (%)	LL	PL	IP	IL	p′	e_o	OCR	γ	G	G/Gmax	D
	(%)	(%)		(%)	(%)	(%)	(%)	(%)	(-)	(-)	(%)	(MPa)	(-)	(%)
count	1227	1227	1227	1227	1227	1227	1227	1227	1227	1227	1227	1227	1227	1227
mean	17.83	35.38	46.79	33.77	15.28	18.49	0.04	193.96	0.440	1.37	0.0117	120.75	0.87	4.57
std	7.26	17.01	21.83	11.74	4.86	7.58	0.11	104.61	0.104	0.85	0.0220	76.70	0.20	2.78
min	10.00	20.00	6.00	17.70	9.10	6.60	−0.19	30.00	0.280	1.00	0.0001	6.45	0.16	1.57
25%	14.00	23.00	17.00	24.50	12.24	12.03	0.01	110.00	0.375	1.00	0.0008	69.82	0.81	2.57
50%	14.00	26.00	60.00	32.75	14.26	16.79	0.02	180.00	0.433	1.00	0.0031	108.61	0.97	3.73
75%	21.00	60.00	62.00	37.25	17.14	25.11	0.07	270.00	0.510	1.37	0.0111	152.26	1.00	5.45
max	50.00	69.00	70.00	70.95	33.11	37.84	0.24	415.00	0.747	6.32	0.2219	431.76	1.02	19.04

Notes: std—standard deviation, 25%—25th percentile, 50%—median, 75%—75th percentile.

After the first look at the data, the next particularly important, but often underestimated, step is to create a test set. The separation of the test set at such an early stage is to prevent the occurrence of a phenomenon called the "data snooping bias". When we look at the test set, we can see some interesting pattern (not always right) that will make us choose a specific machine learning model based on it. When dealing with exceptionally large datasets, it is enough to randomly select some examples and put them back to create a test dataset. This is a satisfactory solution, but it does not work well with small datasets. Next time the program runs, a completely different test set is generated. After some time, the person creating the model or the machine learning algorithm itself will see the entire dataset, and this is what we want to avoid. This is why the random seed was used, which allowed to obtain the same pseudorandom test set every time. The test set accounts for 20% of the total data.

3.3. Exploratory Data Analysis

The objectives of exploratory data analysis (EDA) and exploratory data analysis (EDA) methods are classified according to [34].

3.3.1. Searching for Correlation

At this stage of the project, we only deal with training data. The first step in regression problems should be looking for the correlation between the features and the label (damping ratio, D). To do this, we calculate a linear correlation coefficient (Pearson correlation coefficient) between each pair of values (Table 4).

Table 4. Correlation matrix.

	Cl	Si	Sa + Gr	LL	PL	IP	IL	p'	e_o	OCR	γ	G	G/G_{max}	D
Cl	1.00	0.52	−0.75	0.85	0.84	0.78	−0.48	0.12	0.65	0.41	0.02	−0.02	0.01	−0.13
Si	0.52	1.00	−0.96	0.57	0.61	0.48	−0.25	0.29	0.48	0.11	−0.01	0.09	0.11	−0.31
Sa + Gr	−0.75	−0.96	1.00	−0.73	−0.76	−0.64	0.36	−0.26	−0.60	−0.22	0.00	−0.06	−0.09	0.28
LL	0.85	0.57	−0.73	1.00	0.91	0.96	−0.33	0.14	0.74	0.23	0.05	0.10	−0.02	−0.21
PL	0.84	0.61	−0.76	0.91	1.00	0.77	−0.33	0.15	0.86	0.34	−0.05	0.06	0.09	−0.25
IP	0.78	0.48	−0.64	0.96	0.77	1.00	−0.30	0.12	0.59	0.13	0.11	0.11	−0.10	−0.16
IL	−0.48	−0.25	0.36	−0.33	−0.33	−0.30	1.00	−0.23	0.08	−0.42	−0.04	−0.18	0.03	−0.05
p'	0.12	0.29	−0.26	0.14	0.15	0.12	−0.23	1.00	−0.08	−0.29	−0.19	0.63	0.21	−0.35
e_o	0.65	0.48	−0.60	0.74	0.86	0.59	0.08	−0.08	1.00	0.16	−0.02	−0.19	0.06	−0.18
OCR	0.41	0.11	−0.22	0.23	0.34	0.13	−0.42	−0.29	0.16	1.00	−0.03	−0.11	0.00	0.07
γ	0.02	−0.01	0.00	0.05	−0.05	0.11	−0.04	−0.19	−0.02	−0.03	1.00	−0.42	−0.85	0.83
G	−0.02	0.09	−0.06	0.10	0.06	0.11	−0.18	0.63	−0.19	−0.11	−0.42	1.00	0.48	−0.53
G/G_{max}	0.01	0.11	−0.09	−0.02	0.09	−0.10	0.03	0.21	0.06	0.00	−0.85	0.48	1.00	−0.88
D	−0.13	−0.31	0.28	−0.21	−0.25	−0.16	−0.05	−0.35	−0.18	0.07	0.83	−0.53	−0.88	1.00

Thanks to the correlation matrix, we can determine the degree of correlation of individual features with the damping ratio (Table 5).

Table 5. Degree of correlation (Pearson's correlation coefficient) of individual features with the damping ratio.

D	γ	Sa + Gr	OCR	IL	Cl	IP	e0	LL	PL	Si	p'	G	G/Gmax
1.00	0.83	0.28	0.07	−0.06	−0.13	−0.16	−0.18	−0.21	−0.25	−0.31	−0.35	−0.53	−0.88

The values of the correlation coefficient range from −1 to 1. Values close to 1 indicate a strong positive correlation, values close to −1 say that there is a strong negative correlation, while values close to 0 indicate no linear correlation. However, it should be remembered that the Pearson correlation coefficient measures only a linear relationship. It may completely ignore the nonlinear relationship. The highest positive correlation with the damping factor is shown by the γ, while the highest negative correlation with the G/G_{max} is shown by the G and p'.

Two variables may also be related by a nonlinear relationship or may have a non-Gaussian distribution. In these cases, Spearman's correlation coefficient can be useful to determine the strength and direction between the two data samples. Spearman's rank correlation can also be used if there is a linear relationship between the variables, but will have slightly less power [35]. Table 6 presents the result of the Spearman correlation test.

Table 6. Degree of correlation (Spearman's correlation coefficient) of individual features with the damping ratio.

D	γ	Sa + Gr	OCR	IL	e_o	IP	Cl	LL	p'	Si	PL	G	G/G_{max}
1.00	0.62	0.39	0.11	−0.04	−0.29	−0.30	−0.32	−0.40	−0.40	−0.48	−0.48	−0.65	−0.71

From Spearman's correlation, we see a moderately strong negative correlation of the damping ratio (D) with PL, Si, p′, LL, Cl, and IP. We can observe that the other features are also characterized by a higher degree of correlation than in the Pearson's correlation, which may indicate that the dependence of these features with D is nonlinear. On the other hand, a decrease of degree of correlation of values, such as the G/G_{max} and γ, can be noticed.

Another way to check the correlation between the attributes is a scatter matrix. The scatter matrix of all numerical attributes would be unreadable (and unjustified). Figure 1 shows a scatter matrix for some of the most promising attributes. Both the color and size of the markers relate to the value of the damping ratio (D). The plot shows a matrix of scatter plots where each variable is plotted against all other variables, resulting in a grid of scatter plots. The diagonal of the plot shows the kernel density estimation (KDE) plot for each variable, which provides a visual representation of the distribution of each variable.

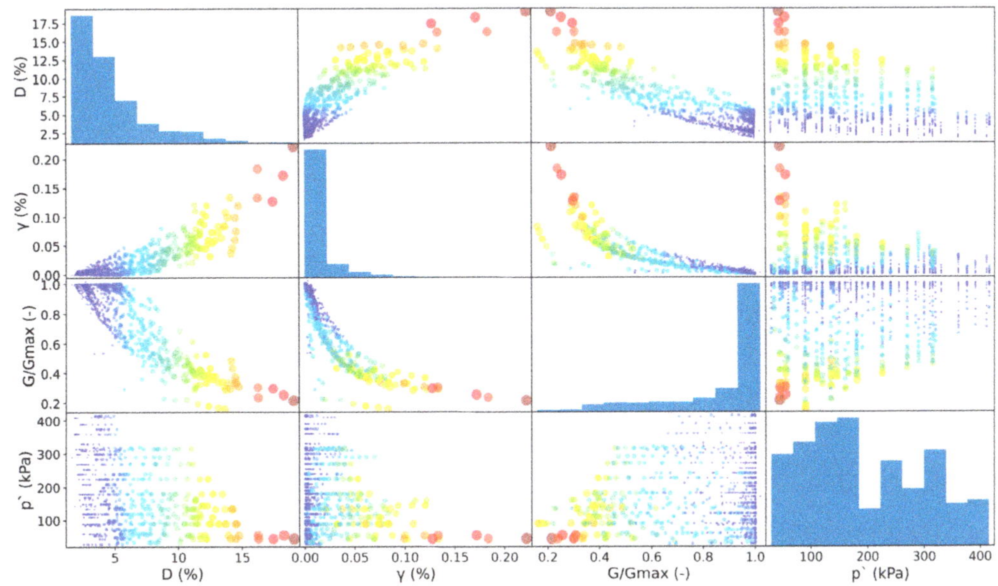

Figure 1. Scatter matrix of the damping ratio (D), shear strain (γ), normalized shear modulus (G/G_{max}), and mean effective stress (p′).

The above charts show that the correlation between the shear strain (γ), normalized shear modulus (G/G_{max}), mean effective stress (p′), and the damping ratio (D) is strong and the individual points are quite close to each other.

3.3.2. Detect Outliers and Anomalies

Some data batches include outliers. An outlier is a data point that differs significantly from other observations. Some outliers may be caused by measuring, recording, or copying errors, or by errors in entering the data into the computer. When such errors occur, it is important to detect and correct them, if possible. If not, excluding the erroneous values from further analysis is another possibility. Not all outliers are erroneous. Some may merely reflect unusual circumstances or outcomes; so, having these outliers called to our attention can help to uncover valuable information [36].

Two methods of the graphical determination of outliers are presented below. The first option is to create histograms of individual features. Histograms, in addition to visualizing outliers, allow researchers to gain insight into data, including the distribution, central tendency, spread, and modality. Figure 2 shows the histograms of the G distribution with outliers (on the left) and without outliers (on the right).

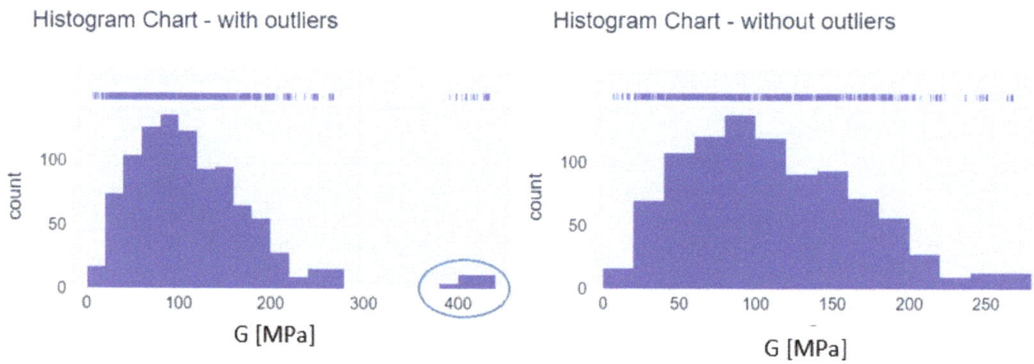

Figure 2. Histogram charts of G with and without outliers.

Another way to determine the outliers is to create boxplots. Boxplots are interesting for representing information about the symmetry, central tendency, skew, and outliers. It is a good EDA technique since it relies on robust statistics such as the median and IQR (interquartile range). Figure 3 presents an annotated boxplot which explains how it is constructed. The central rectangle is limited by Q1 and Q3. The middle line represents the median of the data. The whiskers are drawn in each direction, to the most extreme point that is less than 1.5 IQR beyond the corresponding hinge. Values beyond 1.5 IQR are considered outliers [34]. Figure 4 shows the boxplots of the G feature with outliers (on the left) and without outliers (on the right).

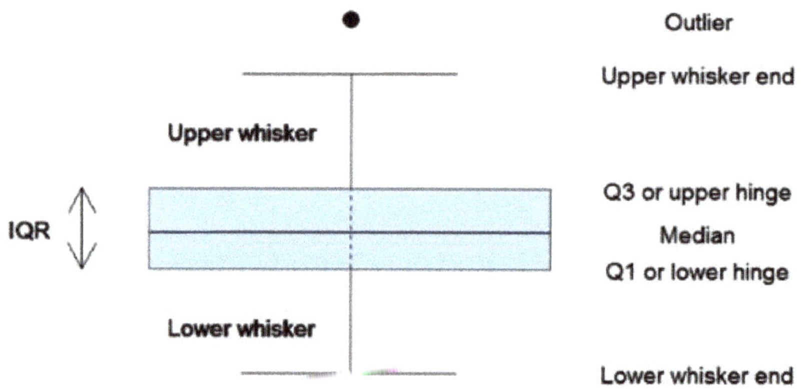

Figure 3. Example of boxplot with annotations.

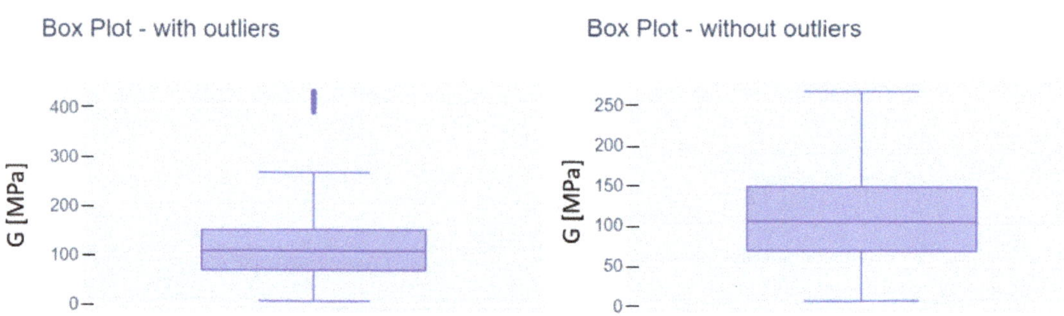

Figure 4. Boxplots of G with and without outliers.

Boxplots are an exploratory data analysis technique, and a researcher should consider designating a boxplot as an outlier as a suggestion that the points might be mistakes or otherwise unusual. Outliers should not be eliminated in the first instance, since depending on their context may provide a lot of information.

3.4. Data Preparation

Based on basic information about the dataset and exploratory data analysis (EDA), we can draw some conclusions which can determine the basic steps for data preparation:

- All data are numeric;
- There are no empty values—no data cleansing is needed;
- Data are of different orders of magnitude—feature scaling may be required;
- γ and G/G_{max} show high linear correlation with the damping ratio—features may be useful in the ANN model;
- LL, p', Si, PL, Cl, and IP show moderately strong nonlinear correlation with the damping ratio—features may be useful in the ANN model;
- Features have outliers—removing outliers should be considered.

Each dataset is different; therefore, there is no one pattern of proceeding in exploratory data analysis or data preparation. When looking at the data, the researcher has to decide what steps should be taken to obtain the best possible machine learning (ML) model. Both exploratory data analysis and data preparation will probably have to be repeated several times during ANN model improvement.

3.4.1. Feature Scaling

Scaling features is one of the most important data transformations. Since the range of values of data may vary widely, it becomes a necessary step in data preprocessing while using machine learning algorithms. Most machine learning algorithms are poor at dealing with numerical attributes that fall within different ranges of the scale. This also applies to the analyzed dataset, where, e.g., the p' feature has values in the range from 30 to 415 and where the values of the G/G_{max} feature range from 0.16 to 1.02.

One of the two most common types of scaling are min–max scaling and standardization. Min–max scaling is commonly known as normalization. Normalization transforms the data in such way that the features are within a specific range, e.g., (0, 1). To do this, it subtracts the minimum value from the given value and divides the result by the difference between the maximum and minimum values. Standardization (also called z-score normalization) transforms data in such way that the resulting distribution has a mean of 0 and a standard deviation of 1. The standardization mechanism is as follows: we subtract the mean value from a given value and then divide it by the standard deviation, thanks to which the resulting distribution has a unit variance. Standardization does not limit the scaled values to a certain range, as is the case, for example, during normalization. It is also much less sensitive to outliers.

The type of value scaling which will be the most appropriate in the analyzed task can be checked by substituting the normalized values, and then the standardized values, into the model. Target values usually do not need scaling. An important issue, that scientists often do incorrectly, is what values are scaled. As with all transformations, the scaling functions should be adjusted only to the training data, not to the entire dataset. Only after fitting against the training data can these functions be used to transform test data (and new cases).

Figures 5 and 6 show the results already obtained from the ANN model. The figures show the MAE and RMSE values of the training and validation set (constituting 20% of the training set) for the data subjected to normalization and standardization.

For the normalized training data: MAE = 0.2809, RMSE = 0.5069, and R^2 = 0.961; for the standardized data: MAE = 0.2032, RMSE = 0.4352, and R^2 = 0.979. For the ANN model aimed at determining the damping ratio based on other soil parameters, the standardization works better, and this scaling method will be applied to data in the final model version.

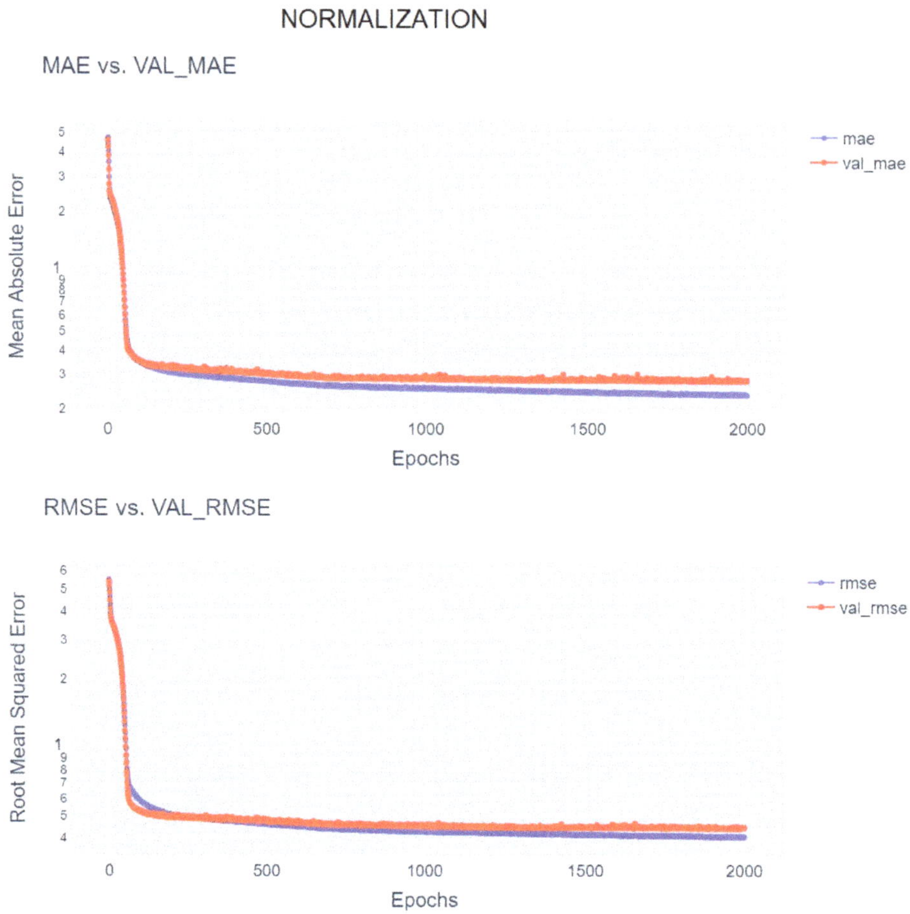

Figure 5. MAE and RMSE from the ANN model for normalized data.

3.4.2. Architecture of Artificial Neural Networks (ANNs)—Model Adjustment

The structure of the designed neural network depends on the complexity of the problem to be solved and the type of independent and dependent variables. In order to define the ANN architecture, the following should be determined: the number of inputs, the number of output neurons, the number of hidden layers, the number of neurons in each hidden layer, the type of activation function in the hidden and output layers, and the weight values of individual neurons. There are also parameters related to training the neural network, such as the batch size, training epochs, learning rate, and momentum. We can operate with all these parameters to improve the accuracy of our model.

This can be done in several ways. One of them is to manually select the hyperparameter values until obtaining the best combination of them. However, this is a very tedious task. Instead, by taking advantage of all the benefits of Python programming, we can outsource the search for the best parameters to the GridSearchCV class (scikit-learn). Grid search is a model hyperparameter optimization technique. For this purpose, it is enough to provide the desired parameters and their proposed values, and all combinations will be assessed using a cross-validation test.

Table 7 below shows the optimized parameters and their values that resulted in the lowest MAE and MSE errors.

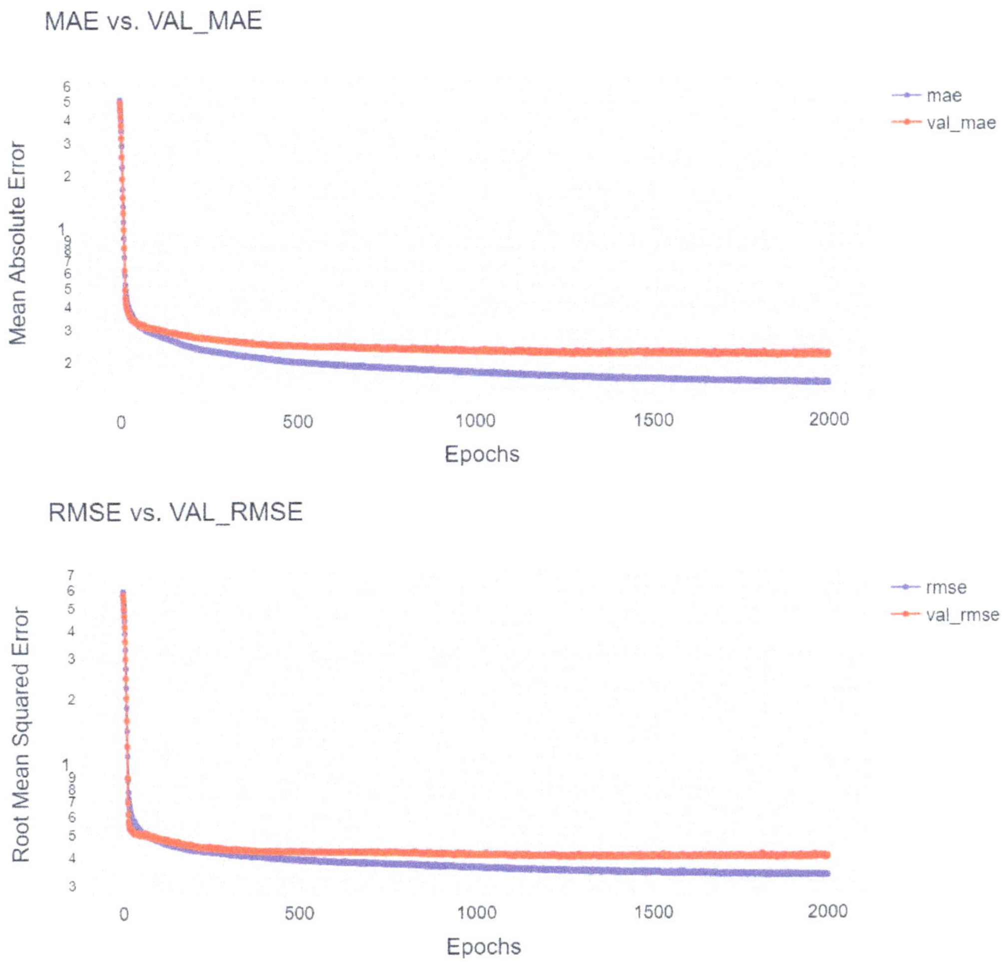

Figure 6. MAE and RMSE from the ANN model for standardized data.

Table 7. Optimized parameter values.

Parameter	Value	Range
Batch Size	10	(1:100)
Number of Epochs	1000	No significant difference when increasing the value
Training Optimization Algorithm	Adam	('SGD', 'RMSprop', 'Adagrad', 'Adadelta', 'Adam', 'Adamax', 'Nadam')
Network Weight Initialization	he_uniform	('uniform', 'lecun_uniform', 'normal', 'zero', 'glorot_normal', 'glorot_uniform', 'he_normal', 'he_uniform')
Neuron Activation Function	relu	('softmax', 'softplus', 'softsign', 'relu', 'tanh', 'sigmoid', 'hard_sigmoid', 'linear')
Dropout Regularization	0	(0.0, 0.1, 0.2, 0.3, 0.4, 0.5, 0.6, 0.7, 0.8, 0.9)
Number of Hidden Layers	2	(0, 1, 2, 3)
Number of Neurons in the 1 Hidden Layer	95	(0:200)
Number of Neurons in the 2 Hidden Layers	80	(0:200)
Learning Rate	0.0001	(0.00001, 0.0001, 0.001, 0.01, 0.1)

4. Results

4.1. ANN Model Results

Using the data from laboratory tests, an artificial neural network was created to predict the damping ratio.

The input parameters included:

- γ—the shear strain;
- G/G_{max}—the normalized shear modulus;
- LL—the liquid limit;
- p'—the mean effective stress;
- Si—the silt content;
- PL—the plasticity limit;
- Cl—the clay content.

The output data of the neural network were the damping ratio determined based on laboratory tests. To determine the value of the damping ratio (D) of clay soils in this study, deep feed-forward (DFF) network was used, which means that we applied more than one hidden layer in the model. The main problem with using only one hidden layer is overfitting. Therefore, by adding more hidden layers, we may achieve (not in all cases) reduced overfitting and improved generalization. The network used has the following architecture: 7-95-80-1, which means an artificial neural network with seven inputs and two hidden layers; the first with 95 neurons, the second with 80 neurons, and an output layer with 1 neuron. The relu function was used as the activation function in the first and second hidden layers. The Adam optimization algorithm was chosen to train the neural network. The input data were divided accordingly for the data used to train the network—80% and 20%—for network testing. From the training set, 20% was used to create a validation set which was used to tune the parameters of a classifier.

The correlation between the damping ratio predicted by the ANN and their values measured in the laboratory is shown in Figure 7. The predicted values of the damping ratio of clay soils are characterized by high convergence ($R^2 = 0.976$) with their values measured in the laboratory, which justifies the use of the ANN to predict their damping ratios.

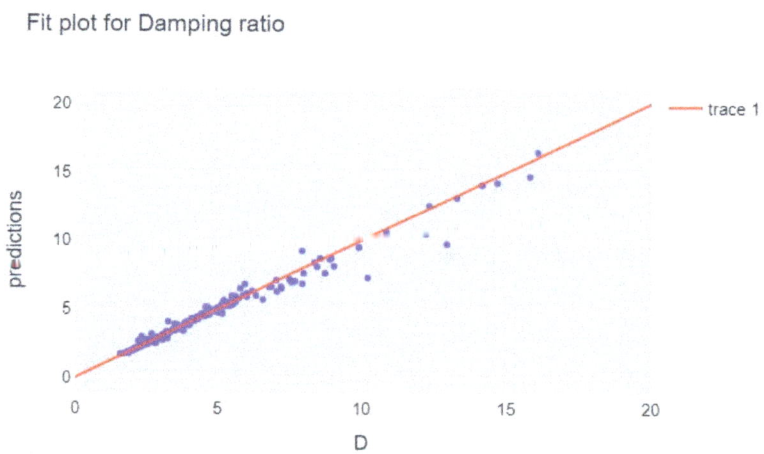

Figure 7. Correlation of the damping ratios of the clay soils predicted by the artificial neural network (test cases) with the values measured in the laboratory.

Figure 8 illustrates the histogram of errors made by a neural network when predicting the damping ratio. The error histogram is the histogram of errors between the target values and predicted values after training a feedforward neural network. These error values indicate how the predicted values differ from the target ones. The mean absolute error (MAE) for the test set was 0.1854, while the mean square error (MSE) was 0.1704.

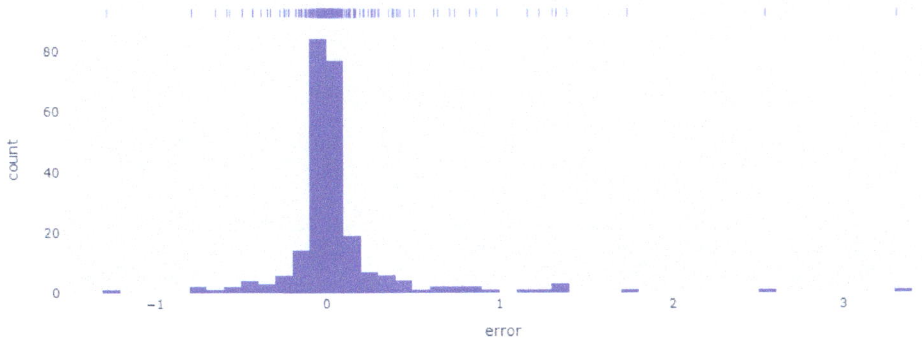

Figure 8. Error histogram.

4.2. Comparison of ANN Model with Other Empirical Formulas Available in the Literature

The obtained results were compared with the results obtained from earlier tests for cohesive soils. The authors chose five formulae for the damping ratio (D) in the wide shear strain range, summarized as follows:

Ishibashi and Zhang [36]:

$$D = \left(\frac{0.333\left(1 + e^{-0.0145 I_P^{1.3}}\right)}{2}\right) \cdot \left[0.586\left(\frac{G}{G_{max}}\right)^2 - 1.547\left(\frac{G}{G_{max}}\right) + 1\right]; \quad (2)$$

Park and Stewart [37]:

$$D = 17.83\left[0.56\left(\frac{G}{G_{max}}\right)^2 - 1.39\left(\frac{G}{G_{max}}\right) + 1\right]; \quad (3)$$

Michaelides et al. [38]:

$$D = 2 + [18 - 0.08(I_P - 15)] \cdot \left[1 - \left(\frac{G}{G_{max}}\right)\right]; \quad (4)$$

Zhang et al. [39]:

$$D = 10.6\left(\frac{G}{G_{max}}\right)^2 - 31.6\left(\frac{G}{G_{max}}\right) + 21 + \left[(0.008 I_P + 0.82)\left(\frac{\sigma'_0}{P_a}\right)^{-0.5k}\right]; \quad (5)$$

Soból et al. [29]:

$$D = \left[a\left(\frac{G}{G_{max}}\right)^2 - b\left(\frac{G}{G_{max}}\right) + c\right] + dI_P + e\left(\frac{p'}{P_a}\right)^f; \quad (6)$$

where a, b, c, d, e, and f are constants, different for both analyzed soil groups, and Pa is the atmospheric pressure equal to 100 kPa. For low- and medium-cohesive soils with an IP < 20%: a = 14.8, b = 34.3, c = 26, d = −0.31, e = 1.36, and f = −0.32; for very cohesive soil with an IP > 20%: a = 6.32, b = 20.36, c = 14.43, d = 0.062, e = 0.75, and f = −1.49.

Each considered model in the literature was suitable only for the tested soils. The equation presented by Ishibashi and Zhang [36] was calibrated for cohesive soil with an IP below 50% and an effective stress below 1000 kPa. Park and Stewart [37] presented an equation for the average damping ratio curve based on nine publications and considered different types of clay, e.g., offshore clayey silt, Edger plastic kaolin, San Old Bay clay, Mexico City clay, etc. Michaelides et al. [38] created an empirical model for cohesive soil with an IP from 0% to 100%. The formula shown by Zhang et al. [39] was appropriated for Tertiary, Quaternary, and older soils with an IP from 0% to 132%, collected from a depth from 0 to 326 m. The equation proposed by Soból et al. [29] was calibrated on the same soil samples as the ANN model. It should be noted, however, that the formula proposed there was different for soils with an IP < 20% and different for those with an IP > 20%. The use of the ANN model allowed to avoid this division, further increasing the accuracy of the damping ratio predictions. Table 8 presents the comparison between the mean absolute error and determination coefficient.

Table 8. The comparison of mean absolute error (MAE) and determination coefficient (R^2) for analyzed models.

	Ishibashi and Zhang [36], Equation (2)	Park and Stewart [37], Equation (3)	Michaelides et al. [38], Equation (4)	Zhang et al. [39], Equation (5)	Soból et al. [29] Equation (6)	Authors ANN Model
MAE	3.63	1.02	1.06	1.77	0.45	0.19
R^2	0.21	0.79	0.78	0.77	0.96	0.98

The conducted comparisons indicate that the best model for cohesive soils taken from the analyzed site is the ANN model, which was proposed by the authors of this article.

The ANN model tested in the presented research can be useful to other researchers in several ways compared to traditional equations, such as Equation (6). Firstly, ANN models have the ability to learn from data and improve their predictive performance over time, which can result in more accurate and reliable predictions. Additionally, ANN models are more flexible than traditional equations and can capture more complex relationships between input and output variables. This can be especially useful in cases where the underlying relationships between variables are not well understood or are nonlinear. Finally, ANN models can be trained on large datasets, allowing for the inclusion of a wide range of input variables, which can lead to more robust predictions. Therefore, ANN models can provide researchers with a powerful tool for predicting the behavior of cohesive soils and can potentially lead to new insights and discoveries in this field.

5. Conclusions

The article presents the procedure for developing an artificial neural network (ANN) model. The potential of using an ANN to increase the confidence level of the prediction of the damping ratio (D) of normally and lightly overconsolidated Quaternary cohesive soil was described. Careful selection of the input variables describing the problem at hand is necessary in order to balance the complexity of the issue at hand with the amount of information available. Minor variables were omitted. The mean absolute error (MAE) for the performed tests was 0.1854, while the mean square error (MSE) was 0.1704. The predicted values of the damping ratio of clay soils are characterized by high convergence (coefficient of determination, $R^2 = 0.976$).

Comparing the ANN model with other published empirical formulas shows an improvement in the prediction accuracy. It is not possible to reliably evaluate the damping ratio for the tested cohesive soil with the use of commonly known empirical equations. A comparison of the measured and calculated value with equations from the literature gave a very large spread of the damping ratio and large error values. The conducted analysis shows that it can be concluded that an empirical model should be applied to the soil, with a similar genesis and mineral composition as the soil for which they were created. Considering the problem, it was shown that artificial neural networks are more flexible compared to formulas and relationships with a predetermined structure. Additionally, another advantage of artificial neural networks is the possibility to further model generalization by inserting newer records and the automatic adaptation of the architecture.

Artificial neural networks are well suited to modeling the complex behavior of most geotechnical engineering materials, which, by their very nature, exhibit extreme variability. The use of artificial neural networks to predict the damping ratio of clay soils is an effective, highly efficient, and easy-to-use method.

Author Contributions: Conceptualization, M.L.-S. and K.Z.; methodology, K.Z. and E.S.; laboratory tests, K.Z.; data analysis, K.Z. and E.S.; writing—original draft preparation, K.Z., A.M. and G.W.; supervision, A.M. and G.W. All authors have read and agreed to the published version of the manuscript.

Funding: This research received no external funding.

Institutional Review Board Statement: Not applicable.

Informed Consent Statement: Not applicable.

Data Availability Statement: The data that were used to support the findings of this study are included within the article.

Conflicts of Interest: The authors declare no conflict of interest.

References

1. Yang, Y.; Rosenbaum, M.S. The artificial neural network as a tool for assessing geotechnical properties. *Geotech. Geol. Eng.* **2002**, *20*, 149–168. [CrossRef]
2. Adeli, H. Neural networks in civil engineering: 1989–2000. *Comput.-Aided Civ. Infrastruct. Eng.* **2001**, *16*, 126–142. [CrossRef]
3. Basheer, I.A. Selection of methodology for neural network modeling of constitutive hysteresis behavior of soils. *Comput.-Aided Civ. Infrastruct. Eng.* **2000**, *15*, 445–463. [CrossRef]
4. Basheer, I.A. Stress-strain behavior of geomaterials in loading reversal simulated by time-delay neural networks. *J. Mater. Civ. Eng.* **2002**, *14*, 270–273. [CrossRef]
5. Basheer, I.A.; Najjar, Y.M. Modeling cyclic constitutive behavior by neural networks: Theoretical and real data. In Proceedings of the 12th Engineering Mechanics Conference, La Jolla, CA, USA, 17–20 May 1998; pp. 952–955.
6. Ellis, G.W.; Yao, C.; Zhao, R. Neural network modeling of the mechanical behavior of sand. In *Engineering Mechanics*; ASCE: Reston, VA, USA, 1992; pp. 421–424.
7. Ellis, G.W.; Yao, C.; Zhao, R.; Penumadu, D. Stress-strain modeling of sands using artificial neural networks. *J. Geotech. Eng.* **1995**, *121*, 429–435. [CrossRef]
8. Ghaboussi, J.; Sidarta, D.E. New nested adaptive neural networks (NANN) for constitutive modeling. *Comput. Geotech.* **1998**, *22*, 29–52. [CrossRef]
9. Gorai, A.K.; Raval, S.; Patel, A.K.; Chatterjee, S.; Gautam, T. Design and development of a machine vision system using artificial neural network-based algorithm for automated coal characterization. *Int. J. Coal Sci. Technol.* **2021**, *8*, 737–755. [CrossRef]
10. Habibagahi, G.; Bamdad, A. A neural network framework for mechanical behavior of unsaturated soils. *Can. Geotech. J.* **2003**, *40*, 684–693. [CrossRef]
11. Haj-Ali, R.; Pecknold, D.A.G.J.; Voyiadjis, G.Z. Simulated micromechanical models using artificial neural networks. *J. Eng. Mech.* **2001**, *127*, 730–738. [CrossRef]
12. Hashash, Y.M.A.; Jung, S.; Ghaboussi, J. Numerical implementation of a neural network based material model in finite element analysis. *Int. J. Numer. Methods Eng.* **2004**, *59*, 989–1005. [CrossRef]
13. Huang, F.; Xiong, H.; Chen, S.; Lv, Z.; Huang, J.; Chang, Z.; Catani, F. Slope stability prediction based on a long short-term memory neural network: Comparisons with convolutional neural networks, support vector machines and random forest models. *Int. J. Coal Sci. Technol.* **2023**, *10*, 18. [CrossRef]
14. Lawal, A.I.; Aladejare, A.E.; Onifade, M.; Bada, S.; Idris, M.A. Predictions of elemental composition of coal and biomass from their proximate analyses using ANFIS, ANN and MLR. *Int. J. Coal Sci. Technol.* **2021**, *8*, 124–140. [CrossRef]
15. Najjar, Y.M.; Ali, H.E. Simulating the stress-strain behavior of Nevada sand by ANN. In Proceedings of the 5th U.S. National Congress on Computational Mechanics (USACM), Colorado, CA, USA, 4–6 August 1999.
16. Najjar, Y.M.; Huang, C. Simulating the stress-strain behavior of Georgia kaolin via recurrent neuronet approach. *Comput. Geotech.* **2007**, *34*, 346–362. [CrossRef]
17. Penumadu, D.; Chameau, J.L. Geo-material modeling using artificial neural networks. In *Artificial Neural Networks for Civil Engineers: Fundamentals and Applications*; ASCE: New York, NY, USA, 1997; pp. 160–184.
18. Penumadu, D.; Zhao, R. Triaxial compression behavior of sand and gravel using artificial neural networks (ANN). *Comput. Geotech.* **1999**, *24*, 207–230. [CrossRef]
19. Romo, M.; Garcia, S.R.; Mendoza, M.J.; Taboada-Urtuzuastegui, V. Recurrent and Constructive-Algorithm Networks for Sand Behavior Modelling. *Int. J. Geomech.* **2001**, *1*, 371–387. [CrossRef]
20. Shahin, M.A.; Indraratna, B. Modeling the mechanical behavior of railway ballast using artificial neural networks. *Can. Geotech. J.* **2006**, *43*, 1144–1152. [CrossRef]
21. Shahin, M.; Jaksa, M.; Maier, H. State of the Art of Artificial Neural Networks in Geotechnical Engineering. *Electron. J. Geotech. Eng.* **2008**, *13*, 1–26.
22. Tutumluer, E.; Seyhan, U. Neural network modeling of anisotropic aggregate behavior from repeated load triaxial tests. *Transp. Res. Rec. Natl. Res. Counc.* **1998**, *1615*, 86–93. [CrossRef]

23. Wrzesiński, G.; Sulewska, M.J.; Lechowicz, Z. Evaluation of the Change in Undrained Shear Strength in Cohesive Soils due to Principal Stress Rotation Using an Artificial Neural Network. *Appl. Sci.* **2018**, *8*, 781. [CrossRef]
24. Wrzesiński, G.; Markiewicz, A. Prediction of Permeability Coefficient k in Sandy Soils Using ANN. *Sustainability* **2022**, *14*, 6736. [CrossRef]
25. Zhou, J.; Lin, H.; Jin, H.; Li, S.; Yan, Z.; Huang, S. Cooperative prediction method of gas emission from mining face based on feature selection and machine learning. *Int. J. Coal Sci. Technol.* **2022**, *9*, 51. [CrossRef]
26. Zhu, J.H.; Zaman, M.M.; Anderson, S.A. Modeling of soil behavior with a recurrent neural network. *Can. Geotech. J.* **1998**, *35*, 858–872. [CrossRef]
27. Zhu, J.H.; Zaman, M.M.; Anderson, S.A. Modeling of shearing behaviour of a residual soil with recurrent neural network. *Int. J. Numer. Anal. Methods Geomech.* **1998**, *22*, 671–687. [CrossRef]
28. Zhu, J.H.; Zaman, M.M.; Trafalis, T.B. Prediction of shear stress-strain behavior of soil with recurrent neural network. *Intell. Eng. Syst. Artif. Neural Netw.* **1996**, *6*, 809–814.
29. Soból, E.; Głuchowski, A.; Szymański, A.; Sas, W. The New Empirical Equation Describing Damping Phenomenon in Dynamically Loaded Subgrade Cohesive Soils. *Appl. Sci.* **2019**, *9*, 4518. [CrossRef]
30. Drozd, W.; Kozioł, P.; Zima, K. *Cyfryzacja W Budownictwie I Architekturze*; PWN: Warsaw, Poland, 2019.
31. Geron, A. *Hands-On Machine Learning with Scikit-Learn, Keras and TensorFlow*, 2nd ed.; Helion: Everett, WA, USA, 2019.
32. *D4015-07*; Standard Test Methods for Modulus and Damping of Soils by Resonant-Column Method. ASTM International: West Conshohocken, PA, USA, 2007.
33. Komorowski, M.; Marshall, D.C.; Crutain, Y. Exploratory Data Analysis. In *Secondary Analysis of Electronic Health Records*; Springer: Cham, Switzerland, 2016.
34. Brownlee, J. Machine Learning Mastery. Available online: https://machinelearningmastery.com/how-to-use-correlation-to-understand-the-relationship-between-variables (accessed on 8 September 2022).
35. Velleman, P.F.; Hoaglin, D.C. *Applications, Basics, and Computing of Exploratory Data Analysis*; Duxbury Press: London, UK, 2004.
36. Ishibashi, I.; Zhang, X. Unified dynamic shear moduli and damping ratios of sand and clay. *Soils Found* **1993**, *33*, 182–191. [CrossRef]
37. Park, D.; Stewart, H. Suggestion of Empirical Equations for Damping Ratio of Plastic and Non-Plastic Soils Based on the Previous Studies. In Proceedings of the International Conferences on Recent Advances in Geotechnical Earthquake Engineering and Soil Dynamics, San Diego, CA, USA, 29 March 2001.
38. Michaelides, O.; Gazetas, G.; Bouckovalas, G.; Chrysikou, E. Approximate non-linear dynamic axial response of piles. *Geotechnique* **1998**, *48*, 33–53. [CrossRef]
39. Zhang, J.; Andrus, R.; Juang, C. Normalized shear modulus and material damping ratio relationships. *J. Geotech. Geoenviron. Eng.* **2005**, *131*, 453–464. [CrossRef]

applied sciences

MDPI

Article

Prediction of Undrained Bearing Capacity of Skirted Foundation in Spatially Variable Soils Based on Convolutional Neural Network

Haifeng Cheng [1], Houle Zhang [2], Zihan Liu [1] and Yongxin Wu [2,*]

[1] Shanghai Investigation, Design & Research Institute Co., Ltd., Shanghai 200434, China; cheng_haifeng@ctg.com.cn (H.C.); liu_zihan@sidri.com (Z.L.)
[2] Key Laboratory of Ministry of Education for Geomechanics and Embankment Engineering, College of Civil and Transportation Engineering, Hohai University, Nanjing 210098, China; ceatczhl@hhu.edu.cn
* Correspondence: yxwuhhu@163.com

Abstract: Skirted foundations are widely used in offshore and subsea engineering. Previous studies have shown that soil undrained shear strength variability has a notable impact on probabilistic analyses of skirted foundation bearing capacity. This study proposes an efficient machine-learning method to predict the uniaxial bearing capacity factors of skirted foundations under pure horizontal and moment loads, without relying on traditional time-consuming random finite element methods. A two-dimensional convolutional neural network is adopted to capture the potential correlation between soil random fields and bearing capacity factors. The proposed CNN-based model exhibits satisfactory prediction performance with regard to coefficients of variation and scale of fluctuations in two directions. Specifically, coefficient of determination (R2) values exceed 0.97, while root mean square error (RMSE) values remain below 0.13 for the surrogate model. In addition, more than 96% of the predictions are associated with a relative error of 5% or less, providing evidence of the proposed 2D-CNN model's satisfactory prediction performance.

Keywords: convolutional neural network; spatial variability; skirted foundation; bearing capacity

Citation: Cheng, H.; Zhang, H.; Liu, Z.; Wu, Y. Prediction of Undrained Bearing Capacity of Skirted Foundation in Spatially Variable Soils Based on Convolutional Neural Network. *Appl. Sci.* **2023**, *13*, 6624. https://doi.org/10.3390/app13116624

Academic Editor: Cheng-Yu Ku

Received: 20 April 2023
Revised: 18 May 2023
Accepted: 26 May 2023
Published: 30 May 2023

1. Introduction

Skirted foundations are widely used in offshore and subsea engineering applications such as the oil and gas industry, wind turbines, and floating structures. Owing to the complex seabed environment, these foundations experience a combination of horizontal (H), vertical (V), and moment (M) loading. Accurate estimation of the foundation's bearing capacity is crucial and has drawn increased attention from geotechnical engineers [1–4]. Previous research has assumed the seabed soil to be homogeneous, ignoring the reality of soil characteristics' spatial variability due to complex deposition history [5]. Thus, soil property uncertainty needs to be considered when analyzing foundation bearing capacity.

Soil spatial variability is simulated by the random field theory [6–9]. The random finite element method (RFEM) and random finite difference method (RFDM) have been widely used in the probabilistic analysis of geotechnical engineering, considering soil spatial variability. These methods have been applied in slope stability [10–13], foundation bearing capacity [14–18], tunnel stability and deformation [19–21], seismic response of structures [22–24], etc. The non-stationary characteristic exists in seabed soil parameters, as shown by Hossain and Randolph [25]. Consequently, the non-stationary random field of soil undrained shear strength has been simulated to investigate the effect of soil spatial variability on the bearing capacity of offshore foundations. For example, Charlton and Rouainia [26] assessed the ultimate capacity of skirted foundation in spatially variable undrained clay under uniaxial and combined loads. Selmi et al. [27] investigated the influence of soil spatial variability and embedment ratio on the capacity of skirted foundations under combined horizontal and moment loading. Ye et al. [16] presented the probabilistic failure envelope of skirted foundations with surrounding soil undrained shear strength linearly increasing with depth. However, these studies are based on Monte Carlo simulation (MCS), which is time-consuming to obtain stochastic results. Usually, more than one hour was required for one MCS, while 500 simulations were necessary for a stochastic analysis [26]. To resolve this problem, there is a need to develop an effective method for probabilistic analysis of foundation bearing capacity.

In recent years, machine learning has attracted significant attention for its unparalleled ability to model non-linear and exceedingly complex functions. Its efficacy in tackling geotechnical issues has been extensively studied and discussed [28–32]. For example, Moayedi and Hayati [33] employed machine-learning techniques to predict the settlement of the ground near sloping terrain due to strip loading. Tran et al. [34] proposed two hybrid machine-learning models that leveraged five input factors to predict the undrained shear strength of Finnish sensitive clays. Wang et al. [35] developed a convolutional neural-network-based global response surface approach to support reliability-based design of a strip footing in soils that exhibit spatial variability. Furthermore, Van et al. [36] used multivariate adaptive regression splines to predict the bearing capacity of conical foundations in clays that are heterogenous and anisotropic. However, the accurate prediction of the bearing capacity of skirted foundations in spatially variable soils has not been studied using machine-learning methods, to date.

This paper proposes an efficient machine-learning-based method to predict the bearing capacity of skirted foundation subjected to pure horizontal and moment loads. The proposed method employs a two-dimensional convolutional neural network (2D-CNN) to establish an efficient surrogate model, which replaces the time-consuming MCS. The input of the surrogate model is the non-stationary random field of soil undrained shear strength, while the output is the bearing capacity factor of the skirted foundation calculated by RFEM. The prediction performance of the proposed model is evaluated by comparing the prediction results with those generated from RFEM.

2. Random Finite Element Method

The bearing capacity of shirted foundation under uniaxial loading is analyzed by utilizing the random finite element method (RFEM), considering the spatial variability of seabed soil. The datasets for the machine-learning model in Section 3 are based on the results of the numerical simulation.

2.1. Simulation of Non-Stationary Soil Random Field

The random field theory extensively delineates the spatial variability of seabed soils [37,38]. Undrained shear strength (S_u) constitutes an essential property of seabed soil, significantly influencing the bearing capacity of offshore foundations in undrained conditions [39,40]. This study examines the spatially variable soil undrained shear strength to investigate the bearing capacity of skirted foundations. Several effective methods for simulating soil property random fields include the spectral representation method (SRM) [16], local average subdivision (LAS) method [41], and Karhunen–Loeve (KL) expansion technique [42,43]. The SRM, characterized by its high efficiency and accuracy in previous research [44], is adopted to simulate the random field of S_u.

Numerous studies illustrate that the mean value of S_u increases linearly with depth for the seabed soil [45–47], which can be expressed by Equation (1):

$$\mu_{S_u} = \mu_{S_{u,0}} + kz \quad (1)$$

where z is the depth of the seabed soil; μ_{S_u} is the mean value of S_u at the depth z; $\mu_{S_{u,0}}$ is the mean undrained shear strength at the mudline; and k is the increasing gradient of strength below the mudline. According to Ye et al. [16], $\mu_{S_{u,0}}$ is set to 10 kPa and k remains fixed at 1 kPa/m, which indicates that the mean undrained shear strength of the soil in the numerical model increases from 10 to 60 kPa as depth ranges from 0 to 50 m.

Determining the scale of fluctuations (SOFs) in horizontal and vertical directions is essential for evaluating the correlation degree between any two points in the random field. When the SOF approaches zero, data from any two points in the soil domain become independent. Phoon and Kulhawy [48] report mean values of horizontal and vertical SOFs (i.e., δ_h and δ_v) for submarine soil as 50.7 m and 3.8 m, respectively. The coefficient of variation (COV), which governs the dispersion degree of S_u in the random field around its mean value, constitutes another important parameter. Generally, Phoon and Kulhawy [48] suggests that the COV of submarine soil ranges between 0 and 0.5. In addition, a lognormal distribution is employed to simulate the variability of S_u to preclude negative values in the random field. The two-dimensional exponential autocorrelation function [49] describes the spatial correlation of S_u, as expressed in Equation (2).

$$\rho(\tau_h, \tau_v) = \exp\left[\left(-\frac{2|\tau_h|}{\delta_h} - \frac{2|\tau_v|}{\delta_v}\right)\right] \tag{2}$$

where ρ is the correlation coefficient; τ_h and τ_v are the distances between any two points in horizontal and vertical directions, respectively.

This paper aims to investigate the predictive performance of the proposed machine-learning model regarding the bearing capacity of the skirted foundations in spatially variable soil. The variation of COV, δ_h, and δ_v are considered separately to provide a comprehensive depiction of the surrogate model's prediction performance. Table 1 lists the specific parameters involved in simulating the random field, and the detailed procedures for simulating the non-stationary random field of S_u are summarized by Ye et al. [16]. A typical series of random field realization is presented in Figure 1.

Table 1. Summary of the parameters of soil random field.

Case	COV	δ_h (m)	δ_v (m)
Ani-1	0.1	50	4
Ani-2	0.2	50	4
Ani-3	0.3	50	4
Ani-4	0.4	50	4
Ani-5	0.5	50	4
Ani-6	0.3	30	4
Ani-7	0.3	40	4
Ani-8	0.3	60	4
Ani-9	0.3	50	2
Ani-10	0.3	50	6
Ani-11	0.3	50	8

2.2. Numerical Model

The finite element method (FEM) was employed to evaluate the bearing capacity of the skirted foundations in homogeneous soil. In accordance with the previous research by Ye et al. [16], a skirted foundation with a diameter (D) of 10 m and length (L) of 10 m was modeled in the commercial finite element software ABAQUS (version 6.14) as a perfectly rigid body. The skirt thickness (t) was set to 0.03 m. Horizontal, vertical, and moment loading (V, H, and M) were applied to the reference point (RP) to determine the corresponding bearing capacity, with RP located at the centre of the skirted foundation lid. Figure 2 presents the details of the skirted foundation and the positive loading and displacement.

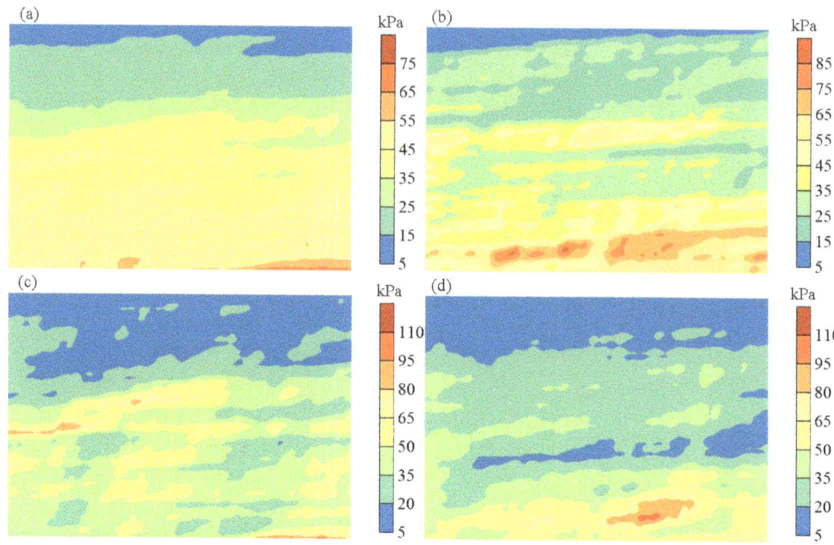

Figure 1. A typical series of the soil random field: (**a**) Case Ani-1; (**b**) Case Ani-3; (**c**) Case Ani-6; (**d**) Case Ani-11.

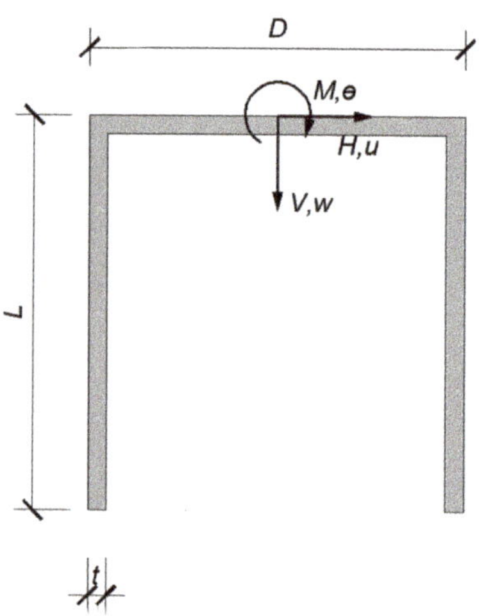

Figure 2. Details of skirted foundation and positive loading and displacement.

In reality, skirted foundation has a circular or square section, but it was considered that plan geometry is secondary to soil strength profile in terms of the effect on bearing capacity by Gourvenec and Barnett [2]. Therefore, a two-dimensional (2*D*) plane strain condition was assumed in this study to allow comparison with existing analytical solution [2], as well as improving calculation efficiency with satisfactory accuracy [16]. The soil domain was modeled as a rectangular region with dimensions 7*D* in length and 5*D* in width (70 × 50 m) to minimize the effect of boundaries. Lateral boundaries were constrained in the horizontal direction, while the bottom boundary was fixed in both horizontal and vertical directions. No constraint was applied in the top boundary. The soil domain was meshed into a square grid with dimensions of 1 × 1 m; however, in the 3*D* × 2*D* area surrounding the skirted foundation, the mesh size was adjusted to 0.5 × 0.5 m to obtain more precise bearing capacity results. A total of 5300 four-node bilinear reduced integration quadrilateral (CPE4R) elements were used to simulate the seabed soil domain in the model, as shown in Figure 3.

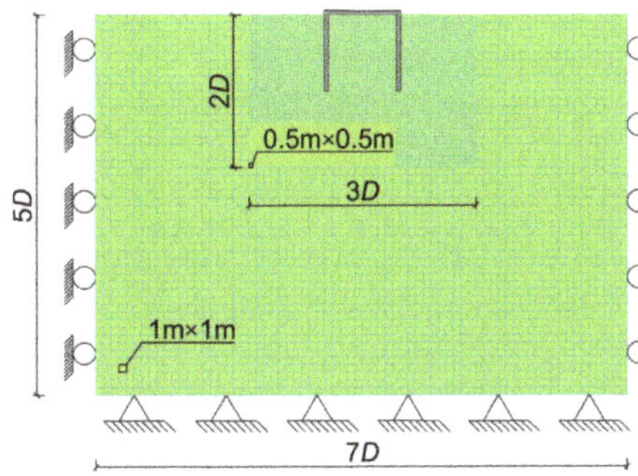

Figure 3. Finite element model of skirted foundation.

The soil was defined as an elastoplastic material that obeys the Tresca yield criterion, in which the maximum shear stress was determined by the undrained shear strength [27]. The soil undrained Young's modulus (E_u) was assumed to be linearly correlated to S_u with the relationship expressed in Equation (3):

$$E_u = K_u \times S_u \tag{3}$$

where K_u is the coefficient. In fact, the ultimate bearing capacity of foundation was confirmed to be barely influenced when K_u was equal to 500 [50]. In addition, the Poisson's ratio was fixed at 0.49 to simulate the undrained condition. As only the ultimate bearing capacities were of interest, the skirted foundation was simulated as a rigid body without any deformation. The skirted foundation and surrounding soil were coupled by the "tie" constraint to avoid the generation of extra interface element.

The bearing capacity factors were introduced in this paper to evaluate the uniaxial bearing capacity normalized by the foundation diameter and the mean undrained shear strength at mudline, as expressed in Equation (4):

$$\begin{aligned} N_{cH} &= H_0/D\mu_{S_{u,0}} \\ N_{cV} &= V_0/D\mu_{S_{u,0}} \\ N_{cM} &= M_0/D^2\mu_{S_{u,0}} \end{aligned} \tag{4}$$

where N_{cH}, N_{cV}, and N_{cM} denote the bearing capacity factors under pure horizontal, vertical, and moment loading, respectively; H_0, V_0, and M_0 denote the ultimate uniaxial bearing capacity. In the deterministic case, without consideration of soil spatial variability, the bearing capacity factors marked by the subscript $_{\text{det}}$ are calculated to make comparison with previous research. The $N_{cH,\text{det}}$ and $N_{cM,\text{det}}$ calculated by the finite element method in this study are 4.20 and 2.54, which are consistent with the results of upper bound analysis by Bransby and Yun [1] (i.e., N_{cH} = 4.00 and N_{cM} = 2.50). The relative errors are no more than 5%, indicating acceptable accuracy of the proposed numerical model.

2.3. Random Finite Element Method

The random finite element method (RFEM) was employed to perform the stochastic analysis of the skirted foundation's bearing capacity, considering spatially variable soil undrained shear strength. The primary concept of RFEM involves incorporating the random field data generated in Section 2.1 into the finite element model simulated in Section 2.2. The general steps are summarized as follows.

(1) A deterministic finite element model, coupling the skirted foundation and soil, is established in ABAQUS software, with the soil elements shown in Figure 1 numbered from 1 to 5300;
(2) The non-stationary random field of S_u is generated using the SRM proposed in Section 2.1. Each value in the random field data is mapped to an element in the finite element model based on its unique position number. Subsequently, each realization of the soil random field is assigned to the deterministic model, replacing the soil undrained shear strength to generate a new calculation document for the subsequent finite element analysis. This process was executed using Python coding;
(3) The number of MCS (n) is determined, and the finite element calculation in Section 2.2 is repeated after generating the random field data. The stochastic analysis comprises n finite element calculations.

Selecting an appropriate MCS number is crucial for obtaining stable and accurate results. The evolution of the mean value and standard deviation of N_{cH} against the required MCS numbers in case Ani-3 are shown in Figure 4. The converging trend suggests that n= 600 is sufficient to provide stable results for the stochastic analysis.

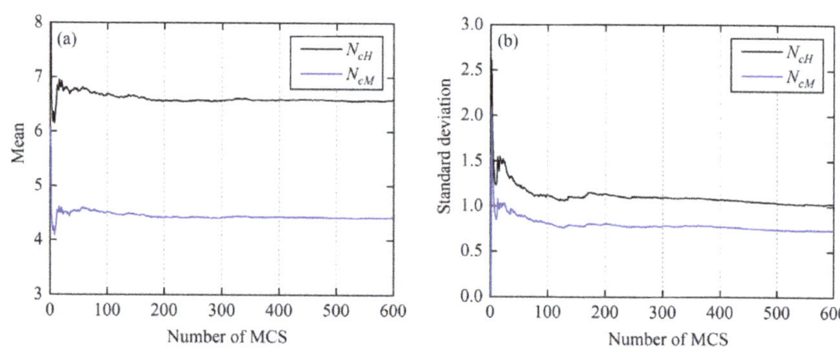

Figure 4. The converging trend of N_{cH} statistics: (**a**) mean; (**b**) standard deviation.

3. Architecture of the Proposed Two-Dimensional Convolutional Neural Network (2D-CNN) Model

The architecture of the proposed 2D-CNN model for predicting the bearing capacity of skirted foundation consists of the input layer, the convolutional layer, the pooling layer, the fully connected layer and the output layer. The input and output of the proposed model are the random field data and bearing capacity factor, respectively. The detailed architecture of the proposed 2D-CNN model is given in Figure 5. The training process runs on the PyTorch platform, which is an open-source deep-learning framework based on Python. The experiments are performed on a standard PC with 8 Intel Core i7-11700 CPUs and two NVIDIA RTX 3060 GPU cards.

3.1. Input and Output Layer

The proposed 2D-CNN model aims to predict the bearing capacity of the skirted foundation in spatially variable soil. As shown in Figure 5, the input is a random field data matrix with size of 70 × 50 mapping to the soil domain in the numerical model. The bearing capacity factor calculated by FEM is the output of the proposed model. A random field matrix and corresponding bearing capacity factor constitute a sample in the dataset, which contains the training, validation, and prediction datasets that are separated from the whole dataset in the ratio of 70%, 10% and 20%, respectively. Taken the case Ani-1 for example, there are 420, 60, and 120 samples in the training, validation, and prediction datasets for the training process of the proposed 2D-CNN model.

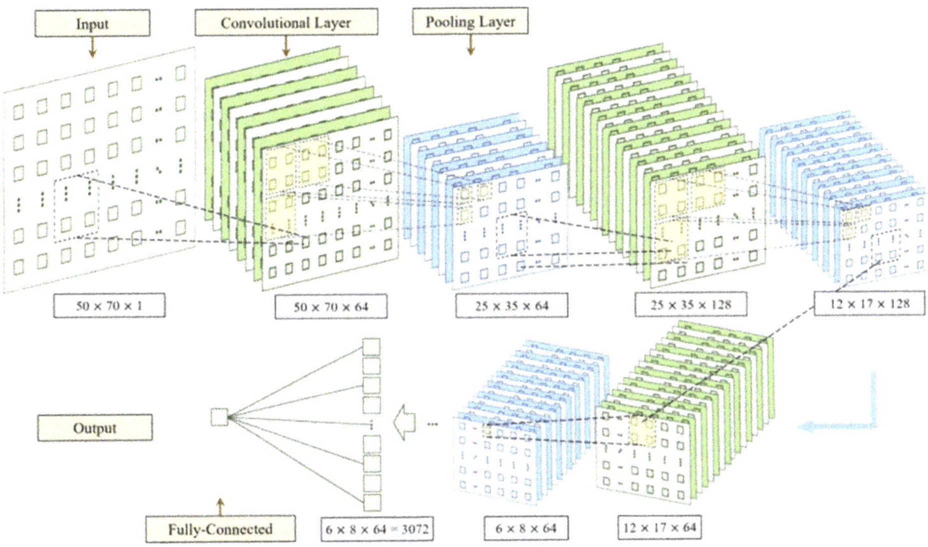

Figure 5. Architecture of the proposed 2D-CNN model.

3.2. Convolutional Layer

The convolutional layer is used to extract the key information from the input data. There are several hyperparameters for a convolutional layer that remain constant in the training process, such as the kernel size, stride, padding, and filters. These parameters were determined by trial and error in this study. The kernel size represents the range of each convolution operation, while the stride controls the crossing length in the horizontal and vertical directions between the two adjacent convolution operations. The zero-padding strategy is adopted to make sure the output has the same size as the input. The filter controls the output channels during the convolutional operation. Generally, a nonlinear activation function is utilized after the convolutional operation, which is beneficial to introduce nonlinearities into the surrogate model. In this study, the rectified linear unit (ReLU) function is employed, as expressed in Equation (5). There are four convolutional layers in the proposed 2D-CNN model and the detailed hyperparameters of these layers are shown in Figure 5.

$$f(x) = \begin{cases} 0\,, x \leq 0 \\ x\,, x > 0 \end{cases} \tag{5}$$

3.3. Pooling Layer

The pooling layer plays significant role in reducing the size of input features. There are two typical patterns of pooling layer. Specifically, the max pooling aims to hold the maximum value of each pooling zone, while the average pooling uses the mean value to present the pooling zone. The average pooling is adopted in this study for a better prediction performance. The pooling size and stride are equal to 2 × 2 to realize a half decreasing in both the horizontal and vertical size of the input data.

3.4. Fully Connected Layer

The fully connected (FC) layer is the final part of a regular CNN model. It works by establishing full connection between the input feature and the output layer. Mathematically, the input feature is transformed to the output layer through a weight matrix and a bias matrix, as illustrated in Equation (6):

$$\mathbf{y}_d = f(\mathbf{W}_d\mathbf{x}_d + \mathbf{b}_d) \tag{6}$$

where $\mathbf{x}_d$ and $\mathbf{y}_d$ denote the input and output vector of the FC layer, respectively; $\mathbf{W}_d$ and $\mathbf{b}_d$ denote the weight and bias matrix, respectively; $f(\cdot)$ denotes the ReLU activation function expressed in Equation (5). Furthermore, the last convolutional layer is flattened to a one-dimensional array before the FC layer.

In the training process, the early stopping method with limited epochs of 2000 is adopted to avoid overfitting of the proposed 2D-CNN model. The adaptive momentum (Adam) algorithm is selected as the optimizer and the learning ratio is set to 8×10^{-5} to reach a reasonable prediction performance. The above-mentioned hyperparameters of the proposed 2D-CNN model are determined by trial and error. Several networks with different parameters of convolutional and pooling layers are examined and the effect on R^2 and RMSE values in Case Ani-5 are summarized in Table 2, where N_{CL} and K_{CL} denote the number and kernel size of convolutional layer, respectively. It is observed from Table 2 that the architecture of three convolutional layers with kernel size at 2×2 and average pooling, as shown in Figure 5, has the maximum R^2 and minimum RMSE value in predicting the uniaxial horizontal and moment bearing capacity factors.

Table 2. Effect of different neural network architecture on R^2 and RMSE values.

N_{CL}	K_{CL}	Pooling Type	R^2		RMSE	
			N_{cH}	N_{cM}	N_{cH}	N_{cM}
2	2 × 2	Maximum	0.9048	0.8996	0.2768	0.2210
2	2 × 2	Average	0.9420	0.9417	0.2501	0.1938
2	3 × 3	Maximum	0.8986	0.8938	0.2911	0.2271
2	3 × 3	Average	0.9189	0.9132	0.2617	0.2051
2	4 × 4	Maximum	0.8968	0.8896	0.3036	0.2459
2	4 × 4	Average	0.9116	0.9077	0.2690	0.2166
3	2 × 2	Maximum	0.9202	0.9166	0.2696	0.2136
3	2 × 2	Average	0.9570	0.9569	0.2433	0.1865
3	3 × 3	Maximum	0.9113	0.9088	0.2836	0.2203
3	3 × 3	Average	0.9345	0.9299	0.2546	0.1987
3	4 × 4	Maximum	0.9078	0.9022	0.2964	0.2384
3	4 × 4	Average	0.9261	0.9212	0.2621	0.2088
4	2 × 2	Maximum	0.9093	0.9066	0.2732	0.2173
4	2 × 2	Average	0.9459	0.9451	0.2476	0.1908
4	3 × 3	Maximum	0.9001	0.8916	0.2877	0.2241
4	3 × 3	Average	0.9215	0.9198	0.2581	0.2029
4	4 × 4	Maximum	0.8988	0.8916	0.3002	0.2429
4	4 × 4	Average	0.9140	0.9112	0.2659	0.2132

4. Results and Discussion

On account of the influence of the number of training samples on the prediction performance of the proposed model, the dataset in case Ani-3 is selected to obtain 16 CNN models with difference size of training samples. The prediction datasets are identical for the 16 models, with 120 samples. Figure 6 displays the variation of R^2 and RMSE values for N_{cH} when the number of training samples increases from zero to 420. It is evident that the R^2 value increases and the RMSE value decreases quickly with the increasing number of training samples. These changing trends gradually flatten out as the number of training samples exceeds 210. The R^2 and RMSE values slightly fluctuate when the training samples arrive at 350, demonstrating that 420 training samples are sufficient to provide stable prediction performance of the surrogate model.

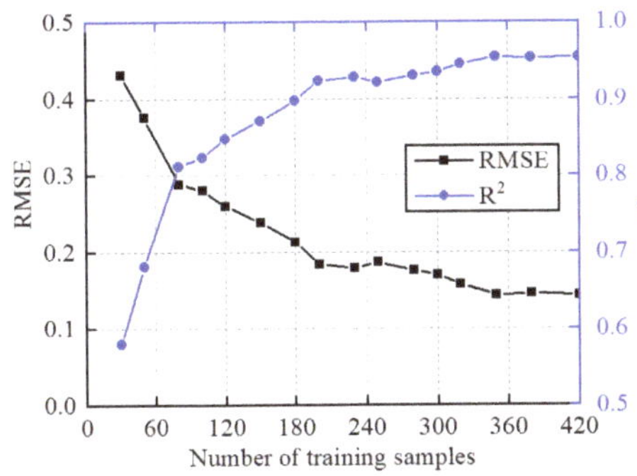

Figure 6. Influence of number of training samples on R^2 and RMSE values.

4.1. *Influence of Coefficient of Variation (COV) in the Random Field*

Figure 7 shows the development of the mean and standard deviation of the uniaxial horizontal and moment bearing capacity factors with the variation of COV in RFEM. It is found that the standard deviations of N_{cH} and N_{cM} increase with COV, for the fluctuation of the undrained shear strength is larger under conditions with higher COV, as illustrated in Figure 1a,b.

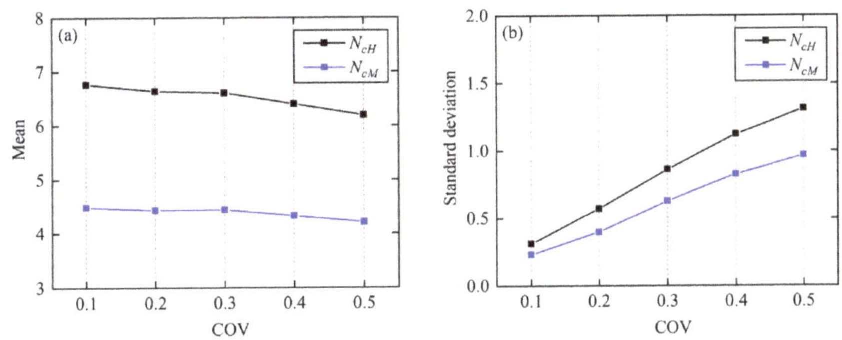

Figure 7. Variation of the bearing capacity factors (N_{cH} and N_{cM}) statistics with COV: (**a**) Mean; (**b**) Standard deviation.

In order to investigate the influence of COV of the soil undrained shear strength random field on the prediction performance of the proposed 2D-CNN model, the datasets consist of the random field data and corresponding bearing capacity factors from case Ani-1 to Ani-are trained separately with 480 samples and tested by the remaining 120 samples. The five-fold cross-validation method [51] is adopted to obtain more persuasive results, in which the total dataset is randomly divided into five subsets and each subset is taken as the prediction dataset while the remaining four subsets are taken as the training dataset to train five CNN models. In other words, the presented prediction performance is the average result of the five models. Figure 8 shows the root mean square error (RMSE) and coefficient of determination (R^2) of the prediction dataset in the five cases. Specifically, the RMSE value reflects the degree to which the predicted result deviates from the actual value, where a smaller RMSE value represents better prediction accuracy. The R^2 value, varying between zero and one, reflects the global performance of the surrogate model. A larger R2 value indicates better prediction performance of the proposed model. It is visible in Figure 8 that the value of RMSE gradually increases when the COV extends from 0.1 to 0.5. Meanwhile, the value of R^2 slightly fluctuates with the increase in COV, which indicates that the performance of the proposed surrogate model is affected by the degree of variation of the random field when one considers the soil spatial variability in predicting the uniaxial bearing capacity of skirted foundation. The 2D-CNN model essentially captures the potential characteristics between the input and output vectors. The output value in the dataset contains a wider distribution range in the case of higher COV, resulting in a decrease in the average density of training data when the total number of samples remains unchanged. This is reflected in the surrogate model, which shows relatively poor prediction performance under higher COV.

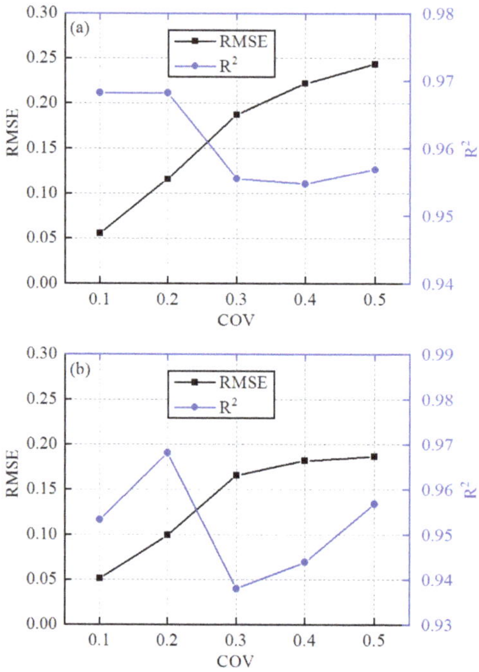

Figure 8. Effect of COV on RMSE and R^2 value for (**a**) N_{cH} and (**b**) N_{cM}.

However, there are five surrogate models for the prediction of bearing capacity factors when COV varies from 0.1 to 0.5. These models are predetermined with the same structure (as shown in Figure 5), which contains the difference in the model parameters trained by different datasets. It is complicated to generate several datasets and train different surrogate models in dealing with the prediction of bearing capacity factors, considering the variation of COV in the random field. It is necessary to train a unified surrogate model in the stochastic analysis under different COV of random field without notably reducing the accuracy of prediction. Therefore, the samples from case Ani-1 to Ani-5 are composed to generate a mixed dataset to train a 2D-CNN model in predicting the bearing capacity factors of the skirted foundation. Specifically, there are 2100, 300 and 600 samples in the training, validation, and prediction datasets for the training process of the unified 2D-CNN model.

Figure 9 presents the uniaxial bearing capacity factors predicted by the unified 2D-CNN model and calculated by the RFEM. It can be seen from Figure 9a,b that both the R^2 values in the prediction datasets of the uniaxial horizontal and moment bearing capacity factor are larger than 0.98, while the RMSE values are smaller than 0.12, indicating that N_{cH} and N_{cM} predicted by the surrogate 2D-CNN model are highly consistent with the results calculated by the RFEM. The unified model performs better than the model trained by the dataset with single value of COV, especially when COV is larger than 0.1, as shown in Figure 8. This is mainly caused by the average effect that the samples with small COV in the dataset are beneficial to the prediction of the unified model. The cumulative density function (CDF) of the predicted and calculated bearing capacity factors under uniaxial horizontal and moment loading are shown in Figure 9c,d, respectively. The satisfactory agreement between the CNN-derived and RFEM-derived CDFs of N_{cH} and N_{cM} demonstrates that the global distribution of the uniaxial bearing capacity factor of the skirted foundation under the soil random field with different COV is well captured by the proposed unified 2D-CNN model. In addition, the mean, minimum and maximum values of the results obtained by the two methods are compared for the quantitative analysis of the prediction performance. As to the uniaxial horizontal bearing capacity factor, the relative errors of the mean, minimum, and maximum values between the predicted and calculated results are 0.15%, 3.16%, and 0.49%, respectively. Meanwhile, the three relative errors of the uniaxial moment bearing capacity factors are 0.23%, 0.42%, and 0.13%, respectively. The relative errors of the three evaluating indicators are all restricted within 4.00%, indicating the distribution of the uniaxial bearing capacity factors calculated by the time-consuming RFEM can be replaced by the proposed 2D-CNN model.

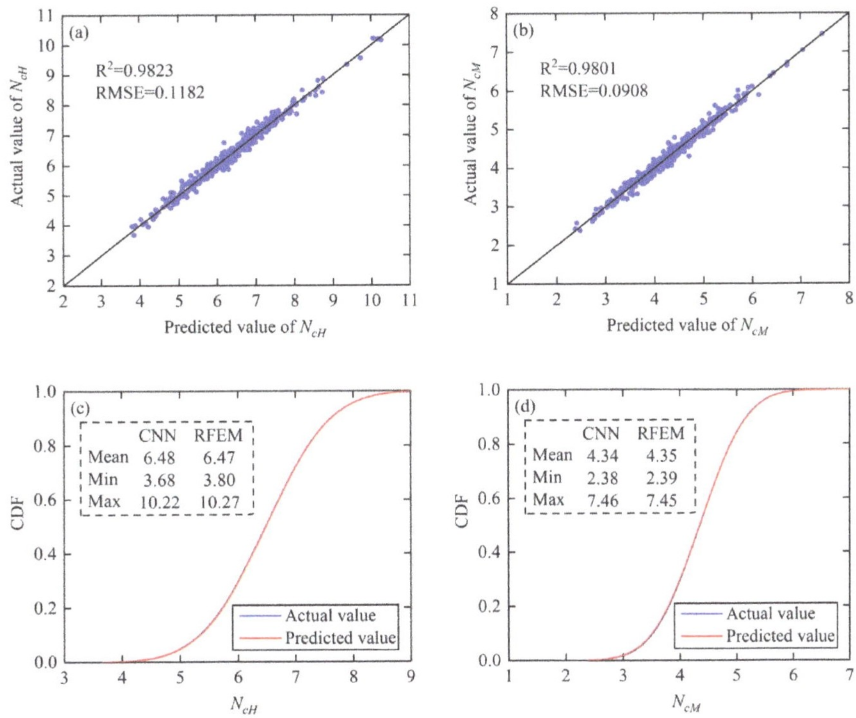

Figure 9. Comparison of the predicted and actual uniaxial bearing capacity factors under mixed COV: (**a**) Scatter of N_{cH}; (**b**) Scatter of N_{cM}; (**c**) CDFs of N_{cH}; (**d**) CDFs of N_{cM}.

The probability density functions (PDFs) of the relative errors between the CNN-predicted and RFEM-calculated bearing capacity factors are given in Figure 10. The relative errors of the 1200 predicted uniaxial horizonal and moment bearing capacity factors are less than 8%. In addition, the confidence intervals (CIs) of the relative errors are 97.7% and 96.3%, with the margin of error at 5%, for the predicted bearing capacity factors under pure horizontal and moment loading, respectively. The unified 2D-CNN model is accurate enough to predict the uniaxial horizontal and moment bearing capacity factors for the probabilistic analysis of the skirted foundation located in spatially variable soil with different COV.

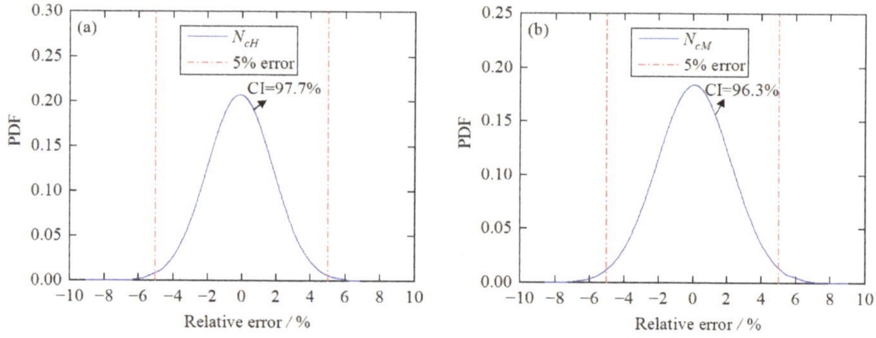

Figure 10. Distribution of the relative errors of the predicted bearing capacity factors under mixed COV: (**a**) PDF of N_{cH}; (**b**) PDF of N_{cM}.

4.2. Influence of Horizontal Scale of Fluctuation (δ_h) in the Random Field

The development of the mean and standard deviation of N_{cH} and N_{cM} with the variation of δ_h is shown in Figure 11. It is indicated that the mean value and standard deviation of N_{cH} and N_{cM} fluctuated slightly with the increase of δ_h from 30 to 60 m (e.g., case Ani-6, 7, 3, and 8 given in Table 1). The samples of the four cases are composed to a mixed dataset in training a 2D-CNN model for the prediction of uniaxial bearing capacity factors of the skirted foundation considering the variation of δ_h. There are 1680, 240 and 480 samples for the training, validation, and prediction datasets, respectively.

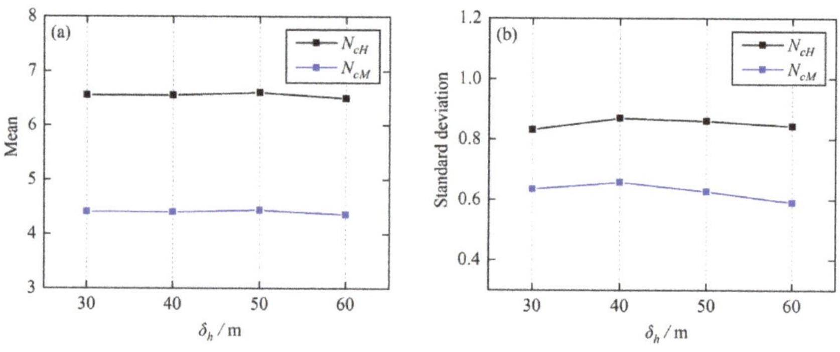

Figure 11. Variation of the bearing capacity factors (N_{cH} and N_{cM}) statistics with δ_h: (**a**) Mean; (**b**) Standard deviation.

Figure 12 displays the comparison of N_{cH} and N_{cM} predicted by the trained 2D-CNN model and calculated by RFEM. As shown in Figure 12a,b, the predicted and calculated uniaxial horizontal and moment bearing capacity factors are located around the solid line, which is predefined as the predicted value equal to the actual value, with the minimum R^2 value at 0.9802 and maximum RMSE value at 0.1204, which indicates a satisfactory prediction performance of the surrogate model. It is observed from Figure 12c that the distribution of the predicted uniaxial horizontal bearing capacity factors is highly consistent with the RFEM-calculated results, in which the relative errors of the mean, minimum, and maximum values are 0.15%, 0.22%, and 0.69%, respectively. Similarly, the relative errors of the three evaluating values are 0.23%, 0.42%, and 0.13% for the uniaxial moment bearing capacity factors predicted by the 2D-CNN model, as presented in Figure 12d, which indicates that the typical distribution of N_{cH} and N_{cM} are well predicted by the surrogate model on account of the variation of δ_h.

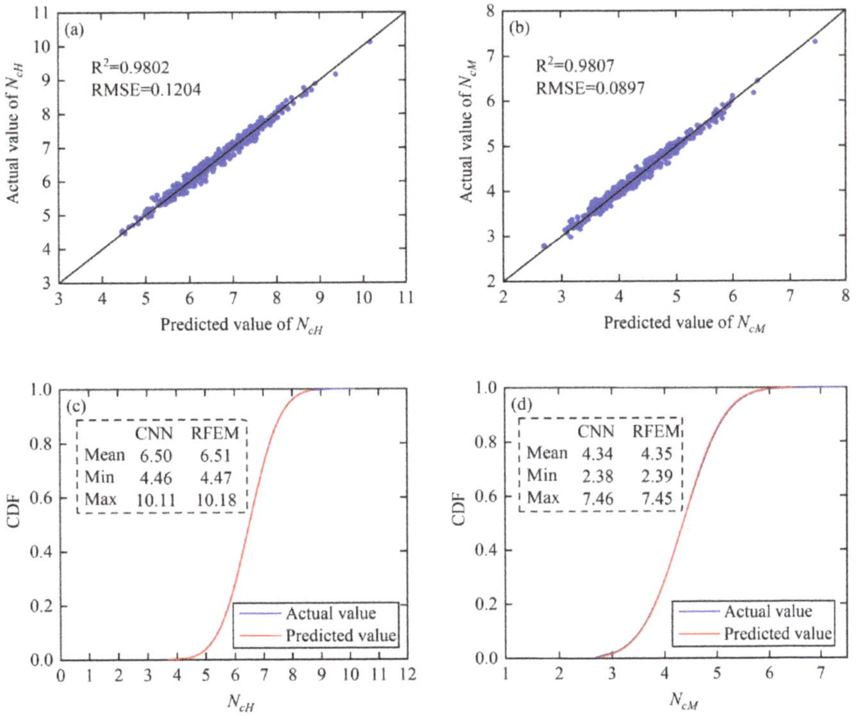

Figure 12. Comparison of the predicted and actual uniaxial bearing capacity factors under mixed δ_h: (**a**) Scatter of N_{cH}; (**b**) Scatter of N_{cM}; (**c**) CDFs of N_{cH}; (**d**) CDFs of N_{cM}.

In order to visualize the accuracy of the CNN-predicted uniaxial bearing capacity factors, the probability distribution of the relative errors between the predicted and calculated results is illustrated in Figure 13. It is obviously that the relative errors are all restricted in 8%, with a CI of 99.2% and 97.3% for the margin of 5% error in the prediction of uniaxial horizontal and moment bearing capacity factors, respectively. It indicates that the architecture of the 2D-CNN model proposed in Figure 5 has the potential to predict the uniaxial bearing capacity factors of the skirted foundation within an acceptable error range when δ_h of the soil random field extends from 30 to 60 m.

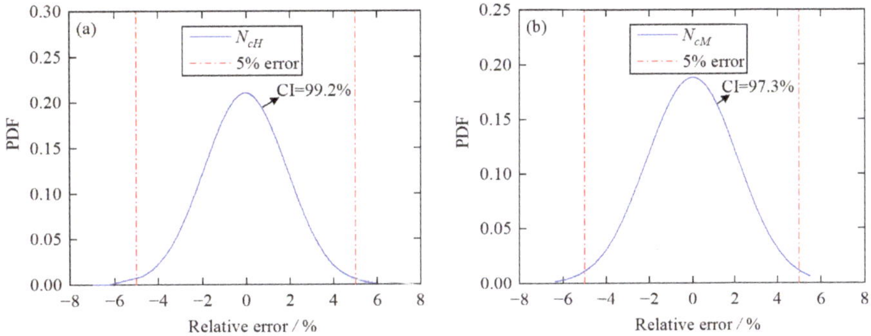

Figure 13. Distribution of the relative errors of the predicted bearing capacity factors under mixed δ_h: (**a**) PDF of N_{cH}; (**b**) PDF of N_{cM}.

4.3. Influence of Vertical Scale of Fluctuation (δ_v) in the Random Field

On account of the increasing of δ_v from 2 to 8 m (e.g., case Ani-9, 10, 3, and 11 shown in Table 1), a mixed dataset consisting of the total samples in the four cases is generated to train a 2D-CNN model, which has the same architecture as shown in Figure 5, to predict the uniaxial bearing capacity factors of the skirted foundation considering the variation of δ_v in the soil random field. A total of 1680 and 240 samples are selected for the training and validation datasets to accomplish the training process of the model, while the remaining 480 samples are used to estimate the prediction performance of the surrogate model.

Figure 14 compares the uniaxial bearing capacity factors predicted by the 2D-CNN model and calculated by RFEM. It can be seen from Figure 14a,b that the CNN-predicted N_{cH} and N_{cM} are close to the calculated results, in which the R^2 value is more than 0.97 and the RMSE value is less than 0.11. According to the cumulative distribution of N_{cH} and N_{cM} shown in Figure 14c,d, it is evident that the proposed 2D-CNN model is feasible in analyzing the uniaxial bearing capacity factors of the skirted foundation in spatially variable soil with different δ_v, for the maximum relative error of the three evaluating indicators is 1.67%.

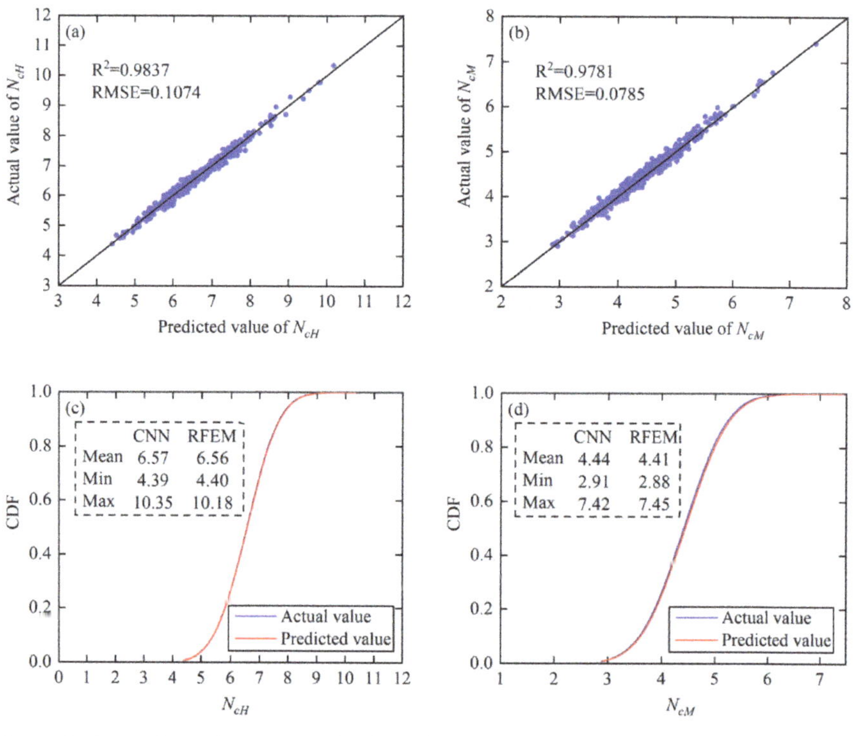

Figure 14. Comparison of the predicted and actual uniaxial bearing capacity factors under mixed δ_v: (**a**) Scatter of N_{cH}; (**b**) Scatter of N_{cM}; (**c**) CDFs of N_{cH}; (**d**) CDFs of N_{cM}.

Furthermore, the probability distribution of the relative errors between the predicted and calculated uniaxial bearing capacity factors is given in Figure 15. With the variation of δ_v from 2 to 8 m, the proposed 2D-CNN model is capable of providing satisfactory prediction performance for the uniaxial bearing capacity factors by restricting the maximum relative error within 9%. More specifically, the Cis of the relative error located within 5% are 99.6% and 97.9% for the prediction of N_{cH} and N_{cM}, respectively. The effect of δ_v in the soil random field on the prediction performance of the uniaxial bearing capacity factors is similar to the influence of δ_h, as shown in Figure 13.

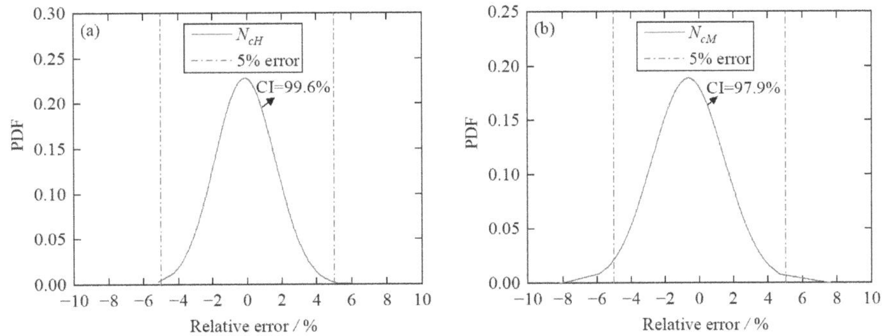

Figure 15. Distribution of the relative errors of the predicted bearing capacity factors under mixed δ_v: (**a**) PDF of N_{cH}; (**b**) PDF of N_{cM}.

The time required for 600 samples in the statistical analysis of tradition RFDM is about 327 min on a standard PC as described in Section 3, while the trained 2D-CNN model takes just 3 s, resulting a 6540-times improvement in calculation efficiency. Only the ultimate bearing capacity of the skirted foundation had been investigated in this study, while more than 600 simulations are essential for further reliability analysis, such as the failure mechanics and failure envelopes of foundation. At this point the trained 2D-CNN model is highly efficient to acquire numerous outputs in several seconds once the random filed data are generated, instead of time-consuming finite element calculations.

5. Conclusions

In summary, this study combines the random field theory and convolutional neural network to develop an effective machine-learning-based model for predicting the uniaxial bearing capacity factors of the skirted foundation in spatial variable soil under pure horizontal and moment loading. The spectral representation method is adopted to simulate the non-stationary random field of soil undrained shear strength with various coefficients of variation and scale of fluctuations, which is acted as the input of the proposed 2D-CNN model. Meanwhile, the uniaxial horizontal and moment bearing capacity factors are defined as the output of the surrogate model. The predicted values are compared with the results calculated by the random finite element method to present the prediction performance of the surrogate model. Several conclusions are summarized as follows:

(1) The proposed 2D-CNN model can replace the time-consuming RFEM in predicting the uniaxial bearing capacity factors of the skirted foundation in spatially variable soil with reasonable accuracy;
(2) There are three 2D-CNN models with the same architecture that are trained to deal with the prediction of skirted foundation bearing capacity considering the variation of COV, δ_h and δ_v in the soil random field, respectively. The minimum R^2 value and maximum RMSE value for the three surrogate models are 0.9781 and 0.1204, indicating satisfactory prediction performance of the proposed model;
(3) The confidence interval of the relative error is more than 96.3% with a margin of 5% for the predicted bearing capacity factor with the variation of COV, while the minimum confidence intervals are 97.3% and 97.9% for the relative errors that are located within 5% on account of the variation of δ_h and δ_v, respectively.

The proposed 2D-CNN model is an effective surrogate model for the prediction of uniaxial bearing capacity factors of skirted foundation. However, it is noted that the bearing capacity prediction of the skirted foundation subjected to the combination of horizontal, vertical and moment loads have not been illustrated. Further study is recommended to develop a reasonable machine-learning model to predict the skirted foundation bearing capacity and failure envelope under different loadings.

Author Contributions: Conceptualization, Y.W. and H.Z.; methodology, H.C.; software, Z.L. and H.Z.; validation, Y.W. and H.Z.; formal analysis, H.Z. and H.C.; investigation, Z.L., H.C. and Y.W.; resources, H.C.; data curation, Z.L. and H.C.; writing-original draft preparation, H.Z.; writing-review and editing, H.C., Y.W. and Z.L.; visualization, H.Z.; supervision, H.Z.; project administration, Y.W. and H.C.; funding acquisition, H.C. All authors have read and agreed to the published version of the manuscript.

Funding: This research was funded by the science foundation of China Three Gorges Corporation grant number 202203004.

Institutional Review Board Statement: Not applicable.

Informed Consent Statement: Not applicable.

Data Availability Statement: The data presented in this study are available on request from the corresponding author.

Acknowledgments: The authors acknowledge the financial support of the science foundation of China Three Gorges Corporation (Grant No. 202203004).

Conflicts of Interest: The authors declare no conflict of interest.

References

1. Bransby, M.F.; Yun, G.J. The undrained capacity of skirted strip foundations under combined loading. *Geotechnique* **2009**, *59*, 115–125. [CrossRef]
2. Gourvenec, S.; Barnett, S. Undrained failure envelope for skirted foundations under general loading. *Geotechnique* **2011**, *61*, 263–270. [CrossRef]
3. Hung, L.C.; Kim, S.R. Evaluation of combined horizontal-moment bearing capacities of tripod bucket foundations in undrained clay. *Ocean. Eng.* **2014**, *85*, 100–109. [CrossRef]
4. Vulpe, C.; Gourvenec, S.M.; Cornelius, A.F. Effect of embedment on consolidated undrained capacity of skirted circular foundations in soft clay under planar loading. *Can. Geotech. J.* **2017**, *54*, 158–172. [CrossRef]
5. Cao, Z.J.; Wang, Y.; Li, D.Q. Quantification of prior knowledge in geotechnical site characterization. *Eng. Geol.* **2016**, *203*, 107–116. [CrossRef]
6. Cho, S.E. Probabilistic assessment of slope stability that considers the spatial variability of soil properties. *J. Geotech. Geoenviron.* **2010**, *136*, 975–984. [CrossRef]
7. Huang, J.S.; Griffiths, D.V.; Fenton, G.A. Probabilistic analysis of coupled soil consolidation. *J. Geotech. Geoenviron.* **2010**, *136*, 417–430. [CrossRef]
8. Bari, M.W.; Shahin, M.A. Probabilistic design of ground improvement by vertical drains for soil of spatially variable coefficient of consolidation. *Geotext. Geomembr.* **2014**, *42*, 1–14. [CrossRef]
9. Sina, J.; Luo, N.; Bathurst, R.J. Probabilistic analysis of simple slopes with cohesive soil strength using RLEM and RFEM. *Georisk* **2017**, *11*, 231–246.
10. Griffiths, D.V.; Fenton, G.A. Probabilistic Slope Stability Analysis by Finite Elements. *J. Geotech. Geoenviron.* **2004**, *130*, 507–518. [CrossRef]
11. Griffiths, D.V.; Huang, J.S.; Fenton, G.A. Influence of spatial variability on slope reliability using 2-D random fields. *J. Geotech. Geoenviron.* **2009**, *135*, 1367–1378. [CrossRef]
12. Li, D.Q.; Qi, X.H.; Phoon, K.K.; Zhang, L.M.; Zhou, C.B. Effect of spatially variable shear strength parameters with linearly increasing mean trend on reliability of infinite slopes. *Struct. Saf.* **2014**, *49*, 45–55. [CrossRef]
13. Liu, L.L.; Deng, Z.P.; Zhang, S.H.; Cheng, Y.M. Simplified framework for system reliability analysis of slopes in spatially variable soils. *Eng. Geol.* **2018**, *239*, 330–343. [CrossRef]
14. Popescu, R.; Deodatis, G.; Nobahar, A. Effects of random heterogeneity of soil properties on bearing capacity. *Probab. Eng. Mech.* **2005**, *20*, 324–341. [CrossRef]
15. Cho, S.E.; Park, H.C. Effect of spatial variability of cross-correlated soil properties on bearing capacity of strip footing. *Int. J. Numer. Anal. Met.* **2010**, *34*, 1–26. [CrossRef]
16. Ye, Z.T.; Gao, Y.F.; Shu, S.; Wu, Y.X. Probabilistic undrained bearing capacity of skirted foundations under HM combined loading in spatially variable soils. *Ocean. Eng.* **2021**, *219*, 108297. [CrossRef]
17. Shen, Z.C.; Pan, Q.J.; Chian, S.C.; Gourvenec, S.; Tian, Y.H. Probabilistic failure envelopes of strip foundations on soils with non-stationary characteristics of undrained shear strength. *Geotechnique* **2022**, *2020*, 1–20. [CrossRef]
18. Chen, H.Z.; Shen, Z.C.; Wang, L.; Tian, Y.H.; Chu, X.X. Undrained capacity of skirted circular foundations under fully three-dimensional loading. *Comput. Geotech.* **2023**, *156*, 105261. [CrossRef]
19. Mollon, G.; Phoon, K.K.; Dias, D.; Soubra, A.H. Validation of a new 2D failure mechanism for the stability analysis of a pressurized tunnel face in a spatially varying sand. *J. Eng. Mech.* **2011**, *137*, 8–21. [CrossRef]

20. Wang, X.R.; Li, Z.; Wang, H.; Rong, Q.G.; Liang, R.Y. Probabilistic analysis of shield-driven tunnel in multiple strata considering stratigraphic uncertainty. *Struct. Saf.* **2016**, *62*, 88–100. [CrossRef]
21. Huang, H.W.; Xiao, L.; Zhang, D.M.; Zhang, J. Influence of spatial variability of soil Young's modulus on tunnel convergence in soft soils. *Eng. Geol.* **2017**, *228*, 357–370. [CrossRef]
22. Wang, S.; Hao, H. Effects of random variations of soil properties on site amplification of seismic ground motions. *Soil. Dyn. Earthq. Eng.* **2002**, *22*, 551–564. [CrossRef]
23. Yue, Q.X.; Ang, A.H.S. 3D reliability evaluation of tunnels under strong-motion earthquakes considering spatial randomness. *Struct. Infrastruct. Eng.* **2017**, *13*, 882–893. [CrossRef]
24. Wang, Y.B.; He, J.J.; Shu, S.; Zhang, H.L.; Wu, Y.X. Seismic responses of rectangular tunnels in liquefiable soil considering spatial variability of soil properties. *Soil Dyn. Earthq. Eng.* **2022**, *162*, 107489. [CrossRef]
25. Hossain, M.S.; Randolph, M.F. Effect of strain rate and strain softening on the penetration resistance of spudcan foundations on clay. *Int. Geomech.* **2009**, *9*, 122–132. [CrossRef]
26. Charlton, T.S.; Rouainia, M. A probabilistic approach to the ultimate capacity of skirted foundations in spatially variable clay. *Struct. Saf.* **2017**, *65*, 126–136. [CrossRef]
27. Selmi, M.; Kormi, T.; Hentati, A.; Ali, N.B.H. Capacity assessment of offshore skirted foundations under HM combined loading using RFEM. *Comput. Geotech.* **2019**, *114*, 103148. [CrossRef]
28. Zhang, J.; Sato, T.; Iai, S. Novel support vector regression for structural system identification. *Struct. Control. Health* **2007**, *14*, 609–626. [CrossRef]
29. Dong, Y.F.; Li, Y.M.; Lai, M.; Xiao, M.K. Nonlinear structural response prediction based on support vector machines. *J. Sound Vib.* **2008**, *311*, 886–897.
30. Lagaros, N.D.; Papadrakakis, M. Neural network based prediction schemes of the non-linear seismic response of 3D buildings. *Adv. Eng. Softw.* **2012**, *44*, 92–115. [CrossRef]
31. Amjad, M.; Ahmad, I.; Ahmad, M.; Wroblewski, P.; Kaminski, P.; Amjad, U. Prediction of pile bearing capacity using XGBoost algorithm: Modeling and performance evaluation. *Appl. Sci.* **2022**, *12*, 2126. [CrossRef]
32. Jolfaei, S.; Lakirouhani, A. Sensitivity analysis of effective parameters in borehole failure, using neural network. *Adv. Civ. Eng.* **2022**, *2022*, 4958004. [CrossRef]
33. Moayedia, H.; Hayati, S. Modelling and optimization of ultimate bearing capacity of stripfooting near a slope by soft computing methods. *Appl. Soft Comput.* **2018**, *66*, 208–219. [CrossRef]
34. Tran, Q.A.; Ho, L.S.; Le, H.V.; Prakash, I.; Pham, B.T. Estimation of the undrained shear strength of sensitive clays using optimized inference intelligence system. *Neural Comput. Appl.* **2022**, *34*, 7835–7849. [CrossRef]
35. Wang, Z.Z.; Goh, S.H.; Zhang, W.G. Reliability-based design in spatially variable soils using deep learning: An illustration using shallow foundation. *Georisk* **2022**, *2022*, 1–15. [CrossRef]
36. Van, C.N.; Keawsawasvong, S.; Nguyen, D.K.; Lai, V.Q. Machine learning regression approach for analysis of bearing capacity of conical foundations in heterogenous and anisotropic clays. *Neural Comput. Appl.* **2023**, *35*, 3955–3976.
37. Cassidy, M.J.; Uzielli, M.; Tian, Y.H. Probabilistic combined loading failure envelopes of a strip footing on spatially variable soil. *Comput. Geotech.* **2013**, *49*, 191–205. [CrossRef]
38. Hentati, A.; Selmi, M.; Kormi, T.; Ali, N.B.H. Probabilistic HM failure envelopes of strip foundations on spatially variable soil. *Comput. Geotech.* **2018**, *102*, 66–78. [CrossRef]
39. Li, J.H.; Zhou, Y.; Zhang, L.L.; Tian, Y.; Cassidy, M.J.; Zhang, L.M. Random finite element method for spudcan foundations in spatially variable soils. *Eng. Geol.* **2016**, *205*, 146–155. [CrossRef]
40. Charlton, T.S.; Rouainia, M. Probabilistic capacity analysis of suction caissons in spatially variable clay. *Comput. Geotech.* **2016**, *80*, 226–236. [CrossRef]
41. Fenton, G.A.; Vanmarcke, E.H. Simulation of random fields via local average subdivision. *J. Eng. Mech.* **1990**, *116*, 1733–1749. [CrossRef]
42. Huang, S.P.; Quek, S.T.; Phoon, K.K. Convergence study of the truncated Karhunen-Loeve expansion for simulation of stochastic processes. *Int. J. Numer. Meth. Eng.* **2001**, *52*, 1029–1043. [CrossRef]
43. Tao, Y.Q.; Sun, H.L.; Cai, Y.Q. Bayesian inference of spatially varying parameters in soil constitutive models by using deformation observation data. *Int. J. Numer. Anal. Met.* **2021**, *45*, 1647–1663. [CrossRef]
44. Wu, Y.X.; Li, R.; Gao, Y.F.; Zhang, N.; Zhang, F. Simple and efficient method to simulate homogenous multidimensional non-Gaussian vector fields by the spectral representation method. *J. Eng. Mech.* **2017**, *143*, 06017016. [CrossRef]
45. Lunne, T.; Andersen, K.H.; Low, H.E.; Randolph, M.F.; Sjursen, M. Guidelines for offshore in situ testing and interpretation in deepwater soft clays. *Can. Geotech. J.* **2011**, *48*, 543–556. [CrossRef]
46. Choo, H.; Lee, W.; Hong, S.J.; Lee, C. Application of the dilatometer test for estimating undrained shear strength of Busan New Port clay. *Ocean. Eng.* **2016**, *115*, 39–47. [CrossRef]
47. Shi, X.S.; Zhao, J.D. Practical estimation of compression behavior of clayey/silty sands using equivalent void-ratio concept. *J. Geotech. Geoenviron.* **2020**, *146*, 04020046. [CrossRef]
48. Phoon, K.K.; Kulhawy, F.H. Characterization of geotechnical variability. *Can. Geotech. J.* **1999**, *36*, 612–624. [CrossRef]
49. Zhang, Y.J.; Zhang, H.L.; Wu, Y.X. Dynamic response of rectangular tunnels embedded at various depths in spatially variable soils. *Appl. Sci.* **2022**, *12*, 10719. [CrossRef]

50. Zhang, Y.H.; Bienen, B.; Cassidy, M.J.; Gourvenec, S. The undrained bearing capacity of a spudcan foundation under combined loading in soft clay. *Mar. Struct.* **2011**, *24*, 459–477. [CrossRef]
51. Zhang, W.G.; Wu, C.Z.; Zhong, H.Y.; Li, Y.Q.; Wang, L. Prediction of undrained shear strength using extreme gradient boosting and random forest based on Bayesian optimization. *Geo. Sci. Front.* **2021**, *12*, 469–477. [CrossRef]

applied sciences

MDPI

Article

Evaluating the Influence of Sand Particle Morphology on Shear Strength: A Comparison of Experimental and Machine Learning Approaches

Firas Daghistani [1,2] and Hossam Abuel-Naga [1,*]

1 Department of Civil Engineering, La Trobe University, Bundoora, VIC 3086, Australia; f.daghistani@latrobe.edu.au
2 Department of Civil Engineering, University of Business and Technology, Jeddah 21448, Saudi Arabia
* Correspondence: h.aboel-naga@latrobe.edu.au

Abstract: Particulate materials, such as sandy soil, are everywhere in nature and form the basis for many engineering applications. The aim of this research is to investigate the particle shape, size, and gradation of sandy soil and how they relate to shear strength, which is an essential characteristic that impacts soil stability and mechanical behaviour. This will be achieved by employing a combination of experimental methodology, which includes the use of a microscope direct shear apparatus, and machine learning techniques, namely multiple linear regression and random forest regression. The experimental findings reveal that angular-shaped sand particles enhance the shear strength characteristics compared to spherical, rounded ones. Similarly, coarser sand particles improve these characteristics compared to finer sand particles, as do well-graded particles when compared to poorly graded ones. The machine learning findings show the validity of both models in predicting shear strength when compared to the experimental results, showing high accuracy. The models are designed to predict shear strength of sand considering six input features: mean particle size, uniformity coefficient, curvature coefficient, dry density, normal stress, and particle regularity. The most important features from both models were identified. In addition, an empirical equation for calculating shear strength was developed through multiple linear regression analysis using the six features.

Keywords: particle size; particle shape; sand; shear strength; machine learning; multiple linear regression; random forest regression

Citation: Daghistani, F.; Abuel-Naga, H. Evaluating the Influence of Sand Particle Morphology on Shear Strength: A Comparison of Experimental and Machine Learning Approaches. *Appl. Sci.* **2023**, *13*, 8160. https://doi.org/10.3390/app13148160

Academic Editor: Wei Gao

Received: 7 June 2023
Revised: 10 July 2023
Accepted: 11 July 2023
Published: 13 July 2023

1. Introduction

Natural particulate materials, such as sandy soil, are found everywhere and are essential to many engineering applications. Various fields, from civil engineering to materials science, require an understanding of the mechanical behaviour of particle-to-particle [1,2] and their interactions with different surfaces [3–6]. Understanding these materials is strongly reliant on particle morphology, which has a significant influence on the mechanical response of granular materials such as sand. The term 'particle morphology' is used to refer to particle shape, size, form, sphericity, or surface roughness. With regards to particle size, the soil size in descending order is boulder, cobbles, pebbles, gravel, sand, silt, and clay. The scope of this paper will be limited to sand, which is a granular material composed of individual particles classified into three sizes: coarse, medium, and fine sand, as specified by the Australian standard [7].

While the particle shape has been a topic that has raised many questions in the literature, its implication on the behaviour of soil is a major area of study with constant research progression. The soil particle shape can be graded on three independent properties: form (sphericity: overall shape), roundness, and roughness, each of which has a different influence on the behaviour of the material [8]. With regards to the sphericity, the soil particles can be bulky, flaky, and needle shaped. Sand particles are considered bulky, and their shape is mostly set during formation. Researchers often use terms such as 'well-rounded', 'rounded', 'sub-rounded', 'subangular', 'angular', and 'very angular' to describe the roundness of bulky particles. However, the sand's surface roughness can change significantly with mechanical and chemical weathering of rocks and minerals over geological time [9]. While sphericity and roundness are macro- and medium-scale particle measurements, particle surface texture is a microscale measurement [10]. With regards to granular sand, the particle shape, including sphericity, roundness, and roughness, affect the sand's stiffness, strength, minimum and maximum void ratio (e_{min} and e_{max}), critical state friction angle (φ_c), dilatancy (ψ), dilation, strain localisation, and the evolution of strength anisotropy [11]. Furthermore, particle shape can significantly influence the compressibility of granular structures. Experimental studies have found that particle roundness and sphericity (particle regularity) can affect both packing density and compressibility [2].

The shape of soil particles, including roundness, angularity, and surface roughness, plays a significant role in determining soil mechanical behaviour. Roundness impacts how particles interact, affecting soil mass packing and stiffness [12–14]. Angular particles, due to their enhanced interlocking, exhibit higher friction angles and shear strength [15]. Li [16] found that as sample convexity decreased, friction angle increased. This statement was supported by an experimental and numerical study by Peng et al. [17], whose results showed that angular particles have more shear strength compared to rounded particles. Surface roughness influences soil stiffness [18] and wave propagation parameters [19]. Angularity affects the undrained response of fine sands, with more angular particles offering increased resistance to movement, thereby boosting soil strength [20,21]. The work of researchers like Miura et al. [22] in studying the impact of these properties on soil behaviour contributes to more accurate predictive models, directly informing engineering practices.

Particle size has an important effect on the behaviour of individual particles and the packaging density. Vangla and Latha [23] investigated the effect of particle size on shear strength characteristics. They attempted to eliminate the effect of morphological characteristics by selecting three sands with different particle sizes (coarse, medium, and fine) but similar particle shapes (angularity, roundness, sphericity, and roughness). The samples were prepared at a similar void ratio, and the test was carried out using direct shear. The results showed that particle size has a slight influence on the peak friction angle but not on the mechanism of shearing, with coarse sand particles taking longer to reach the peak compared to fine sand particles. In contrast, an experiment by Wang et al. [24] investigated the effect of sand and gravel size on shear strength using both direct shear and triaxial tests in the laboratory. The results showed that as the mean particle diameter D_{50} increased, the angle of shearing resistance also increased, leading to higher shear strength. Similar results were reported by [16,25,26], who found that peak and residual shear strength increase as particle size increases, whereas in glass beads, interparticle friction between two glass beads increases as sphere size increases [27]. Interestingly, particle size also affects the compressibility of the granular structure, with smaller particles leading to greater compression compared to larger particles [1].

Researchers in the engineering, geotechnics, as well as the medical field have become more interested in artificial intelligence (AI) techniques over the last two decades. A variety of machine learning algorithms have been utilised with significant success, including multiple linear regression (MLR) and random forest regression (RFR), which we have adopted in our research. In a study by Xie et al. [28], the two models, MLR and RF, were compared for estimating soil extracellular enzyme activities in reclaimed coastal saline land. The authors report that the RF model performed better than the MLR model in predicting the activities of soil amylase and urease, which are important indicators of soil carbon and nitrogen cycling. The article also identifies the main factors affecting soil extracellular enzyme activities, such as soil water content, total nitrogen, and pH. Another study by Zhang et al. [29], who also used MLR and RF models, investigated the prediction of soil organic carbon (SOC) in a coastal reclamation zone of eastern China. The authors compared the effects of different factors on SOC dynamics and found that soil pH, chloride, and silt contents were the most important factors influencing SOC. Results from the study indicated that the RF model also performed better than MLR due to its superiority in handling non-linear relationships between SOC and the predictors. The RF model showed substantially reduced error indices (ME, MSE, and RMSE), as well as a higher R^2. Another interesting technique is the Adaptive Neuro-Fuzzy Inference System (ANFIS) introduced by Jang in 1993. ANFIS integrates the elements of neural networks and fuzzy logic, demonstrating capabilities of learning and generalisation [30]. The system has found diverse applications across various domains. It has been used for predicting skin permeability in drug-delivery scenarios [31], controlling quality and predicting characteristics in food-processing technology [32], determining heavy metal concentrations in water resources [33], and even predicting the security index of ad hoc vehicular networks [34]. Moreover, it has shown efficacy in predicting the higher heating value of biomass [35] and modelling thermal error [36]. In addition, a recent article [37] presents a method to control the cooling of machine tool spindles using ANFIS. The method adjusts the coolant pump frequency based on the spindle speed and thermal state, achieving high accuracy and efficiency in reducing thermal deformation and energy consumption. While MLR and RFR provide robust and interpretable models, the potential of ANFIS, given its successful implementation in various studies, indicates it as an intriguing future direction for predictive modelling research, including predicting the shear strength of cohesionless soil.

The analysis of shear strength of cohesionless soil such as sand can be influenced by granular shape, size, and gradation. However, no comprehensive model taking these parameters into account can be found in the literature. This is because there are many variables that affect it in non-linear ways. In the geotechnical field, machine learning has been used successfully for problems such as slope stability [38], soil mechanics [39,40], soil cracking [41], and soil improvement with recycled materials [42–46]. However, the application of AI methodologies for predicting the shear strength of cohesionless soil, considering the combined influence of particle shape, size, and gradation, has not been sufficiently investigated, indicating a large gap in past research. This research aims to fill this gap by conducting and analysing a series of direct shear tests across different granular sizes and shapes. This is followed by the application of both MLR and RFR, which are based on six input features: mean particle size (D_{50}), coefficient of uniformity (C_u), coefficient of curvature (C_c), dry density (ρ_d), normal stress (σ_n), and particle regularity (ρ_r), the last of which is the average of roundness and sphericity. The research then presents an empirical equation for predicting the shear strength of sand, considering the six input features. Finally, after careful examination of the results derived from the models, the study presents the most effective model and investigates the significance of the inputs involved in each model. This study provides a strong base for a deep investigation into a new area that was not explored before.

2. Materials and Methods

2.1. Material

According to the Australian standard [7], sand sizes range from 2.36 mm to 0.075 mm, with coarse sand ranging from 2.36 to 0.6 mm, medium sand ranging from 0.6 to 0.212 mm, and fine sand ranging from 0.212 to 0.075 mm. Particles larger than 2.36 mm are classified as gravel, while particles smaller than 0.075 mm are classified as silt or clay. Different types of sand were used in the experiments to examine the effect of particle size and shape. The sands used in the study are referred to as L-Sand, M-Sand, P-Sand, and B-Sand.

For the particle shape impact, four types of sand were used, namely L-Sand, B-Sand, M-Sand, and P-Sand, were each sieved and separated into four different sizes (1.18 to 0.6 mm, 0.6 to 0.425 mm, 0.425 to 0.3 mm, and 0.3 to 0.15 mm). Due to the limitation of the microscope lens, which tends to overlook particles larger than 1.18 mm, only particles below this size were selected.

To study the impact of particle size, B-sand was sieved and divided into containers based on their size (Figure 1), and five different sands were selected for testing. According to Australian standards [7], four of the selected sands are poorly graded and are considered to be fine sand (B1-Sand) with D_{50} of 0.11 mm, low medium sand (B2-Sand) with D_{50} of 0.23 mm, high medium sand (B4-Sand) with D_{50} of 0.51 mm, and coarse sand (B6-Sand) with D_{50} of 1.77 mm. The fifth one is a mixture of sand to create a well-graded sand (B-Sand) with D_{50} of 0.58 mm. Therefore, five different sizes were chosen to examine the differences between coarse, medium, and fine sands, as well as to study the effect of poorly graded and well-graded sands.

Figure 1. The coarse soil (B-sand) is sieved and separated into different containers depending on the granular size.

Glass beads were utilised to avoid particle shape influence and concentrate only on particle size impact on mechanical behaviour. The glass beads are made of silica mixed with other minerals melted at high temperature to produce a viscous, thick liquid. The liquid is moulded into spherical shapes and hardens as it cools. The regularity of the particle shape of the glass beads, as observed under the microscope, was found to be almost one. The glass beads were separated into two different sizes: GB5 with a D_{50} of 0.89 mm and GB6 with a D_{50} of 1.77 mm. The specific gravity of the glass beads ranges from 2.45 to 2.50. The specifications of used particulate materials including sand and glass beads are presented in Table 1. The sieve analysis was conducted according to the Australian standard [47], and the results for the used granular material are shown in Figure 2.

Table 1. Specifications of the used particulate materials: sand and glass beads.

Material	Range (mm)	Grade	C_u	C_c	D_{50} (mm)	G_s	R	S	ρ_r
L5-Sand	1.18 to 0.6	PG [1]	1.44	0.96	0.89	2.65	0.288	0.589	0.439
L4-Sand	0.6 to 0.425	PG	1.20	0.97	0.51	2.68	0.421	0.546	0.484
L3-Sand	0.425 to 0.3	PG	1.20	0.97	0.36	2.69	0.302	0.591	0.447
L2-Sand	0.3 to 0.15	PG	1.45	0.96	0.23	2.74	0.288	0.578	0.433
L1-Sand	0.15 to 0.075	PG	1.45	0.96	0.11	-	0.289	0.541	0.415
B-Sand	2.36 to 0.075	WG [2]	6.16	1.24	0.58	2.67	-	-	-

Table 1. *Cont.*

Material	Range (mm)	Grade	C_u	C_c	D_{50} (mm)	G_s	R	S	ρ_r
B6-Sand	2.36 to 1.18	PG	1.45	0.96	1.77	2.66	-	-	-
B5-Sand	1.18 to 0.6	PG	1.44	0.96	0.89	2.67	0.263	0.557	0.410
B4-Sand	0.6 to 0.425	PG	1.20	0.97	0.51	2.66	0.246	0.538	0.392
B3-Sand	0.425 to 0.3	PG	1.20	0.97	0.36	2.69	0.279	0.584	0.432
B2-Sand	0.3 to 0.15	PG	1.45	0.96	0.23	2.69	0.270	0.583	0.427
B1-Sand	0.15 to 0.075	PG	1.45	0.96	0.11	2.70	0.328	0.580	0.454
M5-Sand	1.18 to 0.6	PG	1.44	0.96	0.89	2.68	0.189	0.551	0.370
M4-Sand	0.6 to 0.425	PG	1.20	0.97	0.51	2.69	0.206	0.565	0.386
M3-Sand	0.425 to 0.3	PG	1.20	0.97	0.36	2.67	0.327	0.597	0.462
M2-Sand	0.3 to 0.15	PG	1.45	0.96	0.23	2.72	0.299	0.542	0.421
M1-Sand	0.15 to 0.075	PG	1.45	0.96	0.11	-	0.389	0.499	0.444
P5-Sand	1.18 to 0.6	PG	1.44	0.96	0.89	2.66	0.246	0.559	0.403
P4-Sand	0.6 to 0.425	PG	1.20	0.97	0.51	2.68	0.203	0.575	0.389
P3-Sand	0.425 to 0.3	PG	1.20	0.97	0.36	2.69	0.233	0.523	0.378
P2-Sand	0.3 to 0.15	PG	1.45	0.96	0.23	2.67	0.286	0.565	0.426
P1-Sand	0.15 to 0.075	PG	1.45	0.96	0.11	-	0.330	0.508	0.419
GB6 *	2.36 to 1.18	PG	1.45	0.96	1.77	2.45	1	1	1
GB5	1.18 to 0.6	PG	1.44	0.96	0.89	2.45	1	1	1

* Where GB is glass beads, [1] PG is poorly graded, and [2] WG is well graded sand.

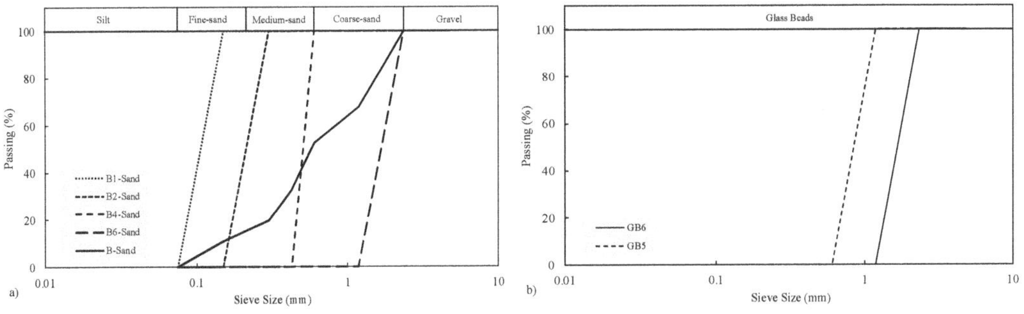

Figure 2. Sieve analysis of the used particulate materials, with (**a**) displaying the sieve analysis for sand, and (**b**) showing the sieve analysis for glass beads.

2.2. Experimental

A total of 1068 tests, including microscopy, direct shear, oedometer, and specific gravity tests, were conducted. Out of these experiments, 1000 involved photographing various types of sand, which include L-Sand, M-Sand, P-Sand, and B-Sand. Each of these sands was sieved and separated into different containers based on their sizes. Subsequently, microscope analysis was performed on uniformly sized specimens. We considered 50 particles in each specimen in order to determine particle regularity. Additionally, 46 direct shear tests were conducted, considering different particle sizes, shapes, and densities. Further, six tests were carried out to measure compressibility across varying particle sizes and densities using an oedometer apparatus. Lastly, 16 tests were conducted to determine the specific gravity of different types of sand of various particle sizes. This was done to investigate the impact of the mean particle size on specific gravity.

2.2.1. Direct Shear Apparatus

A Mateset direct shear apparatus was used to conduct the experiments, which were carried out according to the Australian standard [48]. The dimensions of the mould in the direct shear box were 60 × 60 mm. In each test, different amounts of normal stresses, 25, 50, 100, and 200 kPa, were applied to the sample. Each test was conducted on a dry sample at a shear rate of 1 mm/min, which is the maximum allowable speed according to the standard.

The pluviation technique, sometimes called the rainfall method, is employed in the preparation of granular soil samples, specifically sands, with differing relative densities. Figure 3 provides a schematic diagram of the pluviation technique. By adjusting the height from which the sand particles are dropped, the method allows for the creation of samples with the required density.

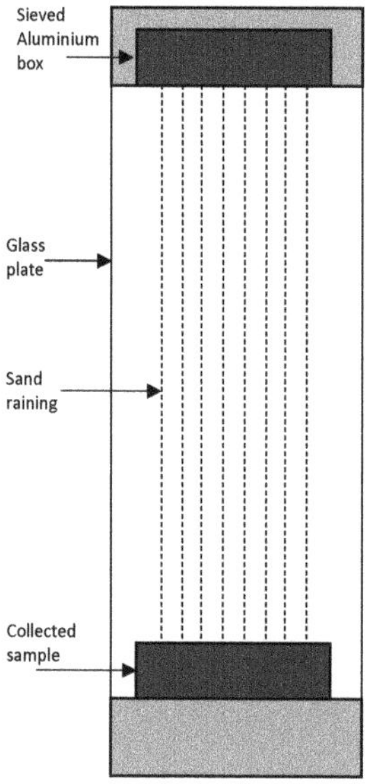

Figure 3. Schematic view of the pluviation technique.

The relationship between drop height versus void ratio and relative density is shown in Figure 4, with the void ratio decreasing as drop height increases. A loose sample is achieved by dropping the soil from a low distance between the cone and mould, reducing the particles' kinetic energy and enabling them to loosely pack. Conversely, a dense sample is formed by dropping the particles from a high distance, increasing their kinetic energy, and causing them to efficiently rearrange and pack densely. Upon dropping the particles from the selected height, the mould is removed, and the sample can be used for shearing testing.

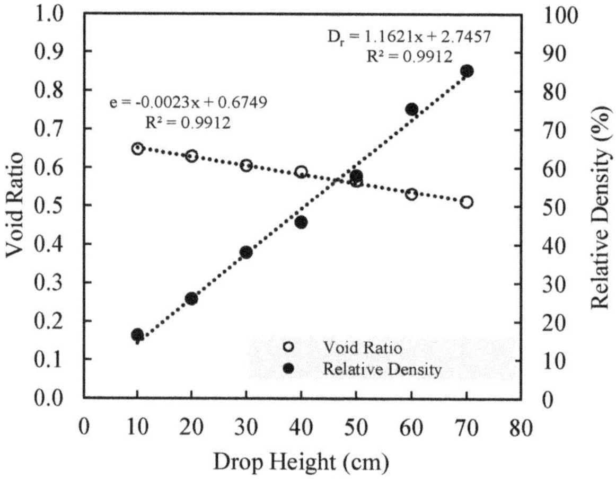

Figure 4. Drop height versus void ratio and relative density of L-Sand.

In the study, two different densities were considered when preparing the sample: the loose and dense density states. In the loose density state, the sand was spooned and dropped from a very low height (zero height). Conversely, in the dense state, the sample was dropped from a cone with a 5.2 mm opening at a height of 83 cm.

2.2.2. Microscope

In this experiment, we used a Nikon Eclipse MA100 microscope, a valuable tool in geotechnical laboratories for identifying particle shapes. The microscope comes with a built-in Progression system, offers high-quality optics that enable accurate and efficient identification of soil particle shapes and sizes. Several parameters are used to characterize sand particle shape and quality, including sphericity and roundness (Figure 5).

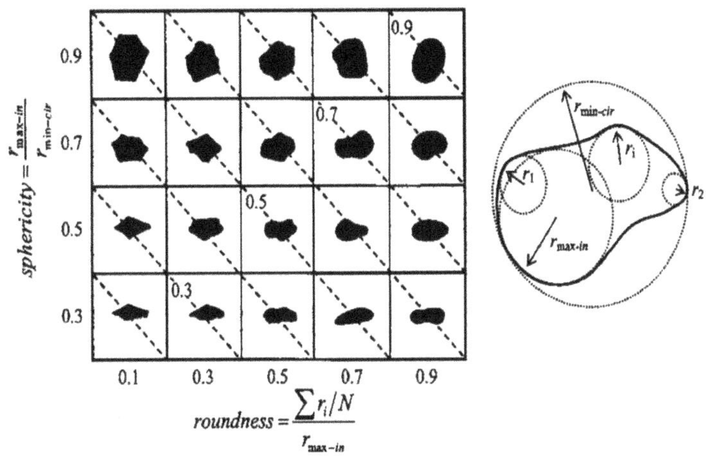

Figure 5. Determining particle shape through sphericity and roundness, with diagonal dotted lines indicating consistent particle regularity $\rho_r = \frac{(R+S)}{2}$ [2,49].

The sphericity of a particle is a measure of how closely it looks like a circle, while roundness determines how curved its corners are. A ceramic proppant and high-quality frac sand are typically both spherical and round, scoring around 0.9 in both metrics. The same high score is observed in silica sand samples with nearly circular particles, where the sphericity measure can reach 0.7 or higher. Nonetheless, sand particles featuring angular edges are expected to have reduced roundness measurements, often falling in the region of 0.2 to 0.5. The schematic representation in Figure 6 shows the method of finding particle shape parameters including roundness, sphericity, and regularity.

2.3. Mathematical Model

The mathematical model was implemented in the Python programming language. The research objectives entailed testing two models: a simple model via multiple linear regression (MLR), and a complex model through random forest regression (RFR). In addition, MLR was specifically applied to model linearity, while RFR was used to navigate nonlinearity. In both implemented models (MLR and RFR) the following libraries were utilised: pandas for data manipulation and analysis, NumPy for numerical computations, scikit-learn for machine learning tasks including data splitting, normalisation, regression modelling, and metric evaluation, and finally matplotlib for data visualisation. The workflow diagram below (Figure 7) outlines the different processes performed for the machine learning algorithm implementation. Further details of these processes are discussed in the following subsections.

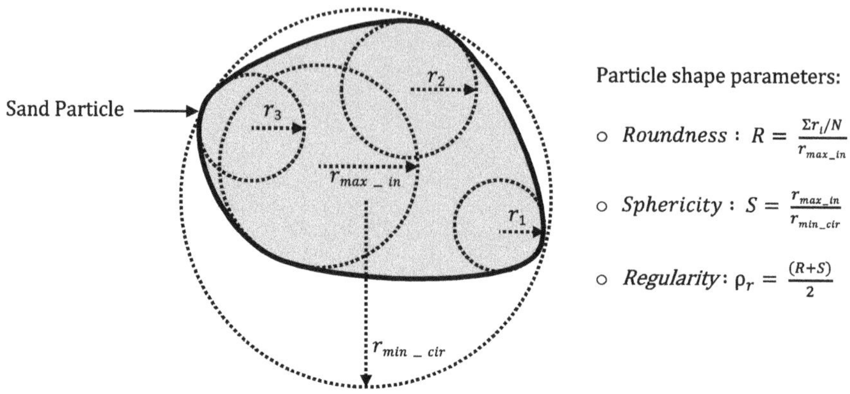

Figure 6. Schematic illustration of determining the particle shape parameters: roundness, sphericity, and regularity.

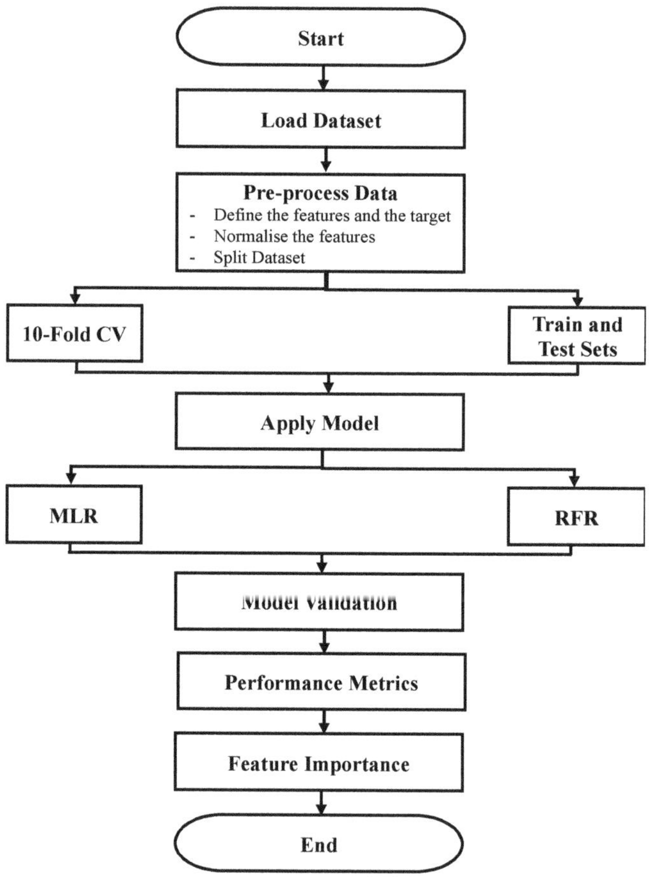

Figure 7. Workflow of the applied machine learning algorithm.

2.3.1. Pre-Process Data

The pre-processing of data involved two steps: normalisation and splitting the dataset. Normalisation in machine learning is a vital process that standardizes numerical data in your dataset, similar to converting measurements from feet, inches, and yards all into metres, so everything is on the same scale. This process ensures that the machine learning models treat all features fairly and do not overvalue one feature while undervaluing another [50]. When all of the features are on the same scale, the model can learn and make predictions more effectively and efficiently. Techniques like min-max normalisation, Z-score normalisation, and robust scaling are commonly used. In other words, normalisation makes the data neat and uniform, helping machine learning models perform at their best.

In the results analysis, the min-max Normalisation method was followed, which rescales the data to a range between 0 and 1. The formula for this is as follows:

$$X' = \frac{(X - X_{min})}{(X_{max} - X_{min})} \quad (1)$$

where X is the original value, and X_{min} and X_{max} are the smallest and largest values in the data.

Following normalisation is the splitting of the dataset. Data splitting is a popular method for model validation in which we divide a given dataset into two distinct sets: training and testing. Following that, the statistical and machine learning models are fitted to the training set and validated using the testing set. By separating a portion of the data for validation purposes, independent of the training process, we can effectively assess and compare the predictive performance of various models. The most used ratio of data splitting is 80:20, where 80% of the data is used for training and 20% for testing. This conventional method relies on a single random split of the data. The 80:20 split draws its justification from the well-known Pareto principle, which states that roughly 80% of the effects come from 20% of the causes or inputs [51]. The train–test split, although commonly used, has been found to have potential biases and limitations in assessing model performance. To overcome these challenges, we implemented the k-fold cross-validation method.

The k-fold cross-validation is a popular statistical method that provides a more comprehensive, robust, and reliable approach to assess the model's performance and reduce computation time without any bias resulting from random dataset splitting [52,53]. This technique enables a more rigorous evaluation of the model's effectiveness compared to the train_test_split approach.

In our own dataset, we incorporated both the train_test_split and the k-fold cross-validation (10 folds) methods. For the k-fold cross-validation, the dataset was divided into 10 sections, with nine sections used for training the model and the remaining section for testing. In each fold, a different section was designated for training, while the remaining sections were used for testing. This process was repeated across all folds until each section was used for both training and testing. The final result obtained from our 10-fold cross-validation was an average of the performance across all folds.

2.3.2. Statistical Parameters

Several metrics, each with its own strengths and limitations, can be used to compare the performance of various AI models. The following are some common metrics:

- Mean absolute error (MAE) is a measure that captures the average absolute disparity between predicted and true values. By focusing solely on the magnitude of the error, irrespective of its direction, it provides an evaluation of the model's effectiveness in accurately forecasting the actual values.
- Root mean square error (RMSE) is a performance metric like MAE, but it considers the square of the errors, thus placing more penalty on larger discrepancies. RMSE is typically employed when substantial errors pose a greater problem than minor ones.
- Root mean square log error (RMSLE) is a useful metric when dealing with a target variable that spans a broad range of values. It employs the logarithms of both predicted and actual values, which lessens the effect of substantial discrepancies between these values. When the distribution of the target variable is skewed, employing this metric can be particularly beneficial.

- R-squared (R^2) is a statistical measure that measures the degree to which the model matches the data, relative to a simple, baseline model. R^2 values can range from 0 to 1, with higher values signifying a better fit. However, it is important to note that R^2 can provide a skewed perspective if the underlying baseline model is unfitting, or the data are contaminated with outliers. Equations (2)–(5) show these metrics:

$$MAE = \frac{\sum_N (X_m - X_p)}{N} \tag{2}$$

$$RMSE = \sqrt{\frac{\sum_N (X_m - X_p)^2}{N}} \tag{3}$$

$$RMSLE = \sqrt{\frac{\sum_N (log(X_m + 1) - log(X_p + 1))^2}{N}} \tag{4}$$

$$R^2 = \left[\frac{\sum_{i=1}^{N} (X_m - \overline{X_m})(X_p - \overline{X_p})}{\sum_{i=1}^{N} (X_m - \overline{X_m})^2 \sum_{i=1}^{n} (X_p - \overline{X_p})^2} \right]^2 \tag{5}$$

where N is the number of datasets, X_m and X_p are actual and predicted values, and $\overline{X_m}$, $\overline{X_p}$ are the average of actual and predicted values, respectively. The model should ideally have an R^2 value of 1 and MAE, $RMSE$, and $RMSLE$ values of 0.

2.3.3. Multiple Linear Regression

In the realm of statistical modelling, multiple linear regression (MLR) is a powerful method that is used to understand the relationship between multiple predictors and a single response variable. This method, which extends the principles of simple linear regression, allows us to uncover complex dependencies and valuable insights hidden within the data. MLR aims to establish a linear relationship between the predictors and the response variable, capturing their combined effect on the result. This method becomes useful in real-life situations where there are multiple factors that simultaneously influence the target variable. Multiple linear regression makes several assumptions to ensure the validity of the regression model. These assumptions include linearity, independence, homoscedasticity (constant variance), and normality of residuals. Any deviations from these assumptions can affect the accuracy and reliability of the regression model and may require additional measures to address them. The MLR code utilises the scikit-learn library with the default hyperparameter values. Furthermore, the numerical hyperparameters that were set for pre-processing data, feature importance estimation, and the visualisation process are displayed on Table 2.

Table 2. Numerical hyperparameters for the multiple linear regression code, including parameters for both with and without the application of 10-fold CV.

Phase	Parameter	Value
Train and Test Sets	test_size	0.2
	random_state	0
K-Fold Cross-Validation	n_splits	10
	random_state	0
	shuffle	True
Feature Importance Estimation	n_repeats	10
Visualisation	start_point	0
	boundary_shift	20%

2.3.4. Random Forest Regression

Random forest regression (RFR) has several advantages that make it a popular choice for regression tasks, including its robustness in dealing with many input features, both numerical and categorical variables, and its ability to deal with outliers and missing values in the data, reducing the need for extensive data preprocessing. Furthermore, RFR is capable of capturing complex non-linear correlations between input data and the target variable, making it appropriate for applications where linear models are insufficient. Random forest regression also provides useful insights on feature importance, which aids in finding the underlying relationships in the data. Because of its versatility, it can be used in a variety of regression tasks and can effectively handle large datasets, making it a useful technique for a wide range of applications. The RFR code utilises the scikit-learn library with the default hyperparameter values. Furthermore, the numerical hyperparameters that were set for pre-processing data, model, and the visualisation process are displayed in Table 3.

Table 3. Numerical hyperparameters for the random forest regressor code, including parameters for both with and without the application of 10-fold CV.

Phase	Parameter	Value
Train and Test Sets	test_size	0.2
	random_state	0
K-Fold Cross-Validation	n_splits	10
	random_state	0
	shuffle	True
Model	n_estimators	100
	random_state	0
Visualisation	start_point	0
	boundary_shift	20%

3. Results

3.1. Experimental Results

3.1.1. Packing Density

The structure of a sand sample (skeleton) plays a crucial role in determining the mechanical behaviour, which can be controlled by density and anisotropy. The packing density of sand can depend on multiple factors, such as the particle shape, size, and gradation along with the arrangement of particles. A sample consisting of particles with high regularity has a higher density and low void ratio compared to a sample with low regularity particles [2].

Sand gradation can be poorly graded, well graded, or gap graded. A poorly graded sand represents sand that has similar grain sizes; in contrast, the well-graded sand has a percentage of each size when the C_u is greater than 6 and when the C_c lies between 1 and 3. The gap-graded sand represents sand that has two different mixed sizes, in other words, two different poorly graded sands mixed together [9]. A well-graded sample will have a high density and a lower void ratio compared to a poorly graded sample.

According to Burmister [54], when the particle size range is coarser, the density increases and the void ratios decrease. In the poorly graded sand used, it was shown that as the mean particle size increases, the density also increases, and the coefficient of volume compressibility decreases, as demonstrated in Figure 8. These findings are consistent with the works of Burmister [54] and Lafata [1].

In terms of shape, there is a strong correlation between particle shape and packing density. A complete sphere shape has the densest possible structure compared to other shapes [1]. Spherical shapes require less compressive force to achieve a dense state because they are easier to reorient compared to less spherical shapes [1]. However, further studies [55] have shown that for particles of similar sizes, the optimal shape for achieving maximum packing fraction is not necessarily a perfect sphere. A comparison between a marble-ball model and M&M candies (which have an elongated and flattened shape) showed that the M&M candy shape has a higher packing fraction of C = 71% compared to a sphere shape with C = 64%. When examining the relationship between particle shape and void ratio, Cho, Dodds and Santamarina [2] found that as particle roundness, sphericity, and regularity approach one (indicating complete rounded and spherical shape), the difference between the maximum and minimum void ratio decreases. Similar results were found by Maroof et al. [56], where the void ratio decreased as regularity increased.

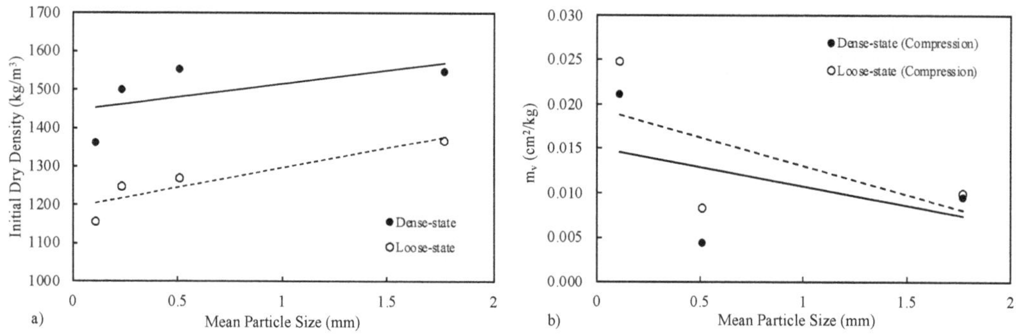

Figure 8. The relationship between the mean particle size and (**a**) initial dry density and (**b**) coefficient of volume compressibility.

The mineral composition of a soil is one of its essential characteristics. Mineralogy influences properties such as specific gravity, Young's modulus, shear modulus, and the Poisson ratio [12,57]. The dry unit weight and specific gravity of sand are important, as they can influence the sample's void ratio, as shown in the following equation:

$$e = \frac{G_S\,\gamma_w}{\gamma_d} - 1 \quad (6)$$

where G_s represents the specific gravity, γ_w indicates the unit weight of water, and γ_d represents the dry unit weight of the sample. According to the equation, when the specific gravity increases, the void ratio also increases. Similarly, when the dry unit weight decreases, the void ratio increases. The specific gravity of sand typically ranges from 2.65–2.67, while that of inorganic clay ranges from 2.70–2.80 [58]. Based on the lab experiment, it was observed that among the four types of sand, as the D_{50} (mean particle size) of the sand increases, the dry unit weight of the sample also increases, while the specific gravity decreases, even within the range of sand particles (2.36 mm to 0.075 mm according to the Australian standard [7]). Consequently, the void ratio decreases as the mean particle size increases, as shown in Figure 9.

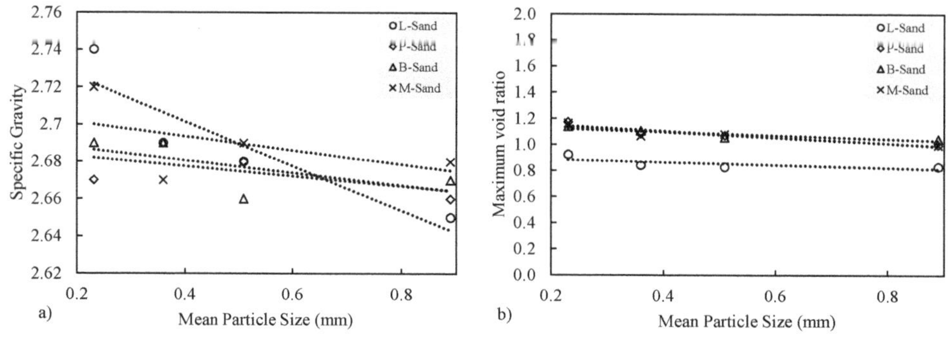

Figure 9. The mean particle sizes of different sands in relation to (**a**) specific gravity and (**b**) maximum void ratio.

3.1.2. Shear Strength

The shear strength of the sand can depend on multiple factors related to the specimen, such as the particle shape, size, and gradation of the sand particles. Upon comparing the poorly graded fine, medium, and coarse sand, it was found that the coarse sand exhibited a higher shear strength compared to the others, as shown in Figure 10.

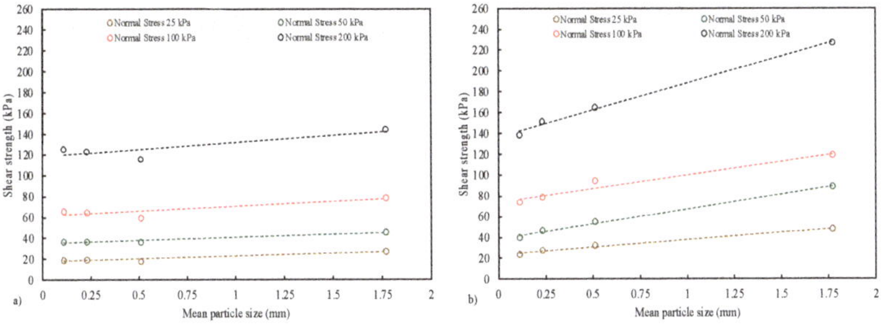

Figure 10. Shear strength versus mean particle size for B-Sand in (**a**) a loose density state and (**b**) a dense density state.

Furthermore, by examining the impact of gradation, the well-graded sand exhibits a higher density and shear strength when compared to fine and medium poorly graded sand, as shown in Figure 11. The well-graded sand had a higher shear strength value, though not as high as the coarse, poorly graded sand. This can be related to the particle shape, size, and surface roughness. Coarse sand particles, particularly those that are angular, can achieve higher shear strength due to particle interlocking. Furthermore, particles with high surface roughness can induce even greater shear strength due to the interlocking of asperities between the particles.

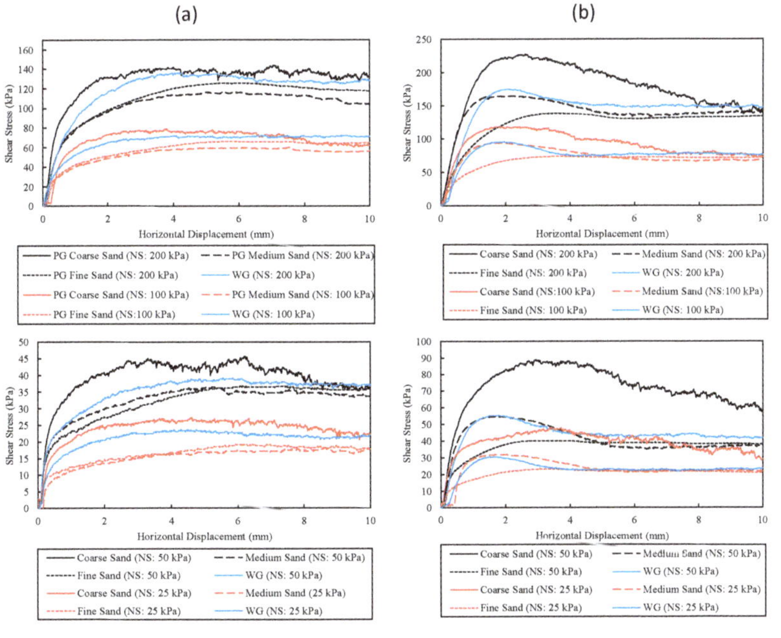

Figure 11. The particle size and gradation impact on the shear strength at different normal stresses (25, 50, 100, and 200) at different densities: (**a**) loose state, and (**b**) dense state.

3.2. *Machine Learning Models*

3.2.1. Multiple Linear Regression

After conducting multiple linear regression (MLR) analyses, the most suitable regression model was identified. The comparison between the predicted values generated by the MLR model for training, testing, and 10-fold CV data and the actual experimental values of shear strength in the direct shear tests is presented in Figure 12. Based on the findings, it can be concluded that the MLR model has a high level of accuracy.

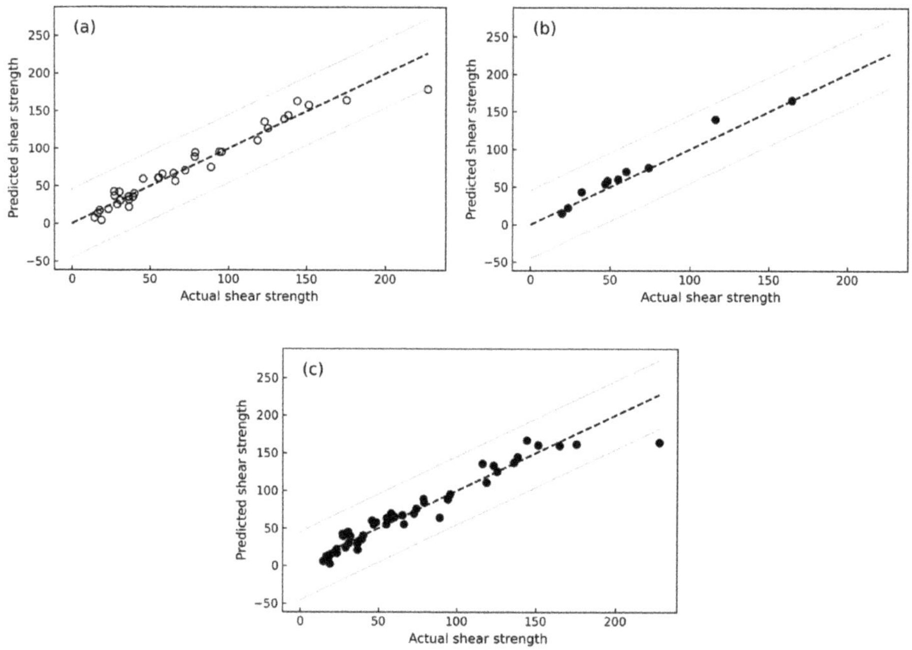

Figure 12. Multiple linear regression was performed to compare actual shear strength with predicted shear strength using (**a**) the training database, (**b**) the testing database, and (**c**) 10-fold cross-validation.

The usage of an MLR model for the objective of predicting the shear strength of sand showed high accuracy, as shown by the varied metrics gathered from the training, testing, and 10-fold CV data (Table 4).

Table 4. The performance of MLR model to predict shear strength.

	Training Database	Testing Database	10-Fold CV
Observations	36	10	46
MAE	8.31	7.67	9.28
RMSE	11.87	10.08	13.57
RMSLE	0.29	0.17	0.35
R^2	0.95	0.94	0.93

The training database included 36 observations, with an MAE of 8.31, RMSE of 11.87, RMSLE of 0.29, and R^2 value of 0.95, indicating a high level of prediction accuracy. The model was then tested on a separate dataset consisting of 10 observations, where it demonstrated a slightly improved MAE of 7.67 and a reduced RMSE of 10.08, and an impressive decrease in RMSLE to 0.17, maintaining a high R^2 value of 0.94. Furthermore, a 10-fold cross-validation (CV) was performed on all 46 observations, yielding an MAE of 9.28, RMSE of 13.57, and RMSLE of 0.35, along with an R^2 of 0.93. The model performance across the training, testing, and cross-validation demonstrates its robust predictive capability for shear strength. This outcome promotes confidence in the predictive capability of the model and its applicability to new data. Thus, an empirical equation was generated to predict the shear strength of sand with high level of accuracy. The empirical equation is as follows:

$$\tau = 15.57 + (7.28 \times D_{50}) + (6.75 \times C_u) - (24.53 \times C_c) + \left(53.90 \times \rho_{dry}\right) + (121.64 \times \sigma_n) - (36.45 \times \rho_r)$$

where D_{50} is the mean particle size, C_u is the coefficient of uniformity, C_c is the coefficient of curvature, ρ_{dry} represents the dry density, σ_n is the normal stress, and ρ_r refers to the sand particle shape regularity.

3.2.2. Random Forest Regression

The comparison between the actual values of shear strength in direct shear testing and the predicted values for training, testing, and 10-fold CV data produced by the RFR model is shown in Figure 13. Based on the results, it can be concluded that the RFR model is highly accurate.

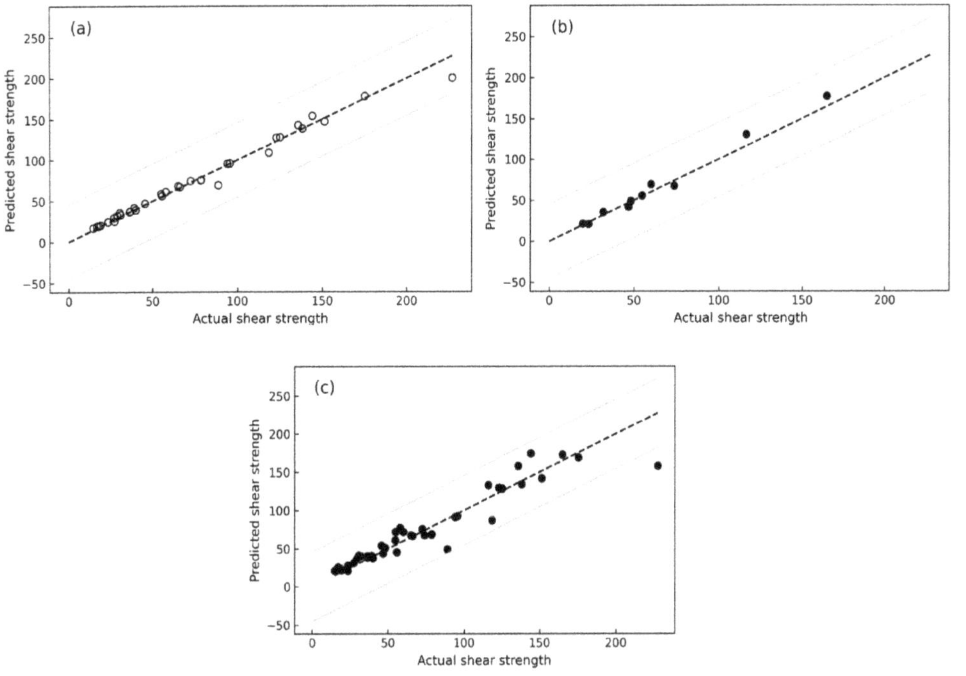

Figure 13. Random forest regression was performed to compare actual shear strength with predicted shear strength using (**a**) the training database, (**b**) the testing database, and (**c**) 10-fold cross-validation.

The Python-based RFR model has demonstrated remarkable accuracy in predicting the shear strength of sand, as evidenced by the metrics calculated for the training, testing, and 10-fold CV data (Table 5).

The training database used contained 36 observations, with an MAE of 3.79, RMSE of 6.55, RMSLE of 0.07, and an impressive R^2 value of 0.98, signifying an excellent fit of the model. In the testing phase, using a distinct database of 10 observations, the model demonstrated slightly higher MAE and RMSE values of 5.68 and 7.37, respectively. The RMSLE also slightly increased to 0.09, yet the R^2 value remained high at 0.97, indicating strong prediction performance. A 10-fold CV performed on the complete dataset of

46 observations resulted in an MAE of 9.83, RMSE of 15.8, RMSLE of 0.19, and an R^2 value of 0.90. Despite the slight increase in error values during cross-validation, the RFR model demonstrated robust and reliable performance in predicting shear strength. These metrics serve as evidence of the model's outstanding predictive performance and its ability to deliver consistent results on new data.

Table 5. The performance of RFR to predict shear strength.

	Training Database	Testing Database	10-Fold CV
Observations	36	10	46
MAE	3.79	5.68	9.83
RMSE	6.55	7.37	15.8
RMSLE	0.07	0.09	0.19
R^2	0.98	0.97	0.90

4. Discussion

4.1. Particulate Shape and Size

Particle morphology can be identified at larger scales, such as that of the particle itself, as spherical, rounded, blocky, bulky, platy, elliptical, elongated, and so on. On a smaller scale, texture is essential because it reflects local roughness properties such as surface smoothness, roundness of edges and corners, and asperities. As shown in Figure 14, there is no direct correlation between particle size and particle shape.

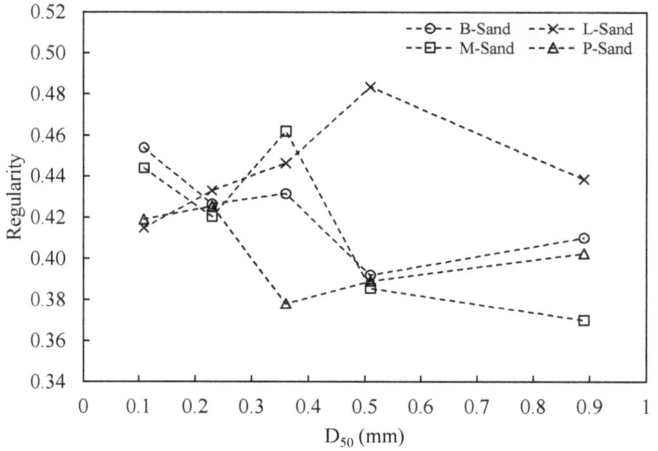

Figure 14. Mean particle size versus the regularity.

Polydispersity, a key concept in materials science and chemistry, refers to the distribution of particles with varying sizes or masses within a sample. Unlike monodisperse systems, where all particles are of the same size, polydisperse systems are characterized by non-uniform particles. It significantly influences the physical properties and behaviour of materials like soil samples, polymers, and colloids [59]. A shear test was conducted on glass beads of two different sizes. Each sample had a monodisperse size. Despite both samples having the same shape regularity, valued at 1, it was observed that the larger beads, with a D_{50} value of 1.77, exhibited higher shear strength at normal stresses of 25, 50, and 100 kPa, as shown in Figure 15. This was in comparison to the finer beads, which had a D_{50} value of 0.89. Therefore, we can conclude that larger particles can induce higher shear strength compared to finer particles.

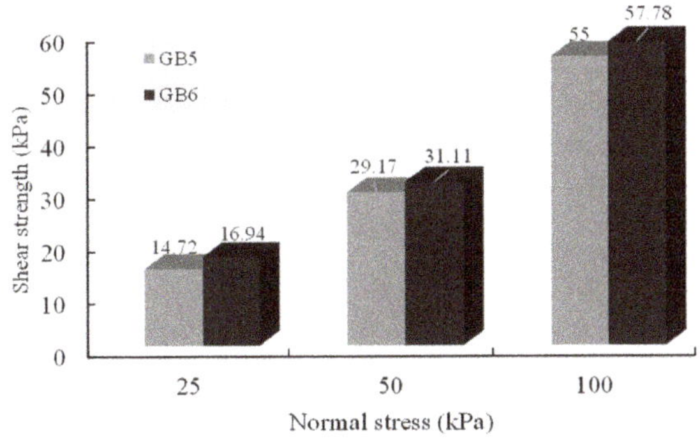

Figure 15. Shear strength versus normal stress for two different mean particle sizes of glass beads.

4.2. Active Lateral Earth Pressure

The effects of particle size, density, and confining pressure on the active lateral earth pressure within a uniform type of sand, sorted into different mean particle sizes, are explored in Figure 16. The active lateral earth pressure increases as the mean particulate size increases. This result is because larger particles will have a higher dry density than smaller particles, which will increase the lateral earth pressure. Also, the active lateral earth pressure increases equally with increasing sample density; this increase is most likely due to an increase in particle content, as the number of particles in compact samples is greater than in loose samples. Therefore, the active lateral earth pressure is greater for denser samples. A similar correlation exists between an increase in normal stress and an increase in active lateral earth pressure. This phenomenon is related to the increased force applied perpendicular to the soil particles, which increase the active lateral earth pressure. In conclusion, the study highlights the importance of mean particle size, density, and normal stress on the active lateral earth pressure, where the active lateral earth pressure increases as the particle size, density, and confining pressure increase.

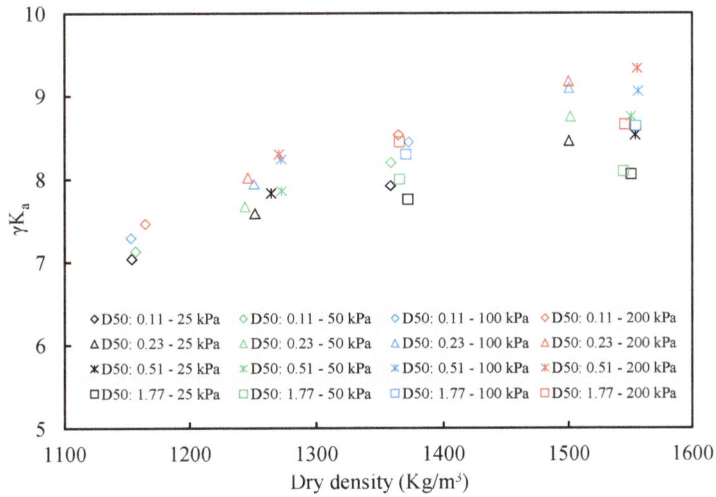

Figure 16. Comparison of active lateral earth pressure and dry density for different particle sizes of various sands under different normal stresses.

4.3. Method Comparison

Multiple linear regression (MLR) is a widely used method in supervised learning, especially for understanding and predicting linear relationships between variables. Owing to its optimal modelling strategy within the linear causal category, MLR often outperforms other standard models [60]. The motivation for using MLR stems from its established reputation as a simple, traditional model. Its strength lies in its capacity to capture linear relationships effectively between multiple predictors (independent variables) and a single response (dependent variable), making it particularly appealing for data analysis [29]. Additionally, MLR can provide an empirical equation for calculating shear strength using multiple inputs, offering utility in geotechnical engineering applications and practices.

Despite these strengths, MLR has its limitations, including its inability to handle nonlinear correlations or complex interactions between input data and the target variable. In response to these limitations, random forest regression (RFR) was employed. This robust and versatile technique navigates these challenges, offering high prediction performance [61]. The RFR model is an ensemble of regression trees, building a large number of these trees before combining them for a final prediction [62]. Moreover, RFR provides insights into feature importance, thereby helping to unravel underlying relationships in the data.

A comparison of the results and performance of both models can offer valuable insights into their respective strengths and weaknesses. This allows for an evaluation of how well each model captures the underlying patterns in the dataset and a determination of which model yields better results.

Table 6 presents a comparative performance analysis of MLR and RFR models applied to training, testing, and 10-fold cross-validation datasets. When comparing MAE values, the RFR model shows superior performance, particularly with the training data, where it achieved an MAE of 3.79, compared to 8.31 with MLR. This trend of enhanced performance continues in the testing data, but not in the cross-validation, where MLR produced a slightly better MAE result. For RMSE and RMSLE, RFR consistently outperforms MLR across all datasets, but not in the cross-validation, where MLR outperformed RFR in RMSE. In terms of R^2 values, which indicate the goodness of fit, RFR shows a slight edge in the training and testing data, but MLR secures a slightly higher value in the 10-fold cross-validation data. Despite some minor variances, both models demonstrate robust predictive capabilities, although RFR generally exhibits stronger performance, particularly on the training and testing datasets.

Table 6. Comparative performance of multiple linear regression and random forest regression on training, testing, and 10-fold cross-validation datasets.

Performance	MLR			RFR		
Metrics	Training Data	Testing Data	10-Fold CV	Training Data	Testing Data	10 Fold CV
MAE	8.31	7.67	9.28	3.79	5.68	9.83
RMSE	11.87	10.08	13.57	6.55	7.37	15.8
RMSLE	0.29	0.17	0.35	0.07	0.09	0.19
R^2	0.95	0.94	0.93	0.98	0.97	0.90

4.4. Importance of Features

The application of machine learning algorithms (MLR and RFR) in our study involves six input features (D_{50}, C_u, C_c, ρ_d, σ_n, ρ_r) and one output which is the shear strength of sand. In Figure 17, both MLR and RFR models, when using the train test splits, identified normal stress as the principal factor, underlining its essential role in governing shear strength, while dry density followed as the second most influential parameter, highlighting its significance in determining the mass per unit volume. However, when using the 10-fold cross-validation method, mean particle size showed the highest feature importance, followed by coefficient of uniformity.

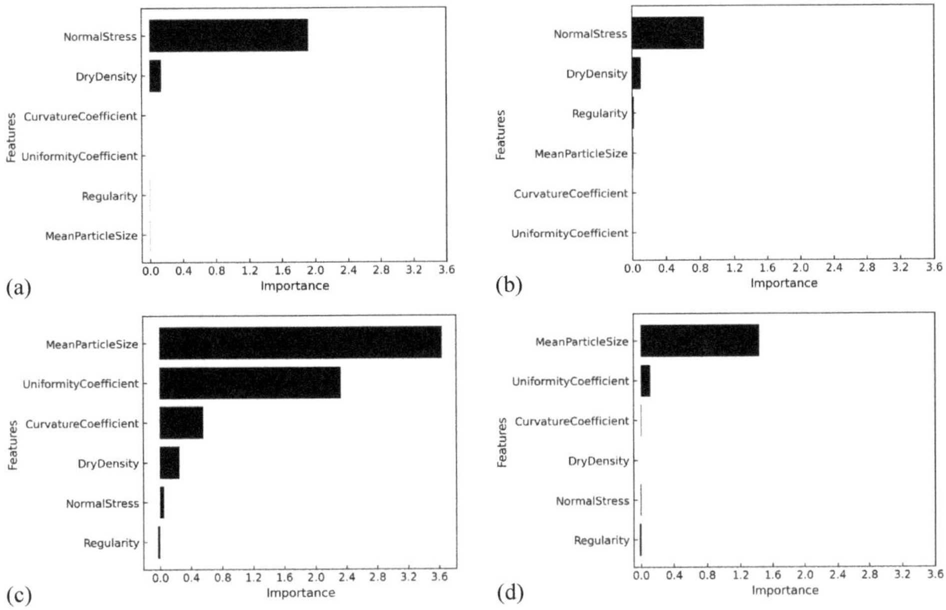

Figure 17. Feature importance analysis comparing multiple linear regression and random forest regression with and without 10-fold cross-validation: (**a**) MLR without 10-fold cross-validation, (**b**) RFR without 10-fold cross-validation, (**c**) MLR with 10-fold cross-validation, (**d**) RFR with 10-fold cross-validation.

5. Conclusions

This study investigated the influence of sand particulate morphology on the shear strength characteristics using experimental and machine learning approaches. The findings can be summarized in the following points.

- Across the range of poorly graded sand sizes, the large sand sample exhibits higher density and shear strength compared to both medium and fine sand.
- The shear strength of well-graded sand is higher than that of poorly graded medium and fine sand, but not as high as that of poorly graded coarse sand.
- The particle shape regularity, including its roundness and sphericity, is not related to the mean particle size.
- In a monodisperse system of glass beads with a similar shape and size, larger particles contribute to greater shear strength compared to their smaller counterparts.
- As the mean particle size of sand decreases, the specific gravity increases and the density decreases, leading to a sample with a higher void ratio. Therefore, finer sand has a higher coefficient of volume compressibility compared to coarse sand.
- The active lateral earth pressure increases as the particle size, density, and confining pressure increases.
- The machine learning models (MLR and RFR) show excellent accuracy in predicting the shear strength of sand based on different particle shapes, sizes, and gradations. In the case of MLR, the R-squared accuracy is 0.95 for the training data, 0.94 for the testing data, as well as 0.93 when using the entire dataset with 10-fold CV method. Similarly, for RFR, the R-squared accuracy is 0.98 for the training data, 0.97 for the testing data, and 0.90 when employing the entire dataset with the 10-fold CV method.
- When using the train–test split, the machine learning models (MLR and RFR) agree on the importance of the following input features in sequence: normal stress and dry density. However, when using the 10-fold CV, the importance of the input features shifts to mean particle size and coefficient of uniformity.

- Future research could address different types of soil (silt and clay), different parameters that could influence the shear strength (moisture content, temperature, strain rate, and stress history), as well as different machine learning algorithms for further exploration.

Author Contributions: Conceptualization, F.D. and H.A.-N.; methodology, F.D.; software, F.D.; writing—original draft preparation, F.D.; writing—review and editing, F.D. and H.A.-N.; supervision, H.A.-N. All authors have read and agreed to the published version of the manuscript.

Funding: The authors declare that no funds, grants, or other financial support were received during the preparation of this manuscript. This paper is part of the PhD of the first author.

Institutional Review Board Statement: Not applicable.

Informed Consent Statement: Not applicable.

Data Availability Statement: Not applicable.

Conflicts of Interest: The authors declare no conflict of interest.

Nomenclature

AI	Artificial Intelligence
MLR	Multiple Linear Regression
RFR	Random Forest Regression
ANN	Artificial Neural Network
SVM	Support Vector Machine
ANFIS	Adaptive Neuro-Fuzzy Inference System
ME	Mean Error
MAE	Mean Absolute Error
MSE	Mean Square Error
RMSE	Root Mean Square Error
RMSLE	Root Mean Square Log Error
CV	Cross-validation
R^2	R-squared
D_{50}	Mean Particle Size
C_u	Coefficient of Uniformity
C_c	Coefficient of Curvature
ρ_d	Dry Density
σ_n	Normal Stress
R	Roundness
S	Sphericity
ρ_r	Particle Regularity
D_r	Relative Density
e	Void ratio
e_{min}	Minimum Void Ratio
e_{max}	Maximum Void Ratio
G_s	Specific Gravity
γ_w	Unit Weight of Water
γ_d	Dry Unit Weight of The Sample
τ	Shear Strength
SOC	Soil Organic Carbon

References

1. Lafata, L. Effect of Particle Shape and Size on Compressibility Behavior of Dredged Sediment in a Geotextile Tube Dewatering Application. 2014. Available online: https://surface.syr.edu/honors_capstone/757/ (accessed on 3 January 2023).
2. Cho, G.-C.; Dodds, J.; Santamarina, J.C. Particle shape effects on packing density, stiffness, and strength: Natural and crushed sands. *J. Geotech. Geoenviron.* **2006**, *132*, 591–602. [CrossRef]
3. Frost, J.; Han, J. Behavior of interfaces between fiber-reinforced polymers and sands. *J. Geotech. Geoenviron. Eng.* **1999**, *125*, 633–640. [CrossRef]

4. Shaia, H. *Behaviour of Fibre Reinforced Polymer Composite Piles: Experimental and Numerical Study*; The University of Manchester: Manchester, UK, 2013.
5. Su, L.-J.; Zhou, W.-H.; Chen, W.-B.; Jie, X. Effects of relative roughness and mean particle size on the shear strength of sand-steel interface. *Measurement* **2018**, *122*, 339–346. [CrossRef]
6. Vaid, Y.; Rinne, N. Geomembrane coefficients of interface friction. *Geosynth. Int.* **1995**, *2*, 309–325. [CrossRef]
7. *AS1289.3.6.1*; Method of Testing Soils for Engineering Purposes—Soil Classification. Australian Standard: Sydney, NSW, Australia, 2009.
8. Barrett, P. The shape of rock particles, a critical review. *Sedimentology* **1980**, *27*, 291–303. [CrossRef]
9. Das, B.M. *Principles of Geotehcnical Engineering*; Cengage Learning: Boston, MA, USA, 2010.
10. Sarkar, D. Influence of Particle Characteristics on the Behaviour of Granular Materials under Static, Cyclic and Dynamic Loading. 2023. Available online: https://www.researchgate.net/profile/Debdeep-Sarkar/publication/370265264_Influence_of_particle_characteristics_on_the_behaviour_of_granular_materials_under_static_cyclic_and_dynamic_loading/links/6448dc28d749e4340e389659/Influence-of-particle-characteristics-on-the-behaviour-of-granular-materials-under-static-cyclic-and-dynamic-loading.pdf (accessed on 3 January 2023).
11. Dodds, J.S. *Particle Shape and Stiffness: Effects on Soil Behavior*; Civil and Environmental Engineering, Georgia Institute of Technology: Atlanta, GA, USA, 2003.
12. Mitchell, J.K.; Soga, K. *Fundamentals of Soil Behavior*; John Wiley & Sons: New York, NY, USA, 2005; Volume 3.
13. Wadell, H. Volume, shape, and roundness of rock particles. *J. Geol.* **1932**, *40*, 443–451. [CrossRef]
14. Powers, M.C. A new roundness scale for sedimentary particles. *J. Sediment. Res.* **1953**, *23*, 117–119. [CrossRef]
15. Schanz, T.; Vermeer, P. Angles of friction and dilatancy of sand. *Géotechnique* **1996**, *46*, 145–151. [CrossRef]
16. Li, Y. Effects of particle shape and size distribution on the shear strength behavior of composite soils. *Bull. Eng. Geol. Environ.* **2013**, *72*, 371–381. [CrossRef]
17. Peng, Z.; Chen, C.; Wu, L. Numerical investigation of particle shape effect on sand shear strength. *Arab. J. Sci. Eng.* **2021**, *46*, 10585–10595. [CrossRef]
18. Otsubo, M.; O'sullivan, C.; Sim, W.W.; Ibraim, E. Quantitative assessment of the influence of surface roughness on soil stiffness. *Géotechnique* **2015**, *65*, 694–700. [CrossRef]
19. Santamarina, C.; Cascante, G. Effect of surface roughness on wave propagation parameters. *Geotechnique* **1998**, *48*, 129–136. [CrossRef]
20. Tsomokos, A.; Georgiannou, V. Effect of grain shape and angularity on the undrained response of fine sands. *Can. Geotech. J.* **2010**, *47*, 539–551. [CrossRef]
21. Menq, F.; Stokoe, K. Linear dynamic properties of sandy and gravelly soils from large-scale resonant tests. In Proceedings of the Deformation Characteristics of Geomaterials, IS Lyon 2003, Lyon, France, 22–24 September 2003; pp. 63–71.
22. Miura, K.; Maeda, K.; Furukawa, M.; Toki, S. Physical characteristics of sands with different primary properties. *Soils Found.* **1997**, *37*, 53–64. [CrossRef]
23. Vangla, P.; Latha, G.M. Influence of particle size on the friction and interfacial shear strength of sands of similar morphology. *Int. J. Geosynth. Ground Eng.* **2015**, *1*, 6. [CrossRef]
24. Wang, J.-J.; Zhang, H.-P.; Tang, S.-C.; Liang, Y. Effects of particle size distribution on shear strength of accumulation soil. *J. Geotech. Geoenviron. Eng.* **2013**, *139*, 1994–1997. [CrossRef]
25. Islam, M.N.; Siddika, A.; Hossain, M.B.; Rahman, A.; Asad, M.A. Effect of particle size on the shear strength behavior of sands. *arXiv* **2019**, arXiv:1902.09079.
26. Alias, R.; Kasa, A.; Taha, M. Particle size effect on shear strength of granular materials in direct shear test. *Int. J. Civ. Environ. Eng.* **2014**, *8*, 1144–1147.
27. Skinner, A. A note on the influence of interparticle friction on the shearing strength of a random assembly of spherical particles. *Geotechnique* **1969**, *19*, 150–157. [CrossRef]
28. Xie, X.; Wu, T.; Zhu, M.; Jiang, G.; Xu, Y.; Wang, X.; Pu, L. Comparison of random forest and multiple linear regression models for estimation of soil extracellular enzyme activities in agricultural reclaimed coastal saline land. *Ecol. Indic.* **2021**, *120*, 106925. [CrossRef]
29. Zhang, H.; Wu, P.; Yin, A.; Yang, X.; Zhang, M.; Gao, C. Prediction of soil organic carbon in an intensively managed reclamation zone of eastern China: A comparison of multiple linear regressions and the random forest model. *Sci. Total Environ.* **2017**, *592*, 704–713. [CrossRef] [PubMed]
30. Jang, J.-S. ANFIS: Adaptive-network-based fuzzy inference system. *IEEE Trans. Syst. Man Cybern.* **1993**, *23*, 665–685. [CrossRef]
31. Keshwani, D.R.; Jones, D.D.; Brand, R.M. Takagi–Sugeno Fuzzy Modeling of Skin Permeability. *Cutan. Ocul. Toxicol.* **2005**, *24*, 149–163. [CrossRef] [PubMed]
32. Al-Mahasneh, M.; Aljarrah, M.; Rababah, T.; Alu'datt, M. Application of hybrid neural fuzzy system (ANFIS) in food processing and technology. *Food Eng. Rev.* **2016**, *8*, 351–366. [CrossRef]
33. Sonmez, A.Y.; Kale, S.; Ozdemir, R.C.; Kadak, A.E. An adaptive neuro-fuzzy inference system (ANFIS) to predict of cadmium (Cd) concentrations in the Filyos River, Turkey. *Turk. J. Fish. Aquat. Sci.* **2018**, *18*, 1333–1343. [CrossRef] [PubMed]
34. Bensaber, B.A.; Diaz, C.G.P.; Lahrouni, Y. Design and modeling an Adaptive Neuro-Fuzzy Inference System (ANFIS) for the prediction of a security index in VANET. *J. Comput. Sci.* **2020**, *47*, 101234. [CrossRef]

35. Akkaya, E. ANFIS based prediction model for biomass heating value using proximate analysis components. *Fuel* **2016**, *180*, 687–693. [CrossRef]
36. Abdulshahed, A.M.; Longstaff, A.P.; Fletcher, S. The application of ANFIS prediction models for thermal error compensation on CNC machine tools. *Appl. Soft Comput.* **2015**, *27*, 158–168. [CrossRef]
37. Hsieh, M.-C.; Maurya, S.N.; Luo, W.-J.; Li, K.-Y.; Hao, L.; Bhuyar, P. Coolant Volume Prediction for Spindle Cooler with Adaptive Neuro-fuzzy Inference System Control Method. *Sens. Mater.* **2022**, *34*, 2447–2466. [CrossRef]
38. Bardhan, A.; Samui, P. Application of artificial intelligence techniques in slope stability analysis: A short review and future prospects. *Int. J. Geotech. Earthq. Eng. (IJGEE)* **2022**, *13*, 1–22. [CrossRef]
39. Inazumi, S.; Intui, S.; Jotisankasa, A.; Chaiprakaikeow, S.; Kojima, K. Artificial intelligence system for supporting soil classification. *Results Eng.* **2020**, *8*, 100188. [CrossRef]
40. Singh, B.; Sihag, P.; Pandhiani, S.M.; Debnath, S.; Gautam, S. Estimation of permeability of soil using easy measured soil parameters: Assessing the artificial intelligence-based models. *ISH J. Hydraul. Eng.* **2021**, *27*, 38–48. [CrossRef]
41. Baghbani, A.; Costa, S.; Choundhury, T.; Faradonbeh, R.S. Prediction of Parallel Desiccation Cracks of Clays Using a Classification and Regression Tree (CART) Technique. In Proceedings of the 8th International Symposium on Geotechnical Safety and Risk (ISGSR), Newcastle, Australia, 14–16 December 2022.
42. Daghistani, F.; Baghbani, A.; Abuel Naga, H.; Faradonbeh, R.S. Internal Friction Angle of Cohesionless Binary Mixture Sand–Granular Rubber Using Experimental Study and Machine Learning. *Geosciences* **2023**, *13*, 197. [CrossRef]
43. Baghbani, A.; Daghistani, F.; Baghbani, H.; Kiany, K. *Predicting the Strength of Recycled Glass Powder-Based Geopolymers for Improving Mechanical Behavior of Clay Soils Using Artificial Intelligence*; EasyChair: Manchester, UK, 2023.
44. Baghbani, A.; Daghistani, F.; Kiany, K.; Shalchiyan, M.M. *AI-Based Prediction of Strength and Tensile Properties of Expansive Soil Stabilized with Recycled Ash and Natural Fibers*; EasyChair: Manchester, UK, 2023.
45. Baghbani, A.; Daghistani, F.; Baghbani, H.; Kiany, K.; Bazaz, J.B. *Artificial Intelligence-Based Prediction of Geotechnical Impacts of Polyethylene Bottles and Polypropylene on Clayey Soil*; EasyChair: Manchester, UK, 2023.
46. Baghbani, A.; Daghistani, F.; Naga, H.A.; Costa, S. Development of a Support Vector Machine (SVM) and a Classification and Regression Tree (CART) to Predict the Shear Strength of Sand Rubber Mixtures. In Proceedings of the 8th International Symposium on Geotechnical Safety and Risk (ISGSR), Newcastle, Australia, 14–16 December 2022.
47. *AS1774.19*; The Determination of Sieve Analysis and Moisture Content. Australian Standard: Sydney, NSW, Australia, 2003.
48. *AS1289.6.2.2*; Soil Strength and Consolidation Tests—Determination of Shear Strength of a Soil—Direct Shear Test Using a Shear Box. Australian Standard: Sydney, NSW, Australia, 2020.
49. Krumbein, W.; Sloss, L. *Stratigraphy and Sedimentation*, 2nd ed.; Friedman, WH and Company: San Francisco, CA, USA, 1963; Volume 660.
50. Liu, Y.-L.; Nisa, E.C.; Kuan, Y.-D.; Luo, W.-J.; Feng, C.-C. Combining deep neural network with genetic algorithm for axial flow fan design and development. *Processes* **2023**, *11*, 122. [CrossRef]
51. Joseph, V.R. Optimal ratio for data splitting. *Stat. Anal. Data Min. ASA Data Sci. J.* **2022**, *15*, 531–538. [CrossRef]
52. Vakharia, V.; Shah, M.; Nair, P.; Borade, H.; Sahlot, P.; Wankhede, V. Estimation of Lithium-ion Battery Discharge Capacity by Integrating Optimized Explainable-AI and Stacked LSTM Model. *Batteries* **2023**, *9*, 125. [CrossRef]
53. Rochman, E.; Rachmad, A.; Fatah, D.; Setiawan, W.; Kustiyahningsih, Y. Classification of Salt Quality based on Salt-Forming Composition using Random Forest. *J. Phys. Conf. Ser.* **2022**, *2406*, 012021. [CrossRef]
54. Burmister, D.M. *Study of the Physical Characteristics of Soils, with Special Reference to Earth Structures*; Department of Civil Engineering Columbia University: New York, NY, USA, 1938.
55. Guyon, E.; Delenne, J.Y.; Radjai, F.; Kamrin, K.; Butler, E. *Built on Sand: The Science of Granular Materials*, MIT Press: Cambridge, MA, USA, 2020.
56. Maroof, M.A.; Mahboubi, A.; Vincens, E.; Noorzad, A. Effects of particle morphology on the minimum and maximum void ratios of granular materials. *Granul. Matter* **2022**, *24*, 1–24. [CrossRef]
57. Terzaghi, K.; Peck, R.B.; Mesri, G. *Soil Mechanics in Engineering Practice*; John Wiley & Sons: Hoboken, NJ, USA, 1996.
58. Bowles, J.E. *Engineering Properties of Soils and Their Measurement*; McGraw-Hill, Inc.: New York, NY, USA, 1992.
59. Voivret, C.; Radjai, F.; Delenne, J.-Y.; El Youssoufi, M.S. Space-filling properties of polydisperse granular media. *Phys. Rev. E* **2007**, *76*, 021301. [CrossRef]
60. Etemadi, S.; Khashei, M. Etemadi multiple linear regression. *Measurement* **2021**, *186*, 110080. [CrossRef]
61. Breiman, L. Random forests. *Mach. Learn.* **2001**, *45*, 5–32. [CrossRef]
62. Pahlavan-Rad, M.R.; Dahmardeh, K.; Hadizadeh, M.; Keykha, G.; Mohammadnia, N.; Gangali, M.; Keikha, M.; Davatgar, N.; Brungard, C. Prediction of soil water infiltration using multiple linear regression and random forest in a dry flood plain, eastern Iran. *Catena* **2020**, *194*, 104715. [CrossRef]

applied sciences

MDPI

Article

Machine Learning Techniques for Soil Characterization Using Cone Penetration Test Data

Ayele Tesema Chala * and Richard P. Ray *

Structural and Geotechnical Engineering Department, Faculty of Architecture, Civil and Transport Sciences, Szechenyi Istvan University, H-9026, Egyetem Ter 1, 9026 Gyor, Hungary
* Correspondence: chala.ayele.tesema@hallgato.sze.hu (A.T.C.); ray@sze.hu (R.P.R.)

Abstract: Seismic response assessment requires reliable information about subsurface conditions, including soil shear wave velocity (V_s). To properly assess seismic response, engineers need accurate information about V_s, an essential parameter for evaluating the propagation of seismic waves. However, measuring V_s is generally challenging due to the complex and time-consuming nature of field and laboratory tests. This study aims to predict V_s using machine learning (ML) algorithms from cone penetration test (CPT) data. The study utilized four ML algorithms, namely Random Forests (RFs), Support Vector Machine (SVM), Decision Trees (DT), and eXtreme Gradient Boosting (XGBoost), to predict V_s. These ML models were trained on 70% of the datasets, while their efficiency and generalization ability were assessed on the remaining 30%. The hyperparameters for each ML model were fine-tuned through Bayesian optimization with k-fold cross-validation techniques. The performance of each ML model was evaluated using eight different metrics, including root mean squared error (RMSE), mean absolute error (MAE), mean absolute percentage error (MAPE), coefficient of determination (R^2), performance index (PI), scatter index (SI), $A10-I$, and U_{95}. The results demonstrated that the RF model consistently performed well across all metrics. It achieved high accuracy and the lowest level of errors, indicating superior accuracy and precision in predicting V_s. The SVM and XGBoost models also exhibited strong performance, with slightly higher error metrics compared with the RF model. However, the DT model performed poorly, with higher error rates and uncertainty in predicting V_s. Based on these results, we can conclude that the RF model is highly effective at accurately predicting V_s using CPT data with minimal input features.

Keywords: shear wave velocity; cone penetration test; machine learning; Random Forests; support vector machine; decision trees; eXtreme gradient boosting; regression

Citation: Chala, A.T.; Ray, R.P. Machine Learning Techniques for Soil Characterization Using Cone Penetration Test Data. *Appl. Sci.* **2023**, *13*, 8286. https://doi.org/10.3390/app13148286

Academic Editor: Wei Gao

Received: 6 June 2023
Revised: 10 July 2023
Accepted: 13 July 2023
Published: 18 July 2023

1. Introduction

Soil characterization plays a vital role in seismic response assessment and interpreting subsurface conditions for large-scale engineering projects. To properly assess seismic response, engineers need accurate and reliable information about subsurface conditions, including soil shear wave velocity (V_s), an essential parameter for evaluating the propagation of seismic waves [1–6]. Seismic-refraction and reflection methods using geophysical signal processing [7–10] measure V_s at various depths and precision to produce a profile (preferably to bedrock) for later analysis. To measure V_s, an active source generates a wave and its travel time to one or more receivers is measured. The velocity results from knowing the time and distance traveled between the source and receiver. There are several V_s measurement methods, including seismic cone penetration testing (SCPT) [7], Multi-Channel Analysis of Surface Waves (MASW) [8], Cross-hole testing [8], and down-hole testing methods. These techniques provide valuable information about subsurface conditions but become more complex with increasing soil layering.

Additionally, laboratory tests such as bender element [11], triaxial test [12], and resonant column tests [13] measure V_s in different ways. These tests are conducted on soil samples collected from the site and offer a controlled environment for testing, providing detailed information on soil behavior under varying stress conditions. However, retrieving high-quality intact soil samples is a challenging task that requires specialist equipment [14]. Furthermore, it is important to note that the properties of collected samples may significantly change over time due to variations in stress conditions, temperature fluctuations, and moisture content.

A viable alternative is to correlate V_s to cone penetration test (CPT) data, a relatively easier approach. Many empirical correlations have been developed over the past couple of decades to estimate V_s from CPT data [15–19]. The CPT test involves pushing a cone-shaped instrument into the ground at a constant rate while measuring the resistance of the soil. Two measurements are typically taken during this test: cone tip resistance (q_c) and sleeve friction (f_s) [20,21]. The CPT test provides continuous and reliable soil data, making it an efficient and cost-effective method in geotechnical engineering practice. This wealth of CPT data has attracted the attention of many geotechnical researchers to further improve the prediction accuracy of V_s employing machine learning (ML) algorithms [22–26]. ML algorithms have shown great promise in accurately predicting V_s from CPT data. The ML algorithms can learn complex relationships between input variables (e.g., q_c and f_s records) and output variables (e.g., soil V_s) from large datasets without the need for explicit mathematical models.

Many ML algorithms, such as gradient boosting, random forest, support vector machine (SVM) artificial neural network (ANN), and decision trees (DT), have been used in various geotechnical applications, including soil classification [27–33], V_s prediction [23–26,34], liquefaction analysis [35–40], stability analysis [41–45], and settlement prediction [46–48]. The application of ML algorithms in geotechnical engineering has shown promising results in terms of efficiency and accuracy. For example, Tsiaousi et al. [25] successfully employed an ANN model to characterize soil stratigraphy and predict V_s. This study demonstrates how ML approaches can be used to improve soil characterization and prediction of important geotechnical parameters. Assaf et al. [24] and Riyadi et al. [49] have also used ML algorithms, including RF and XGBoost, to predict V_s. Their findings confirm that ML models can achieve high accuracy and performance in predicting V_s. Previous research has also shown that SVM performs well in predicting V_s [50,51]. These studies collectively demonstrate the potential of ML algorithms in improving the accuracy of V_s prediction in geotechnical engineering applications.

The aim of this study is to improve the prediction of V_s using various ML algorithms with minimal input features. Four ML algorithms, namely RF, SVM, DT, and eXtreme gradient boosting (XGBoost), are employed to predict V_s from CPT data. The study also aims to minimize the need for expensive and time-consuming fields or laboratory measurements. The development of ML models can lead to higher accuracy and performance in predicting V_s. The improvement in the accuracy of V_s prediction has significant implications for site response assessment and seismic risk reduction. By utilizing ML to predict V_S, this study has the potential to enhance existing knowledge and inspire future research in the field of ML applications for soil characterization.

The rest of this document is organized as follows: Section 2 discusses dataset preprocessing and visualization, Methodology and performance metrics are described in Section 3, Section 4 describes the ML models, and Section 5 presents the results. Finally, Section 6 outlines the main results of the study and concludes by suggesting future research.

2. Datasets Preprocessing and Visualization

The dataset used in this study was obtained from a previously published dataset [52]. This study utilized 61 CPT soundings, each containing over 1000 q_c and f_s recordings. These data sets were collected from various regions of Austria, including the Vienna Basin, Gastein Valley, and Zell Basin. The data is publicly accessible and can be downloaded from the following link: https://www.tugraz.at/en/institutes/ibg/research/computational-geotechnics-group/database/ (accessed on 12 May 2023). The CPT datasets were preprocessed before applying ML training and testing techniques. The preprocessing step involved removing outliers from the data. Specifically, outliers were identified and removed from both the q_c and f_s values in the raw CPT data. Any data point that exceeded twice the interquartile range (IQR), where IQR is the difference between the third quartile ($Q3$) and the first quartile ($Q1$), was considered an outlier. Next, the target variable, which in this case was the shear wave velocity, V_s, was estimated using Equation (1) [16]. Subsequently, the datasets were divided into a training set and a testing set, with a ratio of 0.7:0.3 for training and testing purposes.

$$V_S = \sqrt{\left(\frac{q_c - \sigma_{v0}}{p_a} \times 10^{0.55I_c+1.68}\right)} \quad (1)$$

where q_c represents cone tip resistance, σ_{v0} represents total overburden pressure, p_a represents atmospheric pressure, and I_c represents soil behavioral type index estimated as follows:

$$I_c = \left(3.47 - log((q_c - \sigma_{v0})/\sigma\prime_{v0})^2 + (logF_r + 1.22)^2\right)^{0.5} \quad (2)$$

$$F_r = f_s/(q_c - \sigma_{v0}) \times 100 \quad (3)$$

where f_s represents sleeve friction, $\sigma\prime_{v0}$ is the effective overburden stress, and F_r represents normalized friction ratio.

The statistical summaries of both the training and testing datasets considered in this study are presented in Table 1. To gain further insights into the relationship between the input features and the target variable (V_s), scatter plots are presented in Figure 1. Each scatter plot indicates the correlation between an individual input feature and the target variable. In addition, Figure 2 shows the frequency distribution of the input features and target variable, providing a visual representation of their distribution patterns. Furthermore, box plots of both the input features and the target variable are presented in Figure 3, offering an overview of their distribution.

Table 1. Statistical summary of training and testing datasets.

Features	Unit	Class	Training Dataset					Testing Dataset				
			Mean	SD	Min	Max	Count	Mean	SD	Min	Max	Count
D	m	Input	12.42	8.88	0.01	40	79,579	12.38	8.78	0.01	40	34,104
q_c	MPa	Input	4.89	3.60	0.01	17	79,579	4.87	3.59	0.01	17	34,104
f_s	kPa	Input	42.91	35.35	0.07	142	79,579	42.88	35.39	0.10	142	34,104
R_f	%	Input	1.56	6.11	0.00	1121	79,579	1.57	6.28	0.00	1083	34,104
V_s	m/s	Target	166.76	55.89	10.06	322	79,579	166.55	55.57	9.93	322	34,104

The interdependencies among input features in ML models can lead to overfitting and decreased efficiency. To assess the correlation between each input feature, a Pearson's correlation analysis was conducted. Figure 4 displays the correlation coefficients among the input features in the dataset. The correlation coefficients range from −0.08 to 0.51, indicating a combination of weak to moderate correlations among the features. The absence of highly correlated features in the correlation analysis suggests a lower risk of overfitting, as no redundant features were observed.

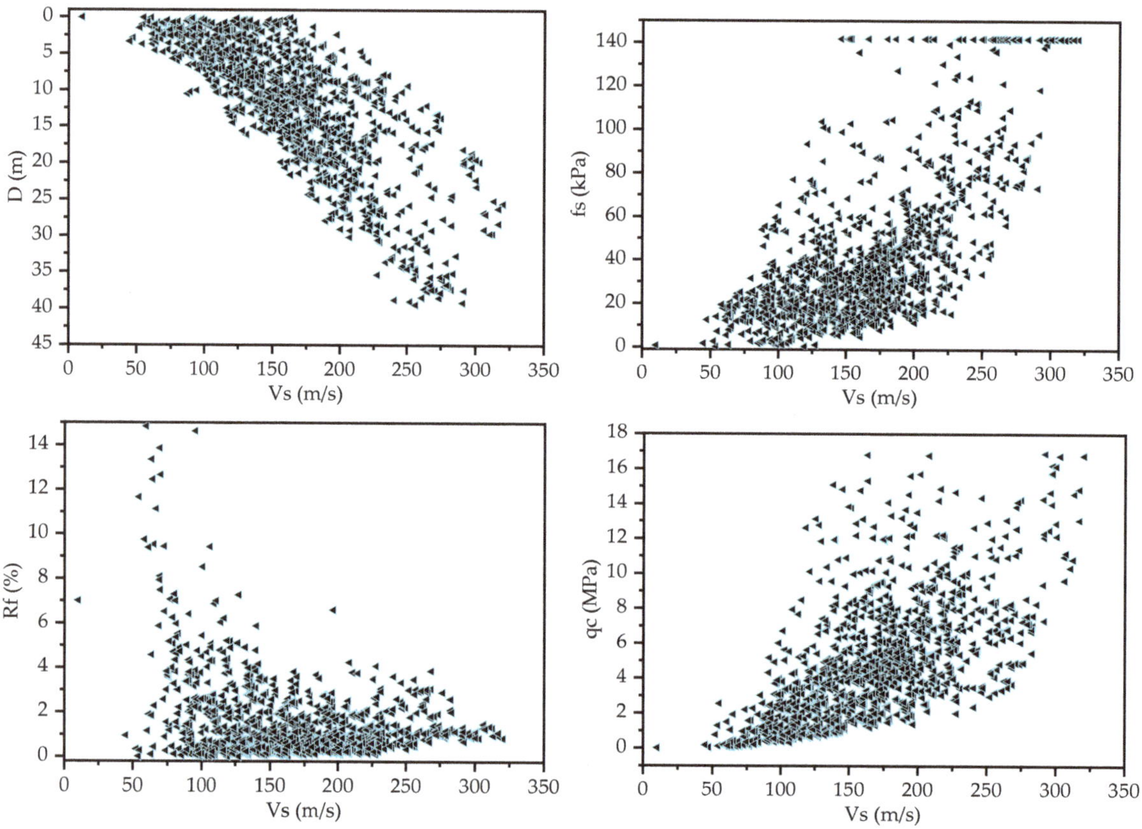

Figure 1. Scatter plots of input features with respect to target variable.

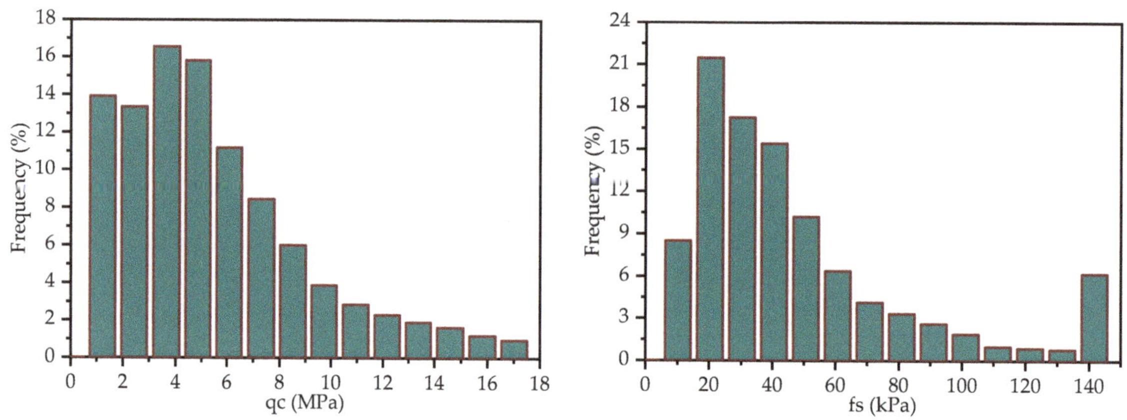

Figure 2. *Cont.*

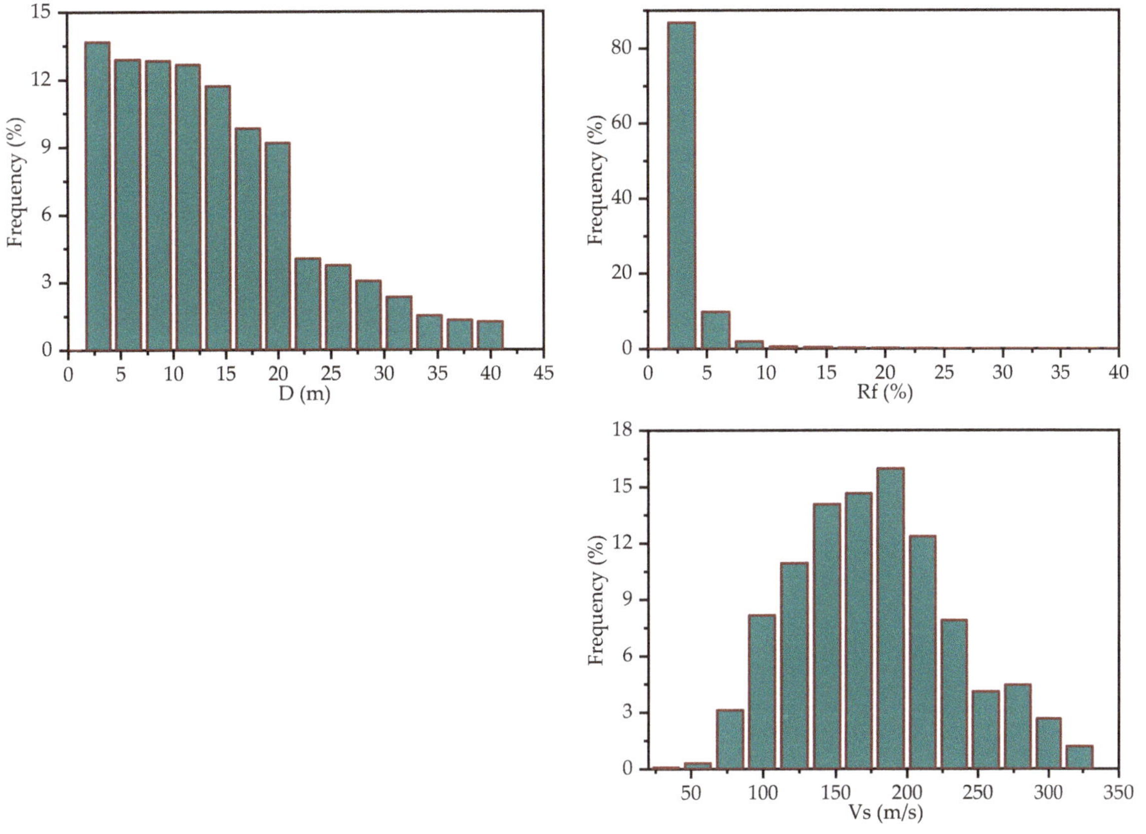

Figure 2. Frequency distribution of input features and the target variable.

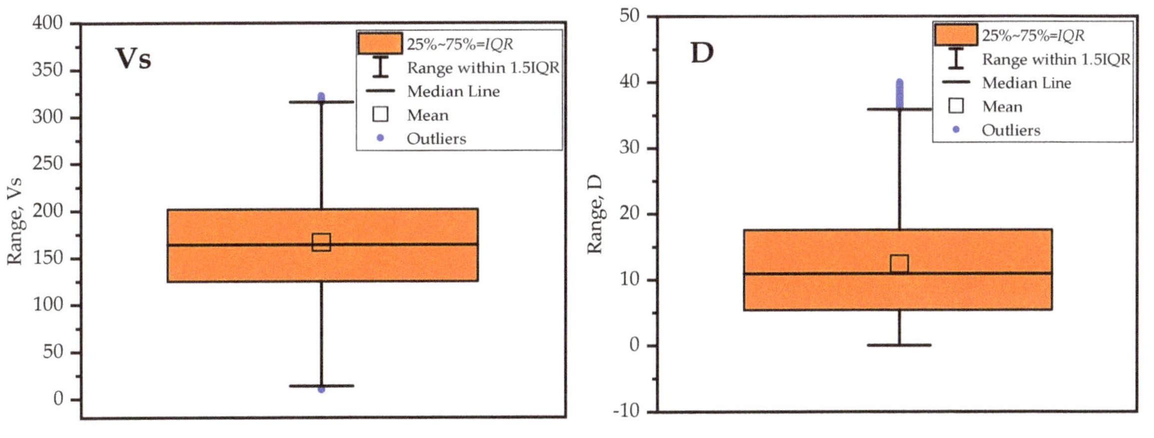

Figure 3. *Cont.*

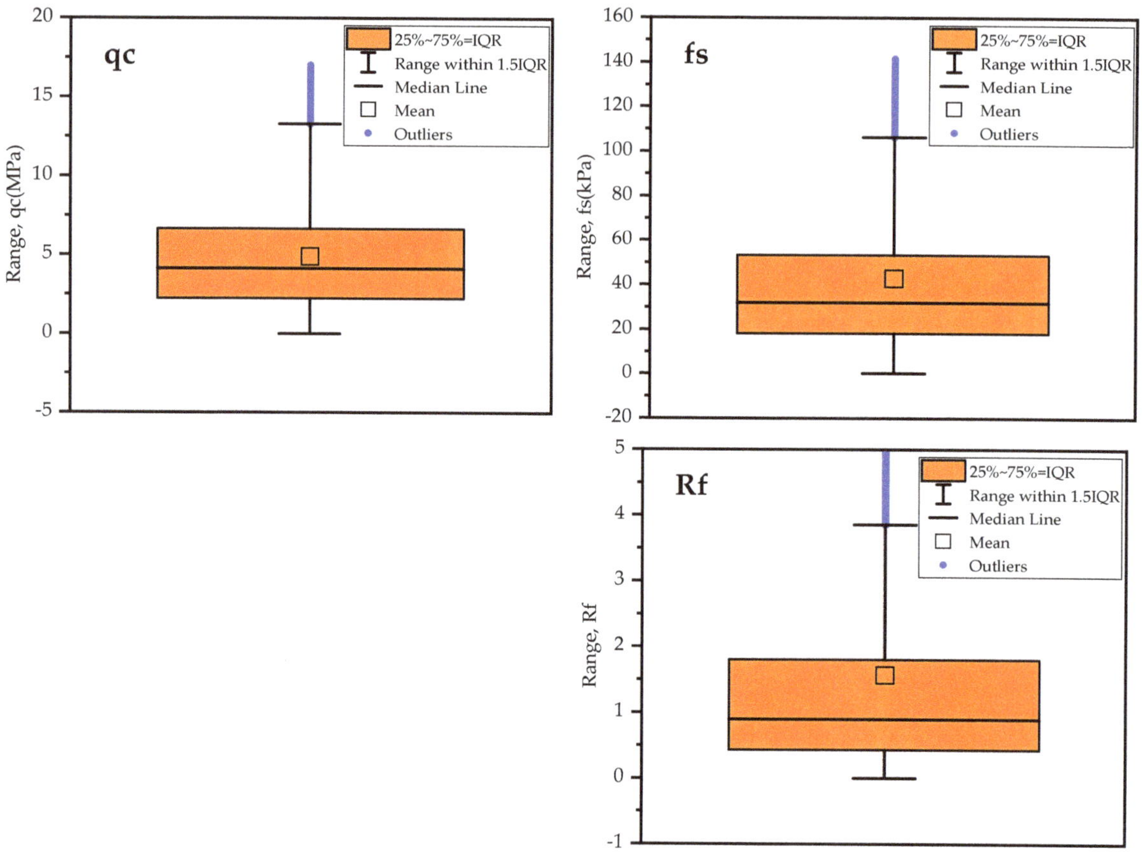

Figure 3. Box plot of input features and target variable.

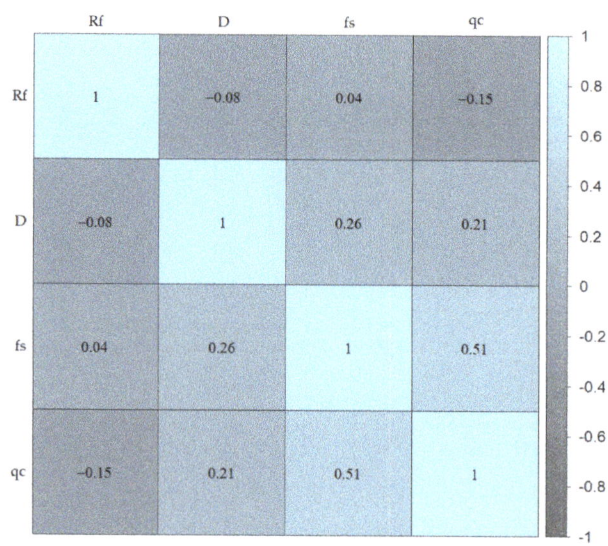

Figure 4. Table of feature correlations.

3. Methodology

This section outlines the training and testing procedures for the ML models used to predict V_S. Four ML models, namely RF, SVM, DT and XGBoost, were trained for this purpose. Each ML model was trained on the training datasets, with q_c, f_s, friction ratio (R_f), and soil depth (D) serving as input features and V_s as target variable (output). To optimize the performance of these models, the hyperparameters of each model were fine-tuned using a model-based Bayesian optimization technique. To ensure that the models can generalize well to new data, the commonly used k-fold cross-validation techniques were employed. This involves dividing the data into k subsets, training the model on k-1 subsets, and evaluating their performance on the remaining subset. The root mean squared error (RMSE) was used as the evaluation metric to assess the models' accuracy. In addition to hyperparameter tuning, permutation feature importance and/or recursive feature elimination techniques were applied using the optimized models. This technique was used for the identification and removal of irrelevant features, if present, in the input features. Once the irrelevant features were eliminated, the hyperparameter tuning process was repeated with the updated features to further enhance the models' performance. Finally, the performances of the optimized models were assessed using the testing dataset. The entire process of training and testing the models is presented in Figure 5, providing a visual representation of the workflow.

The performance of optimized ML models was evaluated using multiple statistical metrics such as root mean squared error (RMSE), mean absolute error (MAE), mean absolute percentage error (MAPE), coefficient of determination $\left(R^2\right)$, scatter index (SI), and performance index (PI) (see Table 2). Many researchers utilized these indices to evaluate the predictive performance of different ML models [53–60]. The RMSE measures the average magnitude of the errors between the predicted and actual values, indicating the model's predictive accuracy. A lower RMSE indicates better model performance. MAE estimates the average absolute difference between the predicted and actual values. Like RMSE, a lower MAE indicates better model performance. MAPE represents the average percentage difference between the predicted and actual values. A lower MAPE signifies better model accuracy. R^2 measures the proportion of the variance in the target variable (V_S) that can be explained by the model, with values closer to 1 indicating a better fit. Furthermore, a newly proposed engineering index ($A10 - I$) was used to evaluate the predictive performance of the models [55,59,61–63]. In an ideal model, the value of $A10 - I$ is expected to be one. The $A10 - I$ has significance in engineering as it represents the proportion of samples that fall within $\pm$10% deviation from the predicted values compared with the target value. Additionally, the efficiency of the models was evaluated using uncertainty analysis at 95% confidence level (U_{95}) [64,65].

Table 2. Performance indices used to evaluate the efficiency of the models.

Metrics	Best Performance	Equations	Equation No.
Root mean squared error	Lower value	$RMSE = \sqrt{\frac{\sum_{i=1}^{n}\left(X_i - \widehat{X}_i\right)^2}{n}}$	(4)
Mean absolute error	Lower value	$MAE = \frac{1}{n} \times \left\lvert \sum\limits_{i=1}^{n} X_i - \widehat{X}_i \right\rvert$	(5)
Mean absolute percentage error	Lower value	$MAPE = \frac{1}{n} \times \left\lvert \sum\limits_{i=1}^{n} \frac{\left(X_i - \widehat{X}_i\right)}{X_i} \right\rvert \times 100\%$	(6)
Coefficient of determination	unity	$R^2 = 1 - \sum\limits_{i=1}^{n} \frac{\left(X_i - \widehat{X}_i\right)^2}{\left(X_i - \overline{X}\right)^2}$	(7)

Table 2. *Cont.*

Metrics	Best Performance	Equations	Equation No.
$A10-I$	unity	$A10-I=\frac{n10}{n}$	(8)
Scatter index	Lower value	$SI=\frac{RMSE}{\overline{X}}$	(9)
Performance index	Lower value	$PI=\frac{RMSE}{\overline{X}\times\sqrt{R^2+1}}$	(10)
Uncertainty at 95% confidence level	Lower value	$U_{95}=\sqrt{SD^2+RMSE^2}$	(11)
SI [64,65]	SI < 0.05: excellent precision (EP), 0.05 < SI < 0.1: good precision (GP), 0.1 < SI < 0.15: fair precision (FP), SI > 0.15: poor precision (PP)		

n is total number of datasets, X_i is the actual value of the ith observation, $\widehat{X}_i$ is the predicted value of the ith observation, and $\overline{X}$ is mean of target variable. $n10$ is the number of samples with actual/predicted value between 0.90 and 1.10, U_{95} is uncertainty with 95% confidence intervals, and SD is standard deviation of residuals (the difference between target V_s and predicted V_s).

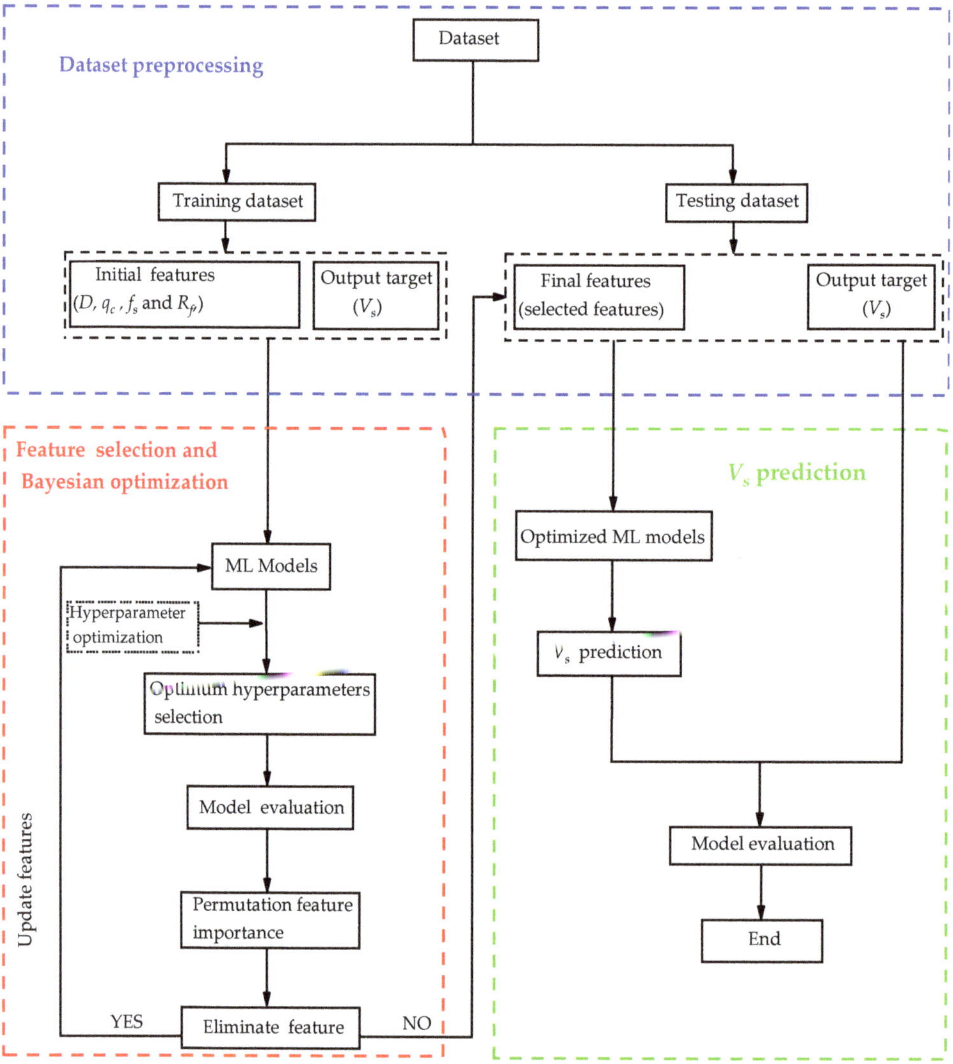

Figure 5. Flow diagram illustrating machine learning models used for predicting V_S.

4. Machine Learning Models

4.1. Random Forests

RF is an ML algorithm that has been widely used for classification and regression tasks [66,67]. It is an ensemble method that combines multiple decision trees to improve predictive accuracy and reduce overfitting. RF has several advantages over other machine learning algorithms, including its ability to handle high-dimensional data, nonlinear relationships between variables, and missing values [68]. In addition, it provides measures of variable importance that can be used for feature selection and interpretation [69].

RF classification and regression can be implemented in R using various packages such as randomForest [68] and ranger [70]. The randomForest package is one of the most widely used packages for RF classification and regression in R. It provides a simple interface for building and evaluating RF models and supports both classification and regression tasks. The ranger package is another popular package for RF classification and regression in R. It is designed to be faster and more memory-efficient than the randomForest package, and supports parallel processing [70].

The ranger package provides several hyperparameters that can be tuned to improve the performance and robustness of RF models. To tune these hyperparameters, one common approach is to use cross-validation. This involves splitting the data into training and test sets, fitting the model on the training set with different combinations of hyperparameters, and evaluating the performance on the validation set. Bayesian optimization is one of the most efficient methods for hyperparameter tuning. It uses a probabilistic model to predict the performance of different hyperparameter configurations based on previous evaluations [71].

4.2. Support Vector Machine

SVMs have gained immense popularity in the field of machine learning due to their ability to solve both classification and regression problems effectively. SVMs work by constructing hyperplanes that can optimally separate data points belonging to different classes or predict target variables with maximum margin. One of the most significant advantages of using SVMs is their ability in handling high-dimensional datasets and nonlinear relationships between variables [72].

In R, e1071 package [73] is commonly utilized to implement SVM models for regression and classification tasks. The package provides options for tuning hyperparameters such as the kernel function, regularization parameter, and cost parameter. One important consideration when using SVMs is their robustness to outliers and noise in the data. Outliers influence the position of the hyperplane and lead to poor generalization performance. To address this issue, Bayesian optimization can be utilized to increase the model's performance and robustness. Bayesian optimization has been shown to be effective at tuning hyperparameters in various machine-learning algorithms, including SVMs [71].

4.3. Decision Trees

The DT algorithm is commonly used for both classification and regression tasks. The DT algorithms recursively partition data into subsets based on the values of input features and then assign labels to each subset based on the majority class or average value of the target variable. The resulting tree structure can be used to make predictions on new data by traversing the tree from the root node to a leaf node that corresponds to a specific class or value. According to Quinlan [74], decision trees are particularly useful for problems with discrete-valued output variables and can handle both categorical and continuous input features. They are also easy to interpret and visualize, making them a popular choice for exploratory data analysis and decision-making tasks. Figure 6 presents a sample decision tree structure to provide insights into the relationships and decision-making process within the data, aiding in understanding and interpreting the model's predictions. The node numbers are depicted within the boxes, while the input features are represented by the variables (see Sections 1 and 3). The green leaves in the figure represent the target value, V_s.

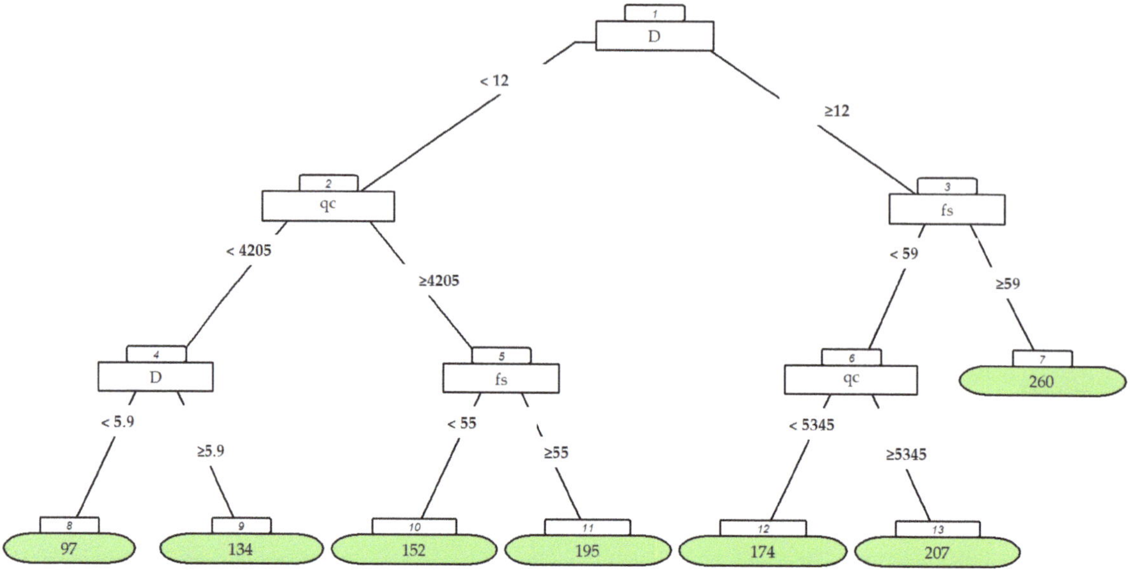

Figure 6. Sample decision tree structure illustrating the splitting criteria for predicting the V_S.

The implementation of the DT algorithm for regression tasks is usually performed using the rpart package [75]. The package provides ranges of DT hyperparameters, including complexity parameters, maximum number of trees, minimum number of splits, etc., that can be tuned through grid search or Bayesian optimization. DT algorithms have been used for a variety of geotechnical applicatons, including classification [76,77] and soil parameters predictions.

4.4. eXtreme Gradient Boosting

Recently, the XGBoost algorithm has gained popularity due to its high accuracy and efficiency. XGBoost is an ensemble method that combines multiple weak learners such as decision trees into a single strong learner [78,79]. The algorithm iteratively adds decision trees to the model, with each tree attempting to correct the errors of the previous trees.

XGBoost package [78] is usually utilized to implement the XGBoost regression model in R. The XGBoost package also offers support for hyperparameter tuning, which can significantly improve the model's performance. It provides a range of options for tuning its hyperparameters, including learning rate (eta), maximum depth of each tree (max_depth), number of trees, and regularization parameters (alpha and gamma). Bayesian optimization can be used for hyperparameter tuning in XGBoost.

5. Results and Discussion

In Section 2, we indicated that the datasets were randomly split into training and testing datasets. The training datasets were used to train the ML models, while the testing datasets were used to evaluate the efficiency of each model in predicting V_s. In this section, we will discuss the results obtained from training and testing ML models. All the ML models were trained and tested using a personal computer with 8GB RAM and Intel(R) Core(TM) i7-1065G7 CPU @ 1.30GHz 1.50 GHz processor (Intel Co., Santa Clara, CA, USA). The performance of each ML model was evaluated using the multiple performance metrics listed in Table 2.

5.1. Hyperparameter Optimization Results

The hyperparameters of each model were fine-tuned using Bayesian optimization with a k-fold cross-validation strategy. Specifically, we used 10-fold cross-validation with the RMSE as the evaluation metric for fine-tuning the hyperparameters. The goal was to minimize the RMSE, as lower values indicate better performance. The maximum number of iterations for the fine-tuning process was set to 100 for each model. Figure 7 illustrates the convergence behaviors of the ML models during the fine-tuning process. It shows how the performance metric (RMSE) changed over the iterations. We observed that all the ML models reached stable results within 100 iterations.

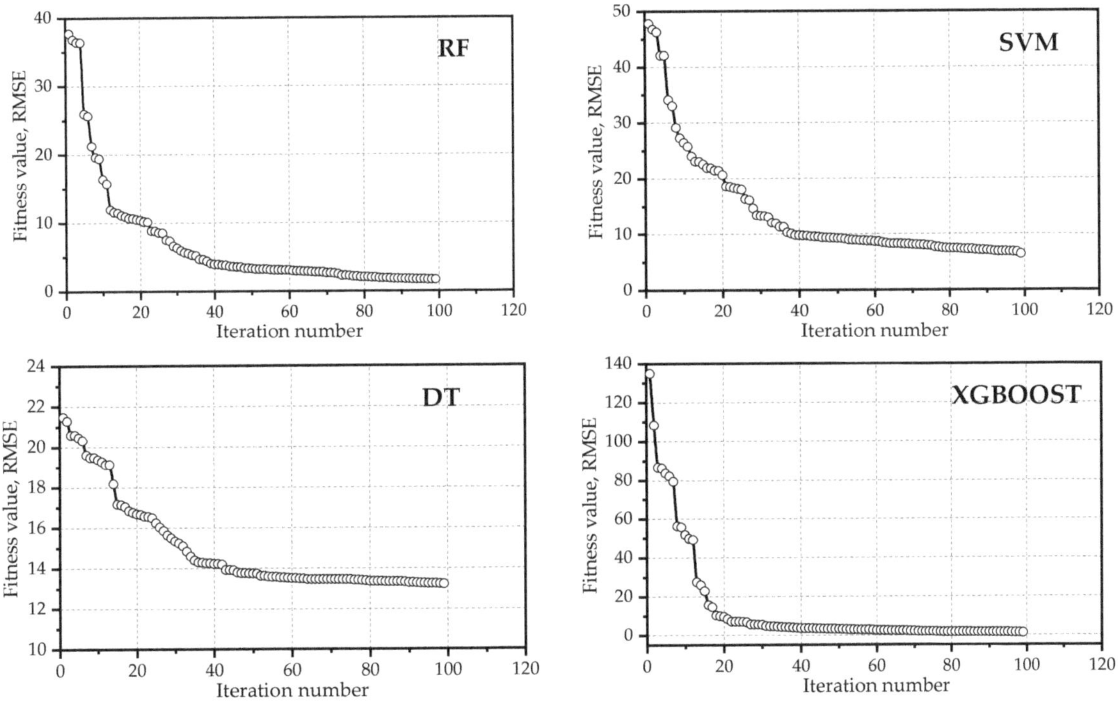

Figure 7. Convergence behavior of ML models.

Table 3 presents the hyperparameters of the ML models along with their optimized values. The optimized values represent the set of hyperparameters that yielded the best average performance according to the RMSE metric.

Table 3. Optimized hyperparameters.

ML Models	Tuned Hyperparameters		
	Names	**Ranges**	**Optimized Values**
RF	Number of variables, mtry	1–4	3
	Minimum node of tree	1–30	2
	Maximum depth of tree	2–100	64
	Number of trees in the forest	1–30	12
SVM	Penalty parameter, Cost	0.1–100	58.75
	Kernel coefficient, gamma	0.01–10	9.44
	Margin of tolerance, Epsilon	0.01–1	0.026
	Kernel type	radial	radial

Table 3. *Cont.*

ML Models	Tuned Hyperparameters		
	Names	Ranges	Optimized Values
DT	Complexity parameter, cp	0.001–1	0.001
	Maximum depth of trees	1–30	20
	Minimum number of splits	2–20	5
	Minimum number of observations at terminal node, minbucket	2–20	6
	Maximum number of splits at node, maxcompete	1–20	9
XGBoost	Learning rate, eta	0.01–1	0.26
	Loss reduction term, gamma	0.01–10	3.79
	L2 regularization term, lambda	0.01–1	0.38
	L1 regularization term, alpha	0.01–1	0.83
	Number of boosting rounds, nrounds	1–100	84
	Maximum depth of trees	2–10	9
	Fraction of samples for each tree, subsample	0.1–1	0.79

5.2. Performance of ML Models

Figure 8 illustrates actual V_s and predicted V_s using the optimized ML models, along with ±10% error lines (red lines). The green lines show a match between actual and predicted V_s values. The results demonstrate that all ML models, except for the DT model, achieved excellent predictive accuracy on both training and testing datasets with high R^2 and $A10 - I$ score values of 1. This shows that the RF, SVM, and XGBoost models can explain all the variance in the V_s using the given features. Furthermore, the scatter plots for these models show that many data points are closer to the error bounds, indicating that the models performed well. In contrast, the DT model achieved lower R^2 and $A10 - I$ values ranging from 0.94 to 0.95 and from 0.77 to 0.78 on the testing and training data, respectively.

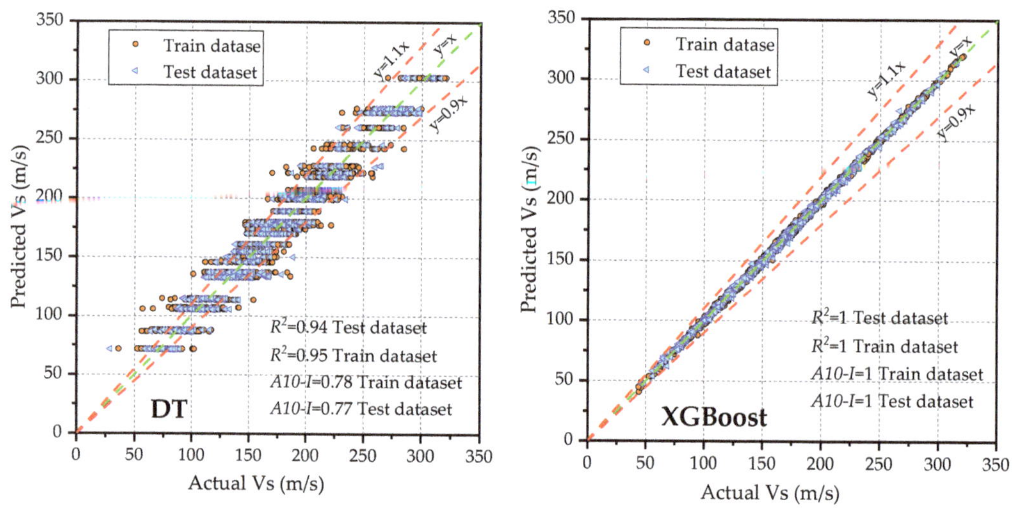

Figure 8. *Cont.*

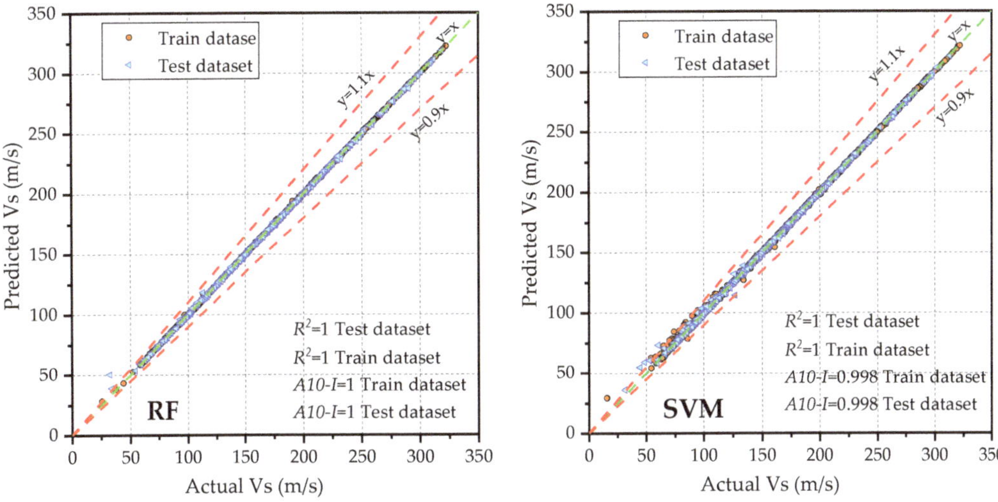

Figure 8. Scatter plot illustrating the correlation between actual V_S and predicted V_S.

The residual plots of ML models are shown in Figure 9, illustrating a random distribution of points around the horizontal orange line at y = 0 (line of zero error). This indicates that the model's predictions are unbiased and have captured the underlying patterns in the data. Additionally, the frequency distributions of residuals (green bars) are shown in the figure. The distribution is approximately symmetric, indicating that the errors are normally distributed, a desirable property. To gain more insight into the performance of the ML models, a further comparison is carried out in the following subsection.

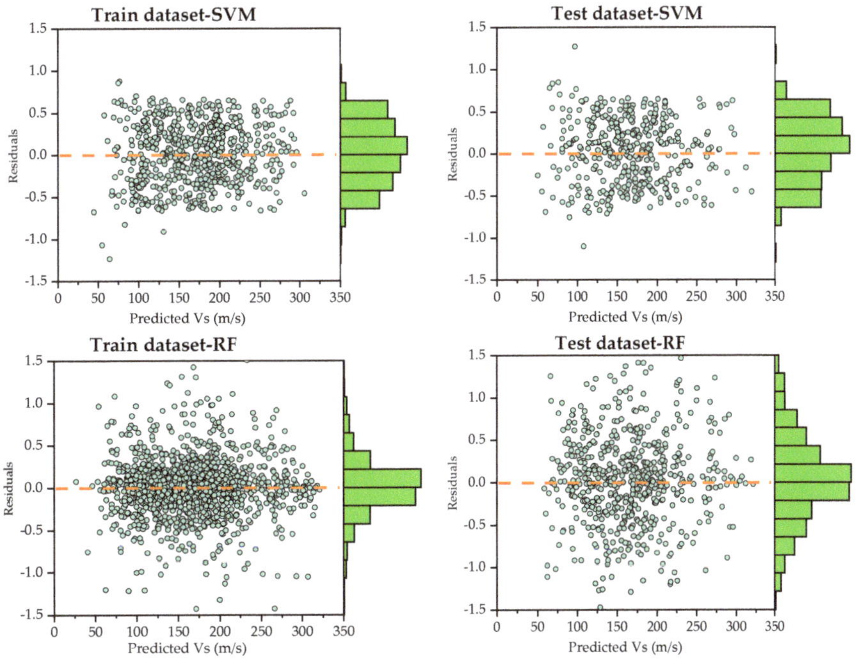

Figure 9. *Cont.*

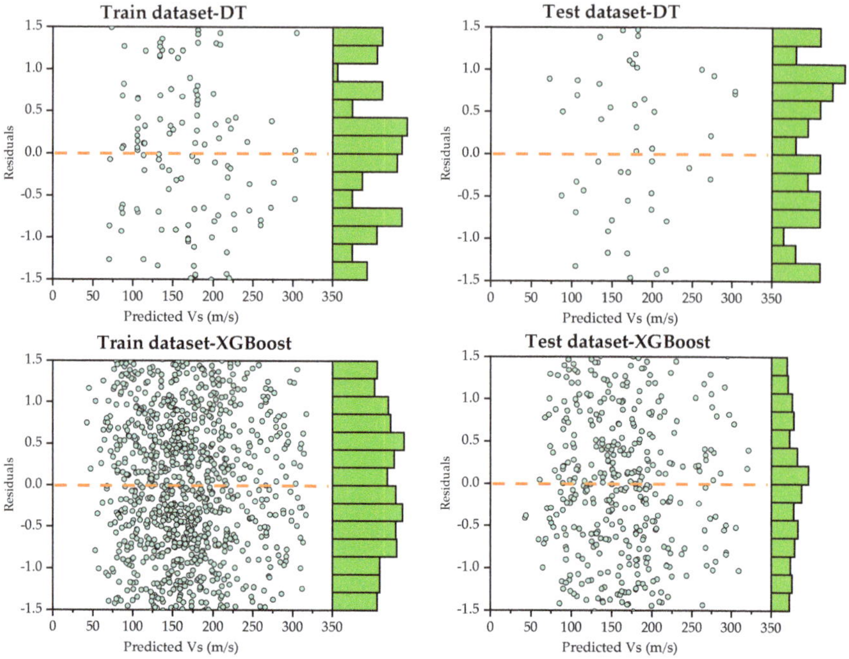

Figure 9. Scatter plots and frequency distributions of residuals.

5.3. Comparisons of ML Models

Table 4 provides a summary of the evaluation results for the ML models. Based on our results, the DT model exhibited lower accuracy, as evidenced by a lower R^2 on both training and testing data. The DT model recorded RMSE of 13.06 and 13.16, MAE of 10.27 m/s and 10.34 m/s, MAPE of 7.27% and 7.31%, and R^2 of 0.95 and 0.94 on the training and testing datasets, respectively. Additionally, Spider charts were utilized to visualize and assess each model's efficiency relative to others (Figure 10). The spider chart shows that the DT model significantly diverged towards higher RMSE, MAPE, and MAE on both training and testing datasets in comparison to other ML models. The RF, SVM, and XGBoost models outperform the DT model in terms of RMSE, MAE, MAPE, and R^2. All RF, SVM, and XGBoost models have lower error values and higher R^2 scores, indicating higher accuracy and better performance in predicting V_S from the input features.

Table 4. Summary of evaluation results for each ML model using training and testing datasets.

Models	Train dataset								Rank
	$A10-I$	RMSE	R^2	PI	SI	MAE	MAPE	U_{95}	
RF	1 (1)	0.46 (1)	1 (1)	0.002 (1)	0.003 (1)	0.24 (1)	0.17 (1)	1.24 (1)	1
SVM	0.998 (2)	1.11 (2)	1 (1)	0.005 (2)	0.007 (2)	0.37 (2)	0.28 (2)	3.07 (2)	2
DT	0.78 (3)	13.1 (4)	0.95 (2)	0.06 (4)	0.08 (4)	10.27 (4)	7.23 (4)	36.20 (4)	4
XGBoost	1 (1)	1.68 (3)	1 (1)	0.007 (3)	0.01 (3)	1.29 (3)	0.87 (3)	4.65 (3)	3
Models	Test dataset								Rank
	$A10-I$	RMSE	R^2	PI	SI	MAE	MAPE	U_{95}	
RF	1 (1)	0.96 (1)	1 (1)	0.004 (1)	0.006 (1)	0.50 (2)	0.36 (2)	2.66 (2)	1
SVM	0.998 (2)	1.36 (2)	1 (1)	0.006 (2)	0.008 (2)	0.38 (1)	0.31 (1)	2.3 (1)	2
DT	0.77 (3)	13.2 (4)	0.94 (2)	0.06 (4)	0.08 (4)	10.34 (4)	7.31 (4)	36.48 (4)	4
XGBoost	1 (1)	1.86 (3)	1 (1)	0.008 (3)	0.01 (3)	1.40 (3)	0.94 (3)	5.16 (3)	3

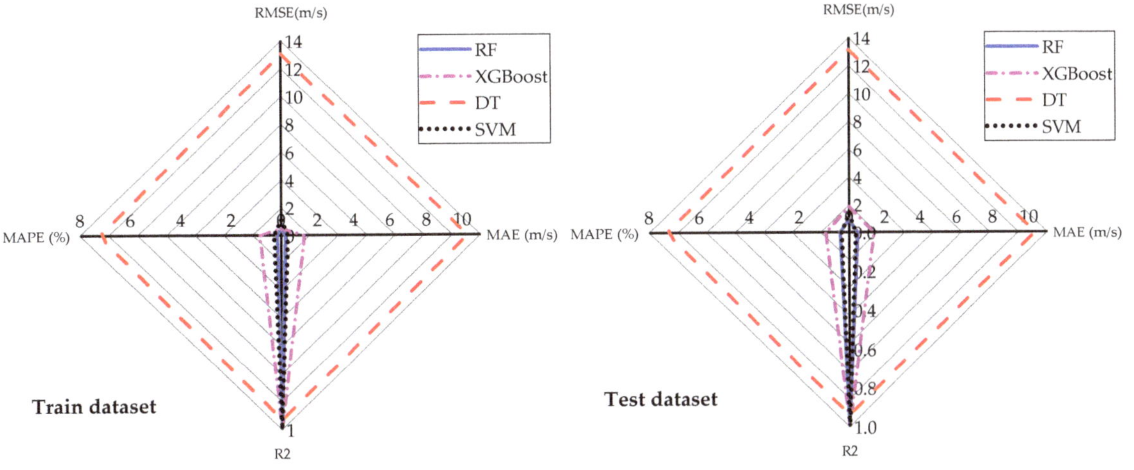

Figure 10. Spider plot showing the performance metrics of the different models.

Figure 11 illustrates the performance of ML models in predicting V_s as evaluated using four performance metrics: U_{95}, $A10-I$, PI, and SI. The results indicate that the DT model achieved lower performance in comparison to other models, as evidenced by its higher U_{95}, PI, and SI scores and lower $A10-I$ values. On the other hand, the RF model demonstrated exceptional performance, outperforming the other models in terms of these performance indicators. In terms of SI, the RF, SVM, and XGBoost models achieved excellent precision (EP), with $SI < 0.05$ on both the training and testing datasets. In contrast, the DT model achieved good precision with $0.05 < SI < 0.1$ on the training and testing datasets. Overall, the RF model ranked first, outperforming the other three ML models, while the DT model ranked fourth. The SVM and XGBoost models ranked second and third, respectively.

To further assess the performance of the ML models, a comparison was made between model-predicted V_S values and estimated V_S values based on existing empirical correlation. A correlation model was selected to estimate V_S from CPT soundings. Equation (12) [80] was utilized for the estimation of V_S from CPT soundings.

$$V_S = 10^{0.31I_c+0.77} \times \sqrt{(q_c - \sigma_{v0})/p_a} \tag{12}$$

where V_S is soil shear wave velocity, I_c is soil behavior type index, q_c is cone tip resistance, σ_{v0} is total overburden pressure, and p_a is atmospheric pressure.

This correlation model served as a benchmark for evaluating the accuracy and reliability of the ML models' predictions. Figure 12 illustrates the models' predictions (red) alongside the profiles of estimated V_S values (black) based on the empirical correlations. The results of this comparison indicate a high level of agreement between the predicted V_S values and the estimated V_S values. This demonstrates that the ML models can produce accurate predictions in line with the established correlations.

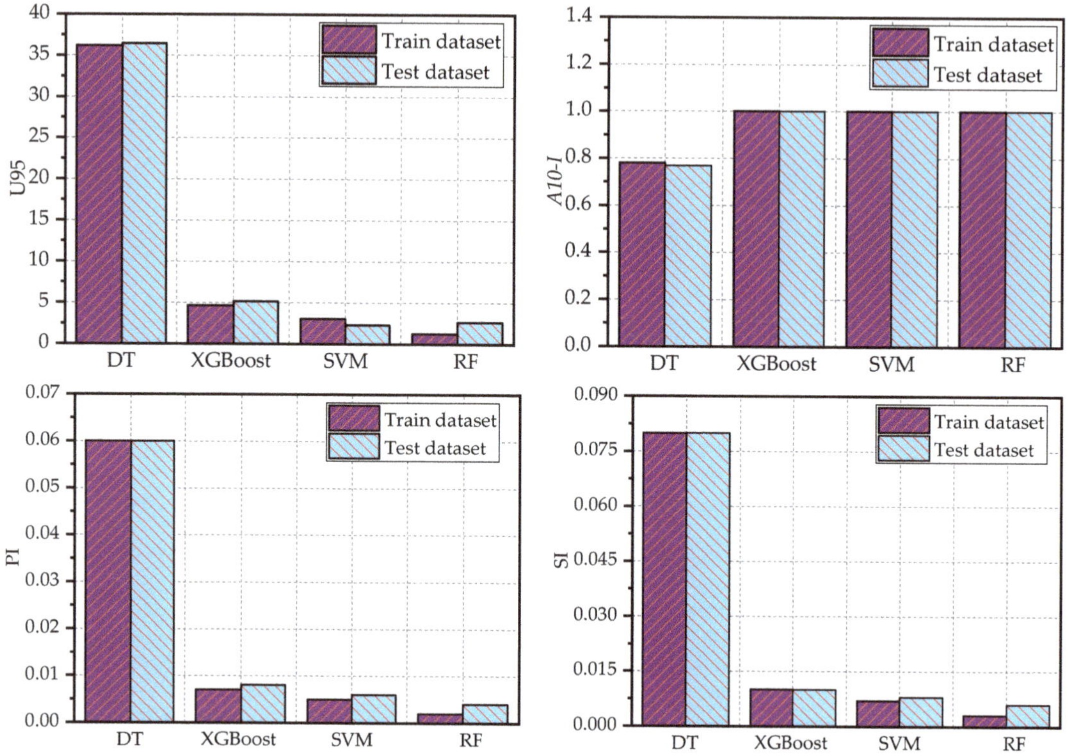

Figure 11. Performance of ML models based on of U_{95} , $A10-I$, PI, and SI indices.

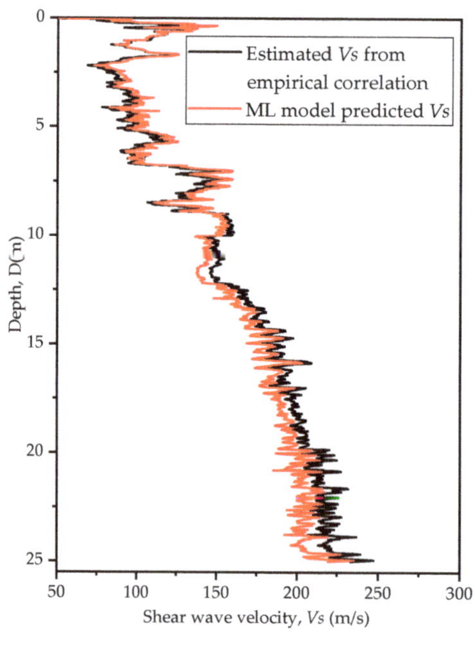

Figure 12. Comparison of predicted and estimated V_S based on empirical correlation.

6. Conclusions

This study utilized various ML algorithms, including RF, SVM, DT, and XGBoost, to predict V_S from CPT data. To train and test these ML models, we used a previously published open-source CPT dataset. The hyperparameters of each ML model were fine-tuned through Bayesian optimization with cross-validation techniques. Eight performance metrics, namely $RMSE$, MAE, $MAPE$, R^2, $A10-I$, SI, PI and U_{95}, provided quantitative evaluation of the models. Based on our results, the following conclusion can be drawn:

- The RF model outperformed the other ML models, achieving the lowest error metrics on both the training and testing datasets. Specifically, it achieved an $RMSE$ of 0.46 and 0.96, an MAE of 0.24 m/s and 0.5 m/s, and an $MAPE$ of 0.17% and 0.36%, respectively. The model also demonstrated low scatter, with SI values of 0.003 and 0.006, and PI values of 0.002 and 0.004 on the training and testing datasets, respectively. Additionally, the RF model achieved R^2 and $A10-I$ values of 1 on both datasets, indicating a perfect fit. Furthermore, the RF model recorded the lowest uncertainty, with a U_{95} value of 1.24 on the training dataset.
- The SVM and XGBoost models also exhibited strong performance, with slightly higher error metrics compared with the RF model. These two models ranked second and third, respectively, following the RF model, which achieved the highest performance. However, the DT model performed poorly, with higher error rates and uncertainty in predicting V_s.
- The RF model demonstrated its overall superior performance and high accuracy in predicting soil V_s, even when trained with minimal input features. Hence, owing to its excellent performance across multiple metrics, the RF model can be integrated into a software package for rapid and accurate prediction of soil V_s.
- In summary, while this study relied solely on CPT data for training ML models, it is important to recognize the limitations of the CPT, particularly its primary suitability for fine-grained soils. To further enhance the application of ML models in soil characterization, future research should consider incorporating experimental results and data for coarse-grained soil types.

Author Contributions: Conceptualization, A.T.C. and R.P.R.; methodology, A.T.C.; software, A.T.C.; validation, A.T.C. formal analysis, A.T.C.; writing—original draft preparation, A.T.C.; writing—review and editing, R.P.R.; visualization, A.T.C.; supervision, R.P.R. All authors have read and agreed to the published version of the manuscript.

Funding: This publication was financially supported by Széchenyi István University.

Institutional Review Board Statement: Not applicable.

Informed Consent Statement: Not applicable.

Data Availability Statement: The CPT data can be downloaded from the following link: https://www.tugraz.at/en/institutes/ibg/research/computational-geotechnics-group/database/ (accessed on 12 May 2023).

Conflicts of Interest: The authors declare no conflict of interest.

Abbreviations

$A10-I$	Engineering index with ±10% deviation	R^2	Coefficient of determination
ANN	Artificial neural network	R_f	Friction ratio
CPT	Cone penetration test	RF	Random forest
D	Depth of soil (m)	$RMSE$	Root mean squared error
DT	Decision trees	$SCPT$	Seismic cone penetration testing
F_r	Normalized friction ratio	SD	Standard deviation
f_s	sleeve friction	SI	Scatter index
I_c	Soil behavioral type index	SVM	Support vector machine

IQR	interquartile range	U_{95}	Uncertainty at 95% confidence interval
MAE	Mean absolute error	V_s	Shear wave velocity
$MAPE$	Mean absolute percentage error	σ_{v0}	Total overburden stress
$MASW$	Multi-Channel Analysis of Surface Waves	$\sigma\prime_{v0}$	Effective overburden stress
ML	Machine learning	Pa	Atmospheric pressure
n	Total number of datasets	$Q3$	Third quartile
Pa	Atmospheric pressure	$\overline{X}$	Mean
PI	Performance index	$\widehat{X}_i$	Predicted value of ith observation
q_c	cone tip resistance	X_i	Actual value of ith observation
$Q1$	First quartile	$XGBoost$	Extreme gradient boosting

References

1. Kalinina, A.V.; Ammosov, S.M.; Bykova, V.V.; Tatevossian, R.E. Effect of the Upper Part of the Soil Profile on the Site Response. *Seism. Instrum.* **2018**, *54*, 499–513. [CrossRef]
2. Kawase, H. Site Effects on Strong Ground Motions. In *International Geophysics*; Kyushu University: Fukuoka, Japan, 2003; Volume 81, pp. 1013–1030.
3. Borcherdt, R.D.; Glassmoyer, G. On the Characteristics of Local Geology and Their Influence on Ground Motions Generated by the Loma Prieta Earthquake in the San Francisco Bay Region, California. *Bull. Seismol. Soc. Am.* **1992**, *82*, 603–641. [CrossRef]
4. Hanks, T.C.; Krawinkler, H. The 1989 Loma Prieta Earthquake and Its Effects: Introduction to the Special Issue. *Bull. Seismol. Soc. Am.* **1991**, *81*, 1415–1423. [CrossRef]
5. Aki, K. Local Site Effect on Ground Motion. *Am. Soc. Civil Eng.* **1988**, *20*, 103–155.
6. Tokimatsu, K. Geotechnical Site Characterization Using Surface Waves. In *Earthquake Geotechnical Engineering, Proceedings of the IS-Tokyo'95, the First International Conference on Earthquake Geotechnical Engineering, Tokyo, Japan, 14–16 November 1995*; A.A. Balkema: Rotterdam, The Netherlands, 1995; pp. 1136–1333.
7. Robertson, P.K.; Campanella, R.G. Interpretation of CPT-Sand&Clay. *Can. Geotech. J.* **1983**, *20*, 718–733.
8. Park, C.B.; Miller, R.D.; Xia, J. Multichannel Analysis of Surface Waves. *Geophysics* **1999**, *64*, 800–808. [CrossRef]
9. Aka, M.; Agbasi, O. Delineation of Weathered Layer Using Uphole and Surface Seismic Refraction Methods in Parts of Niger Delta, Nigeria: Delineation of Weathered Layer. *Sultan Qaboos Univ. J. Sci. SQUJS* **2021**, *26*, 58–66. [CrossRef]
10. Musgrave, A.W. *Seismic Refraction Prospecting*; Society of Exploration Geophysicists: Houston, TX, USA, 1967; ISBN 1560802677.
11. Viggiani, G.; Atkinson, J.H. Interpretation of Bender Element Tests. *Geotechnique* **1995**, *45*, 149–154. [CrossRef]
12. Nishio, S.; Tamaoki, K. Measurement of Shear Wave Velocities in Diluvial Gravel Samples Under Triaxial Conditions. *Soils Found.* **1988**, *28*, 35–48. [CrossRef]
13. Drnevich, V. *Resonant-Column Testing—Problems and Solutions*; ASTM International: Singapore, 1978.
14. Le, T.T.; Skentou, A.D.; Mamou, A.; Asteris, P.G. *Correlating the Unconfined Compressive Strength of Rock with the Compressional Wave Velocity Effective Porosity and Schmidt Hammer Rebound Number Using Artificial Neural Networks*; Springer: Vienna, Austria, 2022; Volume 55, ISBN 0123456789.
15. Andrus, R.D.; Mohanan, N.P.; Piratheepan, P.; Ellis, B.S.; Holzer, T.L. Predicting Shear-Wave Velocity From Cone Penetration Resistance. In Proceedings of the 4th International Conference on Earthquake Geotechnical Engineering, Thessaloniki, Greece, 25–28 June 2007.
16. Robertson, P.K. Interpretation of Cone Penetration Tests—A Unified Approach. *Can. Geotech. J.* **2009**, *46*, 1337–1355. [CrossRef]
17. Wolf, Á.; Ray, R.P. Comparison and Improvement of the Existing Cone Penetration Test Results. Shear Wave Velocity Correlations for Hungarian Soils. *Int. J. Environ. Chem. Ecol. Geol. Geophys. Eng.* **2017**, *11*, 338–347.
18. Mayne, P.W.; Rix, G.J. Correlations Between Shear Wave Velocity and Cone Tip Resistance in Natural Clays. *Soils Found.* **1995**, *35*, 107–110. [CrossRef] [PubMed]
19. Tonni, L.; Simonini, P. Shear Wave Velocity as Function of Cone Penetration Test Measurements in Sand and Silt Mixtures. *Eng. Geol.* **2013**, *163*, 55–67. [CrossRef]
20. Robertson, P.K. Cone Penetration Test (CPT)-Based Soil Behaviour Type (SBT) Classification System—An Update. *Can. Geotech. J.* **2016**, *53*, 1910–1927. [CrossRef]
21. Robertson, P.K.; Campanella, R.G.; Gillespie, D.; Greig, J. Use of Piezometer Cone Data. In *Use of In Situ Tests in Geotechnical Engineering*; ASCE: Reston, VA, USA, 1986; pp. 1263–1280.
22. Chen, J.; Vissinga, M.; Shen, Y.; Hu, S.; Beal, E.; Newlin, J. Machine Learning–Based Digital Integration of Geotechnical and Ultrahigh–Frequency Geophysical Data for Offshore Site Characterizations. *J. Geotech. Geoenvironmental Eng.* **2021**, *147*, 04021160. [CrossRef]
23. Olayiwola, T.; Tariq, Z.; Abdulraheem, A.; Mahmoud, M. Evolving Strategies for Shear Wave Velocity Estimation: Smart and Ensemble Modeling Approach. *Neural Comput. Appl.* **2021**, 33, 17147–17159. [CrossRef]
24. Assaf, J.; Molnar, S.; El Naggar, M.H. CPT-Vs Correlations for Post-Glacial Sediments in Metropolitan Vancouver. *Soil Dyn. Earthq. Eng.* **2023**, *165*, 107693. [CrossRef]

25. Tsiaousi, D.; Travasarou, T.; Drosos, V.; Ugalde, J.; Chacko, J. Machine Learning Applications for Site Characterization Based on CPT Data. In Proceedings of the Geotechnical Earthquake Engineering and Soil Dynamics V, Austin, TX, USA, 10–13 June 2018; American Society of Civil Engineers: Reston, VA, USA, 2018; pp. 461–472.
26. Taheri, A.; Makarian, E.; Manaman, N.S.; Ju, H.; Kim, T.H.; Geem, Z.W.; Rahimizadeh, K. A Fully-Self-Adaptive Harmony Search GMDH-Type Neural Network Algorithm to Estimate Shear-Wave Velocity in Porous Media. *Appl. Sci.* **2022**, *12*, 6339. [CrossRef]
27. Kang, T.H.; Choi, S.W.; Lee, C.; Chang, S.H. Soil Classification by Machine Learning Using a Tunnel Boring Machine's Operating Parameters. *Appl. Sci.* **2022**, *12*, 11480. [CrossRef]
28. Carvalho, L.O.; Ribeiro, D.B. Soil Classification System from Cone Penetration Test Data Applying Distance-Based Machine Learning Algorithms. *Soils Rocks* **2019**, *42*, 167–178. [CrossRef]
29. Eyo, E.; Abbey, S. Multiclass Stand-Alone and Ensemble Machine Learning Algorithms Utilised to Classify Soils Based on Their Physico-Chemical Characteristics. *J. Rock Mech. Geotech. Eng.* **2022**, *14*, 603–615. [CrossRef]
30. Hikouei, I.S.; Kim, S.S.; Mishra, D.R. Machine-Learning Classification of Soil Bulk Density in Salt Marsh Environments. *Sensors* **2021**, *21*, 4408. [CrossRef]
31. Aydın, Y.; Işıkdağ, Ü.; Bekdaş, G.; Nigdeli, S.M.; Geem, Z.W. Use of Machine Learning Techniques in Soil Classification. *Sustainability* **2023**, *15*, 2374. [CrossRef]
32. Carvalho, L.O.; Ribeiro, D.B. A Multiple Model Machine Learning Approach for Soil Classification from Cone Penetration Test Data. *Soils Rocks* **2021**, *44*, 1–14. [CrossRef]
33. Chala, A.T.; Ray, R. Assessing the Performance of Machine Learning Algorithms for Soil Classification Using Cone Penetration Test Data. *Appl. Sci.* **2023**, *13*, 5758. [CrossRef]
34. Akhundi, H.; Ghafoori, M.; Lashkaripour, G. Prediction of Shear Wave Velocity Using Artificial Neural Network Technique, Multiple Regression and Petrophysical Data: A Case Study in Asmari Reservoir (SW Iran). *Open J. Geol.* **2014**, *4*, 303–313. [CrossRef]
35. Demir, S.; Sahin, E.K. An Investigation of Feature Selection Methods for Soil Liquefaction Prediction Based on Tree-Based Ensemble Algorithms Using AdaBoost, Gradient Boosting, and XGBoost. *Neural Comput. Appl.* **2023**, *35*, 3173–3190. [CrossRef]
36. Demir, S.; Şahin, E.K. Liquefaction Prediction with Robust Machine Learning Algorithms (SVM, RF, and XGBoost) Supported by Genetic Algorithm-Based Feature Selection and Parameter Optimization from the Perspective of Data Processing. *Environ. Earth Sci.* **2022**, *81*, 459. [CrossRef]
37. Samui, P.; Sitharam, T.G. Machine Learning Modelling for Predicting Soil Liquefaction Susceptibility. *Nat. Hazards Earth Syst. Sci.* **2011**, *11*, 1–9. [CrossRef]
38. Ozsagir, M.; Erden, C.; Bol, E.; Sert, S.; Özocak, A. Machine Learning Approaches for Prediction of Fine-Grained Soils Liquefaction. *Comput. Geotech.* **2022**, *152*, 105014. [CrossRef]
39. Alobaidi, M.H.; Meguid, M.A.; Chebana, F. Predicting Seismic-Induced Liquefaction through Ensemble Learning Frameworks. *Sci. Rep.* **2019**, *9*, 11786. [CrossRef] [PubMed]
40. Jas, K.; Dodagoudar, G.R. Explainable Machine Learning Model for Liquefaction Potential Assessment of Soils Using XGBoost-SHAP. *Soil Dyn. Earthq. Eng.* **2023**, *165*, 107662. [CrossRef]
41. Wang, L.; Wu, C.; Tang, L.; Zhang, W.; Lacasse, S.; Liu, H.; Gao, L. Efficient Reliability Analysis of Earth Dam Slope Stability Using Extreme Gradient Boosting Method. *Acta Geotech.* **2020**, *15*, 3135–3150. [CrossRef]
42. Zhang, W.; Zhang, R.; Wu, C.; Goh, A.T.C.; Wang, L. Assessment of Basal Heave Stability for Braced Excavations in Anisotropic Clay Using Extreme Gradient Boosting and Random Forest Regression. *Undergr. Space* **2022**, *7*, 233–241. [CrossRef]
43. Bharti, J.P.; Mishra, P.; Moorthy, U.; Sathishkumar, V.E.; Cho, Y.; Samui, P. Slope Stability Analysis Using Rf, Gbm, Cart, Bt and Xgboost. *Geotech. Geol. Eng.* **2021**, *39*, 3741–3752. [CrossRef]
44. Samui, P. Slope Stability Analysis: A Support Vector Machine Approach. *Environ. Geol.* **2008**, *56*, 255–267. [CrossRef]
45. Xiao, L.; Zhang, Y.; Peng, G. Landslide Susceptibility Assessment Using Integrated Deep Learning Algorithm along the China-Nepal Highway. *Sensors* **2018**, *18*, 4436. [CrossRef] [PubMed]
46. Nejad, F.P.; Jaksa, M.B. Load-Settlement Behavior Modeling of Single Piles Using Artificial Neural Networks and CPT Data. *Comput. Geotech.* **2017**, *89*, 9–21. [CrossRef]
47. Nejad, F.P.; Jaksa, M.B.; Kakhi, M.; McCabe, B.A. Prediction of Pile Settlement Using Artificial Neural Networks Based on Standard Penetration Test Data. *Comput. Geotech.* **2009**, *36*, 1125–1133. [CrossRef]
48. Chen, R.; Zhang, P.; Wu, H.; Wang, Z.; Zhong, Z. Prediction of Shield Tunneling-Induced Ground Settlement Using Machine Learning Techniques. *Front. Struct. Civ. Eng.* **2019**, *13*, 1363–1378. [CrossRef]
49. Riyadi, Z.A.; Husen, M.H.; Lubis, L.A.; Ridwan, T.K. The Implementation of TPE-Bayesian Hyperparameter Optimization to Predict Shear Wave Velocity Using Machine Learning: Case Study From X Field in Malay Basin. *Pet. Coal* **2022**, *64*, 467–488.
50. Shooshpasha, I.; Kordnaeij, A.; Dikmen, U.; Molaabasi, H.; Amir, I. Shear Wave Velocity by Support Vector Machine Based on Geotechnical Soil Properties. *Nat. Hazards Earth Syst. Sci.* **2014**, *2*, 2443–2461. [CrossRef]
51. Bagheripour, P.; Gholami, A.; Asoodeh, M.; Vaezzadeh-Asadi, M. Support Vector Regression Based Determination of Shear Wave Velocity. *J. Pet. Sci. Eng.* **2015**, *125*, 95–99. [CrossRef]
52. Oberhollenzer, S.; Premstaller, M.; Marte, R.; Tschuchnigg, F.; Erharter, G.H.; Marcher, T. Cone Penetration Test Dataset Premstaller Geotechnik. *Data Brief* **2021**, *34*, 106618. [CrossRef] [PubMed]
53. Esmaeili-Falak, M.; Benemaran, R.S. Ensemble Deep Learning-Based Models to Predict the Resilient Modulus of Modified Base Materials Subjected to Wet-Dry Cycles. *Geomech. Eng.* **2023**, *32*, 583–600.

54. Harandizadeh, H.; Armaghani, D.J.; Asteris, P.G.; Gandomi, A.H. *TBM Performance Prediction Developing a Hybrid ANFIS-PNN Predictive Model Optimized by Imperialism Competitive Algorithm*; Springer: London, UK, 2021; Volume 33, ISBN 0052102106.
55. Hajihassani, M.; Abdullah, S.S.; Asteris, P.G.; Armaghani, D.J. A Gene Expression Programming Model for Predicting Tunnel Convergence. *Appl. Sci.* **2019**, *9*, 4650. [CrossRef]
56. Li, Z.; Bejarbaneh, B.Y.; Asteris, P.G.; Koopialipoor, M.; Armaghani, D.J.; Tahir, M.M. A Hybrid GEP and WOA Approach to Estimate the Optimal Penetration Rate of TBM in Granitic Rock Mass. *Soft Comput.* **2021**, *25*, 11877–11895. [CrossRef]
57. Abushanab, A.; Wakjira, T.G.; Alnahhal, W. Machine Learning-Based Flexural Capacity Prediction of Corroded RC Beams with an Efficient and User-Friendly Tool. *Sustainability* **2023**, *15*, 4824. [CrossRef]
58. Wakjira, T.G.; Ebead, U.; Alam, M.S. Machine Learning-Based Shear Capacity Prediction and Reliability Analysis of Shear-Critical RC Beams Strengthened with Inorganic Composites. *Case Stud. Constr. Mater.* **2022**, *16*, e01008. [CrossRef]
59. Ahmed, H.U.; Mohammed, A.S.; Faraj, R.H.; Abdalla, A.A.; Qaidi, S.M.A.; Sor, N.H.; Mohammed, A.A. Innovative Modeling Techniques Including MEP, ANN and FQ to Forecast the Compressive Strength of Geopolymer Concrete Modified with Nanoparticles. *Neural Comput. Appl.* **2023**, *35*, 12453–12479. [CrossRef]
60. Skentou, A.D.; Bardhan, A.; Mamou, A.; Lemonis, M.E.; Kumar, G.; Samui, P.; Armaghani, D.J.; Asteris, P.G. Closed-Form Equation for Estimating Unconfined Compressive Strength of Granite from Three Non-Destructive Tests Using Soft Computing Models. *Rock Mech. Rock Eng.* **2023**, *56*, 487–514. [CrossRef]
61. Xu, H.; Zhou, J.; Asteris, P.G.; Armaghani, D.J.; Tahir, M.M. Supervised Machine Learning Techniques to the Prediction of Tunnel Boring Machine Penetration Rate. *Appl. Sci.* **2019**, *9*, 3715. [CrossRef]
62. Mahmood, W.; Mohammed, A. Performance of ANN and M5P-Tree to Forecast the Compressive Strength of Hand-Mix Cement-Grouted Sands Modified with Polymer Using ASTM and BS Standards and Evaluate the Outcomes Using SI with OBJ Assessments. *Neural Comput. Appl.* **2022**, *34*, 15031–15051. [CrossRef]
63. Abdalla, A.; Salih, A. Implementation of Multi-Expression Programming (MEP), Artificial Neural Network (ANN), and M5P-Tree to Forecast the Compression Strength Cement-Based Mortar Modified by Calcium Hydroxide at Different Mix Proportions and Curing Ages. *Innov. Infrastruct. Solut.* **2022**, *7*, 1–15. [CrossRef]
64. Shi, X.; Yu, X.; Esmaeili-Falak, M. Improved Arithmetic Optimization Algorithm and Its Application to Carbon Fiber Reinforced Polymer-Steel Bond Strength Estimation. *Compos. Struct.* **2023**, *306*, 116599. [CrossRef]
65. Behar, O.; Khellaf, A.; Mohammedi, K. Comparison of Solar Radiation Models and Their Validation under Algerian Climate—The Case of Direct Irradiance. *Energy Convers. Manag.* **2015**, *98*, 236–251. [CrossRef]
66. Cutler, D.R.; Edwards, T.C.; Beard, K.H.; Cutler, A.; Hess, K.T.; Gibson, J.; Lawler, J.J. Random Forests for Classification in Ecology. *Ecology* **2007**, *88*, 2783–2792. [CrossRef] [PubMed]
67. Breiman, L. Random Forests. *Mach. Learn.* **2001**, *45*, 5–32. [CrossRef]
68. Liaw, A.; Wiener, M. Classification and Regression by RandomForest. *R News* **2002**, *2*, 18–22.
69. Hastie, T.; Tibshirani, R.; Friedman, J.H.; Friedman, J.H. *The Elements of Statistical Learning: Data Mining, Inference, and Prediction*; Springer: Berlin/Heidelberg, Germany, 2009; Volume 2.
70. Wright, M.N.; Ziegler, A. Ranger: A Fast Implementation of Random Forests for High Dimensional Data in C++ and R. *arXiv* **2015**, arXiv:1508.04409. [CrossRef]
71. Snoek, J.; Larochelle, H.; Adams, R.P. Practical Bayesian Optimization of Machine Learning Algorithms. *arXiv* **2012**, arXiv:1206.2944.
72. Cortes, C.; Vapnik, V. Support-Vector Networks. *Mach. Learn.* **1995**, *20*, 273–297. [CrossRef]
73. Meyer, D.; Dimitriadou, E.; Hornik, K.; Weingessel, A.; Leisch, F. *E1071: Misc Functions of the Department of Statistics, Probability Theory Group (Formerly: E1071), TU Wien*; R Package Version 1.7-13; R Core Team: Vienna, Austria, 2023.
74. Quinlan, J.R. *Induction of Decision Trees*; Springer: Berlin/Heidelberg, Germany, 1986; Volume 1.
75. Therneau, T.; Atkinson, B.; Ripley, B.; Ripley, M.B. *Rpart: Recursive Partitioning and Regression Trees*, R Package version 4.1-10; R Core Team: Vienna, Austria, 2015; pp. 1–9.
76. Song, Y.Y.; Lu, Y. Decision Tree Methods: Applications for Classification and Prediction. *Shanghai Arch. Psychiatry* **2015**, *27*, 130–135. [CrossRef] [PubMed]
77. Bhattacharya, B.; Solomatine, D.P. Machine Learning in Soil Classification. *Neural Networks* **2006**, *19*, 186–195. [CrossRef]
78. Chen, T.; Guestrin, C. Xgboost: A Scalable Tree Boosting System. In Proceedings of the 22nd ACM SIGKDD International Conference on Knowledge Discovery and Data Mining, San Francisco, CA, USA, 13–17 August 2016; pp. 785–794.
79. Fei, Z.; Liang, S.; Cai, Y.; Shen, Y. Ensemble Machine-Learning-Based Prediction Models for the Compressive Strength of Recycled Powder Mortar. *Materials* **2023**, *16*, 583. [CrossRef]
80. Ray, R.P.; Wolf, A.; Kegyes-Brassai, O. Harmonizing Dynamic Property Measurements of Hungarian Soils. In Proceedings of the 6th International Conference on Geotechnical and Geophysical Site Characterization (ISC2020), Budapest, Hungary, 7–11 September 2020.

applied sciences

MDPI

Article

Research on Multi-Objective Optimization Model of Foundation Pit Dewatering Based on NSGA-II Algorithm

Zhiheng Ma [1], Jinguo Wang [1], Yanrong Zhao [1,*], Bolin Li [1,*] and Yufeng Wei [2]

[1] School of Earth Sciences and Engineering, Hohai University, Nanjing 210098, China; zhihengma@163.com (Z.M.); wang_jinguo@hhu.edu.cn (J.W.)

[2] China Energy Engineering Jiangsu Power Design Institute Co., Ltd., Nanjing 211102, China; weiyufeng@jspdi.com.cn

* Correspondence: zhaoyanrong@hhu.edu.cn (Y.Z.); 221309080013@hhu.edu.cn (B.L.)

Abstract: This study focuses on optimizing the foundation pit dewatering scheme using the foundation pit dewatering theory and the principles of multi-objective optimization. It explores the development of a multi-objective optimization model and efficient solution technology for foundation pit dewatering. This research focuses on the foundation pit dewatering project at the inverted siphon section of Xixiayuan canal head, specifically from pile number XZ0+326 to XZ0+500. It establishes an optimized mathematical model for foundation pit dewatering that incorporates three objectives. Additionally, a dewatering optimization program is developed by utilizing the MATLAB optimization toolbox and the multi-objective optimization algorithm program based on the NSGA-II algorithm (Gamultiobj). The multi-objective optimization mathematical model is solved, and a Pareto-optimal solution set with uniform distribution is obtained. The multi-objective optimization evaluation system based on AHP is constructed from the three aspects of dewatering cost, the impact of settlement on the environment, and the safety and stability of the foundation pit. The optimization scheme of the Pareto-optimal solution set is selected as the decision result to provide multiple feasible schemes for the dewatering construction of foundation pits. The optimization scheme is verified by using the GMS software. The simulation results demonstrate that the optimization scheme fulfills the requirements for water level and settlement control. Moreover, the developed optimization program efficiently solves the multi-objective optimization problem associated with foundation pit dewatering. Lastly, an evaluation system incorporating the NSGA-II algorithm and AHP is developed and utilized in the context of dewatering engineering in order to offer multiple viable optimal dewatering schemes.

Keywords: multi-objective optimization; NSGA-II algorithm; Pareto-optimal solution set; evaluation system; foundation pit dewatering

Citation: Ma, Z.; Wang, J.; Zhao, Y.; Li, B.; Wei, Y. Research on Multi-Objective Optimization Model of Foundation Pit Dewatering Based on NSGA-II Algorithm. *Appl. Sci.* **2023**, *13*, 10865. https://doi.org/10.3390/app131910865

Academic Editor: Tiago Miranda

Received: 20 August 2023
Revised: 24 September 2023
Accepted: 25 September 2023
Published: 29 September 2023

1. Introduction

In the construction process of various large-scale projects, the dewatering design of the foundation pit is one of the most important technical and scientific issues. The management objectives of foundation pit dewatering under different working conditions (minimum dewatering cost, minimum land subsidence, maximum drawdown of the foundation pit center water level, etc.), dewatering engineering design (well depth, well diameter, etc.), and geological environment constraints (maximum allowable pumping flow of a single well, allowable value of land subsidence, etc.) should be unified in the optimization model. Taking the strong permeable foundation pit dewatering project in the inverted siphon section of the head of the water diversion project of the Xixiayuan Water Conservancy Project in Henan Province as an example, the multi-objective optimization model of foundation pit dewatering based on the NSGA-II algorithm is established in combination with the multi-objective requirements of groundwater level reduction, settlement deformation control, the groundwater environment, and the economic cost.

In recent years, significant progress has been made in the research and application of optimization algorithms at home and abroad. The genetic algorithm, simulated annealing, the Ant colony algorithm, and particle swarm optimization have made important breakthroughs in theoretical research and practical application [1]. The utilization of optimization algorithms is crucial in enhancing efficiency, reducing costs, and optimizing resource utilization.

Reza introduced a variety of multi-objective optimization algorithms to optimize the power generation efficiency of the hydropower station reservoir [2]. Wang applied the NSGA-II algorithm to power grid optimization planning, established the design model of multi-objective power grid planning, and provided schemes for the trade-off analysis of various objectives. Based on the objective function method [3], Xu optimized and analyzed the dewatering plan for a subway foundation pit with the minimum total water inflow as the objective function and combined with the water level constraints of each control point [4]. Liu applied the genetic algorithm to determine the optimal number of wells for foundation pit dewatering and used a simple and efficient genetic algorithm to control a set of model sets to obtain the overall optimal plan [5]. Yang proposed a NPTSGA algorithm, which combines the genetic algorithm and the tabu search algorithm. The algorithm was used to simulate and optimize the problem of seawater intrusion in coastal areas [6]. Fazli incorporated the crossover operator of the genetic algorithm into the position change phase of the firefly algorithm, and fused the two algorithms to solve the optimization problem [7]. Geng introduced the scatter search algorithm into the computational framework of the particle swarm optimization algorithm, and gave full play to their fast convergence characteristics to study the vehicle scheduling problem with uncertain traffic flow [8]. Li proposed a multi-objective optimization algorithm based on particle swarm optimization, which guided the particle swarm to search more fully, improved the diversity and distribution of its non-inferior solutions, and verified the effectiveness of the particle swarm optimization algorithm by using three multi-objective test functions [9]. Ma systematically summarized the basic principle of the genetic algorithm and introduced the simulated annealing algorithm. The annealing operation was added to the original genetic algorithm, and the algorithm was improved to solve the multi-objective optimization model of subway engineering [10].

Nima proposed a new metaheuristic algorithm, the Crystal structure algorithm, which can effectively handle multi-objective problems [11]. Mohamed helped the metaheuristic algorithm to achieve better results in multi-objective optimization problems based on the marine predators algorithm proposed in recent years, and, compared with other algorithms, achieved remarkable results [12]. Thanh proposed a new Shrimp and Gobi joint search algorithm (SGA) for solving large-scale global optimization problems. This algorithm avoids local optima better than population-based algorithms and has faster convergence speed. Thanh also proposed an improved Grey Wolf optimizer (GWO) algorithm, which improves the speed of the algorithm and can be used to study structural damage identification in high-dimensional problems [13,14]. Matteo proposed that the EPLANopt model developed by the Eurac Research Institute was coupled with the multi-objective evolutionary algorithm of DEAP based on Python, which solved the multi-objective optimization problem of optimizing different energy sources [15]. Zhang proposed an improved particle swarm optimization algorithm to solve the model of multimodal multi-objective problems, and introduced the dynamic neighborhood learning strategy instead of the global learning strategy to enhance the diversity of the population [16]. Srinivas studied the concept of non-dominated sorting called the Goldberg algorithm, while searching for multiple Pareto optimal niche and species formation methods. This method can be extended to higher dimensional and more difficult multi-objective problems [17]. Xu proposed a new multi-objective constraint optimization model, which can normalize the weighted sum of the original objective function and the degree of constraint violation on the basis of minimizing both the objective function and the degree of constraint violation (the degree of violation of each constraint or its sum) [18]. Mirjalili proposed a new multi-objective Grasshopper

Optimization algorithm based on the navigation of locust swarms in nature. The algorithm can estimate the Pareto optimal frontier of multi-objective problems by combining the target selection and archiving technology [19]. Wang proposed an effective coevolutionary multi-group garden balm optimization algorithm (CMGBO) to ensure the convergence of Pareto regions with good diversity [20]. Zhang proposed an evolutionary strategy for solving multi-modal and multi-objective optimization problems, mainly studying the strategy of finding solutions with good convergence and distribution in the decision space. This strategy can effectively solve multiple groups of optimal solutions simultaneously [21]. Gaurav introduced a multi-objective Seagull Optimization Algorithm. This algorithm introduced the concept of dynamic archiving and had the characteristics of caching the non-dominated Pareto optimal solution [22]. Muhammad proposed a search-based software engineering solution by using a multi-objective evolutionary algorithm. The results of the algorithm can be tested under the background of different objectives and two quality indicators. The results reveal the influence of the attribute of the feature model, the implementation environment, and the number of objectives on the performance of the algorithm [23]. Guan established a multi-objective water supply optimization model considering cost, reliability, and water quality for the mountain water distribution network (WDN). The NSGA-II algorithm was used to optimize the WDN design model in the complex terrain of the mountain area, which provided valuable information for the decision makers in the complex terrain WDN [24]. Li has developed a proxy-assisted stochastic optimization inversion algorithm called "dam parameter identification". This algorithm assesses the influence of randomly selected training and testing datasets on the modeling and prediction outcomes of artificial neural networks [25]. Huynh used the dataset collected from the Mekong River test project as an example to train and test a multi-objective dataset by evaluating the results of on-site load tests [26].

Taking the multi-objective optimization model of foundation pit dewatering as the main research objective, this paper carries out research through theoretical research, NSGA-II algorithm design, a multi-objective optimization model MATLAB solving Pareto solution set, and GMS numerical simulation verification, and develops a multi-objective optimization program and quantitative evaluation system based on the Analytic Hierarchy Process. The multi-objective dewatering optimization model solution and numerical simulation verification are carried out for the foundation pit dewatering project of the Xixiayuan canal head inverted siphon section at pile number XZ0+326 to XZ0+500, which provides the dewatering optimization decision scheme for the foundation pit dewatering design and construction.

2. Establishment of Multi-Objective Optimization Model and Evaluation System for Foundation Pit Dewatering

2.1. Establishment of Objective Function, Constraints, and Control Conditions

To construct the optimization model with the objective function method, three basic elements should be determined. The first is the objective function, the second is the constraint condition, and the third is the control condition.

2.1.1. Establishment of Objective Function of Optimization Model

Using the objective function method, three objective functions are constructed, as shown in Equations (1) to (4):

(1) Minimum total cost of dewatering

The objective function is related to the number of wells and the pumping flow of the wells, and the pumping capacity and the number of pumping wells should be reasonably selected to ensure the lowest engineering cost.

$$J_1 = Z_{min} = \alpha_1 \sum_{i=1}^{w} n_i + \alpha_2 \sum_{i=1}^{w} q_i \quad i = 1, 2, \ldots, w \tag{1}$$

Among them, J_1 is the minimum target of the total dewatering cost (Z_{min}); w is the total number of pumping wells; n_i is a binary variable, indicating whether the i-th well is in operation, with a value of 1 indicating operation and 0 indicating no operation; q_i is the flow of the i-th pumping well, m^3/d; α_1 is the construction cost of the pumping well; α_2 is the cost per unit of pumping capacity; and α_1 and α_2 are different coefficients related to local market economic conditions.

(2) The minimum amount of land subsidence caused by dewatering

Considering the influence of dewatering on the premise of meeting the allowable settlement value, it is required that the settlement of important building settlement control points should be the minimum; that is, the settlement influence coefficient should be the minimum. The settlement influence coefficient before optimization is 1, the optimization model can continuously reduce the amount of land subsidence, and the settlement influence coefficient decreases from 1.

$$J_2 = [C]_{min} = \frac{\sum_{g=1}^{n_g} [s]_g}{\sum_{g=1}^{n_g} [s]_y} \quad g = 1, 2, \ldots, n_g \tag{2}$$

Among them, J_2 is the goal of minimizing the settlement influence coefficient ($[C]_{min}$); n_g is the total number of settlement control points; $[s]_y$ is the total allowable settlement, mm; and $[s]_g$ is the optimized settlement of the g-th settlement control point, mm.

(3) The maximum drawdown of the water level in the center of the foundation pit

When implementing the foundation pit dewatering, it is necessary to reduce the groundwater level as much as possible to ensure the safety of the foundation pit structure, but at the same time, it is necessary to ensure that the hydraulic gradient at the bottom of the pit is within the safe allowable range.

$$H^2 - h_j^2 = (2H - S_j)S_j = \sum_{i=1}^{n_w} \frac{q_i}{\pi K} \ln \frac{R_i}{r_{ji}} j = 1, 2, \ldots, n_j \tag{3}$$

$$J_3 = H_{max} = \sum_{j=1}^{n_j} \frac{H - h_j}{n_j} \quad j = 1, 2, \ldots, n_j \tag{4}$$

Among them, H is the initial water level value of aquifer, m; h_j is the water level value of the j-th water level control point after pumping, m; S_j is the water level drawdown value at control point j, m; n_w is the number of pumping wells; q_i is the flow rate of the i-th pumping well, m^3/d; n_j is the number of water level control points; R_i is the influence radius of the pumping well, m; r_{ji} is the distance from the i-th well to the j-th water level control point, m; K is the permeability coefficient, m/d; J_3 is the maximum target of groundwater drawdown at the center of the foundation pit (H_{max}), m.

2.1.2. Determination of Constraints in Optimization Model

The multi-objective optimization model of foundation pit dewatering shall meet the following constraints:

(1) Groundwater level

To meet the construction requirements for foundation pit dewatering, it is necessary to lower the groundwater level in the pit below its bottom and establish water level control points within the pit. The actual drawdown at these control points must exceed the design drawdown.

$$S_j \leq \sum_{i=1}^{w} S_{ji} \quad j = 1, 2, \ldots, n_j \tag{5}$$

where S_j is the design drawdown value at control point j, m; S_{ji} is the drawdown of the i-th well to the water level control point at j, m; w is the number of pumping wells; and n_j is the number of water level control points.

(2) Single well pumping capacity

The pumping capacity of the pumping well is related to the well structure and aquifer permeability. It is required that the maximum pumping capacity of the pumping well shall not exceed its allowable maximum pumping capacity:

$$0 \leq q_i \leq q_{max}(i) \quad i = 1, 2, \ldots, w \tag{6}$$

where $q_{max}(i)$ is the maximum pumping capacity of the i-th pumping well, m^3/d.

(3) Number of pumping wells

In the construction process of pumping wells, the number of pumping wells needs to be restricted. The restriction of the number of pumping wells in operation is as follows:

$$\sum n_i \leq n_{max} \tag{7}$$

where n_i is whether the i-th well exists. If the i-th well exists, then n_i = 1; if the contrary is true, then n_i = 0. n_{max} is the maximum number of pumping wells when all pumping wells exist.

(4) Settlement

In order to ensure the environmental safety around the foundation pit, a settlement control point is set. The ground settlement value at the control point should be less than the allowable settlement at this point:

$$[s]_g \leq [s]_y \tag{8}$$

where $[s]_g$ is the settlement at control point g, mm; variable $[s]_g$ is the function of state variable h_i and decision variable q_i; among them, h_i is the groundwater level, m; q_i is the pumping capacity of a single well, m^3/d; and $[s]_y$ is the allowable settlement, mm.

2.1.3. Determination of Optimal Model Control Conditions

The control conditions are used to determine and control the parameter levels of decision variables and state variables, and the well radius and hydraulic gradient of the pumping well are used as the control conditions.

(1) Well radius

In order to meet the installation of dewatering equipment, the radius of the pumping well is generally required to be greater than or equal to 0.2 m, as follows:

$$0.2 \leq r_w(i) \tag{9}$$

where $r_w i$ is the radius of the i-th well, taken as 0.2 m.

(2) Hydraulic gradient

The water level drawdown at the bottom of the foundation pit is the largest. This is the area with the largest hydraulic gradient. There are potential problems of piping or soil flow. The risk level is the highest, which needs to be considered. In order to consider the safety of foundation pit design, the hydraulic gradient is taken as the control condition:

$$K_s = \frac{i_{cr}}{i_{max}} \tag{10}$$

$$i_{max} = \frac{\Delta h}{\Delta L} \tag{11}$$

$$i_{cr} = \frac{G_s - 1}{e + 1} \tag{12}$$

where K_s is the safety factor, and the safety factor is usually greater than 1.5~2.5 in the dewatering design; i_{max} is the maximum hydraulic gradient; i_{cr} is the critical hydraulic gradient; h is the difference between the internal and external water head, m; ΔL is the seepage path, m; G_s is the specific gravity of aquifer soil particles; and e is the void ratio.

2.2. Establishment of Multi-Objective Optimization Evaluation System

There is a large number and wide distribution range of the Pareto optimal solution set of the multi-objective optimization model. Each Pareto solution can meet the requirements of foundation pit dewatering. Therefore, the Analytic Hierarchy Process [27] (AHP) combined with the evaluation scoring method can be used to establish the evaluation system, so as to select the candidate set with high score in the Pareto solution set as the decision basis. The most important step for evaluation and decision making is to determine the weight and scoring standard of the sub-target layer (total dewatering cost, settlement influence coefficient, and safety and stability of the foundation pit structure).

The evaluation system based on the multi-objective optimization model of dewatering is structured into two layers. The first layer, denoted as A, represents the main objective of optimizing the foundation pit dewatering. The second layer, referred to as the sub-target layer, includes three components: total dewatering cost (A1), settlement influence coefficient (A2), and safety and stability of the foundation pit structure (A3).

According to the dewatering optimization experience, combined with the Saaty scale table [28], the weight judgment among sub-target layers is carried out, and the judgment matrix A is constructed. The maximum eigenvalue and corresponding eigenvector of the judgment matrix are calculated according to Equation (13):

$$AW = \lambda_{max} W \tag{13}$$

where W is the weight corresponding to the sub-target layer, and λ_{max} is the maximum eigenvalue of the judgment matrix.

In order to avoid the influence of human subjective factors, it is necessary to carry out the consistency test on the constructed judgment matrix and calculate the consistency index $C.I.$ of the judgment matrix and the consistency ratio $C.R.$ Generally, when $C.R. < 0.1$, the consistency of the constructed judgment matrix can pass the test.

$$C.I. = \frac{\lambda_{max} - n}{n - 1} \tag{14}$$

$$C.R. = \frac{C.I.}{R.I.} \tag{15}$$

where n is the order of the constructed judgment matrix; $C.I.$ is a consistency index; $C.R.$ is the consistency ratio; and $R.I.$ is the average random consistency index.

Therefore, it is seen from Table 1 that the judgment matrix of the total dewatering cost, the settlement influence coefficient, and the safety and stability of the foundation pit structure on the total target A in the sub-target layer is constructed, and the weight and normalized weight corresponding to the sub-target layer are calculated according to Equation (13).

Table 1. A judgment matrix.

A	A1	A2	A3	W_0	W_0 Normalization
A1	1	1/3	1/5	0.492	0.109
A2	3	1	1/2	1.39	0.309
A3	5	2	1	2.617	0.582

The maximum eigenvalue of the judgment matrix λ_{max} is 3.004. $C.R.$ = 0.002 < 0.1 is calculated by the formula, and the consistency of the matrix can pass the test. From the normalized weights of the three sub-objectives, it can be seen that the decision makers are most interested in the safety and stability of the foundation pit structure, followed by the settlement influence coefficient, and, finally, the total cost of dewatering.

In combination with the actual situation and relevant experience of the project, the total cost of dewatering in the sub-target layer, the settlement influence coefficient, and the safety and stability of the foundation pit structure are evaluated and scored, and the score values are shown in Table 2. Therefore, the solution with the highest score in the Pareto optimal solution set in the multi-objective optimization model is selected as the optimization decision solution.

Table 2. Sub-target layer parameter evaluation score.

Sub-Target Layer	Parameter Range	Scoring Value (0–100)
Total cost of dewatering (Minimum total pumping flow, m^3/d)	$<1.45 \times 10^5$	95
	1.45×10^5–1.5×10^5	85
	1.5×10^5–1.55×10^5	75
	$>1.55 \times 10^5$	65
Impact of settlement on environment (Settlement influence coefficient)	<0.53	90
	0.53–0.55	80
	0.55–0.57	70
Safety and stability of foundation pit structure (The central water level is lower than the foundation pit bottom plate, m)	>1.8	90
	0.6–1.8	80
	0–0.6	70

3. Solution of Multi-Objective Optimization Model

3.1. NSGA-II Algorithm

In 1975, the genetic algorithm appeared in the optimization problem, which was proposed by J Houand and systematically summarized by Goldberg, which realized the development of the population and the continuous improvement of the individual level. The genetic algorithm is based on Darwin's genetic evolution theory and embodies the idea of "natural selection and survival of the fittest" [29]. The genetic algorithm has the characteristics of self-selection and self-adaptation in evolutionary engineering. It has the ability of global search and it can quickly search all solutions. NSGA-II (non-dominated sorting genetic algorithm) is one of the most widely used and effective multi-objective genetic algorithms [17]. It is an algorithm based on the non-dominated sorting algorithm, which has a small amount of calculation and is an elite algorithm. It was originally proposed for 2–3 objective problems [30]. Compared with the original NSGA algorithm, its computational complexity is greatly reduced, and the computation time is greatly reduced. At the same time, it can ensure the diversity of individuals in the population. Therefore, the NSGA-II algorithm has improved in terms of optimization and computation time compared to the original NSGA algorithm, so it is more excellent as a multi-objective algorithm [31].

The flow chart of the NSGA-II algorithm is shown in Figure 1.

3.2. MATLAB Optimization Toolbox

The MATLAB optimization toolbox is an extension toolbox of the MATLAB software's numerical calculation. The toolbox has a variety of conventional functions and heuristic algorithms. It has powerful functions, which can be visualized, and it has high solving efficiency [32]. The optimization functions in the optimization toolbox are shown in Table 3.

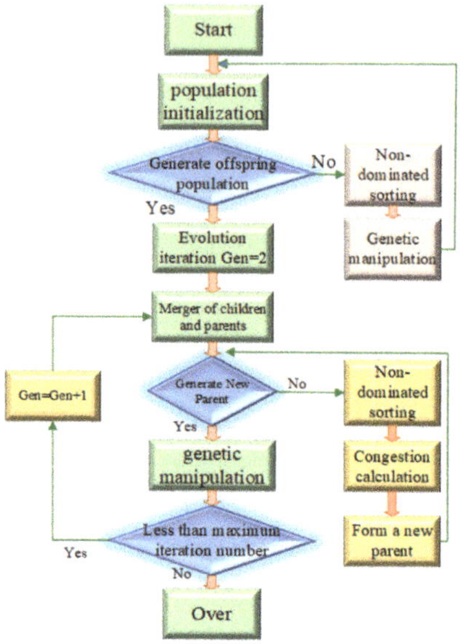

Figure 1. NSGA-II algorithm flow chart.

Table 3. MATLAB main optimization functions.

Function	Description
Fgoalattain	Multi-objective achievement problem
Fmincon	Constrained nonlinear minimization
Fminimax	Minimization and maximum
Linprog	Linear program
Quadprog	Quadratic programming
Gamultiobj	Multi-objective nonlinear minimization

The Gamultiobj program is capable of solving optimization problems with multiple sub-objectives, and it can be utilized either by inputting code or using the graphical toolbox in the MATLAB optimization toolbox. This function incorporates most of the operations from the NSGA-II algorithm, and it is optimized based on it. The Gamultiobj program is consistent with the NSGA-II algorithm in terms of dominance level, non-inferior solution, Pareto ranking, and congestion. However, it introduces the optimal front-end individual coefficient to provide a more precise representation of the Pareto solution. This coefficient has a maximum value of 1 and a minimum value of 0, expressing the ratio of the optimal individual to the total population. A larger ratio results in obtaining more sets of Pareto solutions. The following are the fundamental steps to call the Gamultiobj program, based on the NSGA-II algorithm, for solving practical engineering problems:

(1) According to the actual engineering conditions in the study area, the expression of the objective function is determined, the decision variables are determined, the constraints and control conditions are established according to the construction requirements, and the optimal mathematical model for solving the problem is established;
(2) Launch the MATLAB optimization toolbox and utilize the Gamultiobj program to input the established multi-objective optimization mathematical model into the toolbox, following the specific format guidelines;
(3) Combined with the optimization code, the results of the optimization mathematical model are solved and output.

3.3. Program Design for Solving Multi-Objective Optimization Model

The technical route of the optimization solution using the NSGA-II algorithm is shown in Figure 2.

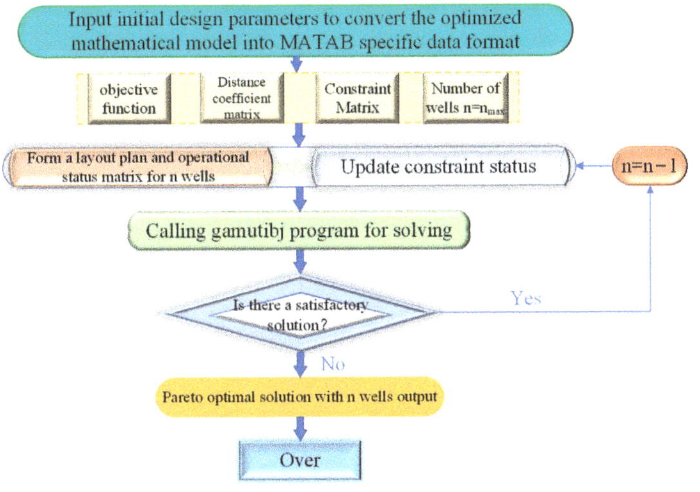

Figure 2. Technical route for optimal solution of multi-objective optimization model of foundation pit dewatering based on NSGA-II algorithm.

The designed dewatering optimization program has the following advantages:

(1) The program can simultaneously optimize the number of pumping wells and the pumping capacity of a single well, and it can obtain the Pareto optimal solution set under different working conditions;
(2) The Pareto solution set is obtained based on the NSGA-II algorithm, with a uniform distribution of Pareto frontiers and while retaining more excellent solutions;
(3) The program avoids complex programming steps and uses the MATLAB optimization solver to input parameters and output results in a visual form.

4. Engineering Background

4.1. General Situation

The water conservancy and irrigation area engineering of the Xixiayuan water conservancy project is located on the North Bank of the Yellow River in Henan Province, China. It is one of the 172 major water conservancy construction projects in the country. The overview of the study area is shown in Figure 3. This paper focuses on the foundation pit project located at the inverted siphon section of the canal head with the specified pile range (XZ0+326 to XZ0+500). This particular section is situated within the Yellow River wetland protection zone, spanning a total length of 174 meters. Notably, the groundwater is found in a highly permeable pebble layer, approximately 11 meters above the foundation surface. Consequently, the excavation of the foundation pit presents challenges pertaining to dewatering and drainage. The constructed multi-objective optimization model and evaluation system are used to optimize the design of the tube well dewatering scheme, providing a reference for similar projects.

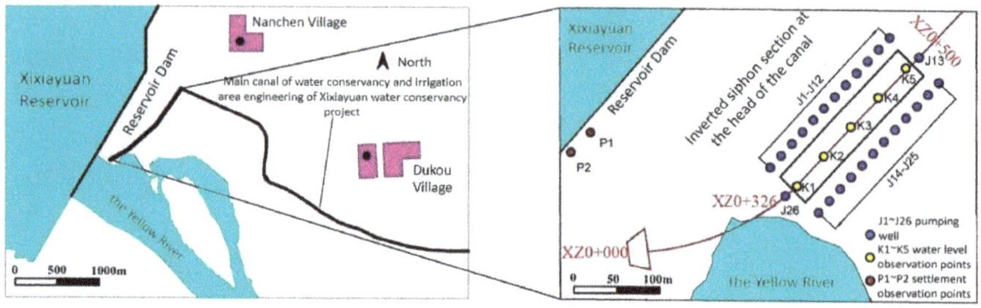

Figure 3. Location distribution map of study area.

4.1.1. Engineering Geological Conditions

The geomorphic unit of the site belongs to the Yellow River beach area. The terrain is flat and open. It is generally high in the north and low in the south. The ground elevation is 122.19–127.75 m, and the bottom elevation of the foundation pit of the inverted siphon section of the canal head is 111.10–111.30 m.

The stratum is quaternary Holocene alluvium, and the lithology is mainly sandy loam and pebble.

Sandy loam (Q_4^{2al}) in the first layer: brown yellow, dry or slightly wet, loose, uneven soil, with high sand content in some parts. The thickness is 1.7~2.0 m, and the bottom elevation is 122.02~122.89 m.

Pebble (Q_4^{2al}) in the second layer: grayish white, purplish red, mixed with a small amount of quartzite and andesite. The particle size is generally 3~6 cm, a small amount of 15~20 cm, and the maximum particle size is more than 20 cm, mostly in sub round shape, accounting for about 55~65%, filled with argillaceous sand and not cemented.

4.1.2. Hydrogeologic Condition

The groundwater in the study area is Quaternary pore phreatic water, which mainly occurs in the pores of the second pebble layer. The buried depth of the groundwater level is 2.60–3.85 m, and the groundwater level is 121.37–121.99 m. Groundwater primarily receives recharge from atmospheric precipitation and the Yellow River, while discharge occurs through evaporation, artificial exploitation, and lateral runoff. The first layer of sandy loam is generally weakly or moderately permeable, with a permeability coefficient of 4×10^{-4} cm/s. The second layer of pebbles is highly permeable, and the permeability coefficient is generally about 3×10^{-1}–1.0 cm/s.

4.2. Initial Scheme of Foundation Pit Dewatering in the Study Area

To achieve dewatering and maintain a dry excavation in the foundation pit of the inverted siphon section of the canal head, this study implements the tube well dewatering method instead of utilizing a waterproof curtain. Moreover, a circular arrangement approach is employed to uniformly position dewatering wells at a distance of 1 m outside the pit to effectively lower the groundwater level. In this paper, the foundation pit with a pile number of XZ0+326–XZ0+500 in the inverted siphon section of the canal head is selected for dewatering design in the pebbles stratum. The length of the foundation pit is 174 meters, and the width is 39 m. The ground elevation of the project area is about 122.5 m, the first layer of sandy loam is about 3 m thick, and the second layer of pebbles is about 20 m thick. The initial groundwater level is about 121.5 m, the bottom elevation of the phreatic aquifer is about 99.5 m, the thickness of the aquifer is about 22 m, the bottom elevation of the foundation pit is 110.3 m, the average excavation depth of the foundation pit is 12 m, the groundwater level is calculated as 1 m below the foundation pit bottom plate, and the drawdown of the water level in the project area is 11~12 m. The engineering geological longitudinal section and foundation pit location of the canal head inverted siphon XZ0+326–XZ0+500 section in the project area is shown in Figure 4. The red line in the figure shows the location of the foundation pit.

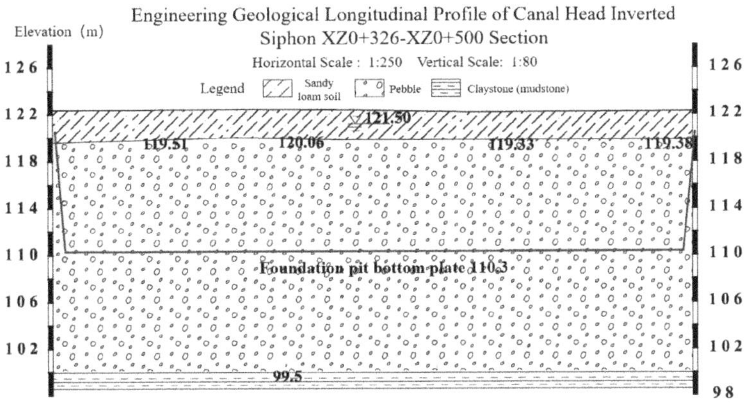

Figure 4. Engineering geological longitudinal section and foundation pit location of canal head inverted siphon section XZ0+326–XZ0+500.

The initial design scheme uses a group of phreatic completely penetrating wells for dewatering, which needs to lower the groundwater to 0.5–1 m below the bottom of the foundation pit. According to the technical specification for building a foundation pit (JGJ/120-2012) [33], calculate the total water inflow of the foundation pit:

$$Q = \pi k \frac{(2H - s_d)s_d}{\ln\left(1 + \frac{R}{r_0}\right)} \tag{16}$$

Among them, Q is the total water inflow, m^3/d; k is the permeability coefficient, m/d; H is the thickness of the phreatic aquifer, m; s_d is the design drawdown of the groundwater level of the foundation pit, m; r_0 is the equivalent radius of foundation pit, m, which can be calculated according to $r_0 = \sqrt{\frac{A}{\pi}}$; A is the area of foundation pit, m^2; and R is the influence radius of dewatering, m.

The influence radius of dewatering can be calculated according to the drawdown value of the observation well arranged in the pumping test, combined with the graphical method. If there is no water level observation well, it can also be solved by using the phreatic aquifer empirical formula method with reference to the parameters obtained from the pumping test:

$$R = 2s_w\sqrt{kH} \tag{17}$$

where s_w is the drawdown value of the well water level, m. The meaning of the other symbols is the same as above.

In order to ensure the dewatering effect, the number of tube wells can be calculated using the following formula when tube wells are arranged at equal intervals for dewatering:

$$n = (1.1 \sim 1.2)\frac{Q}{q_{max}} \tag{18}$$

where n is the number of wells and q_{max} is the maximum water yield of a well, m^3/d.

The maximum allowable pumping capacity of the pumping well is generally obtained from the pumping test. In the absence of a pumping test, it can be solved according to the following empirical formula:

$$q_{max} = 120\pi r l \sqrt[3]{K} \tag{19}$$

where $q_{max}(i)$ is the maximum pumping capacity of the i-th pumping well, m^3/d; r is the screen radius of the pumping well, m; and l is the effective working length of the screen of the pumping well, m.

According to the construction quality acceptance code for building foundation engineering (GB50202-2018) [34], the allowable settlement value caused by foundation pit dewatering is 15 mm, which is related to the height of the dam.

Then, based on survey data and on-site slug test techniques [35–38], the permeability coefficient of the aquifer was obtained. The initial design scheme calculation parameters are shown in Table 4.

Table 4. Calculation parameters of initial design scheme.

Foundation Pit Section	K/(m/d)	Influence Radius R/(m)	q_{max}/ (m^3/d)	Number of Pumping Wells	Water Level Control Point	Settlement Control Points	Total Q/(m^3/d)
XZ0+326–XZ0+500	432	2242.5	6000	26	5	2	156,000

According to the calculation parameters of the initial design scheme, 26 phreatic completely penetrating wells are planned to be arranged, 5 water level control points K1–K5 are arranged along the longitudinal axis of the foundation pit, and 2 settlement control points P1 and P2 are arranged next to the dam of the Xixiayuan water conservancy project in the southwest, as shown in Figure 3.

5. Solving Pareto Optimal Solution Set and Analysis Decision

The multi-objective optimization model of foundation pit dewatering based on the NSGA-II algorithm optimizes the preliminary design scheme, inputs the initial design parameters, calls the Gamultiobj program, and sets the parameters. The initial population size of the Gamultiobj program is set to 100, the genetic iteration is set to 30 generations, the cross ratio is set to 0.7, the function tolerance is set to 1×10^{-6}, and the Pareto set ratio is set to 0.6; that is, the Pareto solution set is generated after 3000 operations and 30 iterations of the function. Through the optimization calculation, in order to meet the program convergence, constraints, and control conditions, at least 24 pumping wells need to be operated. Therefore, the Pareto optimal solution set under three working conditions (24–26 pumping wells in operation) is obtained. The constructed dewatering multi-objective optimization evaluation system assigns scores and evaluates the Pareto optimal solution set. Based on the higher evaluation score, the Pareto solution is analyzed to serve as the foundation for decision making. To assess the feasibility of the proposed scheme, a numerical simulation was conducted to verify its feasibility in Section 6.

5.1. Pareto Optimal Solution Set and Analysis of Running 26 Pumping Wells Schemes

The multi-objective optimization model for foundation pit dewatering allows for the identification of Pareto optimal solutions and the generation of multiple dewatering schemes tailored to different objectives. The first working condition is set with 26 running wells, and the Gamultiobj program, based on the NSGA-II algorithm, is employed to obtain the Pareto optimal solution set and the operational status of the pumping wells. This information is depicted in Figure 5.

It can be seen from Figure 5 that the optimization model maintains the running of all of the initial 26 pumping wells. After 30 iterations of the algorithm, the Pareto optimal solution set is obtained. The light green color ball in the solution set is the Pareto optimal solution. The red dot indicates the relationship between the total water inflow of the foundation pit and the settlement influence coefficient, the green dot indicates the relationship between the central water level drawdown of the foundation pit and the total water inflow, and the blue dot indicates the relationship between the central water level drawdown of the foundation pit and the settlement influence coefficient. Under the condition of meeting constraints and control conditions, the total water inflow ranges from 1.42×10^5 to 1.56×10^5 m^3/d. The settlement influence coefficient is between 0.51 and 0.57, which is converted into the settlement value of 7.65–8.55 mm, which is less than the allowable settlement value of

15 mm. The drawdown of the water level in the center of the foundation pit is 0.2–10 m, meeting the drawdown requirements of more than 0 m. All Pareto solutions of 26 pumping wells under all running conditions are substituted into the dewatering multi-objective optimization evaluation system to calculate the evaluation score, which is arranged in descending order according to the weighted scores of the three objectives. The score results of Pareto solutions obtained by the scheme are shown in Table 5.

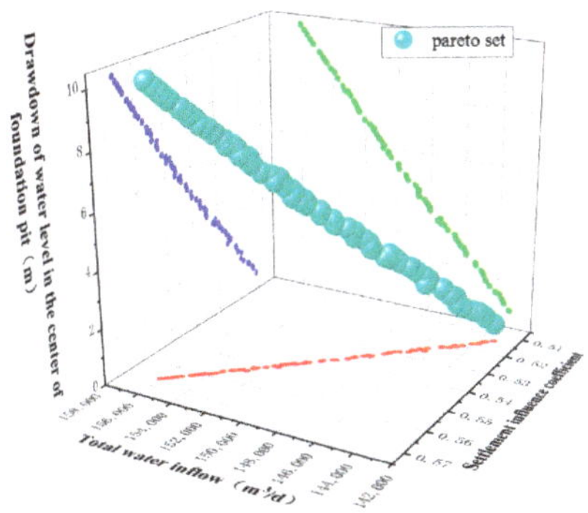

(**a**) Pareto solution set distribution diagram

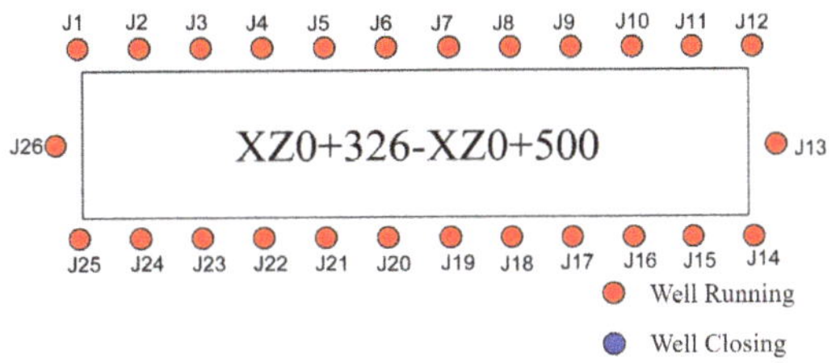

(**b**) Working state of pumping well

Figure 5. Pareto set distribution diagram and pumping well working state under the first optimized working condition. (In the figure (**a**), the red dot indicates the relationship between the total water inflow and the settlement influence coefficient, the green dot indicates the relationship between the central water level drawdown and the total water inflow, the blue dot indicates the relationship between the central water level drawdown and the settlement influence coefficient).

Table 5. Evaluation table for optimal scheme of foundation pit dewatering for running 26 pumping wells.

Pareto Solution No.	Score for Objective I	Score for Objective II	Score of Objective III	Weighted Scores of Three Objectives of Dewatering Optimization
1	95	90	90	90.545
2	85	90	90	89.455
3	85	80	90	86.365
4	75	80	90	85.275
5	95	90	80	84.725
6	75	70	90	82.185
7	65	70	90	81.095

It can be seen from Table 5 that the solutions in the Pareto optimal solution set are evaluated by AHP combined with the scoring method, and the Pareto solution with a high evaluation score is selected. So, the dewatering scheme with a weighted evaluation score of 90.545 is selected. The comparison of the optimization results of the three objectives is shown in Table 6, and the optimal results of decision variables are shown in Figure 6.

Table 6. Three objective optimization results of running 26 pumping wells.

Schemes \ Each Objective	Well Status	Total Pumping Flow/(m^3/d)	Settlement Value (mm)	The Distance Where Water Level in the Center of the Foundation Pit Is Lower than the Bottom Plate/(m)	Hydraulic Gradient Value
Preliminary scheme	26 wells in running	156,000	<15	>0	<0.9
The first optimized working condition	26 wells in running	144,972	7.8	1.84	0.65

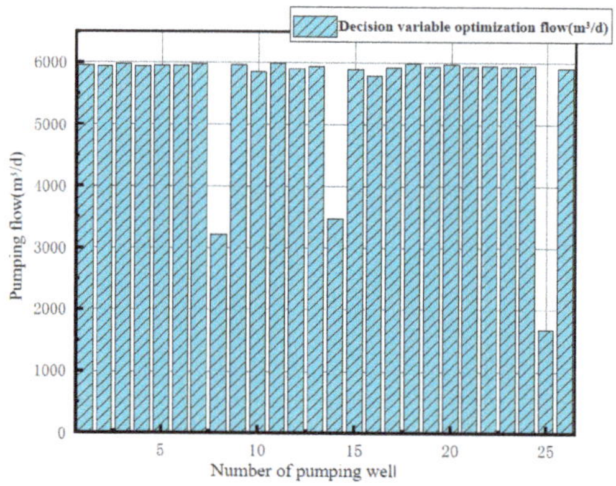

Figure 6. Optimization results of decision variables under the first optimization working condition.

It can be seen from Table 6 that the total water inflow of objective I under the first optimization working condition is about 1.45×10^5 m^3/d, the settlement value of objective II is 7.8 mm, which is less than the allowable settlement value of 15 mm, and the foundation pit central water level of objective III is 0.39 m lower than the foundation pit bottom plate, meeting the design requirements for construction of more than 0 m. In this scheme, if the construction and installation of the pumping well have been completed, the score of the three objectives is the best, but the disadvantage is that the number of the pumping wells is the largest, and it is not necessarily the best scheme in the area with a high cost of construction and installation of the pumping well.

5.2. Pareto Optimal Solution Set and Analysis of Running 25 Pumping Wells Schemes

Similarly, on the premise of meeting the constraints and control conditions, the second optimization working condition is set to arbitrarily close one of the wells, resulting in a total of 26 layout schemes, without changing the location of the wells in the preliminary layout scheme. After optimization by the Gamultiobj program, 26 sets of Pareto solution sets are obtained, and 26 feasible schemes can be optimized for three objectives. Three sets of layout optimization schemes with evenly distributed Pareto solution sets are selected for analysis. The Pareto optimal solution set and pumping well layout state under the second optimization working condition are shown in Figure 7.

It can be seen from Figure 7 that the three schemes under the second optimized working condition are to close the J26, J16, or J3 pumping well, respectively. Similarly, all Pareto solutions of the three schemes under the second optimization condition are substituted into the dewatering multi objective optimization evaluation system, and the evaluation scores are calculated. The weighted scores of the three objectives are arranged in descending order. The score results of the Pareto solutions of each scheme are shown in Table 7.

Select the Pareto solution with the highest evaluation score among the dewatering optimization schemes in Table 7, which are the three dewatering schemes with weighted scores of 89.455, 90.545, and 89.455, respectively. The comparison of the optimization results of the three objectives is shown in Table 8. The optimal results of decision variables are shown in Figure 8.

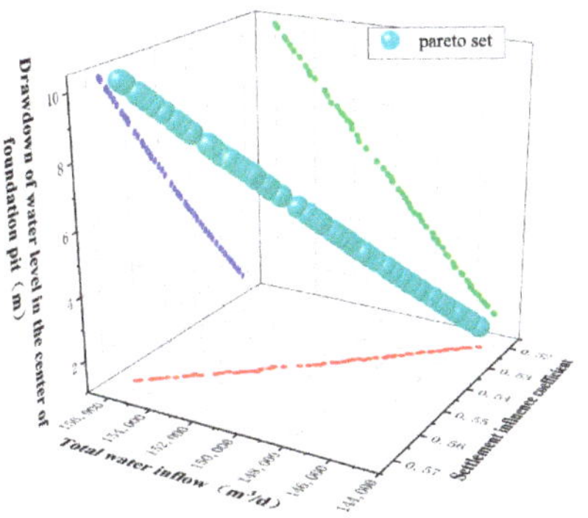

(**a**) Pareto solution set distribution diagram when closing J26 pumping well

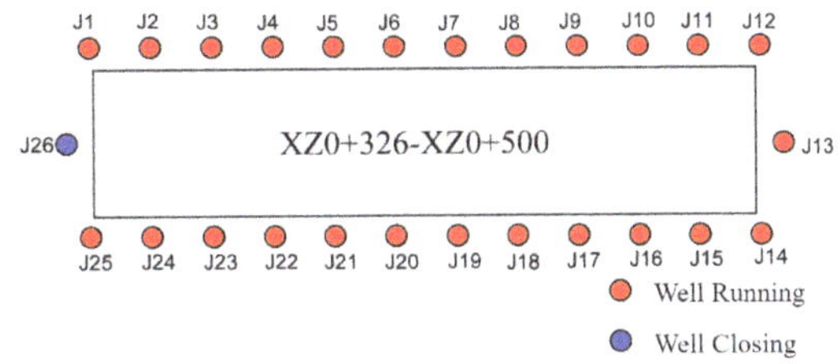

(**b**) Working state of pumping well when closing J26

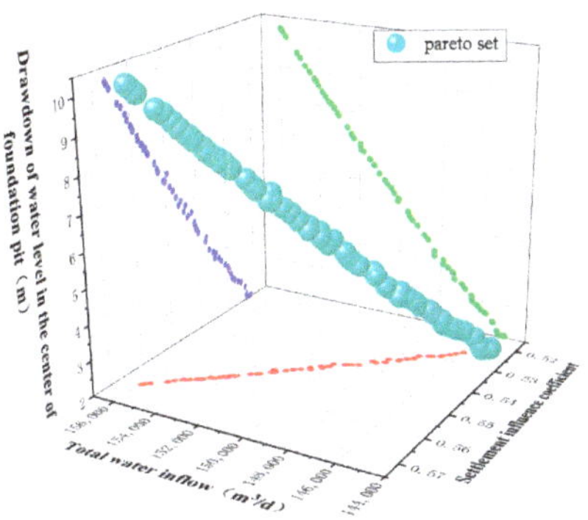

(**c**) Pareto solution set distribution diagram when closing J16 pumping well

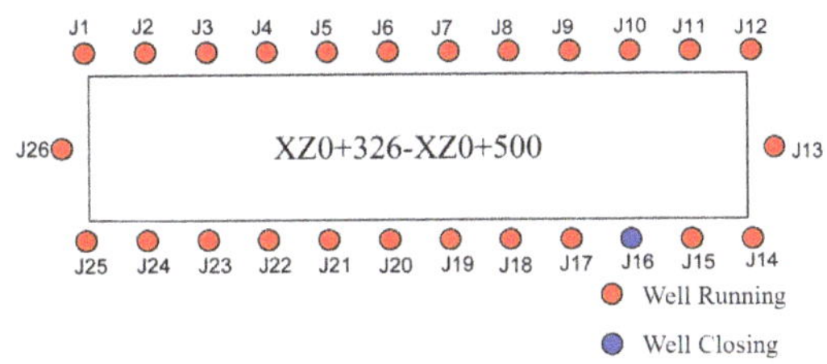

(**d**) Working state of pumping well when closing J16

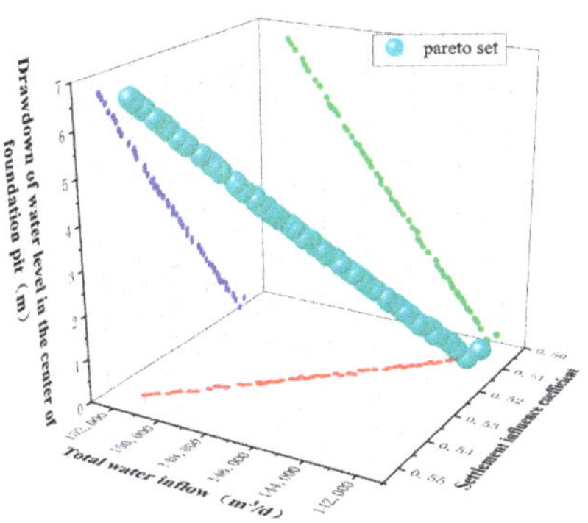

(**e**) Pareto solution set distribution diagram when closing J3 pumping well

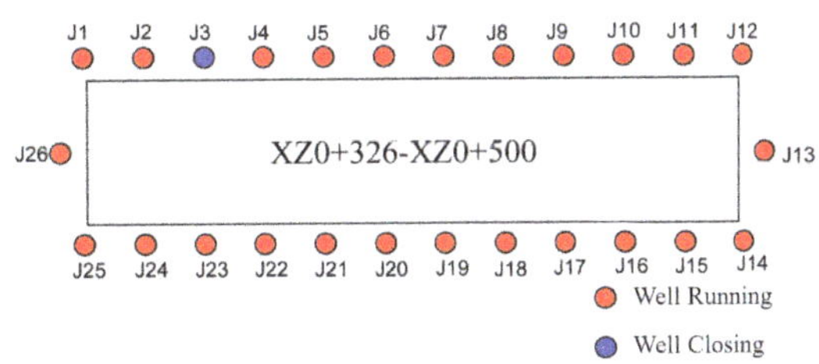

(**f**) Working state of pumping well when closing J3

Figure 7. Pareto set distribution diagram and pumping well working state under the second optimal working condition. (In the figure (**a**,**c**,**e**), the red dot indicates the relationship between the total water inflow and the settlement influence coefficient, the green dot indicates the relationship between the central water level drawdown and the total water inflow, the blue dot indicates the relationship between the central water level drawdown and the settlement influence coefficient).

Table 7. Evaluation table for optimal scheme of foundation pit dewatering for running 25 pumping wells.

Each Scheme	Pareto Solution No.	Score for Objective I	Score for Objective II	Score of Objective III	Weighted Scores of Three Objectives of Dewatering Optimization
Close J26 pumping well	1	85	90	90	89.455
	2	85	80	90	86.365
	3	75	80	90	85.275
	4	95	90	80	84.725
	5	85	90	80	83.635
	6	65	70	90	81.095
Close J16 pumping well	1	85	90	90	89.455
	2	85	90	90	89.455
	3	85	80	90	86.365
	4	75	80	90	85.275
	5	95	90	80	84.725
	6	75	70	90	82.185
Close J3 pumping well	1	85	90	90	89.455
	2	85	90	90	89.455
	3	85	80	90	86.365
	4	75	80	90	85.275
	5	95	90	80	84.725
	6	75	70	90	82.185

Table 8. Three objective optimization results of running 25 pumping wells.

Schemes \ Each Objective	Well Status	Total Pumping Flow/(m^3/d)	Settlement Value (mm)	The Distance Where Water Level in the Center of the Foundation Pit Is Lower than the Bottom Plate/(m)	Hydraulic Gradient Value
Preliminary scheme	26 wells in running	156,000	<15	>0	<0.9
The second optimized working condition	Close J26	145,498	7.9	1.85	0.65
	Close J16	149,608	8.4	1.88	0.67
	Close J3	145,037	7.7	1.83	0.63

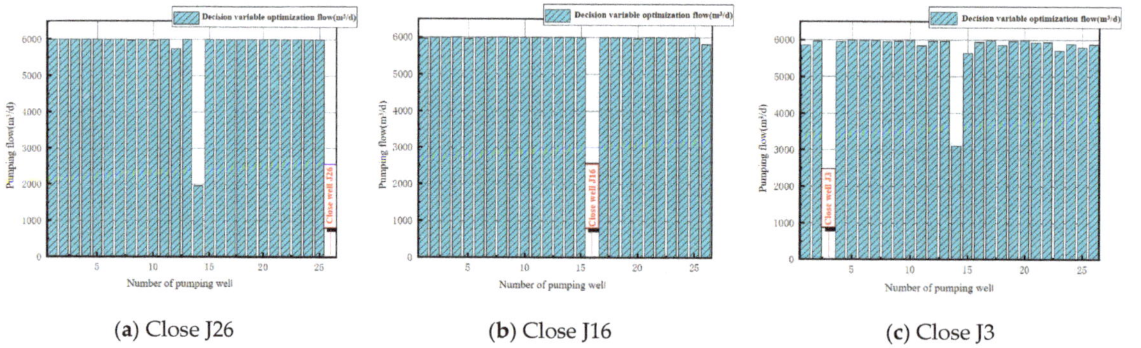

(a) Close J26 **(b)** Close J16 **(c)** Close J3

Figure 8. Optimization results of decision variables under the second optimization working condition.

It can be seen from Table 8 that the total water inflow of objective I under the second optimized working condition is 1.45×10^5–1.49×10^5 m^3/d, and the settlement value of objective II is between 7.85 and 8.4 mm, which is less than the allowable settlement value of 15 mm. The foundation pit central water level of objective III is 1.83–1.88 m lower than the foundation pit bottom plate, which meets the requirements of construction of more than 0 m. In this scheme, the highest evaluation scores for separately closing pumping wells J26, J16, or J3 are basically the same, but according to Table 8, the advantages and disadvantages of each scheme can be analyzed in detail. If the minimum settlement and the lowest hydraulic gradient are considered, pumping well J3 should be shut down; if the maximum water level drawdown value and the safest structure are considered, pumping well J16 should be shut down.

5.3. Pareto Optimal Solution Set and Analysis of Running 24 Pumping Wells Schemes

Similarly, on the premise of meeting the constraints and control conditions, the third optimal working condition is set to arbitrarily close two of the pumping wells, resulting in a total of 325 layout schemes, which does not change the location of the wells in the preliminary layout scheme. After the optimization of the Gamultiobj program, the program converges to 100 sets of Pareto solution sets, and 100 feasible schemes can be optimized for the three objectives. Three sets of layout optimization schemes with evenly distributed Pareto solution sets under the third working condition are selected for analysis. Pareto optimal solution sets and the pumping well layout status are shown in Figure 9.

It can be seen from Figure 9 that under the third optimization condition, the optimization model runs 24 pumping wells. Scheme 1 is to close the J19 and J22 pumping wells, scheme 2 is to close the J7 and J24 pumping wells, and scheme 3 is to close the J1 and J22 pumping wells. Similarly, all Pareto solutions of the three schemes under the third optimization condition are substituted into the dewatering multi-objective optimization evaluation system, and the evaluation scores are calculated. The weighted scores of the three objectives are arranged in descending order from larger to smaller. The score results of Pareto solutions of each scheme are shown in Table 9.

Table 9. Evaluation table for optimal scheme of foundation pit dewatering for running 24 pumping wells.

Each Scheme	Pareto Solution No.	Score for Objective I	Score for Objective II	Score of Objective III	Weighted Scores of Three Objectives of Dewatering Optimization
Close J19 and J22 pumping wells	1	95	90	90	90.545
	2	85	80	90	86.365
	3	75	80	90	85.275
	4	95	90	80	84.725
	5	75	70	90	82.185
	6	95	90	70	78.905
Close J7 and J24 pumping wells	1	95	90	90	90.545
	2	85	80	90	86.365
	3	75	80	90	85.275
	4	95	90	80	84.725
	5	75	70	90	82.185
	6	95	90	70	78.905
Close J1 and J22 pumping wells	1	95	90	90	90.545
	2	85	80	90	86.365
	3	75	80	90	85.275
	4	95	90	80	84.725
	5	75	70	90	82.185
	6	95	90	70	78.905

Select the Pareto solution with the highest evaluation score in the dewatering optimization scheme in Table 9; that is, the three groups of dewatering schemes with the same weighted score of 89.455. The comparison of three objective optimization results is shown in Table 10. The optimization results of decision variables are shown in Figure 10.

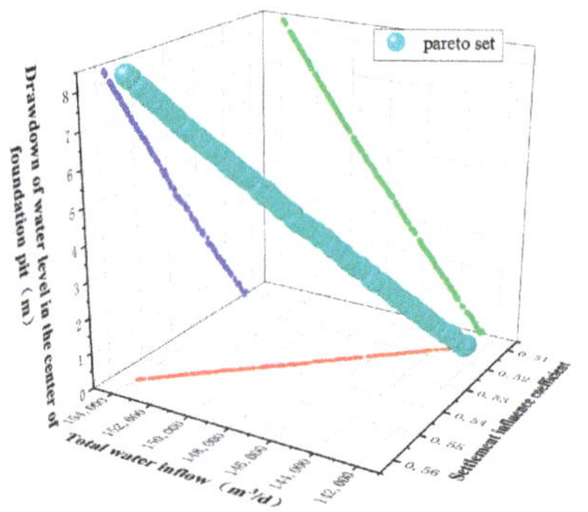

(**a**) Pareto solution set distribution diagram when closing J19 and J22 pumping well

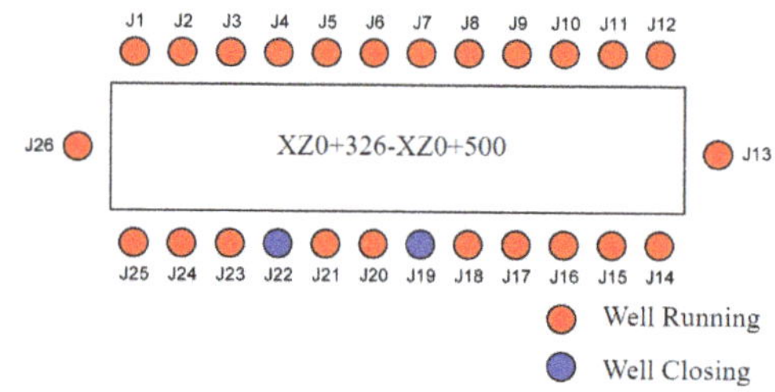

(**b**) Working state of pumping well when closing J19 and J22

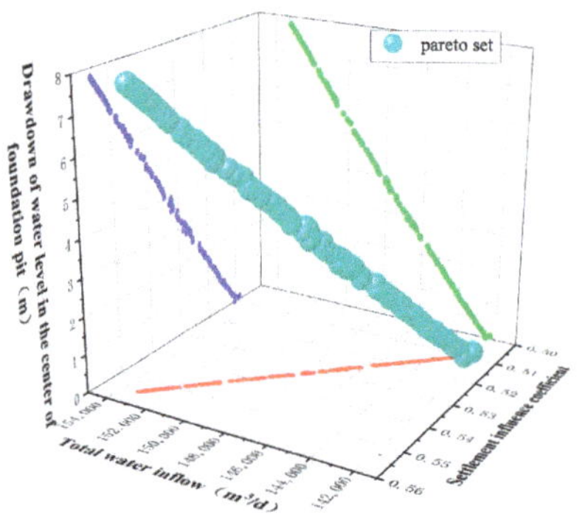

(**c**) Pareto solution set distribution diagram when closing J7 and J24 pumping well

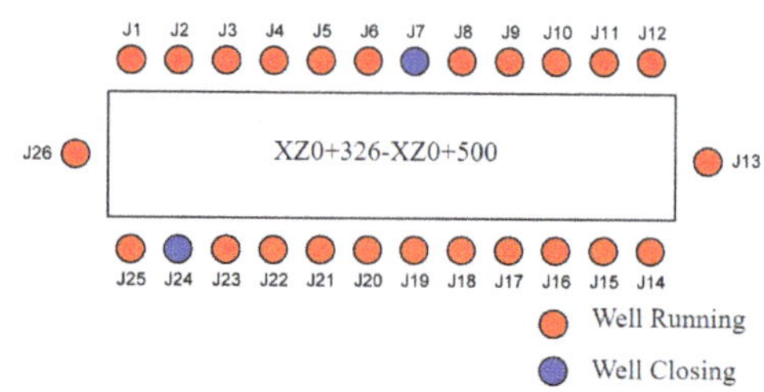

(**d**) Working state of pumping well when closing J7 and J24

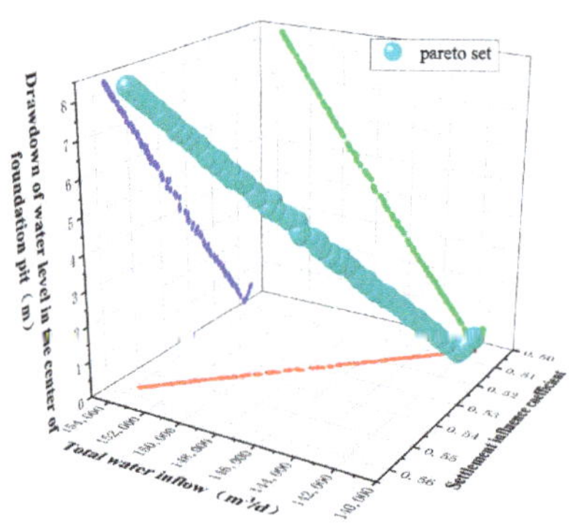

(**e**) Pareto solution set distribution diagram when closing J1 and J22 pumping well

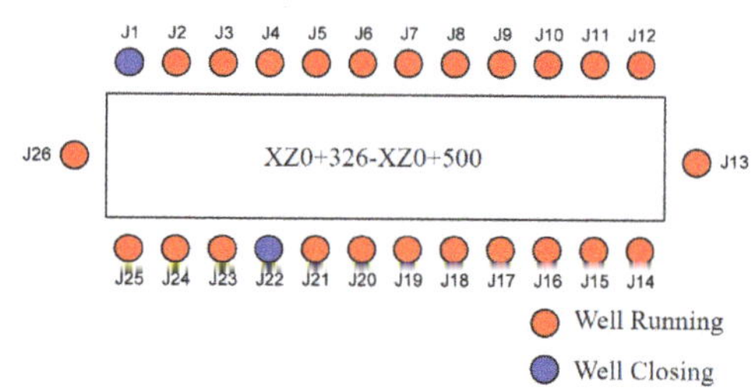

(**f**) Working state of pumping well when closing J1 and J22

Figure 9. Pareto set distribution diagram and pumping well working state under the third optimal working condition. (In the figure (**a**,**c**,**e**), the red dot indicates the relationship between the total water inflow and the settlement influence coefficient, the green dot indicates the relationship between the central water level drawdown and the total water inflow, the blue dot indicates the relationship between the central water level drawdown and the settlement influence coefficient).

Table 10. Three objective optimization results of running 24 pumping wells.

Schemes \ Each Objective	Well Status	Total Water Inflow/(m^3/d)	Settlement Value (mm)	The Distance Where Water Level in the Center of the Foundation Pit Is Lower Than the Bottom Plate/(m)	Hydraulic Gradient Value
Preliminary scheme	26 wells in running	156,000	<15	>0	<0.9
The third optimized working condition	Close J19, J22	143,665	7.95	1.84	0.66
	Close J7, J24	143,121	7.95	1.83	0.65
	Close J1, J22	143,036	7.95	1.83	0.65

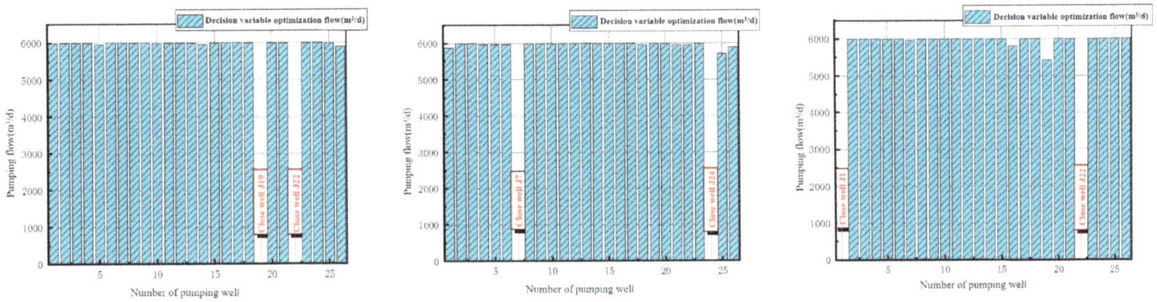

(**a**) Close J19 and J22 (**b**) Close J7 and J24 (**c**) Close J1 and J22

Figure 10. Optimization results of decision variables under the third optimization working condition.

It can be seen from Table 10 that the total water inflow of objective I under the third optimized condition is about 1.43×10^5 m^3/d, the ground settlement value of objective II is about 7.95 mm, which is less than the allowable settlement value of 15 mm, and the foundation pit central water level of objective III is 1.83–1.84 m lower than the foundation pit bottom plate. In this scheme, due to the minimum number of pumping wells and pumping capacity, the cost of dewatering is minimized. However, disadvantages of this scheme are that the settlement of the control point is relatively large, particularly in residential areas where this may not be the optimal choice.

5.4. Pareto Optimal Solution Set Analysis and Decision Making

The multi-objective optimization model of foundation pit dewatering based on the NSGA-II algorithm is solved to obtain the Pareto optimal solution set under three different dewatering conditions. Using the multi-objective optimization evaluation system, the optimization scheme in the Pareto optimal solution set is selected as the decision optimization result from the three objectives of total dewatering cost, settlement influence coefficient, and the safety and stability of the foundation pit structure, which provides a variety of feasible schemes for the dewatering construction of the foundation pit.

The three objectives in the foundation pit dewatering optimization model affect and interact with each other. Therefore, the specific analysis and decision making are as follows:

(1) When giving priority to the total cost of dewatering, it is important to consider the drilling and pumping costs for foundation pit dewatering. Within the optional Pareto solution set, a solution with a smaller number of pumping wells and lower total pumping capacity should be the first choice based on the dewatering multi-objective optimization evaluation system. Specifically, the operation scheme of closing the J1 and J22 pumping wells under the third optimization condition can be preferred;

(2) If the focus is on the impact of dewatering on the surrounding environment, preference should be given to a dewatering scheme with a settlement impact coefficient of less than 0.53. Specifically, the operation scheme corresponding to Pareto solution No. 1 under the first optimized condition in Table 5, and the operation scheme corresponding to closing the J3 pumping well under the second optimized condition

in Table 7, can ensure minimal ground settlement under the conditions of meeting the total cost requirements of dewatering and the safety and stability of the foundation pit structure;

(3) When prioritizing the safety and stability of the foundation pit structure, achieving a significant drawdown of the water level in the center of the foundation pit is crucial for ensuring successful construction and enhancing structural safety. Considering the dewatering multi-objective optimization evaluation system, a preferable dewatering scheme is to maintain the water level approximately 1.5 m below the bottom plate of the foundation pit. Therefore, the operation scheme of running 25 pumping wells while closing the J16 pumping well under the second working condition is recommended to meet the maximum drawdown of the water level.

6. Numerical Simulation Verification of Dewatering Optimization Scheme Based on GMS

6.1. Element Subdivision of Numerical Model and Determination of Boundary Conditions

Based on the hydrogeological conditions in the study area, a numerical model is established using GMS software. The model is divided into two layers, each of which is divided into 22,500 element meshes of 150 × 150. The first layer is a sandy loam layer (119.5–122.5 m), the second layer is a pebble layer (99.5–119.5 m), and the bottom is a claystone layer. The terrain is flat, so it is treated as a horizontal layer. The vertical thickness of the overall model is 23 m. The southwest of the foundation pit in the study area is close to the Yellow River, and the fixed water head boundary is set according to the measured water level of the Yellow River during the simulation period. The Xixiayuan reservoir dam is located in the northwest, which is regarded as the impermeable boundary treatment. The poor permeability of the bottom claystone layer is regarded as the impermeable boundary, and the remaining boundaries are determined as the constant water head boundary according to the influence radius of pumping, which is set to be equal to the initial head of the study area. The grid division of the 3D geological model in the study area is shown in Figure 11. Specifically, the water level values of all constant water head boundaries in the numerical model are consistent with those taken in the optimization algorithm.

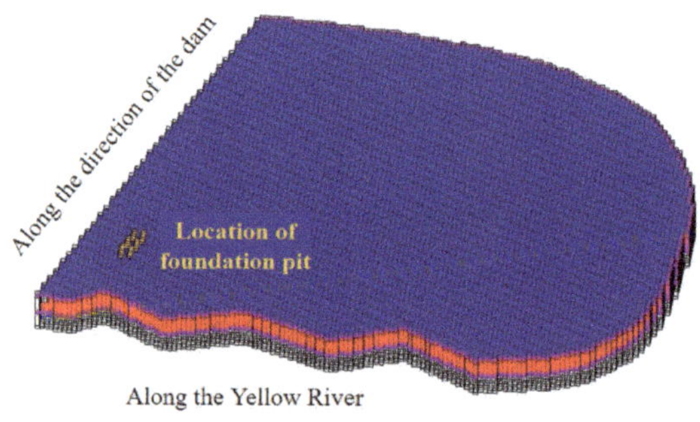

Figure 11. Grid division of 3D geological model of the study area.

The hydrogeological parameters and physical mechanical parameters of each layer are obtained according to the survey reports and on-site slug test results, as shown in Table 11.

Table 11. Hydrogeological parameters and physical and mechanical parameters of each layer.

Aquifer	Permeability Coefficient K/(m/d)	Specific Yield	Porosity	Elastic Water Storage Rate S_{ske} (1/m)	Inelastic Water Storage Rate S_{skv} (1/m)
Sandy loam soil layer	0.35	0.05	0.3	6×10^{-4}	1.2×10^{-3}
Pebble layer	432	0.2	0.4	1.2×10^{-4}	1.6×10^{-4}

6.2. Numerical Simulation Results and Analysis

Based on the MODFLOW module and SUB subroutine package in GMS, according to the requirements of foundation pit dewatering and settlement control, and comprehensively considering three objectives, the dewatering scheme of running 24 pumping wells and closing the J19 and J22 pumping wells under the third optimal condition is numerically simulated and analyzed to verify its feasibility and accuracy. Therefore, the groundwater level and settlement contour maps in the study area under this dewatering scheme are shown in Figures 12 and 13.

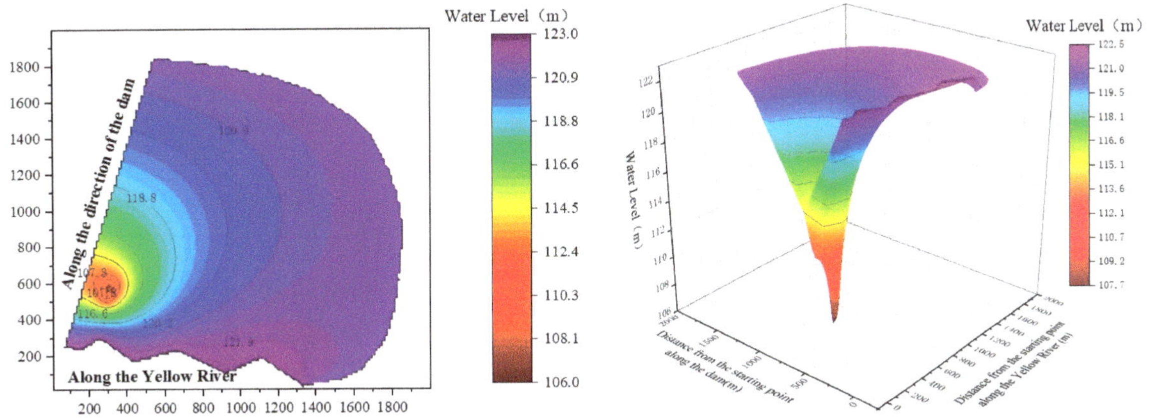

(a) Plane contour map of groundwater level in the study area

(b) Three-dimensional contour map of groundwater level in the study area

Figure 12. Contour map of groundwater level in the study area.

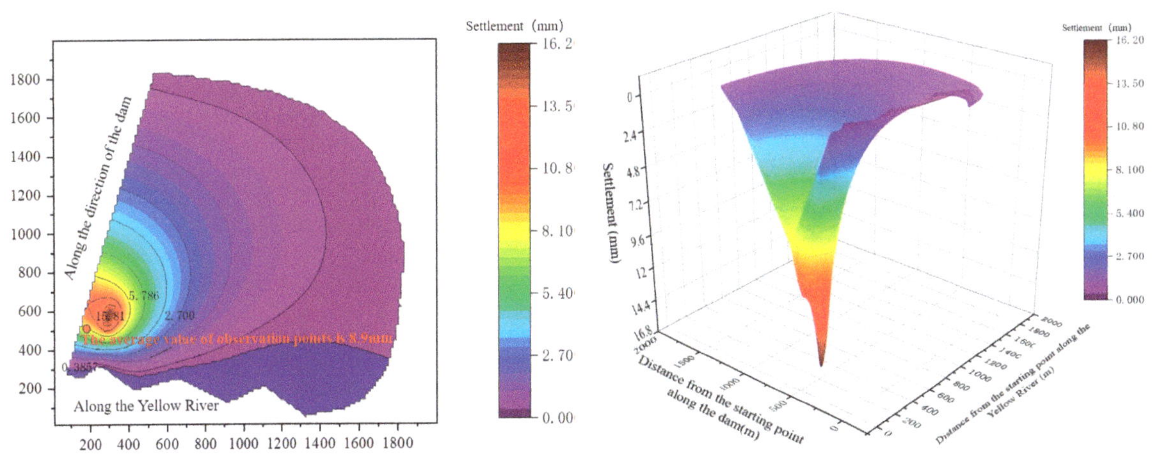

(a) Plane contour map of settlement in the study area

(b) Three-dimensional contour map of settlement in the study area

Figure 13. Contour map of settlement in the study area.

It can be seen from Figure 12 that the water level at the groundwater level control point in the foundation pit is lower than the elevation 110.3 m of the foundation pit bottom, and the water level in the pit is stable between 108.9 and 109.6 m, which can lower the water level to 1.4 m below the foundation pit bottom plate and meet the construction requirements.

It can be seen from Figure 13 that the settlement in the foundation pit settlement observation points is up to 8.8–9.0 mm, and the average settlement is about 8.9 mm. Compared with the allowable settlement of 15 mm calculated by the layered summation method, the simulation results show that the settlement is far less than the allowable settlement, which can reduce about 41% of the settlement. It can be verified that the optimized dewatering scheme of running 24 pumping wells and closing the J19 and J22 pumping wells under the third optimal working condition will not cause settlement damage to the foundation pit and surrounding buildings, and this can ensure construction safety.

The numerical simulation results and optimization results under this dewatering scheme are compared and analyzed, and the comparison results are shown in Table 12. The comparison results show that the numerical simulation results can lower the water level in the center of the foundation pit to 1.4 m below the bottom of the foundation pit and meet the requirements of water level drawdown. The observed value of settlement is 8.9 mm, which is less than the allowable value of 15 mm of settlement, which verifies the feasibility of the optimization results.

Table 12. Verification and comparison of different results under the third optimal working condition.

Types	The Distance Where Water Level in the Center of the Foundation Pit Is Lower than the Bottom Plate/(m)	Mean Value of Ground Settlement Observation Points/(mm)	Total Pumping Wells Flow/(m^3/d)
Allowable value	>0	<15	<144,000
Optimal results	1.84	7.95	143,665
Numerical simulation results	1.40	8.9	143,815

7. Conclusions

(1) The objective function method is used to establish the multi-objective optimization mathematical model of foundation pit dewatering. Combined with the non-dominated sorting genetic algorithm NSGA-II and MATLAB optimization toolbox, the Gamultiobj program is called to develop an iterative program to optimize the pumping capacity of a single well and the number of pumping wells, and the solving process is given. The advantages of multi-objective optimization based on NSGA-II are that the uniformly distributed Pareto optimal solution set can be obtained, and multi-objective optimization problems for foundation pit dewatering can more quickly and efficiently be handled based on the fast-elite selection strategy. Using the Analytic Hierarchy Process (AHP) combined with the evaluation scoring method to establish an evaluation system, the candidate set with high scores in the Pareto solution set is used as the decision-making basis. The construction of an evaluation system combining NSGA-II and AHP applied in foundation pit dewatering engineering represents an innovative technology and method.

(2) Using the NSGA-II algorithm and MATLAB optimization toolbox programming, the dewatering optimization of the foundation pit project of the inverted siphon section of the canal head (pile No. XZ0+326–XZ0+500) in the water conservancy and irrigation area engineering of the Xixiayuan water conservancy project was carried out, and the Pareto optimal solution set under three optimal conditions (24 to 26 pumping wells in running) was obtained. By incorporating the dewatering multi-objective optimization evaluation system based on the Analytic Hierarchy process, the optimization scheme within the set of Pareto optimal solutions is chosen as the ultimate decision for optimization. This scheme takes into account three objectives: the total dewatering cost, the settlement influence coefficient, and the safety and stability of the foundation pit structure. Consequently, it offers a range of workable plans for the construction of the foundation pit dewatering.

(3) The study area's numerical model is created using the MODFLOW module and SUB subroutine package within GMS. The optimization outcomes for the decision variables

in the dewatering scheme, which involves operating 24 pumping wells and closing the J19 and J22 pumping wells based on the third optimal condition, are applied to the numerical model. The numerical simulation of this optimized scheme validates the scientific nature and accuracy of the multi-objective optimization model for foundation pit dewatering. Importantly, the established multi-objective optimization model and evaluation system offer numerous viable dewatering optimization plans.

Author Contributions: Conceptualization, Z.M. and Y.Z.; methodology, Z.M. and Y.Z.; validation, Y.W. and B.L.; formal analysis, Z.M., Y.Z. and Y.W.; resources, Z.M. and Y.Z.; data curation, Y.W. and B.L.; writing—original draft preparation, Z.M. and Y.Z.; writing—review and editing, J.W., B.L. and Y.W.; visualization, B.L. and Y.W.; supervision, Y.Z. and J.W.; project administration, Z.M., Y.Z. and J.W.; funding acquisition, Y.Z. and J.W. All authors have read and agreed to the published version of the manuscript.

Funding: This research was funded by the National Key Research and Development Program of China, grant number 2019YFC1510802, the Fundamental Research Funds for the Central Universities, grant number B220205006, and the scientific research project of water conservancy and irrigation area engineering construction administration of Xixiayuan water conservancy project in Henan Province, China, grant number XXYSS/GQ-KYXM-02.

Institutional Review Board Statement: Not applicable.

Informed Consent Statement: Not applicable.

Data Availability Statement: Not applicable.

Acknowledgments: The authors would like to thank the School of Earth Sciences and Engineering at Hohai University for partial support of the graduate student involved in this project.

Conflicts of Interest: Author Yufeng Wei was employed by the company China Energy Engineering Jiangsu Power Design Institute Co. The remaining authors declare that the research was conducted in the absence of any commercial or financial relationships that could be construed as a potential conflict of interest.

References

1. Zhang, Y.Q.; Li, M.G.; Wang, J.H.; Chen, J.J.; Zhu, Y.F. Field tests of pumping-recharge technology for deep confined aquifers and its application to a deep excavation. *Eng. Geol.* **2017**, *228*, 249–259. [CrossRef]
2. Sharifi, M.R.; Akbarifard, S.; Madadi, M.R.; Qaderi, K.; Akbarifard, H. Optimization of hydropower energy generation by 14 robust evolutionary algorithms. *Sci. Rep.* **2022**, *12*, 7739. [CrossRef]
3. Wang, X.L.; Li, S.H.; Chen, H.Y.; Wang, X.F.; Mei, Y. Multi-objective and multi region power grid planning based on non dominated genetic algorithm and coevolutionary algorithm. *Proceeding CSEE* **2006**, *12*, 11–15.
4. Xu, Y.; Yan, Y. Optimal Design of Foundation Pit Dewatering Based on Objective Functions and Numerical Analysis. *Adv. Mater. Res.* **2011**, *368–373*, 2495–2499. [CrossRef]
5. Liu, J.; Luo, X. Application of Genetic Algorithm in Deep Foundation Pit Dewatering. *Site Investig. Sci. Technol.* **2000**, *3*, 42–45.
6. Yang, Y.; Wu, J.; Sun, X.; Wu, J.; Zheng, C. Development and application of a master-slave parallel hybrid multi-objective evolutionary algorithm for groundwater remediation design. *Environ. Earth Sci.* **2013**, *70*, 2481–2494. [CrossRef]
7. Wahid, F.; Alsaedi, A.K.Z.; Ghazali, R. Using improved firefly algorithm based on genetic algorithm crossover operator for solving optimization problems. *J. Intell. Fuzzy Syst.* **2019**, *36*, 1547–1562. [CrossRef]
8. Geng, J.; Cui, Z.; Gu, X. Scatter search-based particle swarm optimization algorithm for earliness/tardiness flow shop scheduling with uncertainty. *Int. J. Autom. Comput.* **2016**, *13*, 285–295. [CrossRef]
9. Li, N.; Zou, T.; Sun, D. Multi-objective optimization algorithm based on particle swarm optimization. *Comput. Eng. Appl.* **2005**, *23*, 43–46.
10. Ma, C. Multi-Objective Optimization of Subway Construction Projects Based on Improved Genetic Algorithm. Ph.D. Thesis, Lanzhou Jiaotong University, Lanzhou, China, 2020.
11. Khodadadi, N.; Azizi, M.; Talatahari, S.; Sareh, P. Multi-Objective Crystal Structure Algorithm (MOCryStAl): Introduction and Performance Evaluation. *IEEE Access* **2021**, *9*, 117795–117812. [CrossRef]
12. Abdel-Basset, M.; Mohamed, R.; Mirjalili, S.; Chakrabortty, R.K.; Ryan, M. An Efficient Marine Predators Algorithm for Solving Multi-Objective Optimization Problems: Analysis and Validations. *IEEE Access* **2021**, *9*, 42817–42844. [CrossRef]
13. Sang-To, T.; Le-Minh, H.; Wahab, M.A.; Thanh, C.-L. A new metaheuristic algorithm: Shrimp and Goby association search algorithm and its application for damage identification in large-scale and complex structures. *Adv. Eng. Softw.* **2023**, *176*, 103363. [CrossRef]

14. Sang-To, T.; Le-Minh, H.; Mirjalili, S.; Wahab, M.A.; Cuong-Le, T. A new movement strategy of grey wolf optimizer for optimization problems and structural damage identification. *Adv. Eng. Softw.* **2022**, *173*, 103276. [CrossRef]
15. Prina, M.G.; Cozzini, M.; Garegnani, G.; Manzolini, G.; Moser, D.; Oberegger, U.F.; Pernetti, R.; Vaccaro, R.; Sparber, W. Multi-objective optimization algorithm coupled to EnergyPLAN software: The EPLANopt model. *Energy* **2018**, *149*, 213–221. [CrossRef]
16. Zhang, X.; Liu, H.; Tu, L. A modified particle swarm optimization for multimodal multi-objective optimization. *Eng. Appl. Artif. Intell.* **2020**, *95*, 103905. [CrossRef]
17. Srinivas, N.; Deb, K. Muiltiobjective Optimization Using Nondominated Sorting in Genetic Algorithms. *Evol. Comput.* **1994**, *2*, 221–248. [CrossRef]
18. Xu, T.; He, J.; Shang, C.; Ying, W. A New Multi-objective Model for Constrained Optimisation. In *Advances in Computational Intelligence Systems*; Springer: Cham, Switzerland, 2017; pp. 71–85. [CrossRef]
19. Mirjalili, S.Z.; Mirjalili, S.; Saremi, S.; Faris, H.; Aljarah, I. Grasshopper optimization algorithm for multi-objective optimization problems. *Appl. Intell.* **2018**, *48*, 805–820. [CrossRef]
20. Wang, X.; Li, S. Multi-Objective Optimization Using Cooperative Garden Balsam Optimization with Multiple Populations. *Appl. Sci.* **2022**, *12*, 5524. [CrossRef]
21. Zhang, K.; Chen, M.; Xu, X.; Yen, G.G. Multi-objective evolution strategy for multimodal multi-objective optimization. *Appl. Soft Comput. J.* **2021**, *101*, 107004. [CrossRef]
22. Dhiman, G.; Singh, K.K.; Soni, M.; Nagar, A.; Dehghani, M.; Slowik, A.; Kaur, A.; Sharma, A.; Houssein, E.H.; Cengiz, K. MOSOA: A new multi-objective seagull optimization algorithm. *Expert Syst. Appl.* **2021**, *167*, 114150. [CrossRef]
23. Jamil, M.A.; Nour, M.K.; Alotaibi, S.S.; Hussain, M.J.; Hussaini, S.M.; Naseer, A. Software Product Line Maintenance Using Multi-Objective Optimization Techniques. *Appl. Sci.* **2023**, *15*, 9010. [CrossRef]
24. Guan, Y.; Chu, Y.; Lv, M.; Li, S.; Li, H.; Dong, S.; Su, Y. Application of Strength Pareto Evolutionary Algorithm II in Multi-Objective Water Supply Optimization Model Design for Mountainous Complex Terrain. *Sustainability* **2023**, *15*, 12091. [CrossRef]
25. Huynh, T.Q.; Nguyen, T.T.; Nguyen, H. Base resistance of super-large and long piles in soft soil: Performance of artificial neural network model and field implications. *Acta Geotechica* **2023**, *18*, 2755–2775. [CrossRef]
26. Li, Y.; Hariri-Ardebili, M.A.; Deng, T.; Wei, Q.; Cao, M. A surrogate-assisted stochastic optimization inversion algorithm: Parameter identification of dams. *Adv. Eng. Inform.* **2023**, *55*, 101853. [CrossRef]
27. Wu, J.; Zhu, X. Development trend of groundwater flow numerical simulation software based on MODFLOW. *Eng. Investig.* **2000**, *2*, 12–15.
28. Rong, Y.; Fang, Z. Risk assessment of ground settlement induced by construction dewatering of Taizhou Bridge anchorage caisson foundation. In Proceedings of the 2011 Second International Conference on Mechanic Automation and Control Engineering, Huhhot, China, 15–17 July 2011. [CrossRef]
29. Corne, D.W.; Knowles, J.D.; Oates, M.J. The Pareto envelope-based selection algorithm for multiobjective optimization: International Conference on Parallel Problem Solving from Nature. *Lect. Notes Comput. Sci.* **2000**, *1917*, 839–848.
30. Sharma, S.; Kumar, V. A Comprehensive Review on Multi-objective Optimization Techniques: Past, Present and Future. *Arch. Comput. Methods Eng.* **2022**, *29*, 5605–5633. [CrossRef]
31. Shi, F. *Analysis of 30 Cases of MATLAB Intelligent Algorithms [M]*; Beijing University of Aeronautics and Astronautics Press: Beijing, China, 2011.
32. Furtuna, R.; Curteanu, S.; Leon, F. An elitist non-dominated sorting genetic algorithm enhanced with a neural network applied to the multi-objective optimization of a polysiloxane synthesis process. *Eng. Appl. Artif. Intell.* **2011**, *24*, 772–785. [CrossRef]
33. *JGJ/120-2012*; Technical Regulations for Building Foundation Pit Support. Ministry of Housing and Urban-Rural Development, People's Republic of China: Beijing, China, 2012.
34. *GB50202-2018*; The Construction Quality Acceptance Code for Building Foundation Engineering. Ministry of Housing and Urban Rural Development, People's Republic of China: Beijing, China, 2018.
35. Zhao, Y.; Dong, X.; Wang, H.; Wang, J.; Wei, Y.; Huang, Y.; Xue, R. Comparative Study on the Application of Different Slug Test Models for Determining the Permeability Coefficients of Rock Mass in Long-Distance Deep Buried Tunnel Projects. *Appl. Sci.* **2022**, *12*, 10235. [CrossRef]
36. Zhao, Y.; Wang, H.; Lv, P.; Dong, X.; Huang, Y.; Wang, J.; Yang, Y. Theoretical Model and Experimental Research on Determining Aquifer Permeability Coefficients by Slug Test under the Influence of Positive Well-Skin Effect. *Water* **2022**, *14*, 3089. [CrossRef]
37. Zhao, Y.; Wei, Y.; Dong, X.; Rong, R.; Wang, J.; Wang, H. The Application and Analysis of Slug Test on Determining the Permeability Parameters of Fractured Rock Mass. *Appl. Sci.* **2022**, *12*, 7569. [CrossRef]
38. Zhao, Y.; Zhang, Z.; Rong, R.; Dong, X.; Wang, J. A new calculation method for hydrogeological parameters from unsteady-flow pumping tests with a circular constant water-head boundary of finite scale. *Q. J. Eng. Geol. Hydrogeol.* **2022**, *55*, qjegh2021-112. [CrossRef]

applied sciences

MDPI

Article

Effects of Dry Density and Moisture Content on the Kaolin–Brass Interfacial Shear Adhesion

Rayed Almasoudi [1,2], Hossam Abuel-Naga [1,*] and Firas Daghistani [1,3]

1 Department of Civil Engineering, La Trobe University, Bundoora, VIC 3086, Australia; r.almasoudi@latrobe.edu.au (R.A.); f.daghistani@latrobe.edu.au (F.D.)
2 Department of Civil Engineering, Umm Al-Qura University, Makkah 24381, Saudi Arabia
3 Department of Civil Engineering, University of Business and Technology, Jeddah 23435, Saudi Arabia
* Correspondence: h.aboel-naga@latrobe.edu.au

Abstract: Kaolin clay, with its consistent properties, fine particle size, high surface area, and extensive historical use, stands out as a reliable choice for laboratory research. This study aims to assess the interface shear adhesion behaviour between compacted clay and a metallic surface. For this purpose, a new testing approach was developed. The proposed method is simple, requires neither advanced equipment nor specialised test procedures, and, thus, represents an improvement over existing practices in this field. The experimental program focuses on determining the interface shear adhesion strength between reconstituted kaolin clay and a metallic surface. The kaolin clay testing specimens were dynamically compacted at various energy levels and moisture contents. The results indicate that the optimum moisture content is 30%, which provides the highest density to the sample and divides the compaction curve into dry and wet sides. Furthermore, the results demonstrate that the interface shear adhesion strength increases with the clay's dry density. Conversely, there is a significant decrease in strength as the moisture content specifically rises on the wet side of the compaction curve. The adhesion behaviour was also attributed to matric suction, where high suction enhanced interfacial adhesion, while low suction weakened bonding and diminished adhesion. Additionally, this study presents a unique three-dimensional contour graph illustrating the combined effects of dry density and moisture content on the interfacial adhesion.

Keywords: dry density; moisture content; kaolin; brass; interface; shear adhesion

Citation: Almasoudi, R.; Abuel-Naga, H.; Daghistani, F. Effects of Dry Density and Moisture Content on the Kaolin–Brass Interfacial Shear Adhesion. *Appl. Sci.* **2023**, *13*, 11191. https://doi.org/10.3390/app132011191

Academic Editor: Tiago Miranda

Received: 28 August 2023
Revised: 5 October 2023
Accepted: 8 October 2023
Published: 11 October 2023

1. Introduction

Interfacial adhesion refers to the ability of soil particles to adhere to the interface of other materials in the presence of water [1]. The system of soil adhesion comprises three elements: soil, solid surfaces, and their interface [2]. An example of interfacial adhesion commonly encountered is between clay and metal. It is important to differentiate between soil adhesion and soil cohesion, as shown in Figure 1, where the former involves soil particles adhering to other materials, and the latter refers to soil particles sticking to each other [3].

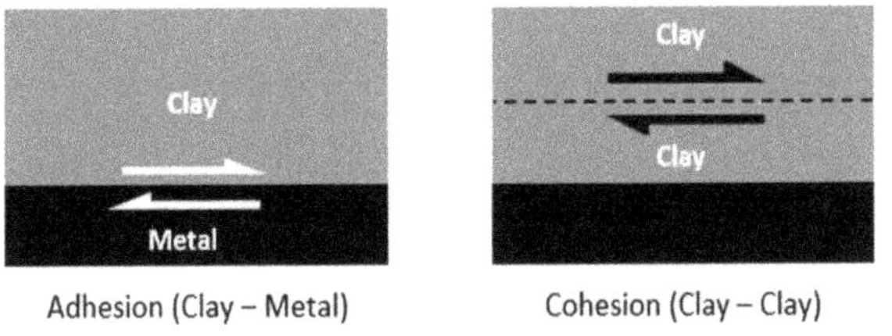

Figure 1. Illustration of adhesion and cohesion of soil.

There are two types of soil adhesion: normal adhesion and shear (tangential) adhesion [2]. The applied load is the factor that determines the adhesion type. The normal adhesion is accommodated with normal tensile load, and the shear adhesion is accommodated with shear load, as shown in Figure 2.

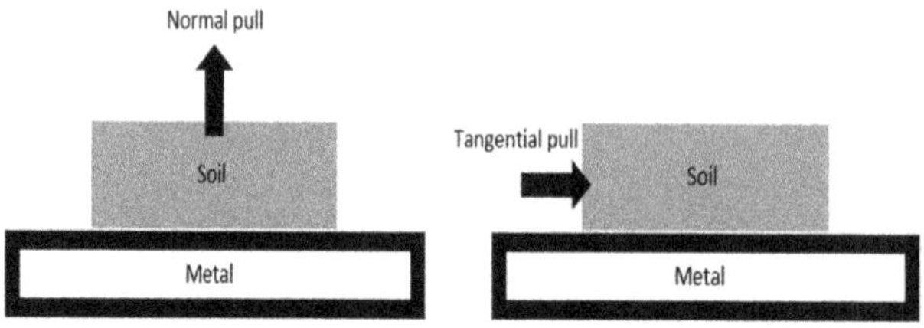

Figure 2. The adhesive tensile strength (normal pull) and the adhesive shear strength (tangential pull).

Adhesion behaviour at interfaces is influenced by various factors, including soil properties, material properties, interface properties and test conditions, as shown in Figure 3 [4,5]. Soil properties affecting interfacial adhesion include soil composition, grain size distribution, porosity, specific surface area (SSA), moisture content, water salinity, matric suction, plasticity, consistency and cohesion [4–6]. According to Zhang and Sang [7], shear adhesion and moisture content exhibit a parabolic relationship, as defined in Equation (1).

$$I_c = \frac{LL - MC}{PI} \tag{1}$$

where I_c is the consistency index of the soil, LL is the liquid limits, MC is the moisture content, and PI is the plasticity index. Thus, there is a relationship between the consistency index and shear adhesion, which assists in understanding the behaviour of clay adhesion [8]. Material properties that influence interfacial adhesion include the contact area, applied normal load, and roughness of the surface area [9]. Testing conditions, such as the contact time at the soil–material interface, rate of loading, humidity, and temperature of the interface surface, also play a significant role in affecting the behaviour and value of interfacial soil adhesion [6,10].

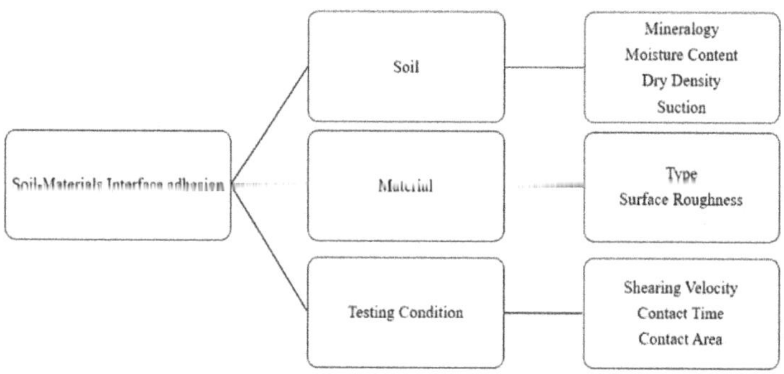

Figure 3. Factors affecting interfacial soil adhesion.

The interfacial soil adhesion derives importance from its applications in geotechnical engineering and agriculture, where it can pose potential challenges. For example, pile foundations, as shown in Figure 4, may encounter stickiness issues during the digging process [11], and the cutter head of a tunnel boring machine (TBM) can experience clogging problems at the interface due to interfacial soil adhesion [12–14].

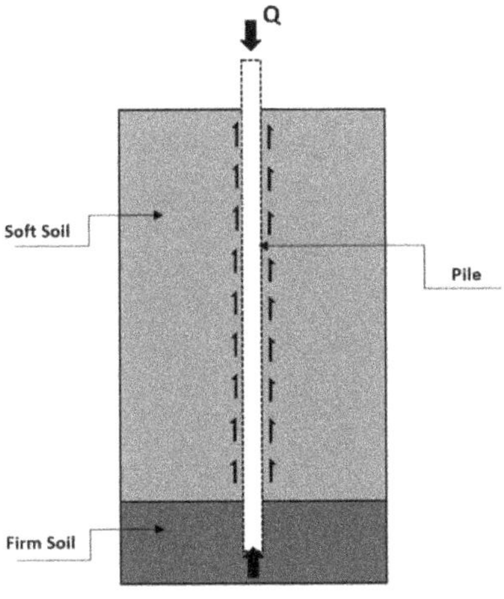

Figure 4. Pile Foundation and stickiness during digging.

These practical applications of interfacial soil adhesion can lead to unexpected cost implications. Additionally, soil sticking to agricultural machines during harvesting can result in high energy consumption and reduced work efficiency [6,8]. Therefore, it is crucial to investigate the behaviour of soil adhesion with other materials.

Several studies have employed different methods to measure soil adhesion; however, there is no specific method or device exclusively designed for soil adhesion measurement [3]. Soil adhesion can be measured in the laboratory using pull-out tests [12], piston separation tests [15] and shear plate tests [16]. Fountaine [1] used a specialised apparatus to measure normal adhesion for loam clay and sandy clay under the influence of water surface tension. Their study revealed that moisture content and material type have an impact on soil adhesion. Thewes and Burger [12] utilised pull-out tests to investigate clogging issues in TBMs, where normal soil adhesion occurs. It was found that clay minerals and soil consistency index (I_c) affect normal adhesion. Azadegan and Massah [17] designed a dedicated instrument to measure normal adhesion between clay and steel. Their study focused on the effect of temperature on adhesion and demonstrated that temperature influences soil adhesion. Mirjavan [15] employed a piston separation device to measure the normal adhesion of soil to metal. The study investigated wetness levels of montmorillonite clay and revealed that wetness is a significant influencing factor. Burbaum and Sass [18] used pull-out load or separation tests to investigate normal adhesion between clay and steel surfaces, finding that adhesion is influenced by the soil consistency index. Zumsteg and Puzrin [16] employed a plate apparatus to examine the clogging issue occurring at the interface between clay and TBM machines during tunnelling. They identified tangential adhesion as the main factor contributing to soil stickiness and clogging issues. This paper aims to investigate the behaviour of shear adhesion at the interface between kaolin clay and the internal surface of a brass mould. This study involves testing thirty compacted kaolin samples with different moisture contents and compaction energy levels (dry density). To conduct the experiment, a new simple test method was developed to measure interfacial shear adhesion between kaolin and brass. The testing methodology introduced in this research study represents a pioneering approach by employing conventional soil testing equipment for evaluating interface shear adhesion. This stands in contrast to previous investigations, wherein specialised testing apparatus, which may not be readily accessible in commercial laboratories, were utilised.

2. Materials and Methods

2.1. Materials

The interface adhesion test conducted in this study involved using kaolin clay as the testing material, sourced from "Scott Chemical Australia Pty Ltd." (Cheltenham, Australia) in Melbourne, Australia. Kaolin clay is commonly used in geotechnical engineering studies due to its availability and representative properties. Different clay compositions can influence various properties, such as shrinkage, plasticity, and strength. These changes in properties can, in turn, impact the experiment results by altering the clay's behaviour under various conditions. In this experiment, we maintained a consistent mineral composition among samples to ensure reliable results. The mineral composition of the kaolin, specified on the clay bag, is shown in Table 1.

Table 1. Mineral composition of the kaolin.

Mineral	Mass Percentage
Kaolinite	83.7
Muscovite	14.0
Quartz	2.3

To further characterize the kaolin clay, Table 2 presents the geotechnical properties that were examined during this study. These properties include parameters such as moisture content, dry density, shear strength, and compaction energy. Understanding the geotechnical properties of the soil is crucial for assessing its behaviour under different loading and moisture conditions.

Table 2. Engineering properties of the used soil (kaolin).

Properties	Values
Specific gravity (G_s)	2.58
LL (%)	74
PL (%)	32
PI (%)	42
Cation exchange capacity (CEC) (meq/100 g)	0.075
Total surface area (m^2/g)	20
Surface charge density ($\mu C/m^2$)	0.36
Silicate SiO_2	45.2
Aluminium Al_2O_3	38.8

Additionally, Figure 5 shows the particle size distribution of the kaolin clay. This distribution illustrates the relative proportions of different particle sizes present within the soil sample. The particle size distribution plays a significant role in determining the mechanical and hydraulic properties of the soil, affecting factors such as compaction behaviour, permeability, and shear strength.

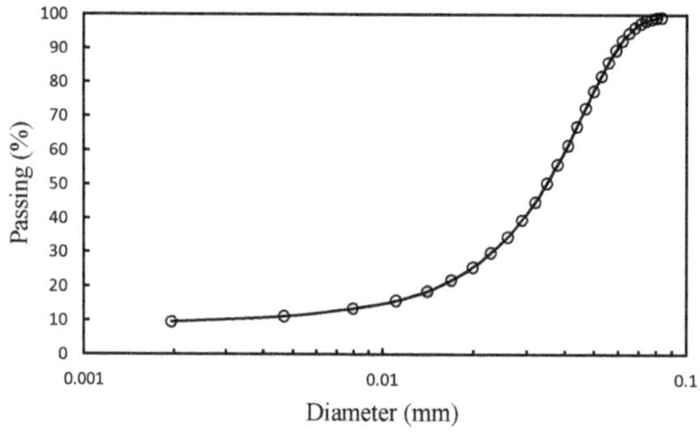

Figure 5. Particle size distribution of kaolin.

2.2. Preparation

To carry out the interfacial adhesion test, 30 different samples were prepared. The prepared samples produced different moisture contents and dry density conditions. The percentage of moisture was between 10 and 50%. The kaolin samples were compacted at different compaction energy levels using a series of blows ranging from 15 to 55 blows to achieve the minimum and maximum dry density at different moisture levels, as listed in Table 3. The soil was compacted in the mould in three equal layers of almost 43 mm. The same amount of energy was applied to each layer. The moisture content used was regular tap water. After adding varying amounts of moisture to the soil samples, each sample was placed in a sealed plastic bag and kept overnight for moisture equalisation.

Table 3. The amount of water content and number of blows used in the experiment.

MC (%)	10	20	25	30	40	50
Number of blows	15	15	15	15	15	15
	25	25	25	25	25	25
	35	35	35	35	35	35
	45	45	45	45	45	45
	55	55	55	55	55	55

2.3. Approach

After the soil had been compacted in the compaction mould, the next step was to test the interfacial shear adhesion with a loading machine. This approach enabled the interfacial shear strength between the compacted soil and the brass mould to be evaluated.

To prepare the compaction mould for testing, it was inverted, and the mould base was removed. This allowed the compacted soil to be extruded from the mould, exposing the surface that was in contact with the brass mould during the shear adhesion test. This setup ensured a consistent interface between the soil sample and the mould and allowed accurate measurements of shear adhesion.

The loading machine was then utilised to apply a displacement-controlled load to the compacted soil, compressing it out of the mould. The loading machine's mechanism and setup are illustrated in Figure 6, providing a visual representation of the process. The displacement rate employed in this study was 5.0 mm/min. This rate ensured a controlled and uniform compression of the compacted soil, allowing precise measurements and reliable data acquisition.

By subjecting the compacted soil to a displacement-controlled load, the loading machine simulated the conditions under which the soil is subjected to external forces or loading in real-world applications. This enabled the shear adhesion strength between the soil and the mould to be evaluated, providing valuable insights into the stability of the soil and its interaction with surrounding materials.

The utilisation of a displacement-controlled load and a consistent loading rate ensured standardised testing conditions, enabling accurate comparisons between different soil samples and variations in compaction parameters. The data obtained from these tests contributed to a better understanding of soil behaviour and its response to external loads and helped in the planning and implementation of various engineering projects.

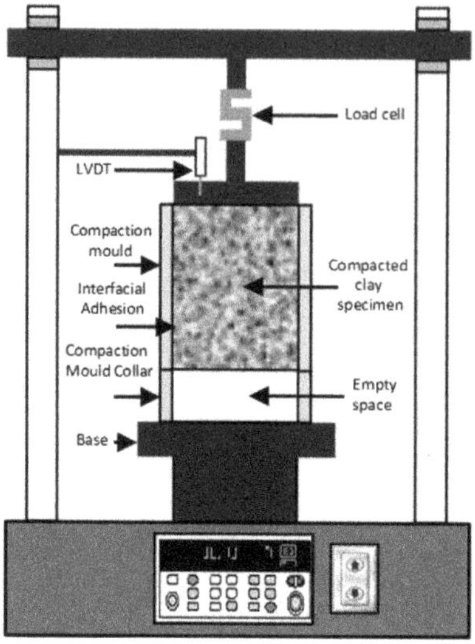

Figure 6. Schematic diagram of the apparatus and the test setup.

The testing process involved shearing the soil against the internal surface of the mould by applying a vertical load. To ensure that the machine load did not cause additional compression on the compacted specimen, which could have potentially damaged it by excessive squeezing, it is crucial for the movement of the compacted specimen to precisely match the loading displacement. Throughout the loading process, both the displacement and the load are accurately recorded and plotted. The peak of the load-displacement curve shows the maximum adhesion capacity of the interface between the compacted kaolin and the inner wall of the brass mould, as shown in Figure 7. The soil–mould interface adhesion can be determined using Equation (2):

$$\alpha = \frac{P}{A} \quad (2)$$

where α is the interfacial adhesion (kPa), P is the peak load (kN), and A is the internal surface of the mould (m^2).

Figure 7. Kaolin adhesion at 30% optimum moisture and varying compaction levels.

3. Results

3.1. Compaction Curve

The compaction curve displayed in Figure 8 provides valuable insights into the behaviour of kaolin under different compaction energies and moisture contents. By examining the curve, the relationship between dry density, compaction energy, and moisture content can be observed.

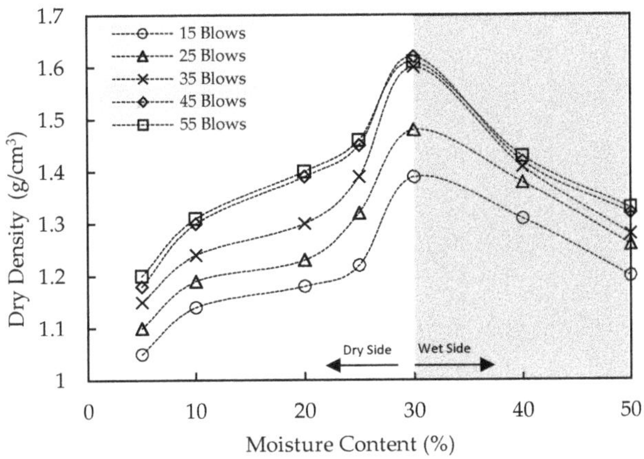

Figure 8. Compaction curve of kaolin at different compaction levels.

The compaction curve shows the variations in dry density as a function of compaction energy, represented by the number of blows (15, 25, 35, 45 and 55 blows) and the moisture content in the range from 10 to 50%. Each combination of compaction energy and moisture content corresponds to a specific point on the compaction curve. At 15 blows, the dry density increases from 1.14 at 10% MC to a peak of 1.39 at 30% MC, then descends to 1.20 at 50% MC. At 55 blows, the dry density ranges from 1.31 at 10% MC to 1.61 at 30% MC, decreasing to 1.33 at 50% MC. The peaks observed on the compaction curve indicate the optimum moisture content (OMC) of the kaolin, which is at 30% moisture content in all cases studied. The OMC represents the moisture content at which the kaolin reaches its maximum dry density during the compaction process. This optimum moisture content is important to achieving the desired compaction properties and overall soil performance. The OMC effectively divides the compaction curve into two distinct sides: the dry side and the wet side. The basis for dividing the compaction curve is primarily to categorize soil adhesion behaviour into two stages: one strongly affected and the other weakly affected. The moisture content ranging from 10 to 30% represents the dry side, while the range of 30 to 50% represents the wet side.

3.2. Dry Density

The relationship between dry density and interfacial shear adhesion at the dry side is presented in Figure 9. It is observed that there is a linear relationship between dry density and shear adhesion, wherein an increase in dry density leads to an increase in shear adhesion at each moisture content value. This finding highlights the importance of compaction in influencing interfacial adhesion behaviour.

Detailed analysis revealed that the impact of dry density on shear adhesion is more pronounced at the dry side of the compaction curve compared to the wet side. When kaolin clay is compacted to a greater extent within the dry side, the soil particles come closer to each other, resulting in creating a denser surface area. At an MC of 10%, adhesion values range significantly from approximately 23.6 to 124 kPa across varied dry densities. This densification of the soil enhances the contact area between kaolin and the brass mould. Consequently, the interfacial shear adhesion is strengthened due to the increased contact between the soil particles and the mould surface. It should be noted that the extent of the increase in dry density is dependent on the moisture content. In other words, the effect of the dry density on interfacial shear adhesion varies with the moisture content.

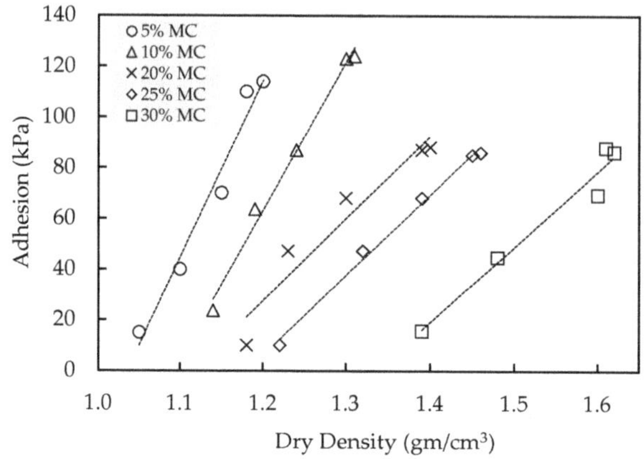

Figure 9. Dry density versus interfacial adhesion on kaolin at different moisture content levels.

3.3. Moisture Content

Figure 10 provides an insight into the behaviour of interfacial shear adhesion in compacted kaolin under the influence of moisture content. It has been observed that with a low moisture content ranging from 5 to 10% MC, the adhesion increases. Specifically, at 10% MC, it shows the highest adhesion among the different levels of compacted kaolin on the dry side. However, when the moisture content approaches the OMC, especially between 20 and 30%, increasing the moisture content has a slight effect on adhesion, which can be considered negligible.

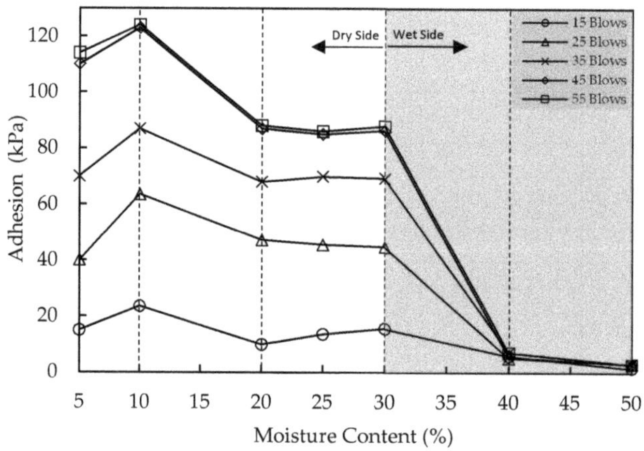

Figure 10. Interface adhesion of kaolin at different moisture contents and densities.

The behaviour of the unaffected area on the dry side can be explained by considering the maximum dry density of the soil at that point. At the OMC, the soil reaches its maximum dry density, resulting in a dense contact area between the kaolin and the mould surface. This dense interface contributes to the formation of the strongest interfacial adhesion. Consequently, there is a balanced effect between dry density, which typically improves adhesion, and moisture content, which often decreases adhesion. On the other hand, on the wet side, which is characterised by extreme moisture content and a high degree of saturation, a significant decrease in interfacial shear adhesion is observed. At the highest moisture content of 50%, the interface adhesion becomes null. This behaviour can be attributed to the detrimental effects of excessive moisture content and saturation on the interfacial adhesion properties of the kaolin.

3.4. Coupling Effect of Dry Density and Moisture Content

Figure 11 presents a novel three-dimensional graph showing the coupling effect of dry density and moisture content on the interfacial shear adhesion between kaolin and a brass mould. This graph provides a comprehensive visualisation of the relationship between these two parameters and their influence on shear adhesion. For the studied case, the light grey shaded area represents the possible shear adhesion values in relation to moisture content and dry density. The diagram of the contour lines provides valuable insights. It shows that at constant dry density, shear adhesion tends to decrease as moisture content increases. This relationship suggests that higher moisture content generally leads to a reduction in interfacial adhesion between kaolin and the brass mould.

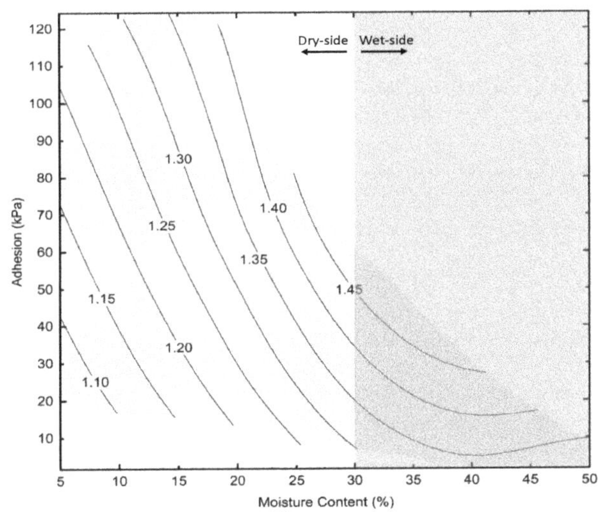

Figure 11. Three-dimensional contour graph of the effects of dry density and moisture content on shear adhesion of kaolin.

Figure 12 represents the adhesion change rate (*ACR*), which is the rate at which the adhesion value changes with variations in dry density at different water content levels. As dry density is a function of the *ACR*, it is important to note that the *ACR* of shear adhesion reduction varies with dry density and can be calculated using Equation (3):

$$ACR = \frac{\alpha_2 - \alpha_1}{\rho_{d2} - \rho_{d1}} \quad (3)$$

where *ACR* is the adhesion change rate; α_2 and α_1 are the final and initial adhesion values, respectively; and ρ_{d2} and ρ_{d1} are the final and initial dry density values, respectively. The information provided by these contour lines allows a more detailed understanding of the interaction between dry density, moisture content, and interfacial shear adhesion. The graph shows that changes in these parameters affect the adhesion behaviour of the kaolin. This understanding can help to optimize compaction processes and control moisture content to achieve desired shear adhesion properties.

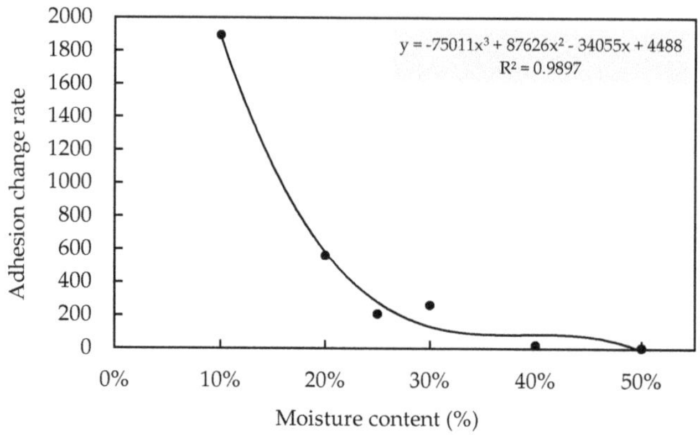

Figure 12. The relationship between the adhesion change rate and moisture content.

4. Discussion

The interfacial adhesion behaviour between kaolin and brass can vary depending on whether it is observed on the dry or wet side of the compaction curve. The compaction curve is a graph showing the relationship between compaction pressure and the resulting density or porosity of a material. An analysis of the interfacial adhesion behaviour on either side of the compaction curve can provide insights into the mechanisms that control the interaction between kaolin and brass.

In Alshameri's [19] experiment, he used a mixture of sand and kaolin at different percentages. In his study, a mixture of 80% sand and 20% kaolin yielded the highest dry density, with an OMC of 12%. In contrast, this experiment exclusively used kaolin clay and achieved the highest dry density with an OMC of 30%. This paper presents a wide range of dry density samples, achieved by considering various numbers of blows at different moisture contents, as shown in Table 4.

Table 4. Summary of interfacial adhesion behaviour for dry and wet sides.

Number of Blows	MC (%)	ρ_d (g/cm^3)	Su (pF)	α (kPa)
15	5	1.05	5.00	15.00
	10	1.14	4.70	23.59
	20	1.18	4.40	9.92
	25	1.22	4.32	13.66
	30	1.39	3.67	15.62
	40	1.31	3.05	6.02
	50	1.20	2.00	1.63
25	5	1.10	5.00	40.00
	10	1.19	4.71	63.60
	20	1.23	4.43	47.17
	25	1.32	4.41	45.55
	30	1.48	3.82	44.73
	40	1.38	3.19	5.37
	50	1.26	2.17	3.25

Table 4. *Cont.*

Number of Blows	MC (%)	ρ_d (g/cm^3)	Su (pF)	α (kPa)
35	5	1.15	5.00	70.00
	10	1.24	4.72	86.86
	20	1.30	4.45	68.00
	25	1.39	4.44	69.95
	30	1.60	3.84	69.30
	40	1.41	3.40	7.32
	50	1.28	2.29	3.09
45	5	1.18	5.00	110.00
	10	1.30	4.75	123.00
	20	1.39	4.48	87.00
	25	1.45	4.20	85.00
	30	1.62	3.89	86.21
	40	1.42	2.83	5.69
	50	1.32	2.20	3.20
55	5	1.20	5.00	114.00
	10	1.31	4.66	124.00
	20	1.40	4.50	88.00
	25	1.46	4.45	86.00
	30	1.61	4.10	87.84
	40	1.43	3.60	7.32
	50	1.33	2.10	3.00

4.1. Mechanism of Adhesion on the Dry Side

It has been observed that on the dry side of the compaction curve, interfacial adhesion increases, with higher dry density values at any moisture content. Conversely, adhesion decreases with increasing moisture content. On the dry side of the compaction curve, where the pressure is relatively low, the interfacial adhesion behaviour between kaolin and brass is mainly influenced by physical interlocking and mechanical friction. When pressure is applied, the kaolin particles tend to encounter the brass surface, creating points of contact and interlocking. This physical interlocking creates a certain level of adhesion between the two materials. In addition, the roughness and surface irregularities of the brass surface can contribute to adhesion. The kaolin particles can fill in the gaps and irregularities on the brass surface, resulting in a larger contact area and stronger adhesion.

4.2. Mechanism of Adhesion on the Optimum Moisture Content

At the OMC point on the compaction curve, where pressure is at its peak, the adhesion behaviour between kaolin and brass undergoes marked changes. Moisture content plays a crucial role in influencing the underlying adhesion mechanisms in that specific region. However, despite the observed changes, it should be noted that the interfacial adhesion at the OMC is not the highest. The adhesion between kaolin and brass weakens with increasing moisture content, demonstrating the negative influence of increased moisture levels on the strength of interfacial adhesion.

4.3. Mechanism of Adhesion on the Wet Side

On the wet side of the compaction curve, the presence of excessive moisture content can adversely affect the bonds between individual kaolin particles. The water essentially acts as a lubricant, reducing friction and interlocking between the particles. This weakens the bond between the particles, resulting in a significant decrease in the overall adhesion strength at the interface. However, low moisture content (5–10% MC) in the sample, as indicated by the results, shows an increase in adhesion. This can be explained by the fact that a slight increase in the moisture content increases the dry density, thereby increasing the contact surface, which leads to higher interfacial adhesion. This is in compliance with Li and Zhang [20], who showed that adhesion initially increases with moisture content and then decreases.

4.4. Matric Suction Role

This complex behaviour is further explained by the matric suction effect. Figure 13 illustrates the role of matric suction in binding the kaolin particles to the inner surface of the compaction mould. On the dry side, which is characterised by low moisture content and high soil suction, interfacial bonding is strong, resulting in increased interfacial adhesion. On the other hand, on the wet side, which is characterised by high moisture content and low soil suction, interfacial bonding becomes weak and has a negligible impact on interfacial adhesion.

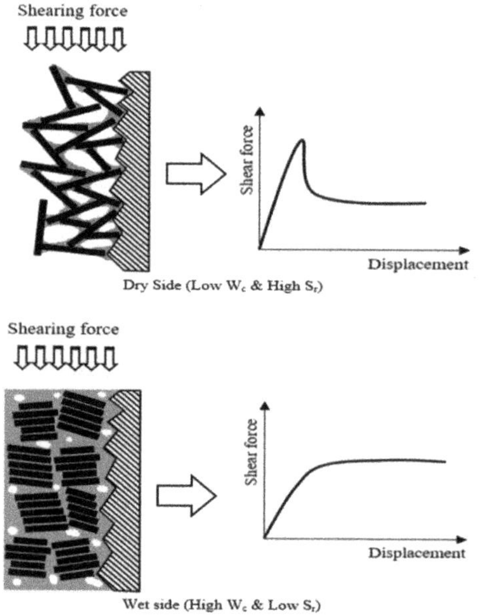

Figure 13. Interface adhesion behaviour for dry side versus wet side.

Kaolin exhibits high suction power on the dry side when moisture content is low and low suction power on the wet side when moisture content is high, as shown in Figure 14.

Figure 15 illustrates the influence of clay suction on the shear adhesion on both sides. On the dry side, where the moisture content is low and the clay is not fully saturated, the suction of clay is high and leads to high interface adhesion. Hence, the significant reduction in soil suction as moisture content increases from 10 to 30% on the dry side accelerates the loss of adhesion. On the wet side, beyond the OMC of 30%, where the moisture content is high and the clay is nearly fully saturated, the suction of clay is becoming low and leads to low interface adhesion. Hence, the low suction due to the high degree of saturation leads to a sudden failure in the interfacial adhesion. This adhesion failure is due to the combined effect of high moisture content, where the soil is fully saturated, and low suction.

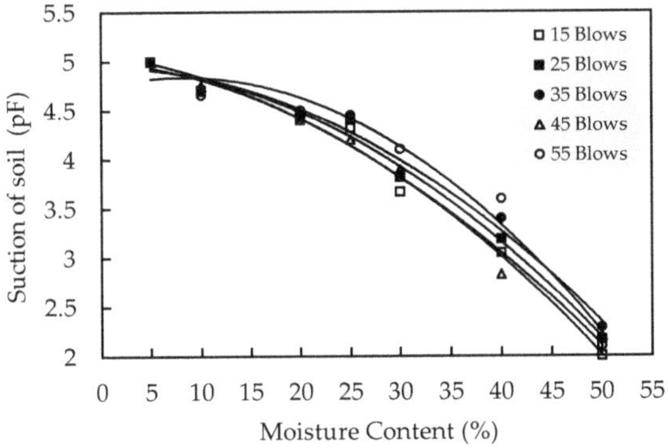

Figure 14. Behaviour of suction against the moisture content for varied compacted kaolin.

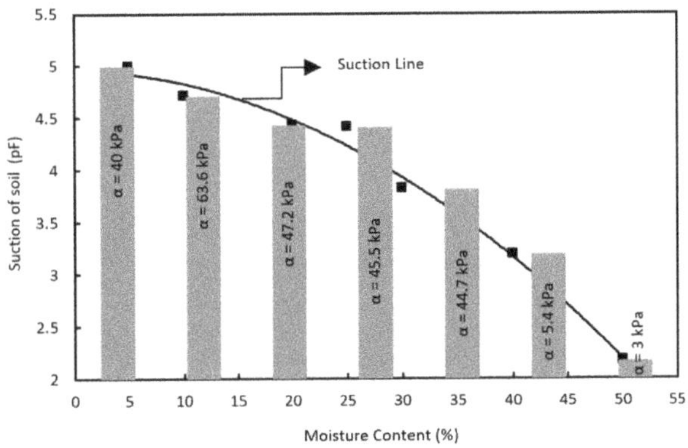

Figure 15. Influence of suction on interface adhesion of kaolin at a standard compaction of 25 blows.

4.5. Limitation

This study has some limitations that should be acknowledged and addressed in future research. Firstly, this study only used kaolin clay as the testing material, which may not represent the behaviour of other types of clay or soil mixtures. Future studies could consider different types of clay mixed with sand or recycled materials [21,22] to investigate the effect of soil composition on interfacial shear adhesion. Secondly, this study only performed the test at one displacement rate of 5.0 mm/min, which may not capture the dynamic loading conditions encountered in real-world applications. Future studies could consider different displacement rates and check the impact of loading rate on the adhesion behaviour. Thirdly, this study conducted the test at a constant room temperature of 23 °C, which may not reflect the environmental variations affecting soil properties and adhesion mechanisms. Future studies could consider the impacts of different temperatures [23] on the interfacial shear adhesion between kaolin and brass. This study focused on the impacts of dry density and moisture content on the adhesion of kaolin clay, which are important factors influencing soil performance and stability. However, other factors such as contact area, surface roughness, and contact time could also affect the interfacial adhesion behaviour and should be explored in future research.

5. Conclusions

This paper investigated the coupling effect of dry density and moisture content on the interfacial shear adhesion between kaolin clay and brass. A simple method was developed to measure the shear adhesion using a loading machine. The main findings of this study are as follows:

- With an increase in moisture content in kaolin, the dry density increases until it reaches a maximum value and then decreases. The maximum dry density of the used kaolin is observed at an OMC of 30%.
- There is a linear relationship between dry density and shear adhesion on the dry side of the compaction curve, where higher dry density leads to higher shear adhesion due to the larger area of contact with the mould surface.
- As moisture content initially increases up to 10% MC, the shear adhesion also increases. Beyond that point, the increase in MC results in a slight decrease in adhesion on the dry side of the compaction curve up to the OMC. Beyond the OMC, on the wet side of the compaction curve, there is a significant decrease in shear adhesion.
- The adhesion behaviour can be further explained by considering the role of matrix suction. High suction on the dry side improves adhesion, while low suction on the wet side weakens adhesion.
- Future research can consider different types of soil and mixtures, as well as tests performed under various shear rates and temperatures.

Author Contributions: Writing—Original draft, R.A.; Writing—Review & editing, H.A.-N. and F.D. All authors have read and agreed to the published version of the manuscript.

Funding: This research received no external funding.

Institutional Review Board Statement: Not applicable.

Informed Consent Statement: Not applicable.

Data Availability Statement: Not applicable.

Conflicts of Interest: The authors declare no conflict of interest.

Nomenclature

I_c	Consistency index
PI	Plasticity index
PL	Plastic limit
LL	Liquid limit
α	Interfacial adhesion
ρ	Dry density
P	Peak load
A	Internal surface area of the mould
Sr	Saturation degree of clay
Su	Suction of clay
G_s	Specific gravity
OMC	Optimum moisture content
MC	Moisture content
ACR	Adhesion change rate
CEC	Cation exchange capacity
TBM	Tunnel boring machine

References

1. Fountaine, E. Investigations into the mechanism of soil adhesion. *J. Soil Sci.* **1954**, *5*, 251–263. [CrossRef]
2. Tong, J.; Ren, L.; Chen, B.; Qaisrani, A. Characteristics of adhesion between soil and solid surfaces. *J. Terramech.* **1994**, *31*, 93–105. [CrossRef]
3. Alberto-Hernandez, Y.; Kang, C.; Yi, Y.; Bayat, A. Clogging potential of tunnel boring machine (TBM): A review. *Int. J. Geotech. Eng.* **2018**, *12*, 316–323. [CrossRef]
4. Jia, X. Theoretical analysis of the adhesion force of soil to solid materials. *Biosyst. Eng.* **2004**, *87*, 489–493. [CrossRef]
5. Soni, P.; Salokhe, V.M. Theoretical analysis of microscopic forces at soil-tool interfaces: A review. *Agric. Eng. Int. CIGR J.* **2006**, *8*, 1–25.

6. Sass, I.; Burbaum, U. A method for assessing adhesion of clays to tunneling machines. *Bull. Eng. Geol. Environ.* **2009**, *68*, 27–34. [CrossRef]
7. Zhang, J.; Sang, Z.; Gao, L. Adhesion and friction between soils and solids. *Nung Yeh Chi Hsieh Hsueh Pao Trans. Chin. Soc. Agric. Mach.* **1986**, *17*.
8. Liu, P.; Wang, S.; Shi, Y.; Yang, J.; Fu, J.; Yang, F. Tangential adhesion strength between clay and steel for various soil softnesses. *J. Mater. Civ. Eng.* **2019**, *31*, 04019048. [CrossRef]
9. Bhushan, B. Adhesion and stiction: Mechanisms, measurement techniques, and methods for reduction. *J. Vac. Sci. Technol. B* **2003**, *21*, 2262–2296. [CrossRef]
10. Satomi, T.; Nihei, H.; Takahashi, H. Investigation on characteristics of soil adhesion to metallic material surface and soil animal's cuticle. In Proceedings of the 15th International Conference on Experimental Mechanics, Online, 7–9 September 2021; pp. 1–9.
11. Sladen, J. The adhesion factor: Applications and limitations. *Can. Geotech. J.* **1992**, *29*, 322–326. [CrossRef]
12. Thewes, M.; Burger, W. Clogging of TBM drives in clay–identification and mitigation of risks. In *Underground Space*; CRC Press: Boca Raton, FL, USA, 2005.
13. Atkinson, J.H.; Fookes, P.G.; Miglio, B.F.; Pettifer, G.S. Destructuring and disaggregation of Mercia Mudstone during full-face tunnelling. *Q. J. Eng. Geol. Hydrogeol.* **2003**, *36*, 293–303. [CrossRef]
14. Jancsecz, S.; Krause, R.; Langmaack, L. Advantages of soil conditioning in shield tunnelling: Experiences of LRTS Izmir. In *Challenges for the 21st Century*; A.A. Balkema: Rotterdam, The Netherlands, 1999; pp. 865–875.
15. Basmenj, A.K.; Mirjavan, A.; Ghafoori, M.; Cheshomi, A. Assessment of the adhesion potential of kaolinite and montmorillonite using a pull-out test device. *Bull. Eng. Geol. Environ.* **2016**, *76*, 1507–1519. [CrossRef]
16. Zumsteg, R.; Puzrin, A.M. Stickiness and adhesion of conditioned clay pastes. *Tunn. Undergr. Space Technol.* **2012**, *31*, 86–96. [CrossRef]
17. Azadegan, B.; Massah, J. Effect of temperature on adhesion of clay soil to steel. *Cercet. Agron. Mold.* **2012**, *45*, 21–27. [CrossRef]
18. Burbaum, U.; Sass, I. Physics of adhesion of soils to solid surfaces. *Bull. Eng. Geol. Environ.* **2017**, *76*, 1097–1105. [CrossRef]
19. Alshameri, B. Maximum dry density of sand–kaolin mixtures predicted by using fine content and specific gravity. *SN Appl. Sci.* **2020**, *2*, 1693. [CrossRef]
20. Li, H.; Zhang, Z.; Zhai, J.; Yang, L.; Long, H. Correlation between Soil Structural Parameters and Soil Adhesion Based on Water Film Theory. *Coatings* **2022**, *12*, 1743. [CrossRef]
21. Daghistani, F.; Baghbani, A.; Abuel Naga, H.; Faradonbeh, R.S. Internal Friction Angle of Cohesionless Binary Mixture Sand–Granular Rubber Using Experimental Study and Machine Learning. *Geosciences* **2023**, *13*, 197. [CrossRef]
22. Daghistani, F.; Baghbani, A.; Abuel Naga, H. Shear strength characteristics of binary mixture sand-carpet fibre using experimental study and machine learning. *Int. J. Geotech. Eng.* **2023**, 1–15. [CrossRef]
23. Li, C.; Kong, G.; Liu, H.; Abuel-Naga, H. Effect of temperature on behaviour of red clay–structure interface. *Can. Geotech. J.* **2019**, *56*, 126–134. [CrossRef]

applied sciences

MDPI

Article

Utilizing Multivariate Adaptive Regression Splines (MARS) for Precise Estimation of Soil Compaction Parameters

Musaab Sabah Abed [1], Firas Jawad Kadhim [1], Jwad K. Almusawi [1], Hamza Imran [2], Luís Filipe Almeida Bernardo [3,*] and Sadiq N. Henedy [4]

1 Department of Civil Engineering, Faculty of Engineering, University of Misan, Amarah 62001, Iraq; musciveng@uomisan.edu.iq (M.S.A.); firasjk@uomisan.edu.iq (F.J.K.); eng.jawad@uomisan.edu.iq (J.K.A.)
2 Department of Environmental Science, College of Energy and Environmental Science, Alkarkh University of Science, Baghdad 10081, Iraq; hamza.ali1990@kus.edu.iq
3 Department of Civil Engineering and Architecture, GeoBioTec-UBI, University of Beira Interior, 6201-001 Covilhã, Portugal
4 Department of Civil Engineering, Mazaya University College, Nasiriyah City 64001, Iraq; sadiq_naama@yahoo.com
* Correspondence: lfb@ubi.pt

Abstract: Traditional laboratory methods for estimating soil compaction parameters, such as the Proctor test, have been recognized as time-consuming and labor-intensive. Given the increasing need for the rapid and accurate estimation of soil compaction parameters for a range of geotechnical applications, the application of machine learning models offers a promising alternative. This study focuses on employing the multivariate adaptive regression splines (MARS) model algorithm, a machine learning method that presents a significant advantage over other models through generating human-understandable piecewise linear equations. The MARS model was trained and tested on a comprehensive dataset to predict essential soil compaction parameters, including optimum water content (w_{opt}) and maximum dry density (ρ_{dmax}). The performance of the model was evaluated using coefficient of determination (R^2) and root mean square error (RMSE) values. Remarkably, the MARS models showed excellent predictive ability with high R^2 and low RMSE, MAE, and relative error values, indicating its robustness and reliability in predicting soil compaction parameters. Through rigorous five-fold cross-validation, the model's predictions for w_{opt} returned an RMSE of 1.948%, an R^2 of 0.893, and an MAE of 1.498%. For ρ_{dmax}, the results showcased an RMSE of 0.064 Mg/m^3, an R^2 of 0.899, and an MAE of 0.050 Mg/m^3. When evaluated on unseen data, the model's performance for w_{opt} prediction was marked with an MAE of 1.276%, RMSE of 1.577%, and R^2 of 0.948. Similarly, for ρ_{dmax}, the predictions were characterized by an MAE of 0.047 Mg/m^3, RMSE of 0.062 Mg/m^3, and R^2 of 0.919. The results also indicated that the MARS model outperformed previously developed machine learning models, suggesting its potential to replace conventional testing methods. The successful application of the MARS model could revolutionize the geotechnical field through providing quick and reliable predictions of soil compaction parameters, improving efficiency for construction projects. Lastly, a variable importance analysis was performed on the model to assess how input variables affect its outcomes. It was found that fine content (C_f) and plastic limit (PL) have the greatest impact on compaction parameters.

Keywords: multivariate adaptive regression splines; soil compaction; optimum water content; maximum dry density; machine learning

Citation: Abed, M.S.; Kadhim, F.J.; Almusawi, J.K.; Imran, H.; Bernardo, L.F.A.; Henedy, S.N. Utilizing Multivariate Adaptive Regression Splines (MARS) for Precise Estimation of Soil Compaction Parameters. *Appl. Sci.* **2023**, *13*, 11634. https://doi.org/10.3390/app132111634

Academic Editor: Wei Guo

Received: 8 October 2023
Revised: 21 October 2023
Accepted: 23 October 2023
Published: 24 October 2023

1. Introduction

Soil compaction, a foundational process in geotechnical engineering, refers to the method of tightening soil particles and reducing air voids, thereby increasing density while maintaining the existing water content [1]. This technique not only enhances soil's mechanical behavior but also provides a stable base for various infrastructures, solidifying its significance in the field of civil engineering. Initially suggested by Proctor in 1933 [2], soil compaction involves applying specified compactive effort in a controlled laboratory setting, allowing for various water contents. This process enables engineers to pinpoint the optimal moisture content and maximum dry density—two essential parameters obtained from the compaction curve [3–5]. In field compaction, the key parameters are the optimum water content (w_{opt}) and maximum dry density ($\rho_{\rm dmax}$). These are crucial in determining the maximum dry density of soil, which indicates an enhanced soil bearing capacity. The optimum water content is the specific moisture level at which soil can attain its highest dry density following a certain compaction effort. It is the minimum water necessary to form a thin film on soil particles, aiding their sliding movement for compaction. Compressing soil to its greatest theoretical density involves eliminating gases within the soil via expelling them through voids in the soil matrix. This process of gas expulsion is fundamental to soil densification, allowing soil to reach its maximum potential density. Hence, w_{opt} and $\rho_{\rm dmax}$ are vital in achieving the desired soil densification and stability [6]. These parameters form the backbone of long-term performance for a myriad of geotechnical structures, such as landfill liners, highway embankments, railway track-beds, earth dams, nuclear waste disposal facilities, and airfield pavements [7–17]. For example, study [7] found that w_{opt} and $\rho_{\rm dmax}$ are crucial factors affecting the compression of compacted Oklahoma soils, influencing collapse potential in embankments. Another study [8] used w_{opt} and $\rho_{\rm dmax}$ as critical parameters to understand the behavior and performance of MX80, which is a versatile clay for nuclear waste barriers. Furthermore, study [14] investigated the microstructure and hydraulic properties of high-speed railway subgrade soil using different values of w_{opt} and $\rho_{\rm dmax}$ to understand its behavior. Thus, accurately predicting and understanding the compaction parameters of diverse soil types is paramount for the successful construction and maintenance of these geotechnical structures.

To find compaction parameters, two main compaction approaches emerged: the standard Proctor compaction and the modified Proctor compaction [1]. The standard method is typically adopted for regular traffic loading situations, while the modified approach is used when dealing with heavy unit weights like those considered for airfield pavements. Outcomes derived from laboratory tests conducted using either the standard Proctor compaction method (ASTM D698) [18] or the modified Proctor compaction method (ASTM D1557) [19] are visually represented via an inverted 'V' curve. This graphical representation illustrates the peak, known as $\rho_{\rm dmax}$, and the corresponding moisture content, termed as the w_{opt} of the soil. Reaching the optimum moisture content and maximum dry density in a laboratory setting, using both standard and modified compaction tests, demands significant resources. It involves a considerable amount of time (typically 2–3 days), substantial effort, and a large volume of soil (around 20 kg per individual test) [1]. Efforts have been made to overcome this challenge through proposing several prediction models for determining the soil compaction parameters more efficiently. Empirical equations, created via multiple linear regression, can predict w_{opt} and $\rho_{\rm dmax}$ of soil [20–25]. These models offer the benefits of quick calculations and easy application, which speed up decision-making in geotechnical projects. They are user-friendly, making them accessible to a variety of users, not just experts. However, these models come with several limitations. One issue is the existence of numerous equations for the same factors, leading to potential confusion and inconsistency in results. This vast array of options makes the selection of the appropriate model challenging, possibly affecting the reliability of predictions. Another significant limitation is the assumption of linear behavior, which often does not reflect the complex, often non-linear nature of soil behavior. These models, while providing an estimation, might not fully encapsulate the range of outcomes in different conditions due to their simplified assumptions.

The complex nature of soil compaction parameters, influenced by numerous interrelated parameters, necessitates the use of robust, data-driven methodologies such as artificial intelligence. Traditional models struggle to capture these intricate relationships, but artificial intelligence and other data-driven methods have shown proficiency in navigating the inherent non-linearity and successfully estimating soil compaction parameters. Continued research should focus on employing precise and reliable techniques, like artificial intelligence, that can effectively handle these complexities. In the area of predicting soil compaction parameters, several studies have demonstrated noteworthy advancements. The use of artificial neural networks (ANNs) has been successfully employed in various models. For instance, one investigation applied an ANN to model w_{opt} and ρ_{dmax} in soil stabilized with nanomaterials. The model exhibited impressive prediction accuracy of over 97%, surpassing conventional statistical models [26]. Extending this approach to different soil types, another study harnessed the potential of deep neural networks to predict soil compaction parameters specific to Egyptian soil. The study achieved remarkable accuracy, with R^2 values of 0.864 for ρ_{dmax} and 0.924 for w_{opt}, thereby surpassing the performance of a simple artificial neural network with just one hidden layer [27]. In the context of highway projects, a multi-layer perceptron neural network model was developed to predict modified compaction parameters for both coarse and fine-grained soil samples. This model, similar to its predecessors, delivered highly accurate results [28]. Stepping beyond traditional ANNs, a novel hybrid intelligence paradigm known as the ANFIS-IGWO model was proposed in a different study. This model demonstrated superior precision in predicting soil compaction parameters, boasting correlation values of 0.9203 and 0.9050 for w_{opt} and ρ_{dmax}, respectively [29]. This accuracy was found to surpass other hybrid ANFIS models. In yet another study, an ANN was utilized to develop predictive equations for the Proctor compaction parameters of fine-grained soil [30]. The resulting ANN model demonstrated high efficiency and accuracy, particularly in predicting unseen datasets, thus setting a new standard compared to existing models in the literature. Moving towards the integration of empirical correlations with machine learning algorithms, one paper investigated the use of the support vector machine algorithm to predict soil compaction properties [31]. This approach aimed at significantly reducing the time and effort required in the laboratory, achieving R^2 values of 0.86 and 0.91 for w_{opt} and ρ_{dmax}, respectively. Lastly, in a significant stride towards machine learning models, a novel model called "ComPara2021" was proposed [17]. This model, based on machine learning, demonstrated superior performance in predicting soil compaction parameters compared to its contemporaries, proving the value of such models in saving both time and cost.

Another form of artificial intelligence is genetic programming approaches such as gene expression programming (GEP) and multigene expression programming (MEP). GEP and MEP are advantageous frameworks that integrate the benefits of both genetic programming and genetic algorithms, enabling them to evaluate more complex functions that express the relationship between input and output data [32,33]. The GEP and MEP models have notably demonstrated their effectiveness in predicting a variety of parameters related to compacted soil [15,34,35].

The multivariate adaptive regression splines (MARS) technique, a form of artificial intelligence, offers potential benefits to geotechnical engineers. Besides being adept at tackling prediction problems, MARS can identify key input parameters that significantly influence output parameters. It also allows for the exploration of intricate nonlinear relationships between a response variable and various predictive variables—a crucial aspect in analyzing and designing soil compaction parameters. This technique has seen successful implementations in diverse fields of geotechnical engineering, including ultimate pile bearing capacity [36], the elastic modulus of rocks [37], Slope reliability analysis [38], the compressive strength of soil [39], liquefaction [40], settlement prediction [41], and penetration resistance in clay [42]. Yet, despite its proven effectiveness in these areas, its application in compaction-related studies remains surprisingly limited.

In this research, MARS is utilized to estimate the optimal moisture content (w_{opt}) and maximum dry density (ρ_{dmax}) of compacted soil. MARS addresses the limitations of the multiple linear regression (MLR) method, as it does not necessitate an a priori form of equation between dependent and independent variables and can precisely capture complex and nonlinear relationships between inputs and outputs [43]. Unlike techniques like the ANN, which often operate as a "black box," MARS can provide an explicit equation linking inputs and outputs [44–46]. This transparency has led to its frequent application in modeling intricate engineering issues. Finally, MARS successfully mitigates one of the key drawbacks of GEP and MEP, which tend to produce highly nonlinear mathematical relations between input and output data [46].

2. Research Significance

This research carries significant implications in the field of soil engineering and artificial intelligence applications. It provides a robust foundation for understanding and predicting soil compaction parameters across a variety of soil classifications. Firstly, this study is rooted in comprehensive literature-based data collection, which has amassed an extensive database of different soil types, including gravel, sand, silt, and clay. This broad spectrum of data offers a detailed and diverse basis for analysis and prediction. Secondly, we employed the MARS model to predict soil compaction parameters. To enhance the accuracy and reliability of predictions, we further fine-tuned the MARS model's performance via hyperparameter tuning. Thirdly, we conducted a comparative study to evaluate the performance of the MARS model against other existing empirical and artificial intelligence models. This comparison seeks to identify the most effective and accurate model for predicting soil compaction parameters, contributing to the optimization of methodologies in this area of research. Lastly, we carried out a sensitivity analysis to determine the relative influence of each input variable on the scour depth. Through identifying the most impactful variables, this study aids in refining the prediction model and enhances the overall understanding of the relationships between different factors.

3. Materials and Methods

3.1. Research Methodology

The comprehensive research approach employed in this study is depicted in Figure 1, encompassing five fundamental phases: gathering and assimilating data, visualizing and statistically assessing the data, carrying out data modeling and analysis, evaluating the performance of machine learning (ML) models, and comparing the findings with earlier research. The data modeling and analysis are conducted using R-project, a programming framework extensively used for data mining and ML applications [47].

3.2. Multivariate Adaptive Regression Splines (MARS)

In the year 1991, the concept of multivariate adaptive regression splines (MARS), a method for non-parametric regression, was proposed by Friedman [48]. This approach aims to model a relationship between a set of predictor variables, represented as X $(n \times p)$, and a target variable, denoted as y $(n \times 1)$. Where n is the number of observations or samples, p is the number of predictor variables or features, X is the matrix of predictor variables with size $n \times p$, and y is the column vector of the target variable with size $n \times 1$. The MARS model's mathematical representation can be expressed through the equation below:

$$y = f(X) + e \tag{1}$$

The term $f(X)$ represents the prediction or approximation of the target variable y given the predictor variables X by the MARS model. The equation also gives us a residual vector, e $(n \times 1)$, which forms an integral part of this model. MARS can be considered as a sophisticated extension of the Classification and Regression Trees (CART) [49] technique, but it stands apart from conventional parametric methods. While parametric methods necessitate a predefined functional correlation between predictor and target variables, MARS refrains from such presumptions. Instead, MARS utilizes the predictor variables present in the specified dataset to produce a set of coefficients and piece-wise polynomials that carry the power 'q'. These polynomials are specifically generated to encapsulate the relationship among variables, which may exhibit either a linear or a non-linear nature. These polynomials, referred to as 'splines', are seamlessly connected to assemble the complete MARS model. Moreover, the process of fitting these splines involves segmenting the data points of the independent variables into discrete regions, with the knot locations symbolized by 't'. The power 'q' of these splines essentially determines their linearity or non-linearity. For any variable 'x', the MARS model calculates these splines through applying the particular equations below:

$$[-(x-t)]_+^q = \begin{cases} (t-x)^q, & \text{if } x < t \\ 0, & \text{otherwise} \end{cases} \tag{2}$$

$$[+(x-t)]_+^q = \begin{cases} (t-x)^q, & \text{if } x \geq t \\ 0, & \text{otherwise} \end{cases} \tag{3}$$

In the given equations, 'q' is not less than zero, and it is a key factor in shaping the smoothness of the ultimate MARS model. In this particular investigation, 'q' is set to 1, signifying the analysis of linear splines. Figure 2 illustrates two splines related to an individual variable 'x', having 'q' set to 1 and a defined knot 't'. The left spline provides positive outcomes, while the right spline produces zero for 'x' values on the left side of the knot. On the flip side, when 'x' values lie on the right side of the knot, the right spline gives positive results, and the left spline is zero. These splines, also called as the foundation functions for 'x', play a significant role in the MARS technique. The MARS model related to the aim variable 'y' is formed via including a series of 'M' basis functions, as demonstrated in the following equation:

$$\hat{y} = \hat{f}_M(x) = a_o + \sum_{m=1}^{M} a_m B_m(x) \tag{4}$$

Equation (4) depicts $\hat{y}$ as the target variable, which the MARS model predicts. The term 'a_o' signifies the intercept term, which is a fixed constant. '$B_m(x)$' denotes the mth basis function, which could be an individual basis function or a combination of multiple basis functions, while 'a_m' symbolizes the coefficient that corresponds to the mth basis function. The estimation of these coefficients is achieved via the least squares method.

The MARS model, as represented by Equation (4), is formulated using a biphasic methodology that encompasses a forward pass and a backward pass. The initial phase, known as the forward pass, involves integrating the predictor variables into the model and fine-tuning the positions of their corresponding knots. This process gives rise to the creation of double-sided basis functions, one for each predictor. In the context of an X matrix, composed of 'n' samples and 'p' predictors, '$n \times p$' pairs of basis functions emerge, as exhibited in Equations (2) and (3). These functions are associated with knots, denoted by x_{ij}, where 'i' ranges from 1 to 'n' and 'j' spans from 1 to 'p'. Every single data point affiliated with a predictor variable is taken into account as a possible knot for that variable's pair of basis functions. However, the forward pass process often results in the formulation of a convoluted model that tends to overfit, hence reducing its predictive capability. To address this issue, the second phase, or the backward pass, is implemented. During this phase, the model is systematically simplified through sequentially eliminating the least contributing, and therefore redundant, basis functions. The backward pass continues until an optimized sub-model is achieved, characterized by its lowest Generalized Cross Validation (GCV). This enhancement substantially boosts the predictive prowess of the MARS model. The computation for determining the GCV follows the formula provided below:

$$GCV(M) = \frac{\frac{1}{n}\sum_{i=1}^{n}(y_i - f_M(\hat{x}_i))^2}{\left(1 - \frac{M + d \times (M-1)/2}{n}\right)^2} \tag{5}$$

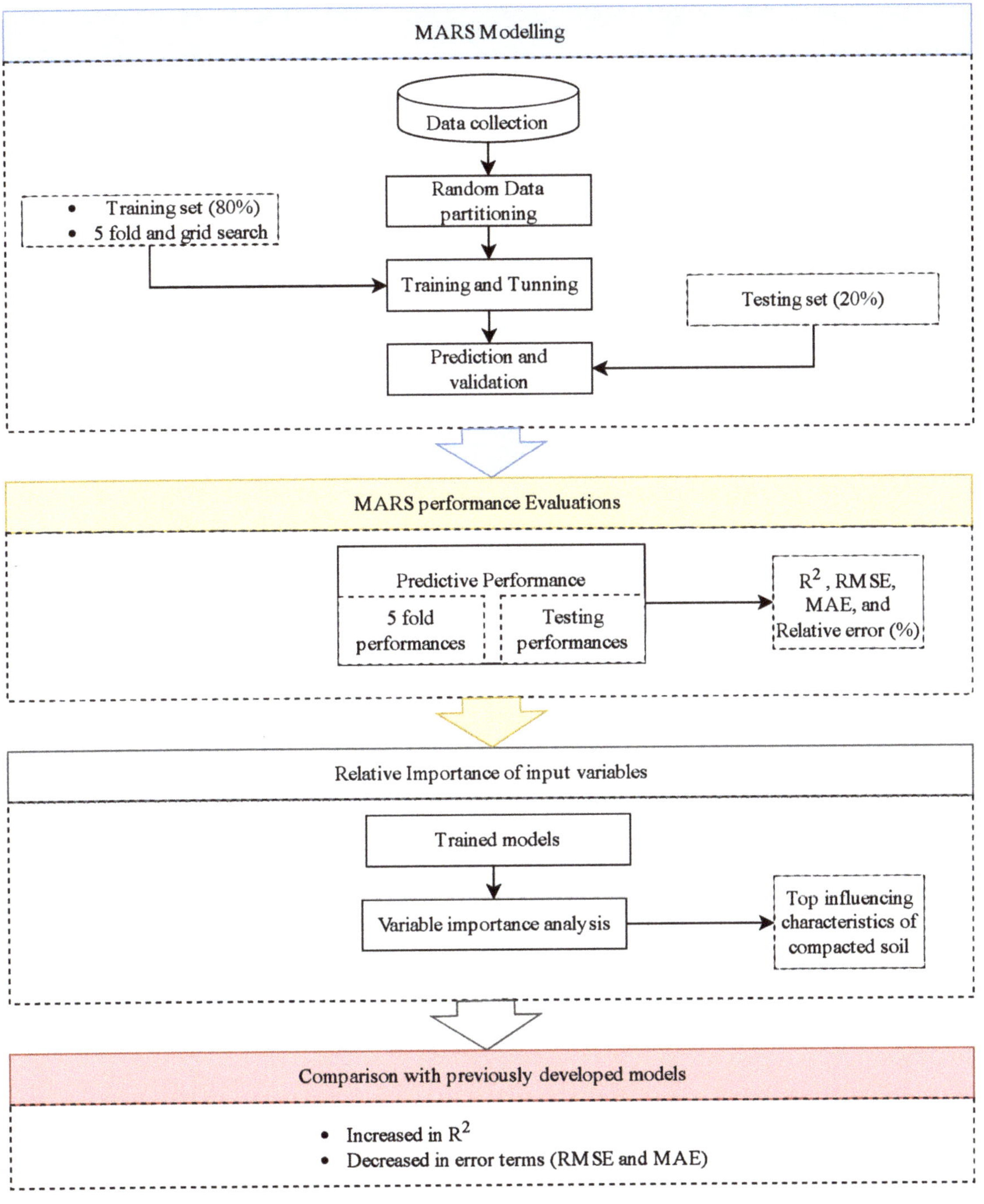

Figure 1. Research methodology.

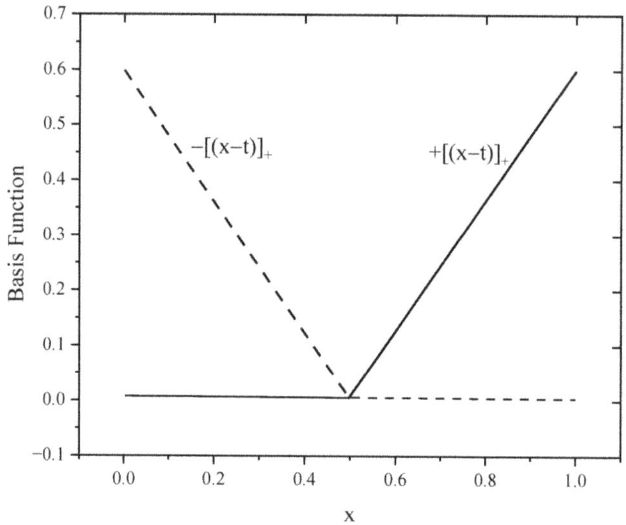

Figure 2. Graph representing two splines associated with the variable x: the left spline, represented by a dashed line (for x less than t, the formula is $-(x - t)$), and the right spline, represented by a solid black line (for x greater than t, the formula is $+(x - t)$).

3.3. Hyperparameter Tuning Procedure

In this research, we utilized a combination of hyperparameter tuning, taking advantage of the grid search approach [50], along with cross-validation (CV) strategies. The grid search methodology essentially provides an ideal setting for parameters through arranging all variable grids within the space of the parameters. Each grid's axis corresponds to a parameter of the algorithm, with a unique set of parameters at each grid point that requires optimization. The primary benefit of combining grid search with CV is to pinpoint the most suitable hyperparameters for the model based on the selected evaluation metric. Furthermore, this integration guarantees that the model's performance is evaluated across various data subsets, resulting in a more robust and dependable assessment of its efficacy. A potential drawback of this methodology is that it is computationally intensive. The model needs to be trained '*k*' times, which can be time-consuming for large datasets or complex models. However, since our data only has 226 points, the computational demand in our specific case is considerably reduced, making the method more manageable and feasible for our dataset size. To ensure the unbiased selection of data, we employed a prevalent validation method known as k-fold CV [51–55] during the tuning process of hyperparameters for MARS models. K-fold CV, a standard type of CV, is extensively used in machine learning. Although there is no definitive rule for choosing the value of K, in practical machine learning scenarios, K is commonly set at 5 or 10. According to Rodriguez [56], the bias in an accurate estimation is likely to be reduced when the fold count is either 5 or 10. Based on this and the suggestions of Kohavi [57] and Wong [58], we chose 5 for K. This decision also took into consideration the trade-off between computational efficiency and bias. As a result, five different sets of training and validation runs were performed, with the outcomes being averaged to depict the overall performance of the MARS models on the training dataset. All the data-related operations in this study were handled using R software 4.1.3 (R Core Team, Vienna, Austria) [47]. The approach for hyperparameter tuning used in this research for training the model and selecting hyperparameters is depicted in Figure 3, as illustrated in reference [59].

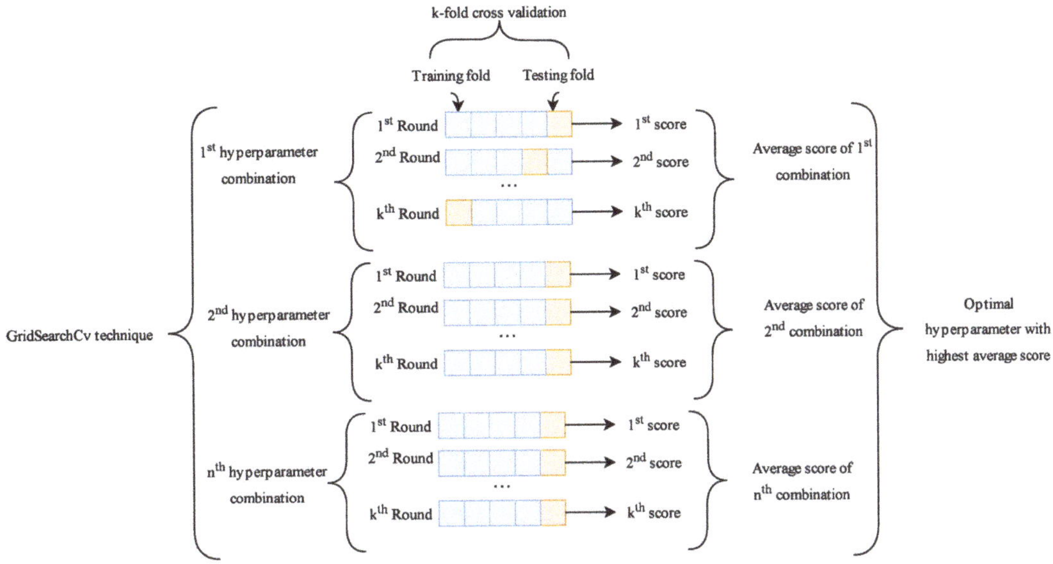

Figure 3. Tuning of hyperparameters using five-fold cross-validation.

Our research made use of the R package "earth" to perform MARS modeling. This implementation called for the user's careful consideration of two particular tuning parameters: "degree" and "nprune". The "degree" parameter indicates the highest degree of interaction allowed in the model. By default, this value is set to 1, leading to the creation of a MARS model that is additive and lacks interaction terms. However, it is noteworthy that MARS is capable of building models that encompass interactions of a degree of two or even higher. Despite this capability, a guideline suggested by Hastie et al. [60] recommends maintaining an upper limit on the degree of interaction. The benefit of keeping a lower degree of interaction lies in its ability to assist in interpreting the final model in a more straightforward manner. On the other hand, a higher degree of interaction could potentially cause occasional volatility in the model's predictive capabilities, possibly causing predictions to deviate significantly from actual values. Hence, to evaluate the impact of varying degrees of interaction, we experimented with three different values: 1, 2, and 3. The second parameter, "nprune", signifies the maximum number of terms (inclusive of the intercept) that can be present in the model. The "nprune" value should be equal to or greater than 2 and less than "*nk*". Here, "*nk*" stands for the maximum number of terms in the model before any pruning activity is initiated. This value of "*nk*" is determined by the formula $nk = \min(200, \max(20, 2 \times \text{ncol}(x))) + 1$, where ncol($x$) is the total number of predictive variables. For our specific study, which involved a total of 6 predictive variables, we looked into 20 potential values for nprune, ranging from 2 to 20. The process of constructing and evaluating these 57 MARS models, featuring degrees of 1, 2, 3 and nprune values ranging from 2 to 20, was accomplished through the implementation of a five-fold cross-validation procedure.

3.4. Performance Metrics

Three error evaluation metrics, namely root mean squared error (RMSE), mean absolute error (MAE), and the coefficient of determination (R^2), are chosen to assess the accuracy of MARS model predictions. The R^2 metric represents the linear relationship between the predicted and actual values, with its value ranging from zero to one. A value of R^2 nearing one signifies a more accurate model prediction. RMSE provides insight into the dispersion between predicted and real data and is sensitive due to its relation to the

Euclidean distance. Meanwhile, MAE is a measure of the average magnitude of errors between predicted and observed values. Lower values of RMSE and MAE suggest superior prediction performance by the algorithm. The three metrics are outlined as follows:

Root mean squared error (RMSE):

$$\mathrm{RMSE} = \sqrt{\frac{\sum_{i=1}^{n}\left(G_{pi} - G_{Ti}\right)^2}{n}} \tag{6}$$

Correlation of determination (R^2):

$$R^2 = 1 - \frac{\sum_{i=1}^{n}\left(G_{pi} - G_{Ti}\right)^2}{\sum_{i=1}^{n}\left(\bar{G}_{Ti} - G_{Ti}\right)^2} \tag{7}$$

Mean absolute error (MAE):

$$\mathrm{MAE} = \frac{1}{n}\sum_{i=1}^{n}\left|G_{pi} - G_{Ti}\right| \tag{8}$$

where G_{Ti} and G_{pi} are the true and predicted values for the observation, respectively. $\bar{G}_{Ti}$ is the mean of all the true values. n is the number of samples.

4. Database Used

In the current research, we utilized data (comprising 226 entries) sourced from the study by [61], which span various soil categories such as fat clay (CH), silty clay (CL-ML), lean clay (CL), clayey gravel (GC), silty gravel (GM), elastic silt (MH), well-graded sand with clay (SW-SC), poorly graded sand with clay (SP-SC), silt (ML), clayey sand (SC), silty sand (SM), poorly graded gravel with clay (GP-GC), and well-graded gravel with clay (GW-GC). The physical attributes of these soil samples are presented in Table 1. This dataset encompasses a diverse array of soils, highlighted by the extensive variety of index properties. Within this dataset, it is posited that the compaction metrics are influenced by factors like liquid limit (LL), plastic limit (PL), compaction energy (E), sand content (C_S), fines content (C_F), gravel content (C_G), optimum water content w_{opt}, and maximum dry density ρ_{dmax}, all of which are detailed in Table 1. Figure 4 presents frequency histograms for each variable, inclusive of soil classification. Analyzing Figure 4 reveals that within this database, a significant portion of soils possess gravel and sand contents ranging from 0% to 20% (as seen in Figure 4a,b). Conversely, the content of fines in the soils predominantly falls within 80% to 100% (illustrated in Figure 4c). A large portion of soils register a liquid limit between 0 and 50 (as displayed in Figure 4g), and 32 soils exhibit a liquid limit exceeding 300, indicative of high plasticity. The plastic limit is chiefly found between 10 and 30 (Figure 4d). Notably, this dataset features a comparably larger quantity of clayey sand, fat clay, lean clay, and clayey gravel in contrast to other soil types (depicted in Figure 4f). A significant number of compaction tests were executed under standard Proctor conditions or with diminished compaction energy (not surpassing 600 kJ/m^3 as shown in Figure 4e). Around 30 modified Proctor compaction tests are also part of the study. In terms of maximum dry density and optimum water content, peak frequencies are predominantly within 1.6 Mg/m^3 to 1.8 Mg/m^3 and 10% to 20%, respectively (as captured in Figure 4h,i).

Table 1. Model variable descriptive statistics.

Statistics	C_G (%)	C_S (%)	C_F (%)	LL (%)	PL (%)	E (kJ/m^3)	w_{opt} (%)	ρ_{dmax} (Mg/m^3)
Standard deviation	14.566	23.388	30.016	164.218	7.405	735.435	5.964	0.199
Mean	7.468	29.448	63.088	108.727	22.005	894.066	17.512	1.751
Median	0.000	27.000	70.000	40.650	20.150	593.000	17.000	1.750
Maximum	67.100	89.000	100.000	608.000	48.300	2755.000	43.700	2.330
Minimum	0.000	0.000	8.600	16.000	6.100	155.000	5.300	1.090
Kurtosis	3.482	−0.471	−1.250	3.264	0.748	2.266	2.862	0.977

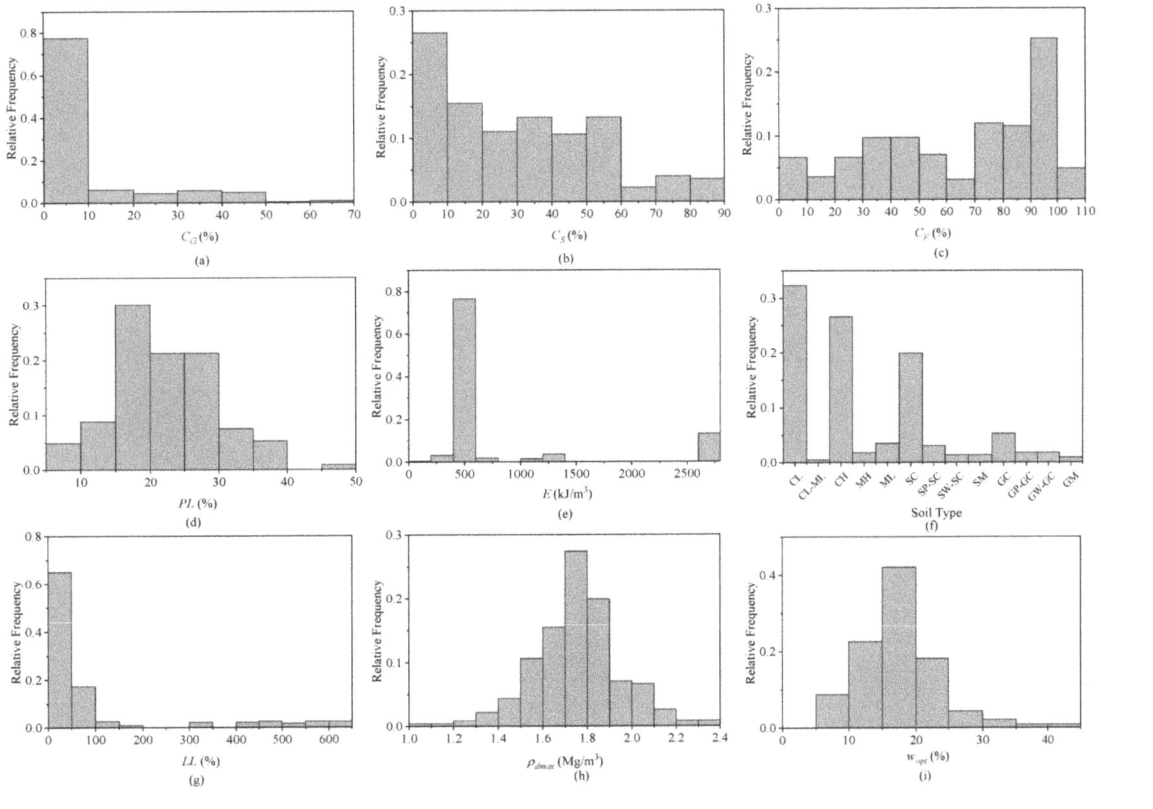

Figure 4. Relative frequency distribution of input and output variables.

5. Model Results

5.1. Hyperparameter Tuning Results for Optimal Model

The hyperparameter tuning procedure, which combined a grid search and k-fold cross-validation, was employed to determine the optimal model parameters for the MARS models, aiming for the lowest RMSE. According to Figure 5, the performance of the MARS model was more sensitive to the 'nprune' parameter than the interaction degree, highlighting the crucial role of 'nprune' in achieving top-tier model accuracy.

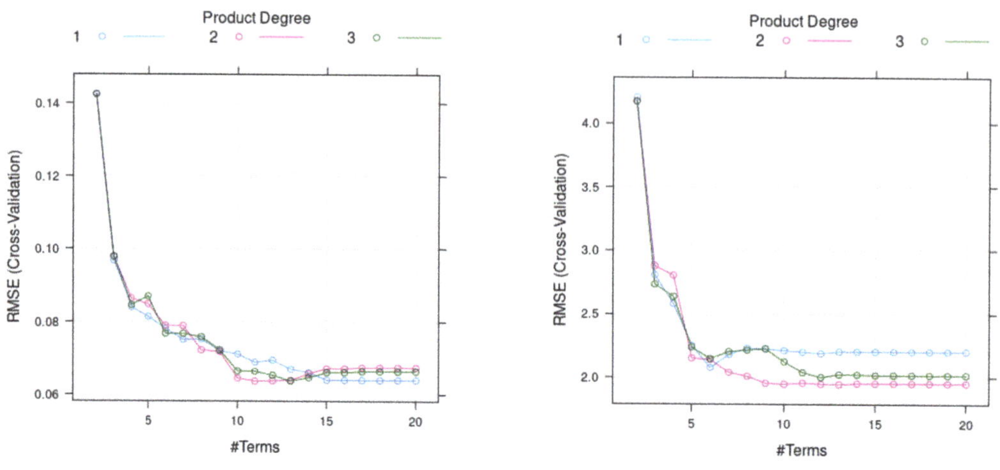

(a)

(b)

Figure 5. Optimization results of hyperparameter tuning for the MARS models: (**a**) ρ_{dmax} and (**b**) w_{opt}.

As detailed in Figure 5, a rigorous five-fold cross-validation process pinpointed 'degree = 2' and 'nprune = 13' as the prime settings for the w_{opt} MARS model. Leveraging these parameters, the model showcased exemplary performance across the 131 training datasets, registering an average RMSE. On the other hand, for the ρ_{dmax} MARS model, the most suitable hyperparameters were 'nprune = 11' and 'degree = 2'. With these settings, the model attained an optimal predictive capacity, reflected by an average RMSE of 0.064 and an R^2 of 0.889, specifically for predicting the maximum dry density in compacted soil.

5.2. Cross-Validation Results after Hyperparameter Tuning

Post tuning, the performance of the final MARS models was assessed using a five-fold cross-validation procedure. The RMSE, R^2, and MAE were computed for each fold to estimate the prediction performance. The results for both models are presented below.

For the w_{opt} model, the five-fold cross-validation results are presented in Table 2:

Table 2. Results derived from a five-fold cross-validation for the w_{opt} model.

Folds	Performance Measures		
	RMSE (%)	R^2	MAE (%)
Fold 1	1.513	0.929	1.182
Fold 2	2.368	0.870	1.867
Fold 3	1.827	0.886	1.351
Fold 4	1.710	0.889	1.361
Fold 5	2.323	0.895	1.728
Average	1.948	0.893	1.498
SD	0.380	0.023	0.287
CoV (%)	0.195	0.026	0.192

Notably, the RMSE, R^2, and MAE values indicate the high predictive performance of the MARS model for the w_{opt} parameter, with high R^2 values over 0.87 for all folds in both cases. This highlights the models' ability to capture a significant portion of the variance in the target variables. The RMSE values were relatively low across all folds for both models, underscoring the models' robustness and their ability to minimize errors in prediction. Moreover, the MAE values were reasonably low, revealing the models' capability to produce accurate predictions with small deviations from the actual values.

The ρ_{dmax} model's cross-validation results are presented in Table 3.

Table 3. Results derived from a five-fold cross-validation for the ρ_{dmax} model.

Folds	Performance Measures		
	RMSE (Mg/m^3)	R^2	MAE (Mg/m^3)
Fold 1	0.068	0.860	0.051
Fold 2	0.067	0.908	0.050
Fold 3	0.048	0.948	0.040
Fold 4	0.071	0.870	0.057
Fold 5	0.065	0.909	0.049
Average	0.064	0.899	0.050
SD	0.009	0.035	0.006
CoV (%)	0.141	0.039	0.120

Once again, the RMSE, R^2, and MAE values for each fold reflect the high predictive performance of the model. The average values highlight the overall accuracy of the model, while the standard deviations suggest that the model's performance is consistent across different folds.

One potential reason some folds perform better than others is due to imbalanced data distribution. If the data are not evenly distributed and one fold ends up with a disproportionate number of certain classes or types of data, the model may perform poorly on that fold. For instance, if one fold contains many outliers or rare events, the model might struggle to generalize. Despite this, the consistently high R^2 values, low RMSE and MAE values, and the low standard deviations for both models across all folds underscore the effectiveness of the MARS models and their robustness in predicting the w_{opt} and ρ_{dmax} parameters of compacted soil.

5.3. Evaluation of MARS Models on Unseen Dataset

The following two equations were derived from the final optimal models after hyperparameter tuning for w_{opt} and ρ_{dmax}. The equations are as follows:

$$
\begin{aligned}
w_{opt} = {} & 19.3773983 \\
& -0.0639987 \times \max(97 - CF, 0) \\
& -5.1538333 \times \max(CF - 97, 0) \\
& -0.0898632 \times \max(64 - LL, 0) \\
& -0.6415284 \times \max(21 - PL, 0) \\
& +0.4288445 \times \max(PL - 21, 0) \\
& +0.0025993 \times \max\left(1347 - \mathrm{EkJ/m^3}, 0\right) \\
& +0.2425873 \times \max(CF - 97, 0) \times PL \\
& -0.0000061 \times LL \times \max\left(\mathrm{EkJ/m^3} - 1347, 0\right) \\
& +0.0578447 \times \max(7.6 - CG, 0) \times \max(21 - PL, 0) \\
& -0.0093695 \times \max(70 - CF, 0) \times \max(PL - 21, 0).
\end{aligned}
\quad (9)
$$

$$
\begin{aligned}
\rho_{dmax} = & 1.39385529 \\
& +0.00261487 \times \max(33.1 - CS, 0) \\
& +0.00621111 \times \max(98 - CF, 0) \\
& +0.00505340 \times \max(LL - 64, 0) \\
& +0.00337354 \times \max(130 - LL, 0) \\
& -0.00526625 \times \max(LL - 130, 0) \\
& -0.01028165 \times \max(PL - 11, 0) \\
& -0.00019175 \times \max(1347 - \mathrm{EkJ/m^3}, 0) \\
& -0.00188667 \times \max(CF - 98, 0) \times PL \\
& -0.00005026 \times \max(CS - 53, 0) \times \max(98 - CF, 0) \\
& +0.00002616 \times \max(CF - 30, 0) \times \max(130 - LL, 0).
\end{aligned}
\tag{10}
$$

The cross plots presented in the subplots (a) and (b) of Figure 6 portray the application of two previous MARS-based models in predicting two crucial parameters for compacted soil: the optimum water content (w_{opt}) and maximum dry density (ρ_{dmax}). These subplots showcase the comparison between the models' predictions and the actual values of w_{opt} and ρ_{dmax} for the testing datasets. The plots illustrate that a significant accumulation of data points clustering near the $Y = X$ line signifies the strength and reliability of the established models. Clearly, the implemented models yield predictions that align remarkably well with the unit slope line. As a result, the proposed models can confidently be considered highly reliable for estimating the w_{opt} and ρ_{dmax} of compacted soil throughout the testing and training phases.

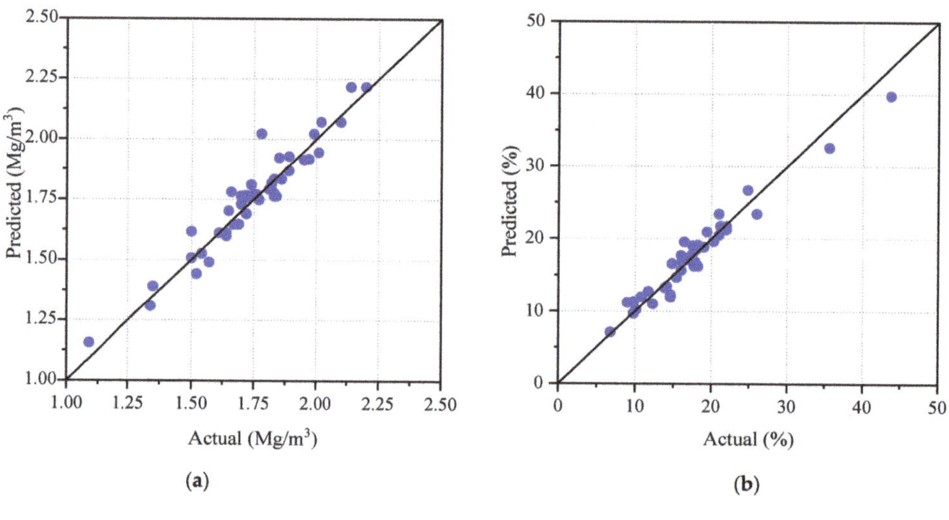

Figure 6. Actual vs. predicted values using the MARS model for the testing dataset: (**a**) ρ_{dmax} and (**b**) w_{opt}.

Figure 7 showcases the relative error of the MARS-based models compared to the actual measurements of w_{opt} and ρ_{dmax}. The subplots in Figure 7 demonstrate that the outputs of the proposed MARS-based models exhibit an acceptable deviation from the real measured data. A significant portion of the data falls within the range of −10% to +10% for w_{opt} and −5% to 5% for ρ_{dmax}.

Additionally, the statistical evaluation of the proposed models is presented in Table 4, providing a comprehensive summary of the assessment. The table includes the MAE, R^2, and RMSE values for the training, testing, and overall datasets. The statistical evaluation clearly indicates that both MARS-based models exhibit outstanding prediction performance. The overall RMSE values for w_{opt} and ρ_{dmax} are 1.428 and 0.052, respectively, highlighting the models' accuracy. The combined findings from Figures 6 and 7, along with the statistical results in Table 4, demonstrate the effectiveness of our newly proposed correlations. They are capable of reliably estimating w_{opt} and ρ_{dmax} for compacted soil across various soil properties and operational conditions, showcasing high integrity and efficiency.

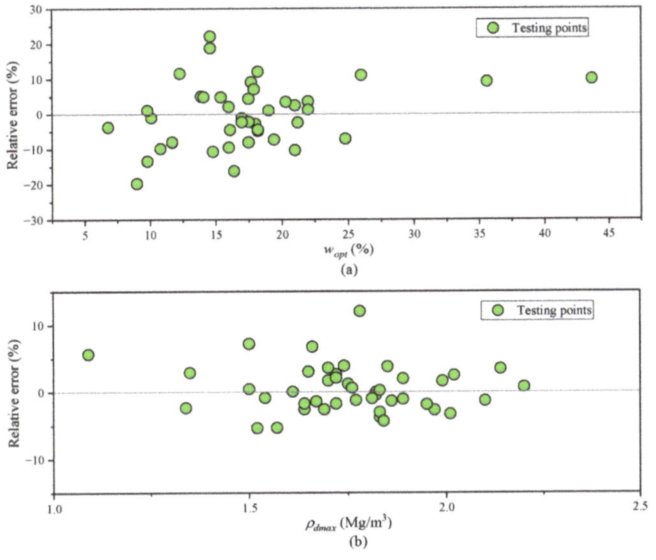

Figure 7. Comparison of relative errors vs. real value: (**a**) MARS-based model for w_{opt}; (**b**) MARS-based model for ρ_{dmax}.

Table 4. Performance metrics of the MARS models for training, testing, and overall datasets.

		Training	Testing	Overall
w_{opt}	MAE (%)	1.12	1.276	1.15
	RMSE (%)	1.392	1.577	1.428
	R^2	0.942	0.948	0.943
ρ_{dmax}	MAE (Mg/m^3)	0.04	0.047	0.041
	RMSE (Mg/m^3)	0.05	0.062	0.052
	R^2	0.936	0.919	0.931

5.4. Comparison between MARS Model with Previously Developed Models

In order to thoroughly assess the reliability of our newly proposed models, we extended the statistical evaluation to compare them with one of the best predictive correlations, namely Multi Expression Programming (MEP) [61]. The comparison is based on various statistical criteria and prediction performance, as presented in Table 5 and visually represented in the bar plots of Figure 8. Analyzing Table 5 and Figure 8 reveals that, despite the satisfactory prediction performance of MEP, our suggested models outperform it in accurately estimating the values of w_{opt} and ρ_{dmax} for compacted soil. Furthermore, MEP frequently creates intricate nonlinear empirical models that can be difficult to handle [62]. In turn, the MARS algorithm offers some distinct benefits. Notably, it effectively captures the intricate interactions between independent and dependent variables. Moreover, it eliminates the need to exert extra effort to confirm any preliminary assumptions regarding their relationship. This advantage becomes increasingly vital as the complexity of the problem grows [63].

Table 5. Performance metrics of the MEP models for training, testing, and overall datasets.

Compaction Parameters	Performance Measures	Overall
w_{opt}	MAE (%)	1.3
	RMSE (%)	1.68
	R^2	0.921
ρ_{dmax}	MAE (Mg/m^3)	0.054
	RMSE (Mg/m^3)	0.073
	R^2	0.867

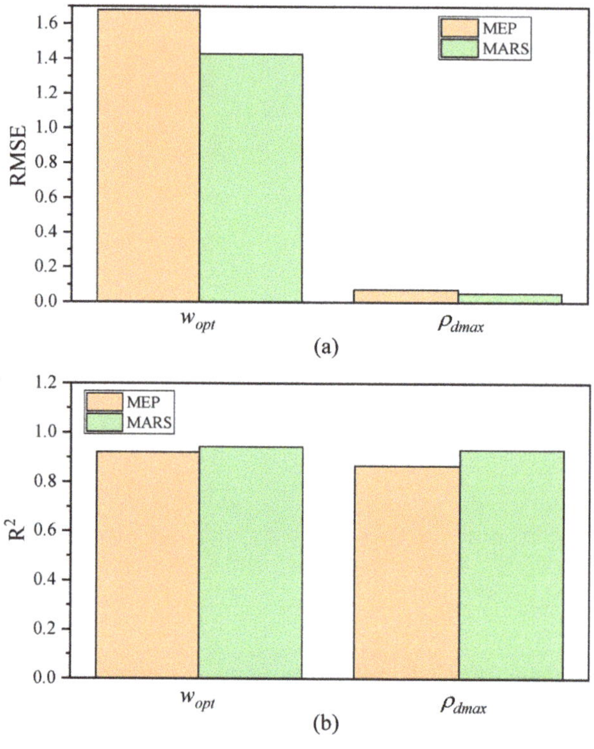

Figure 8. Comparative performance of MARS vs. MEP models: (**a**) RMSE; (**b**) R^2.

5.5. Variable Importance Analysis

A primary strength of the MARS model is its capability to discern if a particular input factor is vital for predicting output values or if it merely offers a slight enhancement to the model's precision.

The varImp function in the caret package [64] produces a matrix that displays the relative importance of features within the model, utilizing two distinct methods to assess feature significance:

- The raw residual sum-of-squares (RSS) method proceeds in two phases. Initially, it assesses the RSS decrease for each subset, contrasting it with the preceding subset's value. Subsequently, for every relevant feature, it accumulates these reductions across all subsets that incorporate that feature. Finally, the overall sum of these reductions is analyzed. Features leading to substantial RSS declines hold greater importance.
- The generalized cross-validation (GCV) method operates analogously to the RSS method, but it employs GCV in place of RSS. GCV assesses feature performance in subsets, pinpointing the most pivotal subset (with smaller GCV values being preferable).

Figure 9 displays the outcomes of the contributions and the selections made by different independent variables on the prediction performance of w_{opt} and ρ_{dmax}.

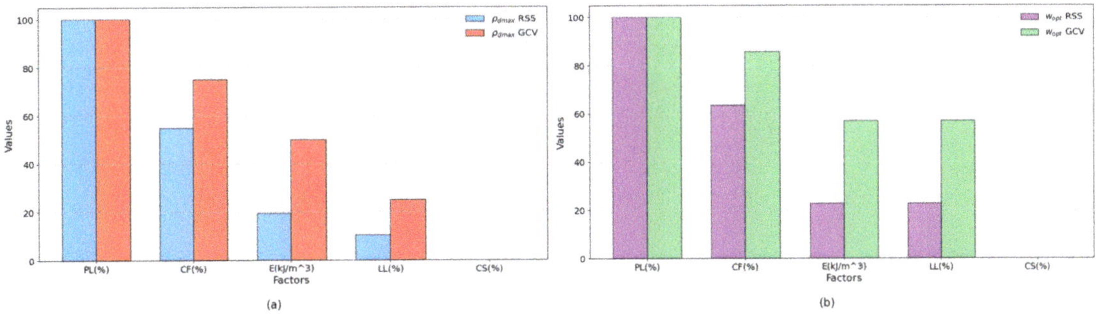

Figure 9. Influence of different independent variables on MARS model: (**a**) for ρ_{dmax}; (**b**) for w_{opt}.

Figure 9a displays the impact of various independent variables on the ρ_{dmax} MARS model. The figure is sorted based on the decreasing influence of these variables. The top two influential factors are PL (%) and C_F (%), with RSS values of 100 and 55.01, and GCV values of 100 and 75, respectively. These statistics suggest that PL (%) holds the most significant influence on the ρ_{dmax} MARS model, followed closely by C_F (%). Other factors listed in decreasing order of influence are E (kJ/m^3), LL (%), and C_S (%).

Figure 9b showcases the influence of different independent variables on the w_{opt} MARS model. The two most impactful factors in this table are C_F (%) and PL (%). C_F (%) takes the lead with RSS and GCV values of 100, indicating its paramount significance in the model. PL (%) follows next with RSS at 63.49 and GCV at 85.71. These results allow us to state that C_F (%) has the most considerable influence on the w_{opt} MARS model, but PL (%) also retains significant sway. Other listed factors, in order of influence, include E (kJ/m^3), LL (%), and C_S (%).

To more deeply assess the reliability of the MARS models, a parametric analysis can be undertaken. This type of analysis examines the alignment of the model's predictions with existing geotechnical knowledge, experimental data, and anticipated outcomes. This study scrutinized how the predicted soil parameters from the suggested models responded to hypothetical data, spanning the lower and upper bounds of the data used in model training. This approach entailed modifying a single input variable while maintaining others at their average. Synthetic data for this individual parameter were produced via raising it step by step. These inputs were then used in the prediction equation to determine soil parameters. This method was iteratively applied for different variables to gauge the model's reaction to each input change.

Figure 10 illustrates how the predicted w_{opt} changes in relation to the C_F (%), the LL (%), the PL (%), and the E (kJ/m^3). Notably, as the C_F (%), the LL (%), and the PL (%) rise, the forecasted w_{opt} also goes up. In contrast, when the E (kJ/m^3) grows, the predicted value goes down. These consistent changes align with the patterns observed for the w_{opt} in relation to each soil property as seen in the figure, indicating that the proposed model is accurate.

Figure 11 displays the parametric analysis of the anticipated ρ_{dmax} in relation to each input variable. Contrary to the w_{opt}, the forecasted ρ_{dmax} goes down as the C_F (%), LL (%), or PL (%) rises and as the E (kJ/m^3) reduces. This trend mirrors the monotonic observations from the actual database illustrated in the figure. Such a parallel between the parametric study and the database in predicting the ρ_{dmax} further confirms the accuracy of the model introduced.

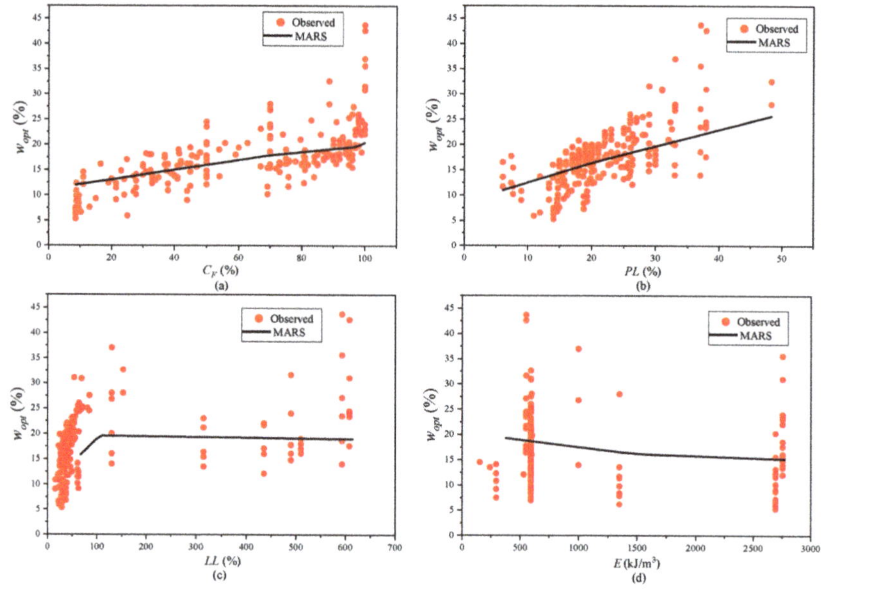

Figure 10. Parametric analysis of the predicted w_{opt} versus: (**a**) C_F (%); (**b**) PL (%); (**c**) LL (%); (**d**) E (kJ/m^3).

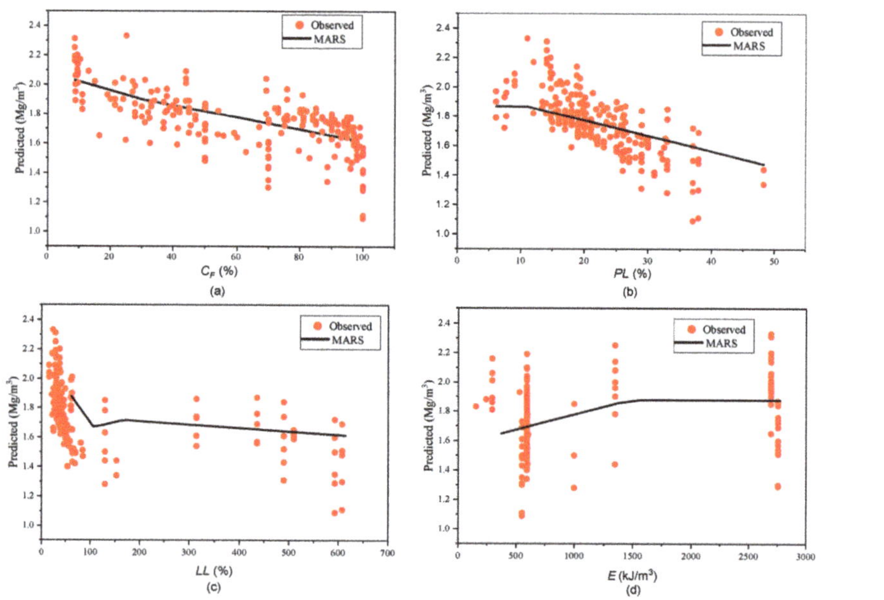

Figure 11. Parametric analysis of the predicted w_{opt} versus: (**a**) C_F (%); (**b**) PL (%); (**c**) LL (%); (**d**) E (kJ/m^3).

The findings from the variable importance analysis are consistent with previously published literature. A respected study explored the influence of C_F on the mechanical response of CDG (completely decomposed granite) subjected to dynamic compaction grouting [65]. Through integrating varying proportions of kaolin clay with CDG, this study aimed to understand the implications of fines concentrations on its compactive behavior. The findings indicated that an increment in C_F correspondingly results in a diminished

ρ_{dmax}, while concurrently elevating the w_{opt}. In another study, sand–kaolin blends were compressed to investigate the connection between ρ_{dmax}, C_F, and specific gravity (Gs) across different moisture levels, with C_F ranging from 20% to 70% [66]. A prominent linear correlation (R^2 = 0.99) was found between ρ_{dmax} and the C_F/Gs^2 ratio. For PL, a respected paper investigates which index properties best correlate with compaction traits. Incorporating both the study's findings and the existing literature, it was determined that the plastic limit correlates strongly with compaction features, such as optimum moisture content and maximum dry unit weight, more effectively than either the liquid limit or plasticity index. For the plastic limit (PL), a notable study examined which index properties have the strongest correlation with compaction characteristics [23]. Through integrating the study's results with the established literature, it was ascertained that the plastic limit shows a significant correlation with compaction attributes, particularly w_{opt} and ρ_{dmax}, surpassing the correlations seen with either the liquid limit or the plasticity index.

6. Conclusions

Determining compacted parameters in geotechnical projects is vital for ensuring soil stability, optimizing foundation designs, mitigating settlement issues, and guaranteeing the safety and longevity of structures built on or within the ground. For this reason, the prediction of those parameters is imperative for proactive planning, cost-effective execution, and minimizing project risks. This research aims to predict those parameters via employing the MARS (multivariate adaptive regression splines) model. Compared to other black-box machine learning models, the MARS model offers advantages such as flexibility in capturing non-linear relationships, automated interaction detection, fewer tuning parameters, and the capability to produce more interpretable results, making it particularly suitable for geotechnical applications where understanding the underlying behavior is as crucial as prediction accuracy.

The results obtained through this research indicate the following:

- Hyperparameter tuning of the MARS model significantly enhanced its predictive performance, reducing the error rate and refining the model's adaptability to complex geotechnical data.
- The final optimal models for parameters w_{opt} and ρ_{dmax} were evaluated using five-fold cross-validation. The findings showed consistency and high accuracy across all folds, demonstrating the model's robustness and reliability in estimating these critical parameters.
- Testing the models on unseen data (testing dataset) revealed that they maintained a commendable level of precision and generalization, further substantiating their efficacy for real-world applications in geotechnical projects.
- Comparing the MARS model with other state-of-the-art models, such as Multi Expression Programming (MEP), indicated that while both models exhibit strong predictive capabilities, MARS demonstrated faster convergence and better handling of non-linear relationships, offering a more efficient and robust alternative for geotechnical parameter prediction.
- Finally, the variable importance analysis revealed that the fine content (C_F) and plasticity limit (PL) are the most influential factors driving the predictive accuracy of the model, highlighting their pivotal role in determining the geotechnical properties under study.

It is worth mentioning that ML research often presents inherent challenges and limitations. In this study, a notable limitation is the data sample size of 226 points. The model formulated in this research delivers the anticipated accuracy when analogous input parameters are subsequently applied. However, errors may arise in the analysis if the same parameters are used but lie outside our established range. Gathering more experimental data in the future is essential to enhance the model's generalization capabilities. Upcoming studies will explore the potential of advanced ML techniques, such as deep learning, in predicting w_{opt} and ρ_{dmax} values.

Author Contributions: Writing—original draft, H.I.; Investigation, F.J.K.; Software, J.K.A. and M.S.A.; Data curation, S.N.H.; Methodology, F.J.K.; Supervision, M.S.A.; Funding acquisition, L.F.A.B.; Validation, J.K.A.; Writing—review and editing, H.I. and L.F.A.B. All authors have read and agreed to the published version of the manuscript.

Funding: This research received no external funding.

Institutional Review Board Statement: Not applicable.

Informed Consent Statement: Not applicable.

Data Availability Statement: Not applicable.

Conflicts of Interest: The authors declare no conflict of interest.

References

1. Verma, G.; Kumar, B. Prediction of compaction parameters for fine-grained and coarse-grained soils: A review. *Int. J. Geotech. Eng.* **2020**, *14*, 970–977. [CrossRef]
2. Proctor, R. Fundamental principles of soil compaction. *Eng. News-Rec.* **1933**, *111*, 245–248, 286–289, 348–351.
3. Hussain, A.; Atalar, C. Estimation of compaction characteristics of soils using Atterberg limits. *Proc. IOP Conf. Ser. Mater. Sci. Eng.* **2020**, *800*, 012024. [CrossRef]
4. Rimbarngaye, A.; Mwero, J.N.; Ronoh, E.K. Effect of gum Arabic content on maximum dry density and optimum moisture content of laterite soil. *Heliyon* **2022**, *8*, e11553. [CrossRef]
5. Spagnoli, G.; Shimobe, S. An overview on the compaction characteristics of soils by laboratory tests. *Eng. Geol.* **2020**, *278*, 105830. [CrossRef]
6. Ren, X.-C.; Lai, Y.-M.; Zhang, F.-Y.; Hu, K. Test method for determination of optimum moisture content of soil and maximum dry density. *KSCE J. Civ. Eng.* **2015**, *19*, 2061–2066. [CrossRef]
7. Lim, Y.Y.; Miller, G.A. Wetting-induced compression of compacted Oklahoma soils. *J. Geotech. Geoenviron. Eng.* **2004**, *130*, 1014–1023. [CrossRef]
8. Delage, P.; Marcial, D.; Cui, Y.; Ruiz, X. Ageing effects in a compacted bentonite: A microstructure approach. *Géotechnique* **2006**, *56*, 291–304. [CrossRef]
9. Rahman, F.; Hossain, M.; Hunt, M.; Romanoschi, S. Soil stiffness evaluation for compaction control of cohesionless embankments. *Geotech. Test. J.* **2008**, *31*, 442–451.
10. Di Matteo, L.; Bigotti, F.; Ricco, R. Best-fit models to estimate modified proctor properties of compacted soil. *J. Geotech. Geoenviron. Eng.* **2009**, *135*, 992–996. [CrossRef]
11. Sun, D.a.; Cui, H.; Sun, W. Swelling of compacted sand–bentonite mixtures. *Appl. Clay Sci.* **2009**, *43*, 485–492. [CrossRef]
12. Zhao, L.-S.; Zhou, W.-H.; Yuen, K.-V. A simplified axisymmetric model for column supported embankment systems. *Comput. Geotech.* **2017**, *92*, 96–107. [CrossRef]
13. Chen, R.-P.; Wang, H.-L.; Hong, P.-Y.; Cui, Y.-J.; Qi, S.; Cheng, W. Effects of degree of compaction and fines content of the subgrade bottom layer on moisture migration in the substructure of high-speed railways. *Proc. Inst. Mech. Eng. Part F J. Rail Rapid Transit* **2018**, *232*, 1197–1210. [CrossRef]
14. Chen, R.-P.; Qi, S.; Wang, H.-L.; Cui, Y.-J. Microstructure and hydraulic properties of coarse-grained subgrade soil used in high-speed railway at various compaction degrees. *J. Mater. Civ. Eng.* **2019**, *31*, 04019301. [CrossRef]
15. Ardakani, A.; Kordnaeij, A. Soil compaction parameters prediction using GMDH-type neural network and genetic algorithm. *Eur. J. Environ. Civ. Eng.* **2019**, *23*, 449–462. [CrossRef]
16. Wang, H.-L.; Chen, R.-P. Estimating static and dynamic stresses in geosynthetic-reinforced pile-supported track-bed under train moving loads. *J. Geotech. Geoenviron. Eng.* **2019**, *145*, 04019029. [CrossRef]
17. Benbouras, M.A.; Lefilef, L. Progressive machine learning approaches for predicting the soil compaction parameters. *Transp. Infrastruct. Geotechnol.* **2023**, *10*, 211–238. [CrossRef]
18. *ASTM D698*; Standard Test Methods for Laboratory Compaction Characteristics of Soil Using Standard Effort (12,400 ft-lbf/ft^3 (600 kN-m/m^3)). ASTM International: West Conshohocken, PA, USA, 2021; p. 13.
19. *ASTM D1557*; Standard Test Methods for Laboratory Compaction Characteristics of Soil Using Modiefied Effort (56,000 ft-lbf/ft^3 (2,700 kN-m/m^3)). ASTM International: West Conshohocken, PA, USA, 2021; pp. 1–10.
20. Blotz, L.R.; Benson, C.H.; Boutwell, G.P. Estimating optimum water content and maximum dry unit weight for compacted clays. *J. Geotech. Geoenviron. Eng.* **1998**, *124*, 907–912. [CrossRef]
21. Omar, M.; Shanableh, A.; Basma, A.; Barakat, S. Compaction characteristics of granular soils in United Arab Emirates. *Geotech. Geol. Eng.* **2003**, *21*, 283–295. [CrossRef]
22. Gurtug, Y.; Sridharan, A. Compaction behaviour and prediction of its characteristics of fine grained soils with particular reference to compaction energy. *Soils Found.* **2004**, *44*, 27–36. [CrossRef]
23. Sridharan, A.; Nagaraj, H. Plastic limit and compaction characteristics of finegrained soils. *Proc. Inst. Civ. Eng. Ground Improv.* **2005**, *9*, 17–22. [CrossRef]

24. Mujtaba, H.; Farooq, K.; Sivakugan, N.; Das, B.M. Correlation between gradational parameters and compaction characteristics of sandy soils. *Int. J. Geotech. Eng.* **2013**, *7*, 395–401. [CrossRef]
25. Farooq, K.; Khalid, U.; Mujtaba, H. Prediction of compaction characteristics of fine-grained soils using consistency limits. *Arab. J. Sci. Eng.* **2016**, *41*, 1319–1328. [CrossRef]
26. Taha, O.M.E.; Majeed, Z.H.; Ahmed, S.M. Artificial neural network prediction models for maximum dry density and optimum moisture content of stabilized soils. *Transp. Infrastruct. Geotechnol.* **2018**, *5*, 146–168. [CrossRef]
27. Othman, K.; Abdelwahab, H. Prediction of the soil compaction parameters using deep neural networks. *Transp. Infrastruct. Geotechnol.* **2021**, *10*, 147–164. [CrossRef]
28. Verma, G.; Kumar, B. Multi-layer perceptron (MLP) neural network for predicting the modified compaction parameters of coarse-grained and fine-grained soils. *Innov. Infrastruct. Solut.* **2022**, *7*, 78. [CrossRef]
29. Bardhan, A.; Singh, R.K.; Ghani, S.; Konstantakatos, G.; Asteris, P.G. Modelling Soil Compaction Parameters Using an Enhanced Hybrid Intelligence Paradigm of ANFIS and Improved Grey Wolf Optimiser. *Mathematics* **2023**, *11*, 3064. [CrossRef]
30. Verma, G.; Kumar, B. Artificial neural network equations for predicting the modified proctor compaction parameters of fine-grained soil. *Transp. Infrastruct. Geotechnol.* **2023**, *10*, 424–447. [CrossRef]
31. Hasnat, A.; Hasan, M.M.; Islam, M.R.; Alim, M.A. Prediction of compaction parameters of soil using support vector regression. *Curr. Trends Civ. Struct. Eng.* **2019**, *4*, 1–7. [CrossRef]
32. Ferreira, C. Gene expression programming: A new adaptive algorithm for solving problems. *arXiv* **2001**, arXiv:pcs/0102027.
33. Raja, M.N.A.; Abdoun, T.; El-Sekelly, W. Smart prediction of liquefaction-induced lateral spreading. *J. Rock Mech. Geotech. Eng.* **2023**. [CrossRef]
34. Jalal, F.E.; Xu, Y.; Iqbal, M.; Jamhiri, B.; Javed, M.F. Predicting the compaction characteristics of expansive soils using two genetic programming-based algorithms. *Transp. Geotech.* **2021**, *30*, 100608. [CrossRef]
35. Jalal, F.E.; Iqbal, M.; Ali Khan, M.; Salami, B.A.; Ullah, S.; Khan, H.; Nabil, M. Indirect estimation of swelling pressure of expansive soil: Gep versus mep modelling. *Adv. Mater. Sci. Eng.* **2023**, *2023*, 1827117. [CrossRef]
36. Samui, P. Determination of ultimate capacity of driven piles in cohesionless soil: A multivariate adaptive regression spline approach. *Int. J. Numer. Anal. Methods Geomech.* **2012**, *36*, 1434–1439. [CrossRef]
37. Samui, P. Multivariate adaptive regression spline (Mars) for prediction of elastic modulus of jointed rock mass. *Geotech. Geol. Eng.* **2013**, *31*, 249–253. [CrossRef]
38. Deng, Z.-P.; Pan, M.; Niu, J.-T.; Jiang, S.-H.; Qian, W.-W. Slope reliability analysis in spatially variable soils using sliced inverse regression-based multivariate adaptive regression spline. *Bull. Eng. Geol. Environ.* **2021**, *80*, 7213–7226. [CrossRef]
39. Ghanizadeh, A.R.; Rahrovan, M. Modeling of unconfined compressive strength of soil-RAP blend stabilized with Portland cement using multivariate adaptive regression spline. *Front. Struct. Civ. Eng.* **2019**, *13*, 787–799. [CrossRef]
40. Zheng, G.; Zhang, W.; Zhou, H.; Yang, P. Multivariate adaptive regression splines model for prediction of the liquefaction-induced settlement of shallow foundations. *Soil Dyn. Earthq. Eng.* **2020**, *132*, 106097. [CrossRef]
41. Zuo, Q. Settlement prediction of the piles socketed into rock using multivariate adaptive regression splines. *J. Appl. Sci. Eng.* **2022**, *26*, 111–119.
42. Sirimontree, S.; Jearsiripongkul, T.; Lai, V.Q.; Eskandarinejad, A.; Lawongkerd, J.; Seehavong, S.; Thongchom, C.; Nuaklong, P.; Keawsawasvong, S. Prediction of penetration resistance of a spherical penetrometer in clay using multivariate adaptive regression splines model. *Sustainability* **2022**, *14*, 3222. [CrossRef]
43. Emamgolizadeh, S.; Bateni, S.; Shahsavani, D.; Ashrafi, T.; Ghorbani, H. Estimation of soil cation exchange capacity using genetic expression programming (GEP) and multivariate adaptive regression splines (MARS). *J. Hydrol.* **2015**, *529*, 1590–1600. [CrossRef]
44. Haghiabi, A.H. Prediction of river pipeline scour depth using multivariate adaptive regression splines. *J. Pipeline Syst. Eng. Pract.* **2017**, *8*, 04016015. [CrossRef]
45. Conoscenti, C.; Ciaccio, M.; Caraballo-Arias, N.A.; Gómez-Gutiérrez, Á.; Rotigliano, E.; Agnesi, V. Assessment of susceptibility to earth-flow landslide using logistic regression and multivariate adaptive regression splines: A case of the Belice River basin (western Sicily, Italy). *Geomorphology* **2015**, *242*, 49–64. [CrossRef]
46. Sharda, V.; Prasher, S.; Patel, R.; Ojasvi, P.; Prakash, C. Performance of Multivariate Adaptive Regression Splines (MARS) in predicting runoff in mid-Himalayan micro-watersheds with limited data/Performances de régressions par splines multiples et adaptives (MARS) pour la prévision d'écoulement au sein de micro-bassins versants Himalayens d'altitudes intermédiaires avec peu de données. *Hydrol. Sci. J.* **2008**, *53*, 1165–1175.
47. R Core Team. *R: A Language and Environment for Statistical Computing*; R Core Team: Vienna, Austria, 2020.
48. Friedman, J.H. Multivariate adaptive regression splines. *Ann. Stat.* **1991**, *19*, 1–67. [CrossRef]
49. Loh, W.Y. Classification and regression trees. *Data Min. Knowl. Discov.* **2011**, *1*, 14–23. [CrossRef]
50. Kuhn, M.; Johnson, K. *Applied Predictive Modeling*; Springer: Berlin/Heidelberg, Germany, 2013; Volume 26.
51. Alhakeem, Z.M.; Jebur, Y.M.; Henedy, S.N.; Imran, H.; Bernardo, L.F.; Hussein, H.M. Prediction of ecofriendly concrete compressive strength using gradient boosting regression tree combined with GridSearchCV hyperparameter-optimization techniques. *Materials* **2022**, *15*, 7432. [CrossRef] [PubMed]
52. Yang, Z.; Gao, W.; Chen, L.; Yuan, C.; Chen, Q.; Kong, Q. A novel electromechanical impedance-based method for non-destructive evaluation of concrete fiber content. *Constr. Build. Mater.* **2022**, *351*, 128972. [CrossRef]

53. Koya, B.P.; Aneja, S.; Gupta, R.; Valeo, C. Comparative analysis of different machine learning algorithms to predict mechanical properties of concrete. *Mech. Adv. Mater. Struct.* **2022**, *29*, 4032–4043. [CrossRef]
54. Tang, F.; Wu, Y.; Zhou, Y. Hybridizing grid search and support vector regression to predict the compressive strength of fly ash concrete. *Adv. Civ. Eng.* **2022**, *2022*, 3601914. [CrossRef]
55. Luo, H.; Paal, S.G. Machine learning–based backbone curve model of reinforced concrete columns subjected to cyclic loading reversals. *J. Comput. Civ. Eng.* **2018**, *32*, 04018042. [CrossRef]
56. Rodriguez, J.D.; Perez, A.; Lozano, J.A. Sensitivity analysis of k-fold cross validation in prediction error estimation. *IEEE Trans. Pattern Anal. Mach. Intell.* **2009**, *32*, 569–575. [CrossRef]
57. Kohavi, R. A study of cross-validation and bootstrap for accuracy estimation and model selection. In Proceedings of the 14th International Joint Conference on Artificial Intelligence (IJCAI), Montreal, QC, Canada, 20 August 1995; Volume 14, pp. 1137–1145.
58. Wong, T.-T. Performance evaluation of classification algorithms by k-fold and leave-one-out cross validation. *Pattern Recognit.* **2015**, *48*, 2839–2846. [CrossRef]
59. Liu, X.; Liu, T.; Feng, P. Long-term performance prediction framework based on XGBoost decision tree for pultruded FRP composites exposed to water, humidity and alkaline solution. *Compos. Struct.* **2022**, *284*, 115184. [CrossRef]
60. Hastie, T.; Tibshirani, R.; Friedman, J.H.; Friedman, J.H. *The Elements of Statistical Learning: Data Mining, Inference, and prediction*; Springer: Cham, Switzerland, 2009; Volume 2.
61. Wang, H.-L.; Yin, Z.-Y. High performance prediction of soil compaction parameters using multi expression programming. *Eng. Geol.* **2020**, *276*, 105758. [CrossRef]
62. Shah, H.A.; Nehdi, M.L.; Khan, M.I.; Akmal, U.; Alabduljabbar, H.; Mohamed, A.; Sheraz, M. Predicting Compressive and Splitting Tensile Strengths of Silica Fume Concrete Using M5P Model Tree Algorithm. *Materials* **2022**, *15*, 5436. [CrossRef]
63. Kaveh, A.; Bakhshpoori, T.; Hamze-Ziabari, S. New model derivation for the bond behavior of NSM FRP systems in concrete. *Iran. J. Sci. Technol. Trans. Civ. Eng.* **2017**, *41*, 249–262. [CrossRef]
64. Kuhn, M. Building predictive models in R using the caret package. *J. Stat. Softw.* **2008**, *28*, 1–26. [CrossRef]
65. Wang, S.; Chan, D.; Lam, K.C. Experimental study of the effect of fines content on dynamic compaction grouting in completely decomposed granite of Hong Kong. *Constr. Build. Mater.* **2009**, *23*, 1249–1264. [CrossRef]
66. Alshameri, B. Maximum dry density of sand–kaolin mixtures predicted by using fine content and specific gravity. *SN Appl. Sci.* **2020**, *2*, 1693. [CrossRef]

applied sciences

MDPI

Article

Prediction of Acceleration Amplification Ratio of Rocking Foundations Using Machine Learning and Deep Learning Models

Sivapalan Gajan

College of Engineering, SUNY Polytechnic Institute, Utica, NY 13502, USA; gajans@sunypoly.edu

Abstract: Experimental results reveal that rocking shallow foundations reduce earthquake-induced force and flexural displacement demands transmitted to structures and can be used as an effective geotechnical seismic isolation mechanism. This paper presents data-driven predictive models for maximum acceleration transmitted to structures founded on rocking shallow foundations during earthquake loading. Results from base-shaking experiments on rocking foundations have been utilized for the development of artificial neural network regression (ANN), k-nearest neighbors regression, support vector regression, random forest regression, adaptive boosting regression, and gradient boosting regression models. Acceleration amplification ratio, defined as the maximum acceleration at the center of gravity of a structure divided by the peak ground acceleration of the earthquake, is considered as the prediction parameter. For five out of six models developed in this study, the overall mean absolute percentage error in predictions in repeated k-fold cross validation tests vary between 0.128 and 0.145, with the ANN model being the most accurate and most consistent. The cross validation mean absolute error in predictions of all six models vary between 0.08 and 0.1, indicating that the maximum acceleration of structures supported by rocking foundations can be predicted within an average error limit of 8% to 10% of the peak ground acceleration of the earthquake.

Keywords: geotechnical engineering; rocking foundations; earthquake engineering; soil-structure interaction; artificial neural network; machine learning

Citation: Gajan, S. Prediction of Acceleration Amplification Ratio of Rocking Foundations Using Machine Learning and Deep Learning Models. *Appl. Sci.* **2023**, *13*, 12791. https://doi.org/10.3390/app132312791

Academic Editors: Rosario Montuori and Wei Gao

Received: 27 September 2023
Revised: 1 November 2023
Accepted: 28 November 2023
Published: 29 November 2023

1. Introduction

Dynamic soil-structure interaction in shallow foundations has generally been modeled using mechanics-based models such as simple spring-dashpot models, beam on nonlinear Winkler foundation models, plasticity-based macro-element models, and continuum-based models. A recent review article summarizes the computational methods generally used to model dynamic soil–foundation–structure interactions during earthquake loading, particularly in the context of geotechnical engineering [1]. As the development of large experimental databases becomes increasingly common, the application of machine learning techniques in geotechnical engineering has been improving and becoming more effective [2]. Machine learning models generalize observed experimental behavior, capture the salient features that may not be captured by mechanics-based models, and can be used with mechanics-based models as complementary measures in engineering applications or can be combined with engineering mechanics using the emerging framework of theory-guided machine learning [3].

Machine learning algorithms such as logistic regression, decision trees, decision tree-based ensemble models, and artificial neural networks have been used in a variety of geotechnical engineering applications that include mechanical properties of soils, strength of soils, soil slope stability, bearing capacity of foundations, and dynamic response of soils during earthquake loading [4–9]. Recently, in dynamic soil–foundation–structure interactions, machine learning algorithms have been used to develop data-driven models for rocking-induced seismic energy dissipation in soil, peak rotation of foundation, and factor of safety for tipping-over failure of rocking shallow foundations [10,11].

The earthquake-induced peak acceleration of structures is one of the key seismic design parameters of buildings and bridges, as the seismic performance of these structures depends heavily on the inertial forces experienced by the structural members and non-structural components induced by the acceleration of structures [12,13]. For instance, base shear force (and bending moment) of structures during earthquake loading, a commonly used seismic design parameter for structures, is directly proportional to the horizontal acceleration at the effective height of the structure [14]. There have been several studies related to floor acceleration demands on structures during seismic loading for structures supported by traditional, fixed-base foundations [15–17].

Rocking shallow foundations is a recent research phenomenon that has been investigated to some extent, particularly using centrifuge and shaking table experiments [18–23]. Research on rocking foundations reveals that they dissipate seismic energy in soil, reduce acceleration, force and flexural drift demands transferred to the structures, and effectively perform as a geotechnical seismic isolation mechanism [24–26]. Numerical modeling methods and empirical methods are available to quantify the moment-rotation response, rotational stiffness, damping ratio and settlement-rotation relationships of rocking foundations [27,28]. This paper presents the development of models to predict the rocking induced, reduced peak acceleration demands on structures using machine learning algorithms that are trained and tested on experimental results from a rocking foundation database that covers a wide range of soil properties, foundation geometry, and structural configurations. Whereas the previously published research on the application of machine learning algorithms to rocking foundations focused on rocking induced seismic energy dissipation, peak rotation, and tipping-over stability of rocking foundations, the current work focuses on the acceleration amplification ratio (AAR) of rocking foundations. The motivation for the current work stems from the importance of reduced acceleration demands transmitted to structures supported by rocking foundations (one of the major, potential beneficial effects of rocking foundations, if adopted in civil engineering practice).

The objective of this study is to develop data-driven models for the prediction of maximum acceleration transmitted to the effective height (center of gravity) of relatively rigid, single degree of freedom type structures founded on rocking shallow foundations during earthquake loading using multiple machine learning and deep learning algorithms. The machine learning algorithms utilized in this study include artificial neural network regression, k-nearest neighbors regression, support vector regression, random forest regression, adaptive boosting regression, and gradient boosting regression. The results of these machine learning model predictions are compared with those of a multivariate linear regression machine learning model (used as the baseline model) and a statistics-based simple linear regression model. A brief background to the problem considered is presented first, along with the experimental data used in this study and input features to machine learning models. It is followed by brief descriptions of the machine learning algorithms utilized and how they are applied to the problem considered. Finally, the results, discussion and conclusions of the study are presented.

2. Rocking Foundations for Seismic Loading

2.1. Rocking Mechanism and Acceleration Amplification Ratio

Figure 1 shows the schematic of a simplified, relatively rigid, shear wall-type rocking structure supported by a shallow foundation, the major forces acting on the structure, and the forces and displacements acting at the soil-footing interface. The key parameters that govern the behavior of a rocking system include the critical contact area ratio of the rocking foundation (A/A_c), slenderness ratio of the rocking structure (h/B), and rocking coefficient of the soil–foundation–structure system (C_r) [26]. C_r is essentially a non-dimensional, normalized form of the ultimate moment capacity of rocking foundations and can be expressed by [29]:

$$C_r = \frac{B}{2 \cdot h} \cdot \left[1 - \frac{A_c}{A}\right] \quad (1)$$

where B is the width of the foundation in the direction of shaking and h is the effective height of the structure. The A/A_c is essentially a factor of safety for rocking foundations and defined as the ratio of total base area of the foundation (A) to minimum foundation area required to be in contact with the soil (A_c) to support the applied vertical load [18].

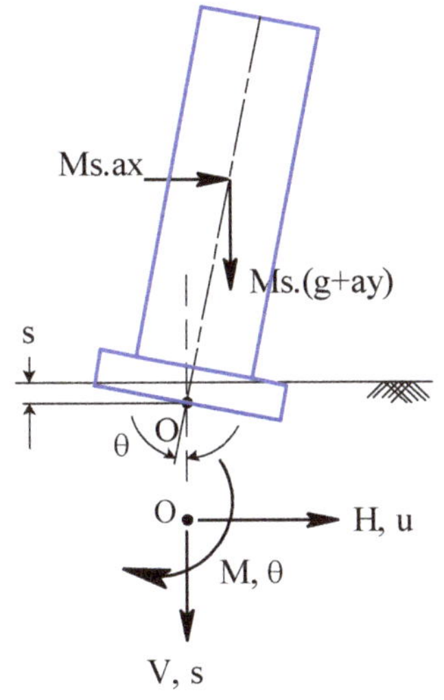

Figure 1. Simplified schematic of a rigid structure-foundation system rocking on soil and the major forces acting on it during earthquake loading.

The output parameter of machine learning models developed in this study is acceleration amplification ratio (AAR) of the rocking foundation and it is defined as:

$$\text{AAR} = \frac{a_{max,str}}{a_{max}} \tag{2}$$

where $a_{max,str}$ is the peak horizontal acceleration at the effective height of the structure and a_{max} is the peak horizontal ground acceleration of earthquake shaking. By comparing the maximum moment experienced by the soil–foundation system due to the inertial forces from the structure with the moment capacity of the rocking foundation, the following approximate relationship can be obtained for a theoretical upper bound for the AAR of a rocking foundation supporting a relatively rigid, SDOF-type structure [26].

$$\text{AAR} \leq \frac{C_r}{a_{max}} \tag{3}$$

Equation (3) implies that the foundation moment capacity limits the maximum seismic force demands transferred to the structure because of nonlinear soil–foundation–structure interaction. Though the relationship given in Equation (3) is approximate, it can be used to obtain simple, statistics-based empirical relationships for AAR.

2.2. Experimental Results

The experimental results utilized in this study are obtained from a total of nine series of centrifuge and shaking table experiments on rocking shallow foundations conducted at the University of California at Davis, the University of California at San Diego, and the National Technical University of Athens in Greece [30]. The details and major results of these individual test series are available in separate publications [18,29,31–36]. A summary of these experimental results and the effects of key rocking system capacity parameters (e.g., A/A_c and C_r) and earthquake demand parameters (e.g., a_{max}) on the performance parameters of rocking foundations (e.g., AAR), derived from the data obtained from these experiments, are available in Gajan et al. (2021) [26].

Altogether, results obtained from 140 experiments on rocking foundations from the abovementioned series of experiments are utilized in this study. Figure 2 presents the variation of AAR with a_{max} (experimental results) for rocking structure-foundation systems with three different clusters of C_r, of which two clusters are for sandy soil foundations and one cluster is for clayey soil foundations. The AAR of rocking foundations are smaller than 1.0 for more than 82% of the experiments considered (for 116 out of 140 experiments), indicating that rocking foundations reduce the accelerations transferred to the structures they support (de-amplifying effect). This effect increases as C_r decreases, indicating that the foundations that have more tendency to rock (smaller C_r) de-amplify the acceleration more. The reduced acceleration demand on the structure during foundation rocking is a consequence of mobilization of bearing capacity and yielding of soil beneath the foundation during rocking. This de-amplifying effect is more pronounced for large amplitude shaking events (greater a_{max}) than for small shaking events, as soil yielding is even more significant during large amplitude shaking (a beneficial consequence of nonlinear soil-foundation interaction). Note that, for the experiments considered in this study, there is no noticeable difference between the AAR values of sandy and clayey soil foundations as long as the a_{max} and the range of C_r remain the same.

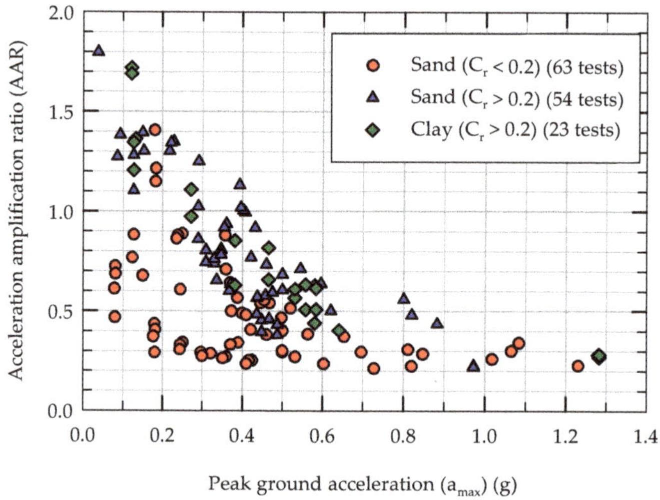

Figure 2. Results obtained from 140 centrifuge and shaking table experiments: Variation of AAR with a_{max} for rocking structure–foundation system with three clusters of rocking coefficient (C_r).

It should be noted that simplified procedures for estimating the peak acceleration demands on traditional fixed-base structures use a trapezoidal distribution, where peak acceleration at roof level could be about 3.0 to 4.0 times the peak ground acceleration of the earthquake [12]. For example, the American Society of Civil Engineers' document Minimum Design Loads for Buildings and Other Structures (ASCE 7-16) indicates that the floor acceleration amplification factor can be as high as 3.0 at roof level [37], while the National Earthquake Hazards Reductions Program's (NEHRP) Building Seismic Safety Council (BSSC) indicates that the floor acceleration amplification factor can be as high as 4.0 at roof level [38]. For the purpose of comparison, the abovementioned values correspond to an equivalent AAR of 3.0 to 4.0. This clearly shows that rocking foundations are much more efficient in de-amplifying the accelerations transferred to the structures during seismic loading (AAR < 1.5 for the vast majority of the experiments plotted in Figure 2).

Figure 3 plots the experimental results of AAR against C_r/a_{max} for all 140 rocking foundation experiments considered in this study (the same data plotted in Figure 2). Also, included in Figure 3 are a 1:1 line and a best fit line obtained from statistics-based simple linear regression (SLR) using log (C_r/a_{max}) and log (AAR) as the independent variable and the dependent variable, respectively. Though C_r/a_{max} is an approximate theoretical upper bound for AAR (Equation (3)), some of the experimental data show AAR values greater than C_r/a_{max}. This could possibly be because of the approximate nature of the upper bound relationship and the assumptions and simplifications involved in the derivations. The best fit SLR relationship yields a coefficient of determination (R^2) value of 0.75 (in log–log scale), indicating that there is room for improvement and better predictive relationships can be obtained by machine learning algorithms. In summary, the experimental results of AAR indicate that rocking foundations reduce the seismic force demands imposed on the structures by decreasing the acceleration transferred to structures, and that this beneficial de-amplifying effect increases as C_r decreases (for foundations that are more prone to rocking) and as a_{max} increases (for relatively larger magnitude earthquakes).

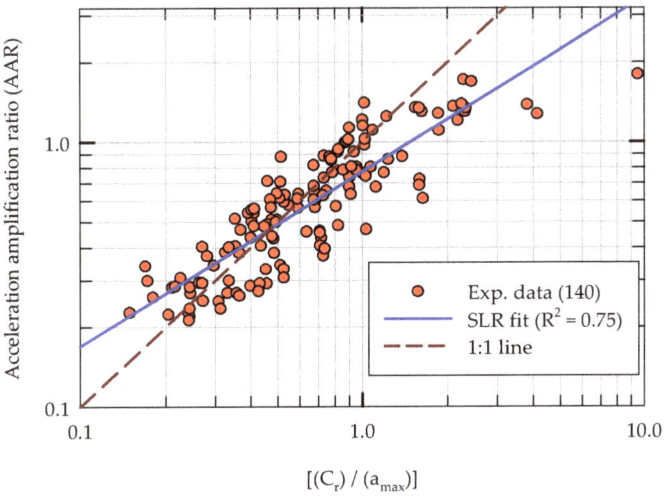

Figure 3. Experimental results for AAR as a function of C_r/a_{max} along with a 1:1 line and a statistics-based simple linear regression (SLR) best fit line in log–log space.

2.3. Input Features for Machine Learning Models

The input features for machine learning models have been selected based on their theoretical and experimentally observed close relationships with AAR, presented in Gajan et al. (2021) [26]. In addition, in order to predict other performance parameters of rocking foundations (namely, seismic energy dissipation, maximum rotation of rocking foundation and factor of safety for tipping over failure), the same set of input features have been found to be appropriate and successful [10,11]. The input features include three non-dimensional rocking system capacity parameters (A/A_c, h/B and C_r), and two earthquake loading demand parameters (a_{max} and Arias intensity of earthquake (I_a)). The a_{max} is the most commonly used ground motion intensity parameter in geotechnical earthquake engineering that characterizes the magnitude of shaking. Arias intensity of earthquake ground motion combines multiple key features of earthquake ground motion through numerical integration of acceleration time history in the time domain. These key features of ground motion include amplitude, duration, frequency content and number of cycles of earthquake loading. All the input feature parameters have been calculated for 140 individual experiments from the abovementioned series of experiments. Figure 4 presents the frequency plots, mean values, and standard deviations of all five input features. For ease of presentation, the frequency plots lump each input feature into five groups; the exact values of each input feature are used in training and testing of machine learning models.

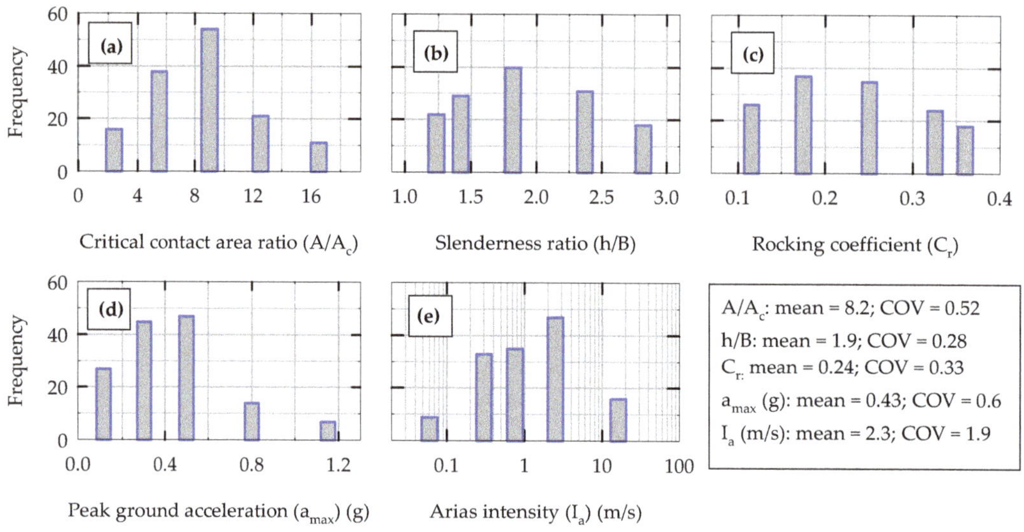

Figure 4. Frequency plots (frequency of occurrence in experiments) of input features for machine learning models developed in this study: (**a**) A/A_c, (**b**) h/B, (**c**) C_r, (**d**) a_{max} and (**e**) I_a.

As shown in Figure 4, the input features used in this study cover a wide range of rocking structure–foundation–soil system parameters (A/A_c, h/B and C_r) and earthquake demand parameters (a_{max} and I_a). As the variation of I_a is relatively high, it is transformed to a log scale (feature transformation). In addition, all the input feature values are normalized to vary between 0.0 and 1.0 (feature scaling). Figure 5 summarizes the research methodology in the form of a flow chart listing the experimental variables, input features, and the machine learning models developed to predict AAR.

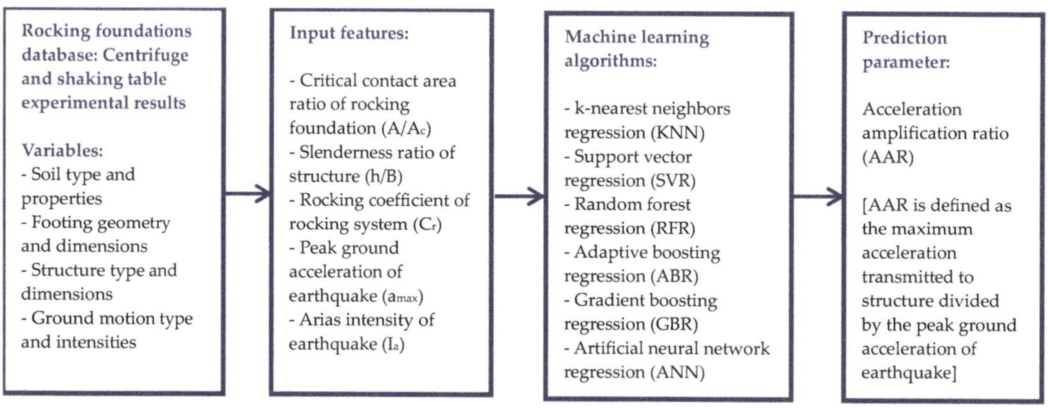

Figure 5. Flow chart showing the research methodology, experimental variables, input features, machine learning algorithms and prediction parameter.

3. Machine Learning Algorithms

3.1. Distance-Weighted K-Nearest Neighbors Regression (KNN)

The KNN algorithm considers data instances as multi-dimensional vectors, with the number of dimensions being equal to the number of input features. The algorithm calculates the distances between data points in this multi-dimensional space and assumes that the data points share similar properties with their close neighbors (and hence similar output values). The Euclidean distance measure is used to calculate the distance between any two data points in 5-D space in this study. During the training phase, the KNN algorithm simply stores the entire training dataset as vectors. During the testing phase, the KNN algorithm goes through the entire training dataset and finds k number of training data points that are closer to the test data point (k nearest neighbors, where k is a hyperparameter of KNN model). The distance weighted KNN model used in this study predicts a weighted average output based on the outputs of the nearest neighbors of the test data and the inverse of the distances between the test data and its nearest neighbors.

3.2. Support Vector Regression (SVR)

Unlike commonly used regression machine learning algorithms (e.g., linear regression) that minimize the error between the predicted values and actual observations, the SVR algorithm fits a hyperplane to represent the training data within a threshold value. This threshold value (called the margin, ϵ) is a hyperparameter of the model. The SVR algorithm uses a kernel function to transform the data instances into multi-dimensional input feature space; the radial basis function (RBF) kernel is used in this study. As highly nonlinear data with multiple input features cannot be completely represented by a hyperplane and a margin, a tolerance is used for the margin. Another hyperparameter (called the penalty parameter, C) of the SVR algorithm controls the magnitude of this tolerance across all dimensions in input feature space. When making a prediction on test data, the SVR model simply uses the hyperplane to make the prediction.

3.3. Decision Tree Regression (DTR)

The DTR algorithm builds an inverted tree-type data structure by going through the training dataset and assigning data instances to branches of the tree using information gain as a measure of reduction in uncertainty in data. While building the tree, the DTR algorithm chooses the best input feature (k) and a threshold value (t_k) for that input feature to decide on the optimum split by minimizing a cost function. The cost function ($J(k, t_k)$) that the DTR algorithm minimizes is given by [39]:

$$J(k, t_k) = \frac{m_l}{m} \cdot E_l + \frac{m_r}{m} \cdot E_r \tag{4}$$

where E and m represent the mean absolute error and the number of data instances, respectively, and the subscripts l and r represent the left and right subsets of that node, respectively ($m = m_l + m_r$). The maximum depth of the tree is the major hyperparameter of the DTR model. When making a prediction on test data, the DTR model finds the appropriate leaf node and makes the prediction using the average value of the prediction parameter (AAR) in that leaf node.

3.4. Random Forest Regression (RFR)

The RFR is a bagging (bootstrap aggregation) ensemble machine learning algorithm that builds multiple DTR models of different depths using random subsets of training dataset (random sampling with replacement). To train individual (and independent) DTR models, the RFR model uses a random number of input features each time (i.e., the maximum number of features to be considered when building a DTR model is a hyperparameter of the RFR model). The idea is that by intentionally introducing randomness in the construction of the RFR model, the accuracy of predictions and the variance in prediction error will be reduced. The number of base DTR models in an RFR model is another hyperparameter of the model. When making a prediction on test data, the RFR model simply outputs the average of predictions of each individual DTR model in the ensemble.

3.5. Adaptive Boosting Regression (ABR)

The ABR algorithm uses a boosting technique, where multiple individual base DTR models are trained sequentially on the entire training dataset. Each successive DTR model attempts to focus more on the "difficult" data instances (i.e., the data instances for which the prediction error of the preceding DTR model is high). Two sets of weights are used by the ABR algorithm: predictor weight for each individual DTR model and instance weight for each training data instance. During the training phase, these weights are adjusted in such a way that, when combined, the final prediction error will be minimum. When making a prediction on test data, the ABR model combines the predictions of all DTR models in the ensemble and weighs them using predictor weights.

3.6. Gradient Boosting Regression (GBR)

The GBR algorithm is similar to the ABR algorithm in that it builds multiple base DTR models in sequence on the entire training dataset with the successive DTR model attempting to correct the error made by its predecessor. The difference between ABR and GBR is that the GBR algorithm trains the successive base DTR models on the residual errors made by its predecessor. When making a prediction on test data, the GBR model simply adds the predictions made by all base DTR models in the ensemble. The optimum value for the learning rate, a hyperparameter of DTR-based boosting ensemble models, is found to be 0.1 for both ABR and GBR models using a trial and error procedure.

3.7. Artificial Neural Network Regression (ANN)

Figure 6 schematically illustrates the architecture of the multi-layer perceptron, deep learning ANN regression models considered in this study. The number of input neurons is equal to the number of input features (five), and the number of hidden layers and the number of neurons in each hidden layer are varied systematically using hyperparameter tuning, grid search and random search techniques. The commonly used feed-forward, back-propagation algorithm is used to propagate the input features and correct the errors during training of ANN models using the stochastic gradient descent (SGD) algorithm.

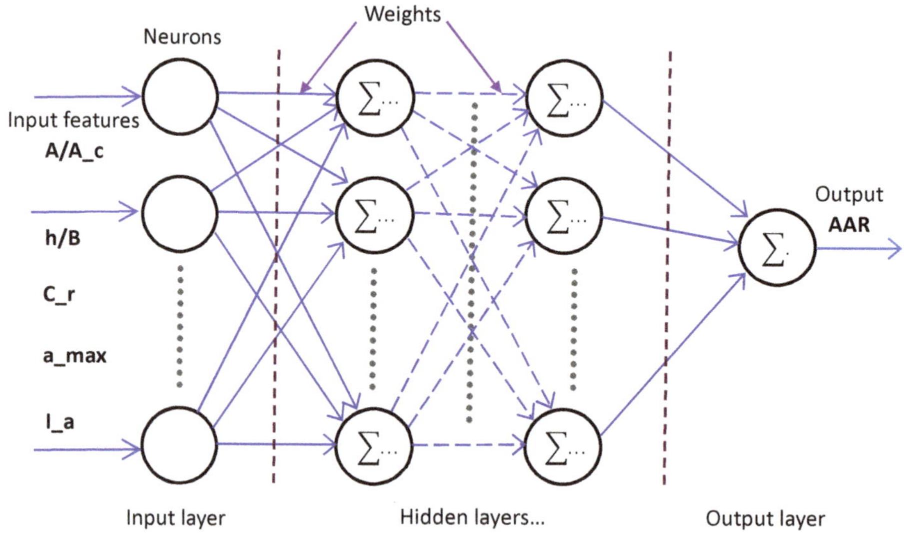

Figure 6. Schematic of the architecture of the multi-layer perceptron artificial neural network (ANN) regression model developed in this study.

In general, the relationship between the inputs and outputs of a neuron in the ANN model can be expressed by [39]:

$$y_i = g\left(\sum_{j=1}^{k}(W_{j,i}X_j) + b_i\right) \quad (5)$$

where y_i is the output of the *i*th neuron in any hidden layer, *j* goes from 1 to the number of neurons (*k*) in the previous layer, *X* are the outputs of neurons from the previous layer, *W* are the connection weights between neurons in the current layer and previous layer, *b* is the bias value, and *g*() is an activation function. The rectified linear unit (ReLU) function is the activation function used in this study. For each training instance, the backpropagation algorithm first makes a prediction using the above relationship and measures the error using the mean squared error loss function. It then goes through each layer in reverse to measure the error contribution from each connection and adjusts the connection weights to reduce the error using the SGD algorithm. During testing, the ANN model simply propagates the input features through the network and calculates the prediction using the optimum connection weights determined in the training phase.

4. Results and Discussion

The performance of machine learning (ML) models developed in this study are evaluated mainly using mean absolute percentage error (MAPE) and mean absolute error (MAE) in predictions. MAPE is defined as:

$$\mathrm{MAPE} = \frac{1}{n} \cdot \sum_{i=1}^{n} \left(\left| \frac{\tilde{y}_i - y_i}{y_i} \right| \right) \tag{6}$$

where *y* is the actual (experimental) value of AAR, $\tilde{y}$ is the output value (AAR predicted by a particular model), and "*i*" goes from 1 up to the number of predictions (*n*). Note that MAE is a similar error measure that calculates the error in terms of the absolute difference between the predicted and experimental values of AAR (i.e., MAE does not normalize the difference between experimental and predicted values). A multivariate linear regression (MLR) ML model is also developed using the same dataset, the same input features, and supervised learning technique. It is used as the baseline model for comparison of performances of the nonlinear ML models developed in this study. All the ML models and deep learning ANN models in this study are developed in the Python programming platform using the implementations of the standard classes available in Scikit-Learn (https://scikit-learn.org/stable/, accessed on 1 June 2023) and TensorFlow and Keras (https://keras.io/, accessed on 1 June 2023) libraries of modules.

4.1. Initial Evaluation (Training and Testing) of Machine Learning Models

The experimental data and results obtained from the abovementioned series of experiments (140 tests) are split into two groups for initial training and testing of ML models using a 70–30% random split of data: training dataset (98 tests) and testing dataset (42 tests). After the initial training of ML models on the training dataset, the models are tested on previously unseen test dataset. Figure 7 presents the comparisons of ML model predictions with experimental results for AAR during the initial testing phase of the models for the KNN and SVR models along with the baseline MLR model. Note that the hyperparameters of the ML models are kept constant at their optimum values (described in Section 4.4): k = 3 in the KNN model, C = 1.0 and ϵ = 0.1 in the SVR model. As seen in Figure 7, both the KNN and SVR models (MAPE = 0.17 and 0.16, respectively) outperform the baseline MLR model (MAPE = 0.21) in terms of accuracy of predictions.

Figure 8 presents the initial testing results of three DTR-based ensemble ML models (RFR, ABR and GBR) along with their MAPE values. As with the previous three models, the hyperparameters for all of these models are also kept at their optimum values: maximum depth of tree = 6 and number of trees in the ensemble = 100 for all three ensemble models. A single DTR model results in a MAPE of 0.17 during the initial testing phase (not shown in the figure). However, as can be seen from Figure 8, when 100 trees are combined together, all three DTR-based ensemble models (MAPE = 0.14 to 0.15) outperform other models presented in Figure 7 in terms of accuracy of predictions. In terms of consistency among different models, the MAPE of all five nonlinear, nonparametric ML models vary between 0.14 and 0.17 and the MAE values vary between 0.08 and 0.11. This shows excellent consistency among the ML models developed for the problem considered in this study. For comparison, the MAPE and MAE resulting from the statistics-based simple liner regression model (SLR) presented in Figure 3 are 0.23 and 0.15, respectively. It should be noted that the SLR model uses the entire dataset for fitting a linear relationship and uses the same dataset for calculating the MAPE and MAE values. Despite that, it is interesting to note that the testing errors of the MLR model (MAPE = 0.21 and MAE = 0.12) are still slightly better than those of the statistics-based (non-ML) SLR model.

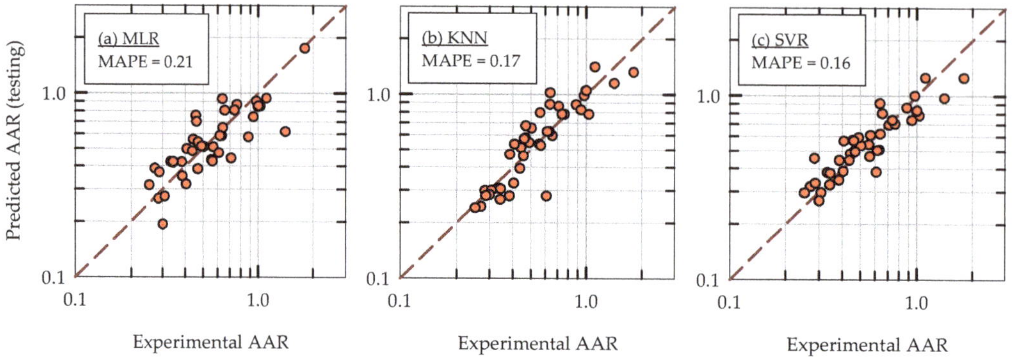

Figure 7. Comparisons of three ML model predictions with experimental results for the acceleration amplification ratio (AAR) during the initial testing phase of models: (**a**) MLR, (**b**) KNN and (**c**) SVR. Note: the dashed lines represent 1:1 lines.

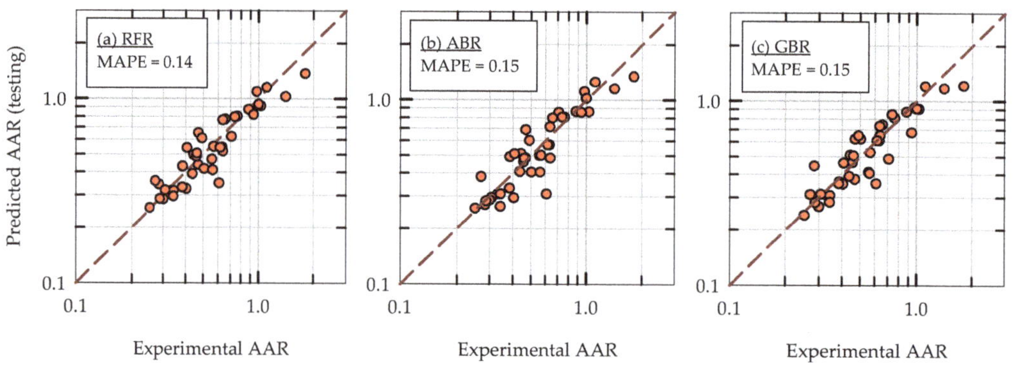

Figure 8. Comparisons of three DTR-based ML model predictions with experimental results for the acceleration amplification ratio (AAR) during the initial testing phase of the models: (**a**) RFR, (**b**) ABR and (**c**) GBR. Note: the dashed lines represent 1:1 lines.

4.2. Significance of Input Features

The significance of input features for the problem considered is quantified using the feature importance scores obtained from the RFR, ABR and GBR models. The feature importance scores are calculated based on how much the base decision-tree nodes that use an input feature reduce uncertainty in the data. The normalized feature importance scores of each input feature are presented in Figure 9 for three DTR-based ensemble ML models after the initial training phase. Figure 9 clearly shows that a_{max} has the highest normalized feature importance score (about 40% to 50%) in the predictions of AAR, followed by C_r (about 20%). This is consistent with the close relationship of AAR with a_{max} and C_r presented in Figure 3 and indicates that the AAR is more sensitive to a_{max} and C_r than the other parameters. The other three input features (A/A_c, h/B and I_a) have approximately 10% of feature importance scores each. These observations are consistent for all three DTR-based ensemble models and confirm that none of the input features considered in this study are redundant. It should be noted that when the type of soil is included as an input feature to ML models, it results in feature importance scores of less than 5%, consistently for all three DTR-based models. In addition, it does not make any significant difference in ML model predictions when the type of soil is included as an input feature, and hence the type of soil is not included as an input feature in this study. However, the effect of soil type on rocking response of foundations is indirectly included in A/A_c and C_r through shear strength and the bearing capacity of the soil.

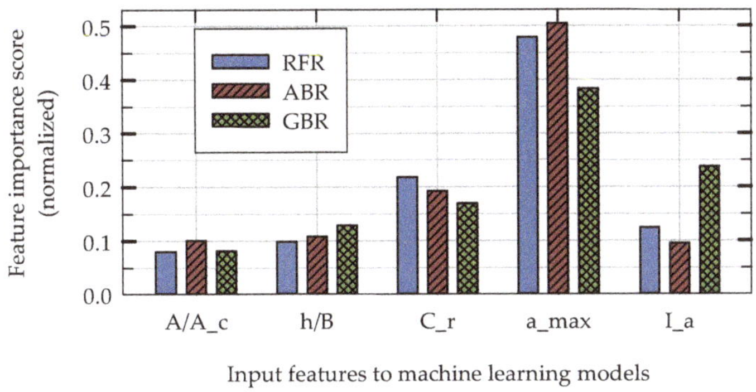

Figure 9. Results of feature importance scores based on the prediction of AAR obtained from three DTR-based ML models (RFR, ABR and GBR).

4.3. K-Fold Cross Validation Tests

In order to evaluate the performance of ML models on multiple, random pairs of training–testing datasets, the k-fold cross validation test is used. In a k-fold cross validation test, one fold of data is used for testing of ML models that are trained on (k − 1) folds of data, and the process is repeated k times using every single fold as the test dataset once. In this study, five-fold cross validation tests with three repetitions (with different randomization of the data in each repetition) are carried out. This repeated cross validation yields 15 different sets of results for AAR and the corresponding MAPE and MAE values. Two types of repeated five-fold cross validations are carried out: (i) considering only the training dataset for hyperparameter turning of each ML model and (ii) considering the entire dataset for final evaluation and comparison of all the ML models developed in this study (in terms of accuracy of predictions and variance in prediction errors).

4.4. Hyperparameter Tuning of Machine Learning Models

The purpose of hyperparameter tuning is two-fold: (i) to determine the optimum values of hyperparameters of ML models for the problem considered and (ii) to ensure that the ML models do not overfit or underfit the training data. The key hyperparameters of ML models are optimized by minimizing the testing MAPE obtained using repeated five-fold cross validation tests on the training dataset. Figure 10 presents the results of hyperparameter tuning of ML models in the form of average testing of MAPE versus the variation of corresponding major hyperparameters of the models. Note that each MAPE value in Figure 10 is the average of 15 different MAPE values resulting from repeated five-fold cross validation tests.

Results presented in Figure 10a show that the average testing MAPE of the KNN model first decreases as the number of nearest neighbors (k) increases, indicating an increase in accuracy. However, when k increases further (k > 3), the accuracy of the model decreases. This indicates that the critical value of k is 3, in order to avoid overfitting (k < 3) or underfitting (k > 3) the training data. Based on this observation, the optimum value for k in the KNN algorithm is chosen to be 3. Similar to the KNN model, the average testing MAPE of the SVR model decreases as the penalty parameter C increases (Figure 10b), indicating that relatively smaller values of C would underfit the training data. Though it is not very apparent from

Figure 10c, relatively larger values of C would overfit the training data. Based on the results obtained and to be consistent with the previously developed ML models related to this topic (performance prediction of rocking foundations), the optimum value for C is chosen to be 1.0.

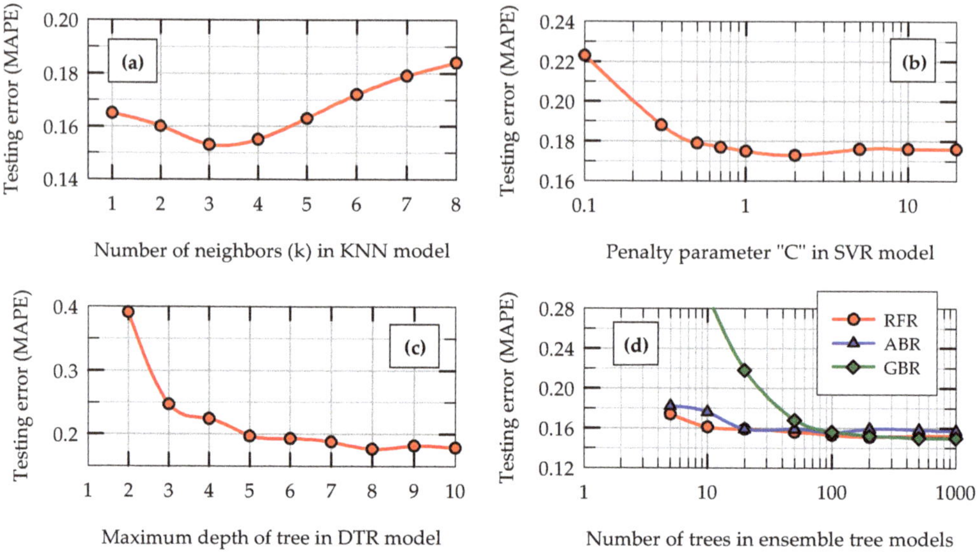

Figure 10. Variation of the average MAPE of machine learning models with their major hyperparameters from k-fold cross validation tests of models using training data for (**a**) KNN, (**b**) SVR, (**c**) DTR and (**d**) DTR-based ensemble models.

It is well known that deep DTR models tend to overfit the training data, while shallow DTR models tend to underfit the training data [39]. Based on the results shown in Figure 10c and to be consistent with the previously developed DTR-based models related to this topic, the optimum value for the maximum depth is set at 6 for base DTR models in the ensembles. Figure 10d shows that, for all three DTR-based ensemble models (RFR, ABR and GBR), the accuracy of the models increases as the number of trees increases (this is more apparent for the GBR model). The number of random features (maximum) to be considered is kept at 4 for the RFR model, and the learning rate is kept at 0.1 for both boosting models (ABR and GBR). When the number of trees in the ensembles increases beyond 100, the average testing MAPE of the models does not decrease any further. This indicates that the minimum number of trees required in DTR-based ensemble models is 100. This is remarkably consistent for all three DTR-based ensemble models. Table 1 summarizes the key hyperparameters chosen for five nonlinear ML models developed in this study.

Table 1. Optimum values chosen for major hyperparameters of machine learning models.

Machine Learning Model	Hyperparameter
k-nearest neighbors regression (KNN)	k = 3 weight = inverse distance
Support vector regression (SVR)	C = 1.0 epsilon = 0.1 mapping function = RBF [1]
Random forest regression (RFR)	max. depth = 6 max. features = 4 number of trees = 100
Boosting models (ABR and GBR)	max. depth = 6 learning rate = 0.1 number of trees = 100

[1] Radial basis function.

4.5. Initial Evaluation of ANN Models

Multiple, multi-layer perceptron, sequential ANN models, with different architectures (varying number of hidden layers and number of neurons in hidden layers) and hyperparameters, are developed and evaluated. The same training dataset and testing dataset are also used for the initial evaluation of ANN models, and the MAPE values of ANN models are calculated using the same procedure (same as described in Section 4.1). In addition to the testing error, the ANN models are also tested with the training data after the models are trained to compute the training error. The purpose of this exercise is to quantify how well the ANN models learn from the training data and their ability to generalize the patterns present in training data. The variation of predicted AAR with experimental results for AAR are presented in Figure 11a,b for the training phase and testing phase, respectively, for one particular ANN model.

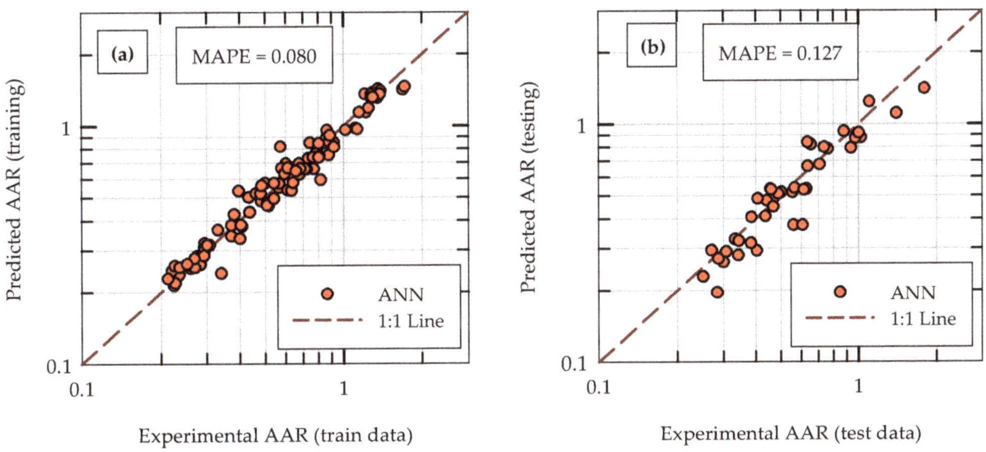

Figure 11. Comparisons of ANN model predictions with experimental results for AAR during initial evaluation of the model: (**a**) training phase and (**b**) testing phase.

The architecture of this particular ANN model consists of four hidden layers (L = 4) with forty neurons (N = 40) in each hidden layer. This is in addition to five neurons in the input layer (one for each input feature) and one neuron in the output layer (for output parameter, AAR). This particular set of hyperparameters turns out to be the optimum for the ANN model architecture for the problem considered (described in Section 4.6). Based on the comparison of predicted versus experimental AAR, with a MAPE of 0.08 and MAE of 0.053 during the training phase (Figure 11a), it is fair to say the ANN model extracts adequate information from data to build a reasonably good neural network structure during the training phase. The ANN model predictions during the initial testing phase are shown in Figure 11b and the resulting MAPE and MAE on test data are 0.127 and 0.082, respectively. This prediction accuracy places the ANN model superior to all other ML models developed in this study during the initial evaluation and testing phase.

4.6. Hyperparameter Tuning of the ANN Model

Similar to the other ML models, the key hyperparameters of the ANN model are optimized by minimizing the average MAPE values obtained from repeated five-fold cross validation tests (number of repeats = 3) carried out on the initial training dataset. The average values of testing MAPE of many different ANN models resulting from the cross validation tests are presented in Figure 12 (each data point represents the average of 15 different MAPE values). Multiple ANN models (with different architecture) are developed to find the optimum number of hidden layers (L) and number of neurons (N) in each hidden layer (Figure 12a,b), while a fixed network architecture is used to tune number of epochs and the learning rate (LR) of the SGD algorithm (Figure 12c,d).

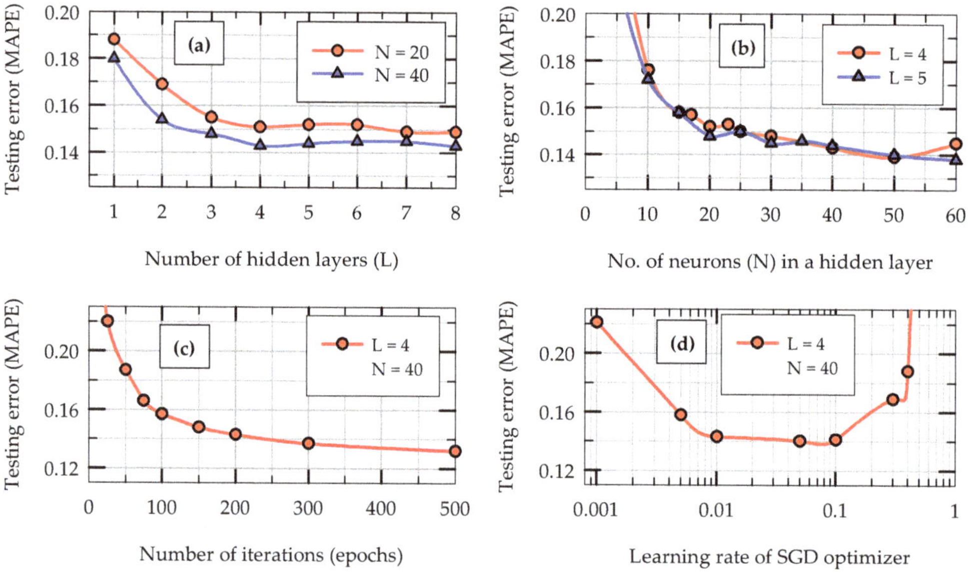

Figure 12. Results of hyperparameter tuning of the ANN model: Variation of average MAPE with (**a**) number of hidden layers, (**b**) number of neurons in each hidden layer, (**c**) number of iterations for each batch of training data, and (**d**) learning rate of SGD algorithm.

Results presented in Figure 12a show that as L increases, the error (average MAPE) in predictions decreases up to when L = 4. When L increases further, the ANN model seems to overfit the training data slightly (increase in testing error). A similar trend is observed for the number of neurons (N) used in each hidden layer (Figure 12b). Based on these observations, the combination of L = 4 and N = 40 is chosen as the optimum combination for the architecture of the ANN model for the problem considered. These observations are confirmed and verified independently by using grid search and random search algorithms going through multiple ANN model architectures with several possible combinations of L and N. As the number of iterations (epochs) increases, as expected, the average MAPE in predictions decreases (Figure 12c); however, once the number of iterations reaches around 200 to 300, no further significant improvement in MAPE is observed with the number of iterations. As for the learning rate (LR) of the SGD algorithm, the optimum learning rate is found to be between 0.01 and 0.1 (Figure 12d). The optimum values chosen for the number of iterations and the learning rate are 300 and 0.01, respectively. Table 2 summarizes the optimum values chosen for the key hyperparameters of the ANN model developed in this study.

Table 2. Optimum values chosen for major hyperparameters of the ANN model.

Hyperparameter of the ANN Model	Value
Number of hidden layers (L)	4
Number of neurons in each hidden layer (N)	40
Activation function	ReLU [1]
Optimizer	SGD [2]
Learning rate	0.01
Batch size for training	2
Number of epochs	300

[1] Rectified linear unit function. [2] Stochastic gradient descent.

4.7. Comparison of Overall Accuracy of Model Predictions and Variance in Prediction Error

Multiple k-fold cross validation tests (k = 5 and number of repeats = 3) are carried out considering the entire dataset to evaluate the overall performance (average testing MAPE and MAE, and the variance in testing MAPE and MAE) of all ML models. Note that for hyperparameter training, the k-fold cross validation tests are performed using the training dataset only, while this final k-fold cross validation test uses the entire dataset. Figure 13 presents the results of the MAPE of predictions of AAR obtained using six nonlinear machine learning and deep learning models (KNN, SVR, RFR, ABR, GBR, and ANN) along with the baseline MLR model. The hyperparameters of all models are kept constant as obtained from the hyperparameter tuning phase of each model. For each model, the testing MAPE results are plotted in the form of boxplots, showing the average MAPE, median MAPE, and the 10th, 25th, 75th and 90th percentile values of MAPE (obtained from 15 values of MAPE for each model).

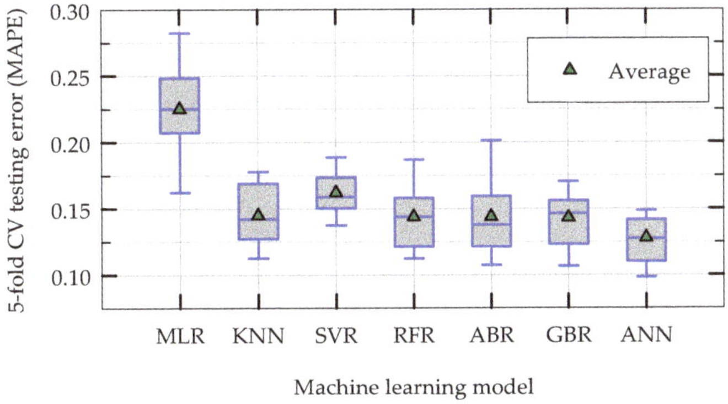

Figure 13. Boxplots of MAPE in the predictions of AAR of machine leaning models during final five-fold cross validation tests of models.

The first observation from Figure 13 is that the average testing MAPE of all six nonlinear models are better (smaller) than that of the baseline MLR model. Among the six nonlinear models, except for the SVR model, the average testing MAPE of the models are smaller than 0.15, and the results for the average MAPE are remarkably consistent across different ML models. The ANN model has the smallest average MAPE value (0.128) and relatively lower variance in MAPE (0.05), indicating that the ANN model outperforms all other models developed in this study for the problem considered. Based on the overall average MAPE in predictions, the ANN model improves the prediction accuracy by 43% compared to the MLR model (MAPE of 0.128 versus 0.225). As the difference in overall model performance in terms of average accuracy among five nonlinear models is relatively small (average MAPE varies from 0.13 to 0.15), and if one prefers simpler ML models, KNN and all three DTR-based ensemble models are almost equally effective for the prediction of AAR. Figure 14 presents the results obtained from the same k-fold cross validation tests (same as the one presented in Figure 13), in the form of MAE in predictions of AAR. As can be observed from Figure 14, the results for testing MAE show a very similar trend that is observed for testing MAPE (Figure 13). Except for the SVR model, the overall average MAE of the other five nonlinear ML models varies between 0.083 and 0.092, once again indicating a remarkable consistency across different ML models. The overall average MAE in predictions of all six nonlinear ML models varies between 0.08 and 0.1, indicating that the maximum acceleration transmitted to structures supported by rocking foundations can be predicted within an average error limit of 8% to 10% of peak ground acceleration of the earthquake.

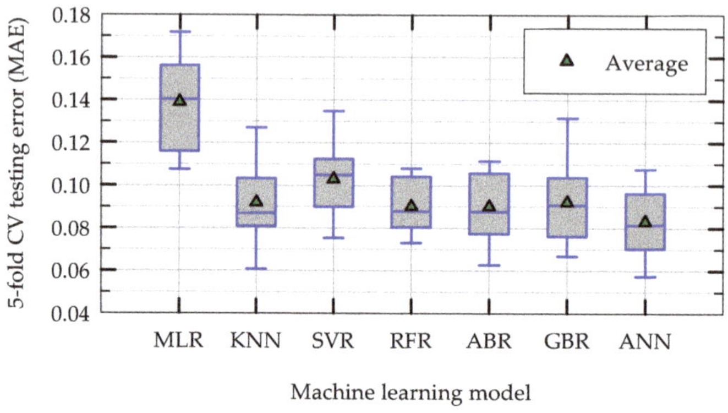

Figure 14. Boxplots of MAE in the predictions of AAR of machine leaning models during final five-fold cross validation tests of models.

Table 3 summarizes the average MAPE and MAE of predictions of all seven ML models in repeated five-fold cross validation tests. Also included in Table 3 is the MAPE and MAE of the statistics-based simple linear regression (SLR) best fit model presented in Figure 3 (using the relationship between log (C_r/a_{max}) and log (AAR)). As can be seen from Table 3, from the statistics based SLR model to MLR machine learning model, the results do not seem to vary much. However, the six nonlinear ML models developed in this study for the prediction of AAR show a significant difference in accuracy. In order to compare the performance of the models using a different error criterion, which is not used in the training phase of models nor in the hyperparameter tuning, a third error measure is also considered. For this purpose, root mean squared error (RMSE), a commonly used error measure in machine learning, is selected. The last column of Table 3 presents the results obtained for average RMSE of predictions of all the models in repeated five-fold cross validation tests. As can be seen from Table 3, the trend in RMSE values is consistent with the trends observed in MAPE and MAE, and it leads to the same conclusion: among the six nonlinear ML models, the ANN model turns out to be the most accurate, the second most accurate model is the RFR, and it is followed by the ABR, GBR, and KNN models.

Table 3. Summary of average MAPE, MAE and RMSE (testing errors) of models in final five-fold cross validation tests.

Model	Ave. MAPE	Ave. MAE	Ave. RMSE
Simple linear regression (SLR) *	0.228	0.148	0.232
Multivariate linear regression (MLR)	0.225	0.139	0.185
Support vector regression (SVR)	0.162	0.103	0.145
k-nearest neighbors regression (KNN)	0.145	0.092	0.137
Random forest regression (RFR)	0.144	0.090	0.124
Adaptive boosting regression (ABR)	0.144	0.090	0.125
Gradient boosting regression (GBR)	0.143	0.092	0.133
Artificial neural network regression (ANN)	0.128	0.083	0.113

* Statistics-based (non-ML) model.

4.8. Parametric Sensitivity Analysis of Models

In order to study the sensitivity of ML model predictions to variations in input feature values, a parametric sensitivity analysis is carried out. For this exercise, the input feature values are systematically varied and are fed into the ML models. As a baseline case, all input feature values are kept at their mean values and the predicted AAR corresponding to this scenario is the most likely value (MLV) of prediction for a particular model. In addition, each input feature is varied to include two other values: mean minus standard deviation and mean plus standard deviation. The predictions of the ML models are obtained by this method using these two extreme values for a certain input feature, while all other input feature values are kept at their mean value. As there are five input features, this method results in eleven combinations of input features. The results of this parametric sensitivity analysis are presented in Figure 15 for four models in the form of "tornado diagrams". In the tornado diagrams presented in Figure 15, the x-axis represents the predicted AAR values by that particular model when the input feature values are varied (mean ± standard deviation). Note that in a tornado diagram, the absolute difference between the prediction values corresponding to the two extreme values of an input feature is called the "swing", and the input feature that has greatest swing is plotted at the top of the plot (the input features are plotted on the y-axis in descending order of their swing values). Also included in these figures are the most likely value (MLV) of predicted AAR (vertical dashed lines), when all the input features are kept at their mean values. Table 4 presents summary results of predicted AAR in parametric sensitive analysis (MLV, minimum and maximum) for all seven ML models.

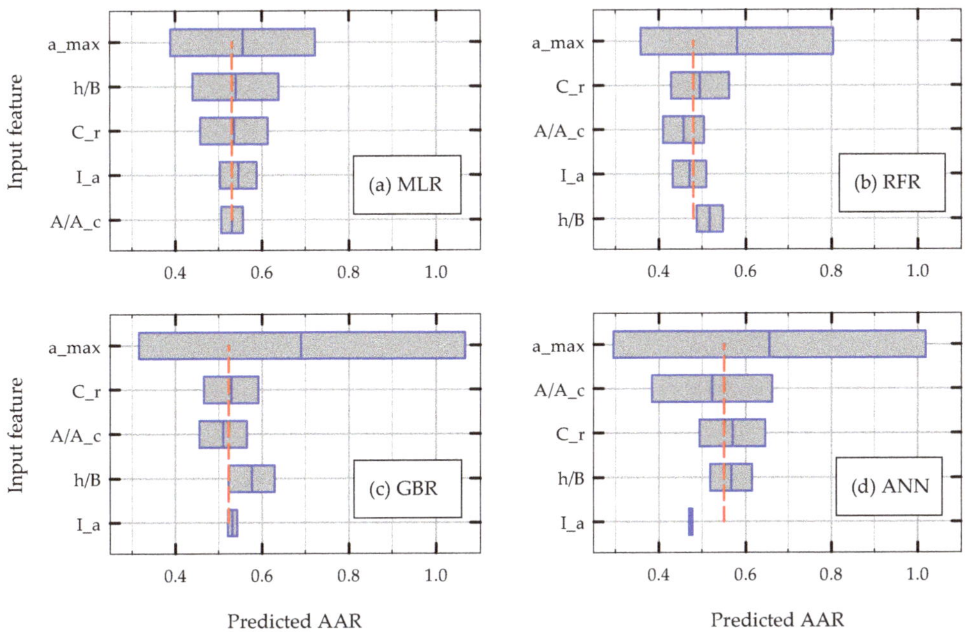

Figure 15. Results of parametric sensitivity analysis of ML models in the form of tornado diagrams when the input feature values are varied one at a time: (**a**) MLR, (**b**) RFR, (**c**) GBR and (**d**) ANN. Note: the dashed vertical lines correspond to the predicted AAR (most likely value) when all input features are at their mean values.

As the results presented in Figure 15 and Table 4 indicate, the predicted AAR is more sensitive to peak ground acceleration (a_{max}) than any other input feature for all the models (i.e., a_{max} produces the maximum swing in predicted AAR). Only the MLR model shows an almost symmetric response around the MLV in tornado diagrams, mostly because it is a linear ML model. The unsymmetric nature of the tornado diagrams of all nonlinear ML models indicates that the relationship between AAR and input features are highly nonlinear. For all six nonlinear ML models, about 45% to 75% of the variance in the prediction of AAR results from the variation in a_{max} (variance in this context is defined as half-swing divided by the most likely value of predicted AAR). Next to a_{max}, C_r and A/A_c have more effect on model predictions in general when compared to h/B and I_a. This trend is consistent with the experimental results plotted in Figure 3, where a_{max} and C_r are identified as the key variables that dictate AAR. This is also consistent with the results presented in Figure 9, where two DTR-based ensemble models (RFR and ABR) identify a_{max} and C_r as the features with the highest and the second highest, respectively, feature important scores to predict AAR. It should also be noted that none of the ML model predictions are extremely high or extremely low when the input feature values are varied. This indicates that the ML models developed in this study do not tend to extrapolate the data beyond a reasonable range of AAR values.

Table 4. Summary results of predicted AAR in parametric sensitivity analysis of ML models when the input feature values are varied one at a time.

Model	MLV *	Minimum	Maximum
Multivariate linear regression (MLR)	0.530	0.389	0.723
Support vector regression (SVR)	0.586	0.332	1.046
k-nearest neighbors regression (KNN)	0.435	0.349	0.634
Random forest regression (RFR)	0.480	0.357	0.804
Adaptive boosting regression (ABR)	0.541	0.341	1.150
Gradient boosting regression (GBR)	0.523	0.316	1.065
Artificial neural network regression (ANN)	0.469	0.295	1.018

* Most likely value.

The interpretability of ML models is often thought to be challenging, as they are agnostic to the underlying scientific principles driving the physical mechanisms of the problem considered. However, apart from the ANN model, all other ML models developed in this study are based on simple, straight forward logic, and they are relatively easy to interpret (i.e., why the model predicts a certain value for AAR given the input feature values). The new data science paradigm of theory-guided machine learning combines the beneficial features of both mechanics-based models and ML models while minimizing or eliminating their adverse effects [3]. This concept forms the basis for future research on this topic.

5. Conclusions

Multiple machine learning (ML) models are developed to predict the maximum acceleration transferred to the center of gravity of structures founded on rocking shallow foundations during earthquake loading. Based on this study, the following major conclusions are drawn.

- Given the five input features representing the key properties of the rocking foundation and earthquake loading (A/A_c, h/B, C_r, a_{max} and I_a), the ML models presented in this paper can be used to predict the maximum acceleration transmitted to structures supported by rocking foundations with reasonable accuracy.
- Based on k-fold cross validation tests, the overall average MAPE in predictions of the KNN, RFR, ABR, GBR, and ANN models are all smaller than 0.145, with ANN being the most accurate and most consistent (MAPE = 0.128). For comparison, the MAPE of the MLR model and statistics based SLR model are around 0.23. This corresponds to an improvement in prediction accuracy of about 43%. Next to the ANN model, the second most accurate model is RFR, and it is followed by ABR, GBR, and KNN. This finding is also supported by another error measure criterion, namely, root mean squared error (RMSE) of model predictions.
- The overall average MAE in predictions of all six nonlinear ML models vary between 0.08 and 0.1, indicating that the maximum acceleration transferred to structures supported by rocking foundations can be predicted within an average error limit of 8% to 10% of the peak ground acceleration of the earthquake.
- Hyperparameter tuning is carried out to obtain the optimum values for hyperparameters and to ensure that the ML models presented in this paper do not overfit or underfit the training data. In terms of the architecture of the ANN model, a relatively simple network (only four hidden layers with 40 neurons in each layer) is found to be

the optimum and most efficient for the problem considered in terms of accuracy of predictions without overfitting the training data.

- Feature importance analysis using the RFR, ABR and GBR models reveals that the chosen five input features capture the maximum acceleration of structures (through AAR) supported by rocking foundations satisfactorily. Parametric sensitivity analysis of all ML models reveals that AAR is more sensitive to peak ground acceleration of the earthquake motion than to other input features.
- The ML models presented in this paper can be used with numerical simulation results as complementary measures in modeling of rocking foundations or can be combined with mechanics-based models using the emerging framework of theory-guided machine learning. This forms the basis for future research on this topic.

Funding: This research was funded by the US National Science Foundation (NSF) through award number CMMI-2138631.

Institutional Review Board Statement: Not applicable.

Informed Consent Statement: Not applicable.

Data Availability Statement: The data presented in this study and more details about the machine learning models developed in this study are available on request from the corresponding author. The data are not publicly available due to ethical.

Conflicts of Interest: The author declares no conflict of interest.

Nomenclature

AAR	Acceleration amplification ratio
ABR	Adaptive boosting regression model
a_{max}	Peak ground acceleration of earthquake
ANN	Artificial neural network regression model
A/A_c	Critical contact area ratio of rocking foundation
C_r	Rocking coefficient of rocking system
GBR	Gradient boosting regression model
h/B	Slenderness ratio of rocking system
I_a	Arias intensity of earthquake
KNN	k-nearest neighbors regression model
MAE	Mean absolute error
MAPE	Mean absolute percentage error
MLR	Multivariate linear regression model
R^2	Coefficient of determination
RFR	Random forest regression model
RMSE	Root mean squared error
SLR	Simple linear regression (non-ML) model
SVR	Support vector regression model

References

1. Bapir, B.; Abrahamczyk, L.; Wichtmann, T.; Prada-Sarmiento, L.F. Soil-structure interaction: A state-of-the-art review of modeling techniques and studies on seismic response of building structures. *Front. Built Environ.* **2023**, *9*, 1120351. [CrossRef]
2. Ebid, A.M. 35 years of AI in geotechnical engineering: State of the art. *Geotech. Geol. Eng.* **2021**, *39*, 637–690. [CrossRef]
3. Karpatne, A.; Atluri, G.; Faghmous, J.H.; Steinbach, M.; Banerjee, A.; Ganguly, A.; Shekhar, S.; Samatova, N.; Kumar, V. Theory-guided data science: A new paradigm for scientific discovery from data. *IEEE Trans. Knowl. Data Eng.* **2017**, *29*, 2318–2331. [CrossRef]
4. Mozumder, R.A.; Laskar, A.I. Prediction of unconfined compressive strength of geopolymer-stabilized clayey soils using artificial neural network. *Comput. Geotech.* **2015**, *69*, 291–300. [CrossRef]
5. Pham, B.T.; Bui, D.T.; Prakash, I. Landslide susceptibility assessment using bagging ensemble based alternating decision trees, logistic regression and J48 decision trees methods: A comparative study. *Geotech. Geol. Eng.* **2017**, *35*, 2597–2611. [CrossRef]
6. Jeremiah, J.J.; Abbey, S.J.; Booth, C.A.; Kashyap, A. Results of application of artificial neural networks in predicting geo-mechanical properties of stabilized clays—A review. *Geotechnics* **2021**, *1*, 144–171. [CrossRef]

7. Amjad, M.; Ahmad, I.; Ahmad, M.; Wroblewski, P.; Kaminski, P.; Amjad, U. Prediction of pile bearing capacity using XGBoost algorithm: Modeling and performance evaluation. *Appl. Sci.* **2022**, *12*, 2126. [CrossRef]
8. Rateria, G.; Maurer, B.W. Evaluation and updating of Ishihara's (1985) model for liquefaction surface expression with insights from machine and deep learning. *Soils Found.* **2022**, *62*, 101131. [CrossRef]
9. Raja, M.N.A.; Abdoun, T.; El-Sekelly, W. Smart prediction of liquefaction-induced lateral spreading. *J. Rock Mech. Geotech. Eng.* **2023**, *in press*.
10. Gajan, S. Modeling of seismic energy dissipation of rocking foundations using nonparametric machine learning algorithms. *Geotechnics* **2021**, *2*, 534–557. [CrossRef]
11. Gajan, S. Data-driven modeling of peak rotation and tipping-over stability of rocking shallow foundations using machine learning algorithms. *Geotechnics* **2022**, *2*, 781–801. [CrossRef]
12. Huang, B.; Lu, W. Evaluation of the floor acceleration amplification demand of instrumented buildings. *Adv. Civ. Eng.* **2021**, *2021*, 7612101. [CrossRef]
13. Wang, T.; Shang, Q.; Li, J. Seismic force demands on acceleration-sensitive nonstructural components: A state-of-the-art review. *Earthq. Eng. Eng. Vib.* **2021**, *20*, 39–62. [CrossRef]
14. Rutenberg, A. Seismic shear forces on RC walls: Review and bibliography. *Bull. Earthq. Eng.* **2013**, *11*, 1727–1751. [CrossRef]
15. Clvi, P.M.; Sullivan, T.J. Estimating floor spectra in multiple degree of freedom systems. *Earthq. Struct.* **2014**, *7*, 17–38. [CrossRef]
16. Vukobratovic, V.; Fajfar, P. Code-oriented floor acceleration spectra for building structures. *Bull. Earthq. Eng.* **2017**, *15*, 3013–3026. [CrossRef]
17. Perrone, D.; Brunesi, E.; Filiatrault, A.; Nascimbene, R. Probabilistic estimation of floor response spectra in masonry infilled reinforced concrete building portfolio. *Eng. Struct.* **2020**, *202*, 109842. [CrossRef]
18. Gajan, S.; Kutter, B.L. Capacity, settlement, and energy dissipation of shallow footings subjected to rocking. *J. Geot. Geoenviron. Eng.* **2008**, *134*, 1129–1141. [CrossRef]
19. Paolucci, R.; Shirato, M.; Yilmaz, M.T. Seismic behavior of shallow foundations: Shaking table experiments versus numerical modeling. *Earthq. Eng. Struct. Dyn.* **2008**, *37*, 577–595. [CrossRef]
20. Gelagoti, F.; Kourkoulis, R.; Anastasopoulos, I.; Gazetas, G. Rocking isolation of low-rise frame structures founded on isolated footings. *Earthq. Eng. Struct. Dyn.* **2012**, *41*, 1177–1197. [CrossRef]
21. Pelekis, I.; Madabhushi, G.; DeJong, M. Seismic performance of buildings with structural and foundation rocking in centrifuge testing. *Earthq. Eng. Struct. Dyn.* **2018**, *47*, 2390–2409. [CrossRef]
22. Khosravi, M.; Boulanger, R.W.; Wilson, D.W.; Olgun, C.G.; Shao, L.; Tamura, S. Stress transfer from rocking shallow foundations on soil-cement reinforced clay. *Soils Found.* **2019**, *59*, 966–981. [CrossRef]
23. Irani, A.E.; Bonab, M.H.; Sarand, F.B.; Katebi, H. Overall improvement of seismic resilience by rocking foundation and trade-off implications. *Int. J. Geosynth. Ground Eng.* **2023**, *9*, 40. [CrossRef]
24. Anastasopoulos, I.; Gazetas, G.; Loli, M.; Apostolou, M.; Gerolymos, N. Soil failure can be used for seismic protection of structures. *Bull. Earthq. Eng.* **2010**, *8*, 309–326. [CrossRef]
25. Pecker, A.; Paolucci, R.; Chatzigogos, C.; Correia, A.A.; Figini, R. The role of non-linear dynamic soil-foundation interaction on the seismic response of structures. *Bull. Earthq. Eng.* **2014**, *12*, 1157–1176. [CrossRef]
26. Gajan, S.; Soundararajan, S.; Yang, M.; Akchurin, D. Effects of rocking coefficient and critical contact area ratio on the performance of rocking foundations from centrifuge and shake table experimental results. *Soil Dyn. Earthq. Eng.* **2021**, *141*, 106502. [CrossRef]
27. Gajan, S.; Raychowdhury, P.; Hutchinson, T.C.; Kutter, B.L.; Stewart, J.P. Application and validation of practical tools for nonlinear soil-foundation interaction analysis. *Earthq. Spectra* **2010**, *26*, 119–129. [CrossRef]
28. Hamidpour, S.; Shakib, H.; Paolucci, R.; Correia, A.A.; Soltani, M. Empirical models for the nonlinear rocking response of shallow foundations. *Bull. Earthq. Eng.* **2022**, *20*, 8099–8122. [CrossRef]
29. Deng, L.; Kutter, B.L.; Kunnath, S.K. Centrifuge modeling of bridge systems designed for rocking foundations. *J. Geot. Geoenviron. Eng.* **2012**, *138*, 335–344. [CrossRef]
30. Gavras, A.G.; Kutter, B.L.; Hakhamaneshi, M.; Gajan, S.; Tsatsis, A.; Sharma, K.; Kouno, T.; Deng, L.; Anastasopoulos, I.; Gazetas, G. Database of rocking shallow foundation performance: Dynamic shaking. *Earthq. Spectra* **2020**, *36*, 960–982. [CrossRef]
31. Deng, L.; Kutter, B.L. Characterization of rocking shallow foundations using centrifuge model tests. *Earthq. Eng. Struct. Dyn.* **2012**, *41*, 1043–1060. [CrossRef]
32. Hakhamaneshi, M.; Kutter, B.L.; Deng, L.; Hutchinson, T.C.; Liu, W. New findings from centrifuge modeling of rocking shallow foundations in clayey ground. In Proceedings of the Geo-Congress 2012, Oakland, CA, USA, 25–29 March 2012.
33. Drosos, V.; Georgarakos, T.; Loli, M.; Anastasopoulos, I.; Zarzouras, O.; Gazetas, G. Soil-foundation-structure interaction with mobilization of bearing capacity: Experimental study on sand. *J. Geot. Geoenviron. Eng.* **2012**, *138*, 1369–1386. [CrossRef]
34. Anastasopoulos, I.; Loli, M.; Georgarakos, T.; Drosos, V. Shaking table testing of rocking—Isolated bridge pier on sand. *J. Earthq. Eng.* **2013**, *17*, 1–32. [CrossRef]
35. Antonellis, G.; Gavras, A.G.; Panagiotou, M.; Kutter, B.L.; Guerrini, G.; Sander, A.; Fox, P.J. Shake table test of large-scale bridge columns supported on rocking shallow foundations. *J. Geot. Geoenviron. Eng.* **2015**, *141*, 04015009. [CrossRef]
36. Tsatsis, A.; Anastasopoulos, I. Performance of rocking systems on shallow improved sand: Shaking table testing. *Front. Built Environ.* **2015**, *1*, 00009. [CrossRef]

37. American Society of Civil Engineers (ASCE). *Minimum Design loads for Buildings and Other Structures*; SEI/ASCE 7-16; American Society of Civil Engineers (ASCE): Reston, VA, USA, 2017.
38. Building Seismic Safety Council (BSSC). *Recommended Provisions for the Development of Seismic Regulations for New Buildings and Structures*; National Earthquake Hazard Reduction Program (NEHRP): Washington, DC, USA, 2015.
39. Geron, A. *Hands-On Machine Learning with Scikit-Learn, Keras, and TensorFlow: Concepts, Tools and Techniques to Build Intelligent Systems*, 2nd ed.; O'Reilly Media, Inc.: Sebastopol, CA, USA, 2019.

applied sciences

Article

Predicting Model of Dual-Mode Shield Tunneling Parameters in Complex Ground Using Recurrent Neural Networks and Multiple Optimization Algorithms

Taihua Yang [1], Tian Wen [1], Xing Huang [2,*], Bin Liu [2,*], Hongbing Shi [3], Shaoran Liu [4], Xiaoxiang Peng [1] and Guangzu Sheng [5]

1 School of Urban Construction, Wuhan University of Science and Technology, Wuhan 430083, China; yangtaihua@wust.edu.cn (T.Y.)
2 State Key Laboratory of Geomechanics and Geotechnical Engineering, Institute of Rock and Soil Mechanics, Chinese Academy of Sciences, Wuhan 430071, China
3 China Construction Civil Infrastructure Corp., Ltd., Beijing 100029, China
4 China Construction South Investment Co., Ltd., Shenzhen 518000, China
5 Wuhan Urban Construction Group Construction Management Co., Ltd., Wuhan 430040, China
* Correspondence: xhuang@whrsm.ac.cn (X.H.); liubin@whrsm.ac.cn (B.L.)

Abstract: Based on the left tunnel of the Liuxiandong Station to Baimang Station section of Shenzhen Metro Line 13 (China), a prediction model for the advanced rate of dual-mode shield tunneling in complex strata was established to explore intelligent tunneling technology in complex ground. Firstly, geological parameters of the complex strata and on-site monitoring parameters of EPB/TBM dual-mode shield tunneling were collected, with tunneling parameters, shield tunneling mode, and strata parameters selected as input features. Subsequently, the Isolation Forest algorithm was employed to remove outliers from the original advance parameters, and an improved mean filtering algorithm was applied to eliminate data noise, resulting in the steady-state phase parameters of the shield tunneling process. The base model was chosen as the Long-Short Term Memory (LSTM) recurrent neural network. During the model training process, particle swarm optimization (PSO), genetic algorithm (GA), differential evolution (DE), and Bayesian optimization (BO) algorithms were, respectively, combined to optimize the model's hyperparameters. Via rank analysis based on evaluation metrics, the BO-LSTM model was found to have the shortest runtime and highest accuracy. Finally, the dropout algorithm and five-fold time series cross-validation were incorporated into the BO-LSTM model, creating a multi-algorithm-optimized recurrent neural network model for predicting tunneling speed. The results indicate that (1) the Isolation Forest algorithm can conveniently identify outliers while considering the relationship between tunneling speed and other parameters; (2) the improved mean filtering algorithm exhibits better denoising effects on cutterhead speed and tunneling speed; and (3) the multi-algorithm optimized LSTM model exhibits high prediction accuracy and operational efficiency under various geological parameters and different excavation modes. The minimum Mean Absolute Percentage Error (MAPE) prediction result is 8.3%, with an average MAPE prediction result below 15%.

Keywords: shield tunneling; complex strata; EPB/TBM dual-mode shield tunneling; tunneling parameter prediction; recurrent neural network

Citation: Yang, T.; Wen, T.; Huang, X.; Liu, B.; Shi, H.; Liu, S.; Peng, X.; Sheng, G. Predicting Model of Dual-Mode Shield Tunneling Parameters in Complex Ground Using Recurrent Neural Networks and Multiple Optimization Algorithms. *Appl. Sci.* **2024**, *14*, 581. https://doi.org/10.3390/app14020581

Academic Editor: José António Correia

Received: 7 December 2023
Revised: 26 December 2023
Accepted: 29 December 2023
Published: 9 January 2024

1. Introduction

The composite stratigraphy dual-mode shield tunnel boring machine (TBM) is a shield construction technology designed to cope with complex geological conditions in subway tunneling. It can efficiently and stably advance in composite strata, such as alternating soft and hard layers or soft overlying complex layers. Among these, the most common is the EPB/TBM dual-mode shield TBM. Compared to traditional single-mode shield TBMs, the composite stratigraphy dual-mode shield TBM can better adapt to geological conditions, enhancing construction efficiency and safety. When using a dual-mode shield TBM for variable excavation modes in complex geological conditions [1–3], it is necessary to comprehensively consider factors such as engineering geology and excavation mode and propose reasonable adjustments to the equipment excavation parameters. Optimizing the shield TBM excavation parameters is crucial for ensuring stable and effective tunneling [4]. Improper excavation parameter settings may lead to various issues, such as over-excavation, deviation of the TBM axis, instability of the face support [5], excessive tool wear, and in severe cases, may even result in TBM accidents. Therefore, in the shield construction process, careful attention must be paid to the setting and adjustment of excavation parameters, and strict adherence to relevant construction regulations and standards is essential to ensure the safety and efficiency of shield construction [6–8]. During the construction of composite stratigraphy shield tunnels, the tunneling speed of the TBM may be influenced by various factors, including geological conditions, shield construction parameters, and environmental factors [9–11]. The relationship between these factors and the TBM's tunneling speed involves complex mathematical relationships, requiring further in-depth research and analysis. Precise control and adjustment of these factors are necessary to ensure smooth construction.

Early tunneling speed prediction models are primarily based on empirical, theoretical, or a combination of both approaches. These models explore the operational patterns of tunneling speed and other characteristics via formula derivation and simulated experiments. Zhao Bojian et al. [12] utilized statistical methods to establish relationships between shield tunneling parameters and strata, conducting a thorough analysis. In the context of composite strata, Li Jie et al. [13] employed orthogonal experiments combined with nonlinear regression analysis to develop a mathematical model for the tunneling speed of Earth Pressure Balance (EPB) shield tunneling. Sapigni et al. [14], via a study of monitoring data from three tunnels, found a close correlation between excavation rate and rock mass classification, which could be fitted with a quadratic regression equation. Based on on-site measured data, Kahrama et al. [15] established a regression model for excavation rate, with statistical analysis results indicating a close correlation with rock properties. Hassanpour et al. [16], using data from the Nowsood Tunnel No. 2, established an empirical formula for excavation rate about different geological parameters, finding a particularly close correlation with rock cuttability, especially the field penetration index. Wang Hongxin et al. [17], based on model test results, successfully developed a structural model to study EPB shield tunneling. They derived specific mathematical expressions for the total thrust, soil chamber pressure, screw conveyor speed, and tunneling speed. Zhang Zhiqi et al. [18] conducted multivariate regression analysis and discovered a certain robustness between shield tunneling speed and cutterhead torque. Xu Qianwei [19], via experiments, identified key shield construction parameters and studied their adaptive relationships with soil properties. However, these methods rely on linear relationships between data, while shield tunneling data often exhibit nonlinear relationships. Additionally, shield tunneling data are exceptionally voluminous, making conventional calculation methods unable to calculate the connections between data precisely.

In recent years, due to the advancement of artificial intelligence technology, many data-driven models have been successfully applied to tunneling speed prediction. Xu et al. [20], based on on-site and laboratory experiment data from a tunnel in Malaysia, proposed five different machine-learning methods for predicting tunneling speed. By comparing the predicted results of each method with actual values, they found that the K-Nearest Neighbors (K-NN) algorithm achieved the best prediction accuracy. Zhang Zheming et al. [21] employed the uniform extraction of samples to establish a model for predicting cutterhead torque, cutterhead thrust, and tunneling speed in the stable section. The model used a radial basis function neural network kernel and ten-fold cross-validation for the training of the Least Squares Support Vector Machine (LS-SVM) model, providing accurate predictions. Based on the Shenzhen Metro project, Li Chao et al. [22] utilized backpropagation (BP) artificial neural network technology to establish a prediction model for shield tunneling parameters under complex geological conditions. Hou Shaokang et al. [23] introduced a novel TBM tunneling parameter prediction model that used an improved particle swarm optimization algorithm to optimize a BP neural network. The enhanced algorithm employed adaptive inertia weights, resulting in higher prediction accuracy than traditional BP and PSO-BP models. Qiu Daohong et al. [24], considering the temporal nature of collected TBM tunneling parameters, constructed a Long Short-Term Memory (LSTM) neural network model. Experimental results demonstrated that this model achieved optimal prediction accuracy for net tunneling speed.

There is a greater abundance of research on the tunneling speed prediction of single-mode shield TBMs under different geological conditions. At the same time, there is relatively less research on the tunneling speed prediction of dual-mode shield TBMs. Dual-mode shield TBMs require the selection of different excavation modes based on varying geological conditions during the tunneling process. Simultaneously, factors such as face support balance, cutterhead tool types, and support methods may change, rendering the parameters obtained by the data acquisition system more complex. Therefore, establishing tunneling speed prediction models for dual-mode shield TBMs is inherently more complex than for single-mode shield TBMs. Furthermore, the application of machine learning algorithm models for predicting tunneling speed has been more prevalent in research. The above model data input, lack of data refinement process, data outlier processing, and its importance. In contrast, there is a relatively limited number of studies focusing on prediction models based on optimization algorithms. There is also a scarcity of research that compares the impact of different optimization algorithms on the prediction accuracy of models.

Based on the left line project of Liuxiandong Station–Baimang Station Tunnel of Shenzhen Metro Line 13, this paper obtains a large number of time series characteristic parameters based on the data acquisition system, eliminates abnormal data via the isolated forest algorithm, and optimizes the original shield parameters with the improved mean filtering algorithm. Considering the influence of stratum conditions on tunneling parameters from three dimensions of surrounding rock grade, tunnel depth–span ratio, and soft–hard composite ratio, an LSTM model integrating four super-parameter optimization algorithms is established. Combined with the dropout algorithm and five-fold time series cross-validation, the two shield tunneling modes of EPB and TBM and the propulsion speed under different strata are predicted and analyzed, which provides feasible guidance for intelligent control of the dual-mode shield tunneling process.

2. Project Overview

The Shenzhen Metro Line 13 is a north–south urban rail transit line starting at the Shenzhen Bay Port and traversing the Nanshan District and Bao'an District, totaling 22,434 km. The line is between the central development axis and the western development axis of the city, connecting the Hohai Central Urban Area and the Western High-Tech Industrial Park. It serves as a fast connection between these two regions. This study is based on the left-line project from Liuxiandong Station to Baimang Station, focusing on the section from 0 to 650 m in the tunnel, with a total length of approximately 2036 m. This section represents a typical complex geological formation. The Liuxiandong Station's elevation to Baimang Station's section ranges from 24.26 to 42.85 m above ground, with a slightly undulating terrain. A mid-tableland landscape, with localized gully landscapes between tablelands, characterizes the initial topography. The predominant geological layers in this section include the fourth series of artificial fill (Q4ml) and fluvial deposits (Q4al + pl). The fluvial deposits mainly consist of silty clay, plastic fine-grained clay, and sand layers. The existing geological layers exhibit distinct stratified structures above the fill layer. The underlying bedrock consists mainly of mixed granites from the Jixian Formation to the Qingbaikou Formation (Jx-Qby) and biotite granite from the Yanshan period ($\gamma\beta$5K1). The tunnel body passes through moderately to slightly weathered rocks as the primary geological formation, followed by residual layers and highly weathered rocks. The tunnel also intersects sporadically with intensely weathered and moderately weathered schists.

The shield tunneling machine selected for the Liubai section of Shenzhen Metro Line 13 is the "Xinhe 15" EPB and single-shield TBM dual-mode shield tunneling machine, designed and manufactured by China Railway Construction Equipment Group. The main configuration parameters are detailed in Table 1. The design concept of this shield tunneling machine integrates the functionalities of both EPB and single-shield TBM, allowing for in-tunnel conversion. The EPB mode is suitable for weak geological conditions, employing the EPB tunneling method to ensure the face support's stability and prevent uneven ground settlement. On the other hand, the single-shield TBM mode applies to complex rock formations, enhancing the tunneling speed in such conditions and avoiding risks associated with slow progress and severe tool wear when using the EPB mode in complex rock formations.

Table 1. "Xinhe 15" dual-mode shield main technical parameters index.

Parameter	Design Value
Shield type	Mono-protecting shield
Total weight/T (Host + Supporting)	650
Total length of machine/m	96
Cutterhead speed/(r/min)	0-2.47-5.59
Excavation diameter/(mm)	φ6980
Maximum thrust/T	5060
Rated torque/kN·m	6080
Number of tools	50 (Hob) + 61 (Scraper) + 12 (Side scraper)

3. Predictive Model Algorithm Principle

3.1. LSTM Model

The data of shield tunneling are time-series, and the data are purely dependent. Traditional machine learning algorithms (such as BP neural network, random forest, etc.) cannot capture the time-series value between data. The LSTM network has an internal gating mechanism, which enables it to effectively capture and retain information from past inputs. At the same time, the generalization ability of the LSTM neural network is stronger. It is a deep learning algorithm widely used in the market and more suitable for engineering needs. In predicting TBM tunneling speed, the LSTM model can be employed by inputting parameters such as excavation parameters, TBM mode, and geological parameters. This allows for the establishment of a model capable of predicting the tunneling speed at the next step. Furthermore, by introducing hyperparameter optimization algorithms into the LSTM model, prediction accuracy and operational efficiency can be enhanced, providing better support for TBM excavation projects. When handling sequential data, the LSTM model effectively addresses issues such as gradient vanishing or exploding that exist in traditional Recurrent Neural Networks (RNNs). Additionally, it can model long-term dependencies. In the LSTM architecture, besides the conventional input layer, output layer, and hidden layer, a memory cell and three gating units (Forget Gate, Input Gate, Output Gate) are introduced. Among these, i_t represents the input gate, f_t is the forget gate, g_t denotes the input supply, and o_t represents the output gate. The formulas for computing the forget gate, input gate, and output gate are as follows:

$$f_t = \sigma(W_f \cdot [h_{t-1}, x_t] + b_f), \tag{1}$$

$$i_t = \sigma(W_i \cdot [h_{t-1}, x_t] + b_i), \tag{2}$$

$$C_t = \tanh(W_C \cdot [h_{t-1}, x_t] + b_C), \tag{3}$$

$$o_t = \sigma(W_o \cdot [h_{t-1}, x_t] + b_o), \quad (4)$$

In the formula, W_f, W_i, $W_{c,}$ and W_o are different calculation matrices; b_f, b_i, b_o, and b_c are the bias terms of the three gated units and the cell state, respectively; σ and *tanh* are activation functions, respectively; σ represents the *sigmoid* function, and its output is between 0 and 1; and *tanh* is the hyperbolic tangent function of the mapping interval $[-1, 1]$. At each time step t, the LSTM introduces a hidden state C (cell state) and employs three gates to control the content of the cell state. The first gate, the Forget Gate, determines how much information from the previous time step's cell state C_{t-1} needs to be retained in the current time step's cell state C_t. The second gate, the Input Gate, regulates how much information from the current time step's input x is stored in the cell state C_t. The Output Gate, the third gate, determines how much information will be output from the cell state C_t to the current time step's output h_t. The LSTM network can selectively retain or output helpful information via this mechanism, enabling improved handling of long sequential data.

3.2. Multiple Optimization Algorithm Model

In the operation of LSTM, performing hyperparameter optimization is crucial as it identifies the optimal combination of hyperparameters that enhances model performance. By selecting appropriate hyperparameter combinations, it is possible to improve the model's generalization ability, resulting in superior performance on the test set. This study employs four widely applicable and practical hyperparameter optimization algorithms: Genetic Algorithm (GA), Differential Evolution (DE), Bayesian Optimization (BO), and Particle Swarm Optimization (PSO). Due to constraints in length, the principles of these algorithms are not extensively elaborated. Subsequently, the study will train these four hyperparameter optimization algorithms, compare the performance of evaluation metrics under different algorithms, and employ a ranking method to determine the optimal hyperparameter optimization algorithm for this model.

After selecting the optimal hyperparameter optimization algorithm, in order to enhance the model's generalization ability, a multi-algorithm optimized model incorporating Time Series cross-validation (TSCV) and the Dropout algorithm was established to further improve the predictive performance of the model on tunneling speed.The Dropout algorithm is a widely used regularization technique for deep learning models proposed by Geoffrey Hinton and his team in 2012. Its primary objective is to prevent neural networks from overfitting, thereby improving the model's generalization ability across different data sets. This study incorporated the Dropout layer into the LSTM model, with a dropout probability set to 0.1. This implementation achieved random dropout of a portion of neurons in the neural network, preventing certain neurons from developing excessive dependence on specific features and thereby reducing the complexity of the neural network.

Tunnel boring machine (TBM) excavation parameters constitute time-series data with temporal dependencies. In conventional cross-validation, data is randomly partitioned, making validating past data using future data highly inappropriate due to the temporal dependencies. Therefore, the data sequence should not be arbitrarily shuffled. This study employs time series cross-validation to evaluate prediction models' performance on time-series data. When assessing time-series prediction models, time-series cross-validation effectively measures the model's generalization ability across different periods. The model utilizes a five-fold time series cross-validation, and the operational workflow is depicted in Figure 1:

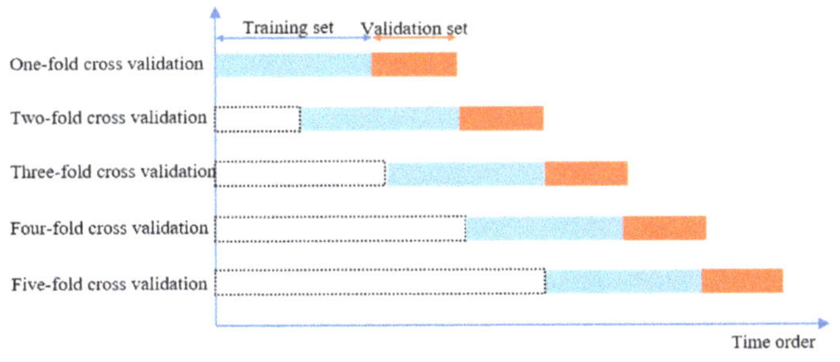

Figure 1. Five-fold time series cross-validation diagram.

The process begins by selecting a window size, which encompasses a specific number of observational values. Subsequently, the window is incrementally moved forward, providing distinct training and validation data subsets for the model. At each step, the model is trained on the data within the window and validated on the data outside the window. This allows for the computation of performance metrics, such as mean squared and absolute percentage errors, for the model on each validation set. Finally, the average of these metrics is calculated to assess the model's overall performance. By combining the above methods, the overall model process is shown in Figure 2.

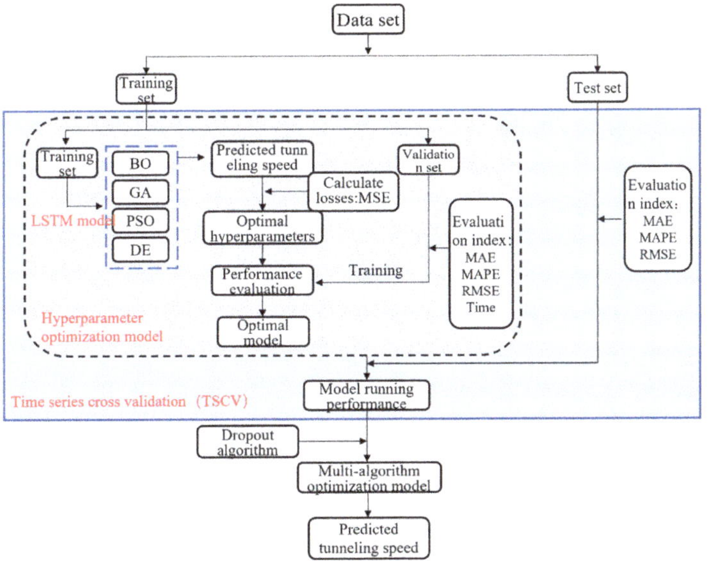

Figure 2. Flow chart of multi-algorithm optimization model.

4. Establishment of a Prediction Model for Dual-Mode Shield Tunneling Parameters

4.1. Filter Input Feature Parameters

The predictive model in this study comprehensively considers the influence of various factors, including geological parameters, shield machine excavation parameters, and shield tunneling modes. Therefore, detailed explanations for the parameters mentioned above are provided in the following sections.

(1) Composite Geological Characteristics Parameters

When considering the geological conditions within the Liubai section's left tunnel range, this study employs rock mass rating and the soft–hard compound ratio as input parameters for composite geological features. According to the national standard 'Code for geotechnical engineering investigation of urban rail transit' (GB50307-2012) Appendix F [25], the geotechnical construction engineering classification of each rock and soil layer revealed by this investigation is carried out. Grade III rock mass predominates along the left tunnel route, accounting for 46.78% of the tunnel length. Grade IV rock mass represents 12.7% of the tunnel length, Grade V rock mass accounts for 38.56%, and Grade VI rock mass covers 1.95%. To better reflect the conditions of alternating soft and hard layers in the composite geological formation, the thickness of the weak soil layer at the excavation face to the total excavation face thickness is defined as the soft–hard compound ratio, representing the composite geological formation. The definition of the soft–hard compound ratio is given by Formula (5):

$$\mu = \frac{H_i}{H_i + H_j}, \tag{5}$$

The equation μ represents the soft–hard compound ratio, within the range of 0 to 1; H_i is the thickness of the weak soil layer, and H_j is the thickness of the hard rock layer.

(2) Operating parameters of shield machine

The original data in the research phase comprises 220-dimensional parameter indicators. Based on the mechanisms affecting shield tunneling speed, this study primarily considers several excavation parameters as input variables for predicting tunneling speed:

(1) Total Thrust: A more significant total thrust reduces resistance encountered by the shield machine during excavation, enabling faster advancement. As the shield machine progresses, the total thrust must overcome the resistance and friction in the geological layers to propel the machine forward. Therefore, the magnitude of the total thrust directly influences the shield tunneling speed.
(2) Thrust Pressure: It represents the force exerted by the shield machine during the excavation process, directly affecting the machine's forward speed in the geological layers. The thrust pressure of the shield machine should be controlled within a specific range to ensure the stability and safety of shield tunneling.
(3) Cutterhead Torque: It represents the force exerted by the shield machine during the excavation process, directly affecting the machine's forward speed in the geological layers. The thrust pressure of the shield machine should be controlled within a specific range to ensure the stability and safety of shield tunneling.
(4) Cutterhead Speed: The higher the rotation speed of the cutterhead on the shield machine, the stronger its cutting ability, resulting in faster advancement of the shield machine in the geological layers.

(3) Shield tunneling mode parameters

To consider variations in the tunneling state of the dual-mode shield machine, it is necessary to select different excavation modes based on distinct geological conditions and excavation performance. They need to be labeled accordingly to differentiate between the EPB and TBM excavation modes. After multiple test runs, the EPB mode is ultimately labeled 1, and the TBM mode is labeled 3. This labeling approach aids the model in learning from the input parameters, reduces the impact on neural network weight training, and consequently enhances the rationality of the model's predictive results.

4.2. Data Preprocessing

4.2.1. Steady-State Segment Data Extraction

During the tunneling process, equipment monitors data on a timely basis. Interruptions in tunneling due to cutterhead tool replacement, segment assembly, and other reasons are within the scope of monitoring. This results in a substantial volume of raw data of lower quality. To extract steady-state data, it is necessary to eliminate empty thrusting data and exclude short-term unstable data [26]. In general processing, cutterhead thrust

(F), cutterhead torque (T), cutterhead speed (RPM), and tunneling speed (V) are considered state-discriminant parameters. Any value of zero parameters is considered blank data recorded during shield machine operations, and the entire row of data is removed. Short-term unstable data in each tunneling cycle, typically occurring during the start-up process of the shield machine, should be selectively excluded to reduce potential errors in subsequent calculations.

Figure 3 shows a shield tunneling process for data segment 140. Although the displayed unstable data in the initial stages of tunneling are minimal, considering that not every start-up process proceeds smoothly, it is advisable to sequentially exclude the first 10% of data from each start-up phase chronologically. This approach aims to optimize the quality of the data.

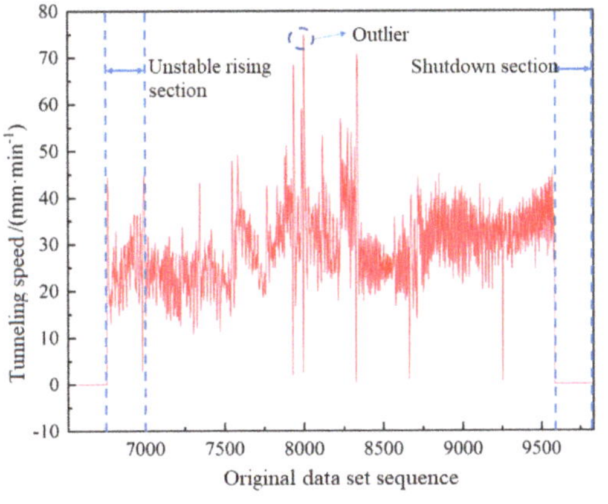

Figure 3. Raw data.

4.2.2. Outlier Handling

In practical scenarios, equipment monitoring generates a large amount of data, and monitoring anomalies may inevitably lead to some outlier values. For such data, commonly used outlier detection methods are employed for exclusion. This study utilizes the Isolation Forest algorithm for identifying and detecting anomalous data. The Isolation Forest algorithm was jointly proposed by Professor Zhou Zhihua and others in 2008 [27] for data mining. It is an unsupervised anomaly detection algorithm suitable for continuous data anomaly detection. Specifically, the Isolation Forest algorithm randomly partitions the data into several subspaces, constructs a set of binary search trees based on random partitions, and inserts data points into the leaf nodes. It determines whether a data point is an outlier by calculating the average path length across all trees for each data point. A shorter path length indicates that the data point is more easily isolated, making it more likely to be an outlier. The isolated forest model is shown in Figure 4.

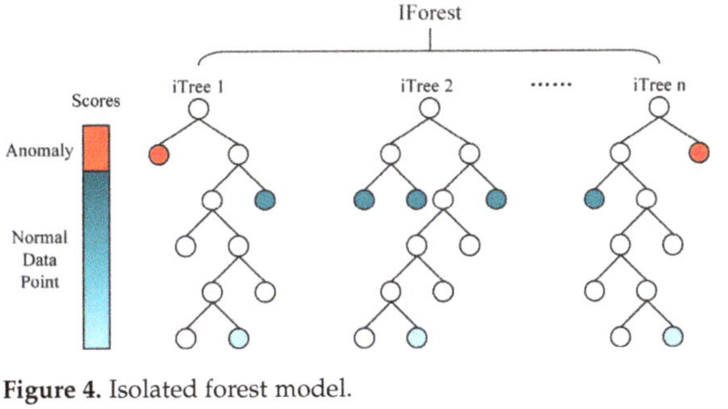

Figure 4. Isolated forest model.

The specific numerical expression is given by the following Formula (6). Firstly, construct $h(x)$ as a metric to measure the "isolation degree" of a data point (sample) from other data points. It is defined as the path length that a data point traverses from the root node to that point on a random tree. For a random tree T and a data point x within it, the number of samples that share the same leaf node with x in T is denoted as $T.size(x)$. Subsequently, for a given $T.size(x)$, a correction term $c(T.size(x))$ can be calculated to represent the average path length of constructing a binary tree with $T.size(x)$ samples.

$$h(x) = e + c(T.size), \tag{6}$$

The second step involves constructing the average path length $c(n)$ of a binary tree using n samples, specifically based on a Binary Search Tree (BST). This metric represents the average distance between any two nodes in a BST with n elements. The average path length $c(n)$ can be calculated using the following formula:

$$c(n) = 2[H(n) - 1] - 2\frac{n - q(n) - 1}{n}, \tag{7}$$

Here, $H(n)$ represents the average path length of a BST with a height of n, and $q(n)$ is an estimation of the number of non-leaf nodes in the BST after randomly constructing and inserting n elements.

$$s(x,n) = 2^{-\frac{E(h(x))}{c(n)}}, \tag{8}$$

Finally, a normalization process is applied to map the range of $h(x)$ to between 0 and 1. Here, $h(x)$ represents the path length of sample point x, and $S(x, n)$ is the anomaly index of the tree, which records the training data of x in n samples. From this formula, it can be observed that as the path length decreases, s approaches 1, and the probability of detecting the sample point as an anomaly increases.

Compared to commonly used calculation methods such as Mahalanobis distance and the 3σ criterion, the Isolation Forest algorithm does not require calculating anomaly standards for data under different geological conditions. Its unsupervised, efficient, and precise advantages make it more suitable for extensive data processing in industries such as tunnel boring machines (TBM). The Isolation Forest algorithm is capable of handling high-dimensional data. However, an increase in data dimensions during processing increases tree depth, resulting in higher time complexity for tree construction and search. Additionally, greater tree depth increases sensitivity to anomalies, making them more prone to be classified as outliers. Considering the focus of this study on predicting the tedious tunnel process's advance rate, the advance rate is paired with the total thrust, thrust pressure, cutterhead torque, and cutterhead speed to form a two-dimensional array. This array is then cyclically fed into the model for training, with all identified anomalies marked after training completion. Finally, the rows containing anomalies are removed. The model is configured with 100 trees, a contamination rate of 0.02 in the data set, and a random seed set to 42. The training results are shown in Figure 5.

For the processing of outliers in the 140th tunneling cycle parameters, a total of 2825 data sets were processed. During the operation, numerous duplicate data points were identified. Via statistical analysis, it was determined that a total of 115 data sets were flagged as outliers. As a result, the final data set comprises 2710 remaining data sets.

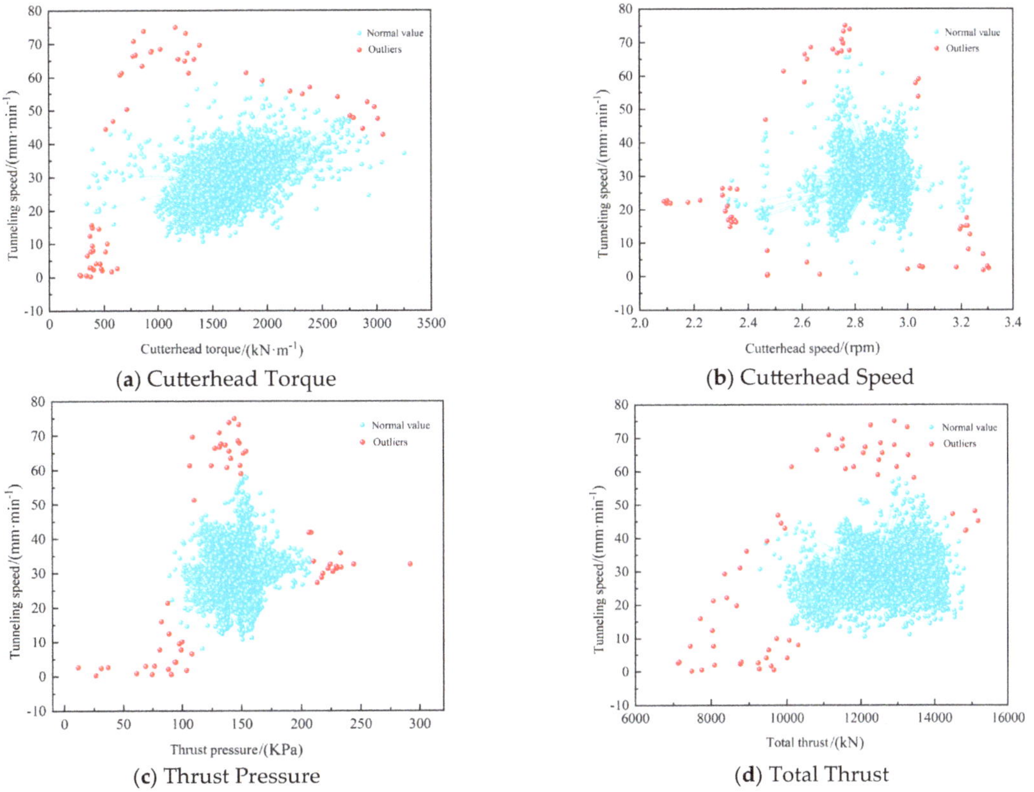

Figure 5. Outlier tag.

4.2.3. Data Denoising

After removing outliers, the data exhibits inevitable fluctuations that can impact the modeling process and even the calculation results. These anomalous fluctuation data in the tunneling parameter sequence are called noise data. This phenomenon is particularly pronounced in geological environments with complex and variable conditions, rendering the temporal parameter data more unstable. Xiao et al. [28] pointed out that denoising tunneling parameters can reduce the spatial variability of these parameters, making it easier for machine learning algorithms to learn the patterns in data changes. An improved mean filtering algorithm is employed to denoise the tunneling parameters to mitigate these irregular variations. The formula for the original mean filtering algorithm is as follows, assuming a non-stationary data set of total length N. To eliminate noise, a window size of 2n + 1 (<N) is set, and as the window slides forward, the average of every 2n + 1 adjacent data $\overline{y}_k$ is taken to represent the measurement result of the midpoint data. This method effectively eliminates noise.

$$\overline{y}_k = \frac{1}{2n+1}\sum_{i=-n}^{2n+1} y_{k+i}, \tag{9}$$

There is significant data fluctuation with sudden rises and falls throughout the shield tunneling process, accompanied by changes in the soft and hard composite strata. An improved mean denoising algorithm is proposed to mitigate the impact of significant differences in window values on denoising effectiveness. In each forward-sliding window process, the data within the window is sorted, and the maximum value y_{kmax} and the minimum number y_{kmin} are excluded. The mean of the remaining data is then calculated to represent the measurement result. The formula for the improved algorithm is as follows:

$$\overline{y}_k = \frac{1}{2n-1}\left(\sum_{i=-n}^{2n+1} y_{k+i} - y_{k\max} - y_{k\min}\right), \tag{10}$$

In order to assess the denoising effect, this paper selects two evaluation metrics: Signal-to-Noise Ratio (SNR) and Peak Signal-to-Noise Ratio (PSNR). Signal-to-Noise Ratio (SNR): It evaluates signal strength and noise level ratio. Peak Signal-to-Noise Ratio (PSNR): It evaluates the ratio between the maximum possible signal and noise power. Excellent *SNR* and *PSNR* depend on the specific application scenarios and requirements. The metrics only need to reach a certain level, and in the context of denoising the same data, higher *SNR* and *PSNR* indicate better denoising effects. The formulas for calculating both metrics are as follows:

$$SNR = 10 * \lg\left(\frac{\sum p^2}{\sum(p-q)^2}\right), \tag{11}$$

$$PSNR = 10 * \lg\left(\frac{{p_{\max}}^2}{\frac{1}{n} * \sum(p-q)^2}\right), \tag{12}$$

where p represents the original data, q is the filtered data, and p_{max} denotes the maximum signal value. In the engineering calculations, considering that the data processing is conducted in a ring-by-ring manner, with varying lengths of the excavation parameter sequences for each ring, a sliding window of size 11 is chosen after multiple tests. Here, the denoising effect after handling outliers is presented. Table 2 compares the evaluation metric values before and after algorithm improvement, while Figure 6 illustrates the denoising effect achieved via the improved mean filtering algorithm.

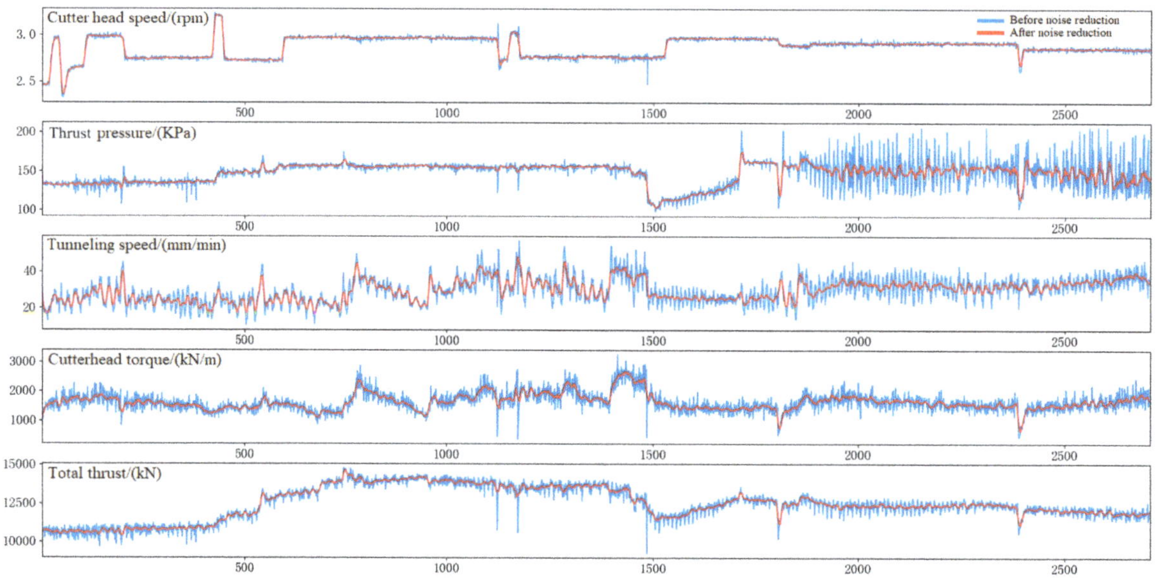

Figure 6. Comparison of variation amplitude of data values before and after denoising.

Table 2. Prediction effect ranking analysis.

	Cutterhead Speed		Thrust Pressure		Tunneling Speed		Cutterhead Torque		Total Thrust	
	SNR	PSNR	SNR	PSNR	SNR	PSNR	SNR	PSNR	SNR	PSNR
Mean filter algorithm	42.10	43.14	22.57	25.49	17.72	23.14	18.96	24.78	33.19	34.65
Improved mean filter algorithm	42.68	43.72	22.58	25.49	17.84	23.27	18.96	24.79	33.19	34.65

The improved algorithm enhances the denoising effect on the cutterhead rotation and tunneling speed. The SNR for the cutterhead rotation and tunneling speed increased by 0.58 and 0.12, respectively, while the PSNR increased by 0.58 and 0.13, respectively. The SNR for tunneling speed and cutterhead torque is below 20, indicating a relatively low signal-to-noise ratio. This suggests the presence of considerable noise in the data signal, likely influenced by changes in the geological conditions and factors such as vibrations and impacts in the mechanical system. This underscores the necessity of data denoising.

4.2.4. Data Normalization

When data features have significantly different magnitudes and exhibit a wide range of values, normalization methods are commonly employed to balance the importance of different features in prediction, thereby enhancing prediction accuracy. Two commonly used normalization methods are min–max normalization and Z-score normalization. The computation formulas for these two methods are as follows:

$$x^* = \frac{x - x_{\min}}{x_{\max} - x_{\min}}, \tag{13}$$

$$x^* = \frac{x - x_{mean}}{\delta}, \tag{14}$$

As shown in Formula (13), the max–min normalization method linearly transforms the original data to the range [0, 1], where x_{max} and x_{min} represent the maximum and minimum values of the column data, respectively. This method is suitable for data distributions with clear boundaries, mainly when scaling the data to a fixed range is necessary. Formula (14) demonstrated that the z-score normalization method normalizes the original data set to have a mean of 0 and a standard deviation of 1. Here, x_{mean} represents the mean of all sample data, and δ represents the standard deviation of all sample data. This method is applicable when data distribution lacks clear boundaries and is suitable for cases where comparing data from different features on the same scale is necessary. In the subsequent sections of this paper, different normalization methods will be applied to the training set, and the same method will then be used to normalize the validation set. The most suitable normalization method will be determined by comparing the model's predictive accuracy on the test set. It is important to note that considering the potential differences in data distribution between the training and validation sets, separate normalization processes for the training and validation sets are necessary to achieve optimal predictive performance.

4.3. Evaluation Index

In the predictive model, various evaluation metrics are employed, each emphasizing different aspects, and relying on a single metric lacks comprehensiveness in model assessment. To evaluate the model's performance, three distinct evaluation metrics are employed: Mean Absolute Error (MAE), Root Mean Square Error (RMSE), and Mean Absolute Percentage Error (MAPE), defined as follows:

$$MAE = \frac{1}{n}\sum_{i=1}^{n}\left|\hat{y}_i - y_i\right|, \tag{15}$$

$$RMSE = \sqrt{\frac{1}{n}\sum_{i=1}^{n}(\hat{y}_i - y_i)^2}, \tag{16}$$

$$MAPE = \frac{100\%}{n}\sum_{i=1}^{n}\left|\frac{\hat{y}_i - y_i}{y_i}\right|, \tag{17}$$

In the formula, y_i is the real value and $\hat{y}_i$ is the predicted value. MAE is a commonly used metric for assessing the accuracy of predictive models, reflecting the degree of closeness between actual and predicted values. Similarly, RMSE is employed to gauge predictive model accuracy. In contrast to MAE, RMSE imposes a higher penalty by squaring the errors, making it more sensitive to significant errors and outliers. MAPE is highly sensitive to extreme values, limiting its ability to handle outliers effectively. However, it expresses errors as a percentage, providing interpretability and comparability.

5. Comparative Analysis of Forecast Results

5.1. Model Parameter Presets

The research process processed 1–680 tunnel boring machine (TBM) shield tunneling parameters. Due to the enormous volume of raw data and variations in data quality, which could impact the model's predictive performance, samples were selected from different tunneling modes. For the TBM mode, the training set comprised rings 120 to 138, with the prediction segment spanning rings 139 to 146. Under the Earth Pressure Balance (EPB) mode, the training set included rings 435 to 444, and the prediction segment covered rings 445 to 450. Approximately 310,000 data points for each parameter—tunneling speed, total thrust, cutterhead torque, tunneling pressure, and cutterhead speed—were selected, totaling around 1.55 million data points. The constructed model is a real-time prediction model with a rectified linear unit (ReLU) as the activation function. The time step was set to 20, utilizing the past 20 time steps of data as input to predict future data. The number of epochs was set to 200. Considering that optimization algorithms can optimize multiple hyperparameters in the LSTM model, such as the number of LSTM layers, the number of LSTM units in hidden layers, learning rate, batch size, etc. Given the limitations of existing computer performance, it is advisable to choose critical hyperparameters. In line with previous research experience, this study selected two commonly optimized hyperparameters before including optimization algorithms: the number of LSTM hidden layer neurons and the learning rate. Prior to incorporating optimization algorithms, preliminary tuning was conducted for other hyperparameters. The pre-training hyperparameter settings are presented in Table 3, with 30 combinations tested during the preliminary tuning, and the relative errors are compared in Figure 7.

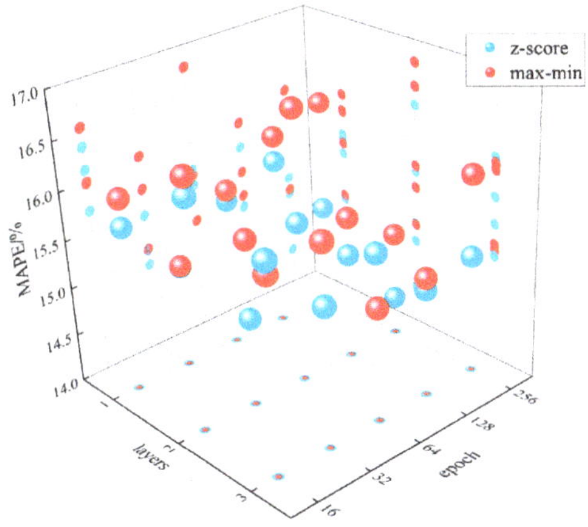

Figure 7. Comparison of relative errors in pre-training.

Table 3. Pre-training setup.

Name	Optimize Parameter Range
Normalization method	max-min normalization, z-score normalization
Batch size	16, 32, 64, 128, 256
Number of LSTM layers	1, 2, 3

After multiple pre-tuning experiments, the minimum average percentage error was 14.4%. Regarding the model's accuracy, using the z-score normalization method for data processing, setting the number of LSTM layers to 2, and configuring the batch size to 256 were more effective in training the model. Considering the dimensional differences in the model's input data, which could influence the model's output, the z-score normalization method was employed to transform data with different dimensions into a unified scale. This ensures relative balance in the importance of these data during model computations, mitigating interference caused by dimensional disparities. The network structure becomes more complex when the number of LSTM layers is excessive. Although increasing the number of layers can enhance model performance and help capture deeper features and relationships, this improvement is dynamic. With the increased model complexity, the training process may face the risk of gradient explosions.

Additionally, computational resources and training time would also increase. After a series of pre-training experiments, it was determined that a two-layer LSTM configuration is optimal for the model. Furthermore, setting the batch size to 256 improves the model's convergence speed and generalization ability. A larger batch size can simultaneously process more data, reducing noise and fluctuations. During training, using a batch size that is too small may lead to overfitting of the model to the training data, resulting in inadequate generalization of new data.

5.2. Excavation Rate Prediction Analysis

The four optimization algorithms set the optimization range for the number of neurons in the LSTM hidden layer between 10 and 100, and the optimization range for the learning rate is set between 0.001 and 0.1. At the same time, other parameters use the model's default values. Figure 8 illustrates the predictive results of the LSTM model on the test set under different optimization algorithms. In regression models, different evaluation metrics focus on different aspects. Evaluating with a single metric may overlook other important factors, so a comprehensive evaluation method using multiple metrics is necessary. Zorlu et al. [29]

proposed the ranking method as a commonly used multi-metric comprehensive evaluation method in 2008. This method involves ranking *N* models under the same evaluation metric, *i*. Then, the rankings of *m* different evaluation metrics for the same model are summed to obtain the multi-metric ranking for that model, as shown in Table 4.

Table 4. Prediction effect ranking analysis.

Predictive Model	Optimization Results		Training Time (s)	MAE		RMSE		MAPE		Overall Ranking	Rank Sort
	Units	Rate		EPB	TBM	EPB	TBM	EPB	TBM		
BO-LSTM	32	0.0094	367(1)	1.63(2)	3.51(1)	2.17(2)	4.49(1)	8.0%(1)	13.8%(1)	9	1
DE-LSTM	36	0.0085	800(3)	1.62(1)	3.93(3)	2.16(1)	4.88(3)	8.0%(1)	15.9%(3)	15	2
GA-LSTM	23	0.0086	1011(4)	1.82(4)	4.22(4)	2.38(4)	5.21(4)	9.3%(3)	16.9%(4)	27	4
PSO-LSTM	19	0.0150	396(2)	1.74(3)	3.77(2)	2.24(3)	4.83(2)	8.7%(2)	14.6%(2)	16	3

Note: ( ) is a single-column ranking.

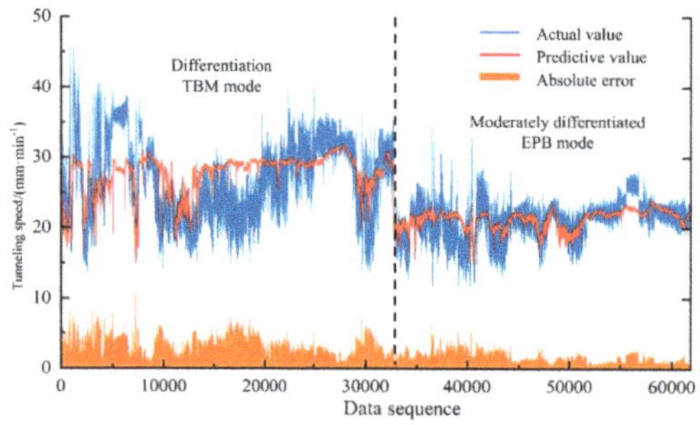

(**a**) BO-LSTM

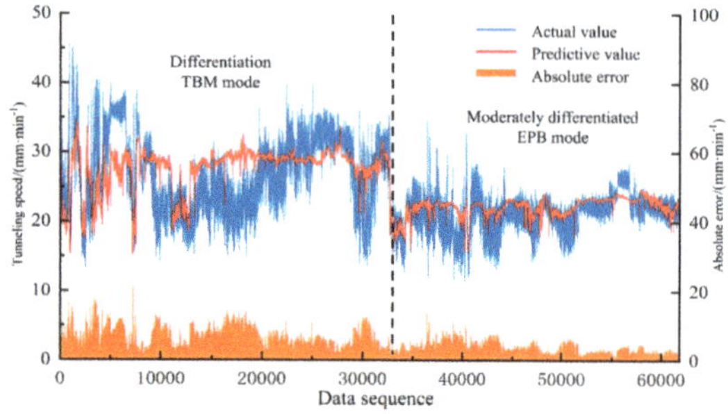

(**b**) DE-LSTM

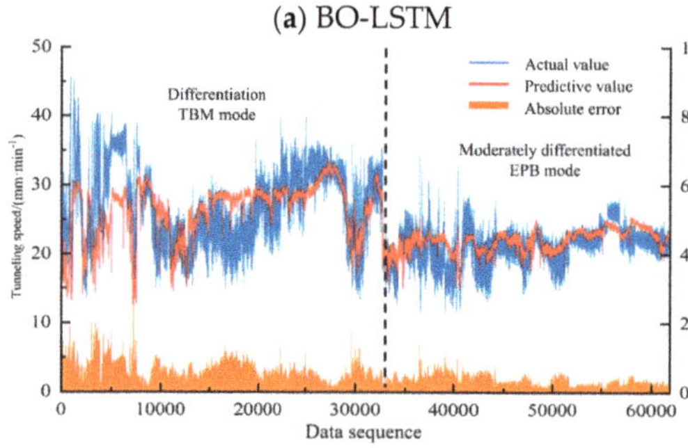

(**c**) GA-LSTM

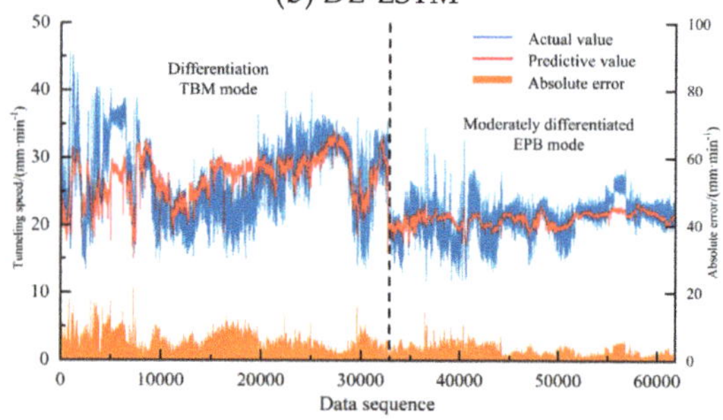

(**d**) PSO-LSTM

Figure 8. Prediction results of different LSTM models for the test set.

After comparing the performance of different optimization models on the test set, the conclusion can be drawn that, regarding the optimization of model hyperparameters, the number of neurons in the LSTM hidden layer or the learning rate should be manageable. Many neurons in the LSTM hidden layer can lead to an overly complex model, making it prone to overfitting. Conversely, more neurons are needed to ensure the model captures complex patterns and relationships, resulting in an insufficient utilization of information from the input data. A learning rate that is too small may slow down network training, requiring a longer time to converge, while a learning rate that is too large may lead to unstable training, skipping the optimal points and preventing convergence. Therefore, it is necessary to use hyperparameter optimization algorithms for multiple training iterations to achieve optimal predictive performance.

Examining the prediction curves for specific tunneling ring numbers reveals that the trends in the predicted data under both tunnel boring machine (TBM) and earth pressure balance (EPB) shield modes generally align with the actual excavation parameter curves. Under the TBM mode, the predicted curve for the weathered zone aligns well with the actual values, demonstrating overall good predictive performance. However, there is significant fluctuation in the local prediction segment from ring 139 to ring 141 due to unfavorable geological conditions characterized by complex rock changes in a poor geological area. This section underwent pre-reinforcement during construction, contributing to relatively poorer predictive results than other segments. In the EPB mode, the predicted curve for the composite soft and hard layer is smoother, reflecting stable changes in tunneling speed due to the softer rock characteristics, resulting in a closer match between predicted and actual values with more minor relative errors. Based on the overall ranking, BO-LSTM performs the best, with PSO-LSTM and DE-LSTM showing similar performance, while GA-LSTM performs the least favorably.

Regarding predictive accuracy, DE-LSTM and BO-LSTM are suitable for predicting relatively stable curve patterns. However, DE-LSTM performs poorly in predicting curves with significant fluctuations and suffers from the drawback of slow operation speed. Overall, BO-LSTM is a preferable model that can be applied to predict different geological environments. In EPB mode, the Mean Absolute Percentage Error (MAPE) prediction result is 8%, while in TBM mode, the MAPE prediction result is 13.8%. PSO-LSTM exhibits good overall performance, providing accurate predictions with a shorter runtime, and can be considered an alternative prediction model.

5.3. Multi-Algorithm Optimization Model Prediction Analysis

The models above exhibit favorable predictive performance on a fixed test set, but this does not necessarily imply the same performance on the global data set. In order to enhance the model's generalization capability, dropout algorithms and five-fold time series cross-validation are introduced to the BO-LSTM model, proposing a multi-algorithm-optimized tunneling speed prediction model. The validation set is partitioned based on the temporal characteristics of the overall data set, and the evaluation metric results are depicted in Figure 9.

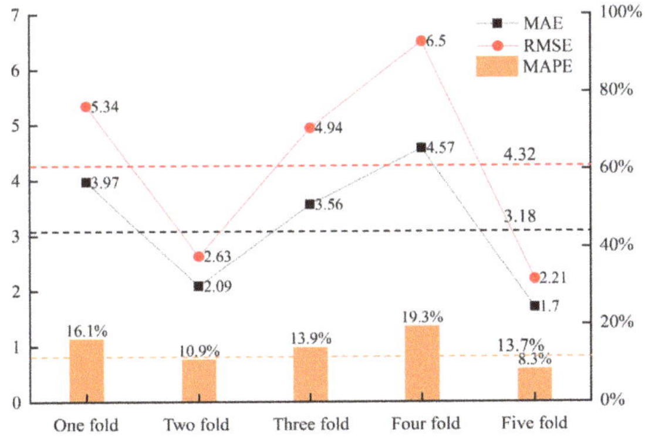

Figure 9. Time series cross-validation results.

In the above pictures, the dotted lines of different colors are the reference lines of the average value of the evaluation index. Via five rounds of time series cross-validation, the model demonstrates strong generalization capabilities, with a mean absolute error (MAE) of 3.18, root mean square error (RMSE) of 4.32, and mean absolute percentage error (MAPE) of 13.7%. In relatively stable geological layers, the MAPE prediction result is 8.3%. Even in challenging geological conditions where the excavation process experiences significant discrete fluctuations, the model maintains an 80% accuracy in predicting tunneling speed.

6. Conclusions

This paper relies on the Shenzhen Metro Line 13 Left Tunnel Project from Liuxiandong Station to Baimang Station. It utilizes geological parameters and real-time monitoring of excavation parameters in the complex strata to establish a prediction model for the excavation rate of EPB/TBM dual-mode tunnel boring machines (TBMs) based on multi-algorithm optimization and recurrent neural networks. The accuracy of the prediction results is analyzed, and the main conclusions are as follows:

(1) The combined impact of the dual-mode TBM excavation mode, TBM operating state parameters, and geological parameter variations was considered in establishing the tunnel boring machine (TBM) excavation data set. The model is more interpretable under actual working conditions, supporting the planning and control of TBM construction.

(2) Isolation Forest and an improved mean filtering algorithm were applied to handle TBM operating state parameters. This reduced the spatial variability of excavation parameters, making it easier for machine learning algorithms to learn the patterns of feature variations. The data set processed via denoising allows the construction of a more accurate prediction model for TBM tunneling speed.

(3) A prediction model for TBM tunneling speed was established using multi-algorithm optimization and recurrent neural networks. The model integrates Bayesian optimization algorithms, dropout algorithms, and time-series cross-validation, demonstrating strong generalization capabilities and operational efficiency. The lowest MAPE prediction result is 8.3%, with an average MAPE prediction result below 15%.

The established model primarily aims to explore the real-time operational patterns of tunneling speed in the dual-mode TBM excavation process under unsupervised training models and provide a foundational model for intelligent decision control. In practice, the data in the tunneling process is input into the model in real time, the model is trained, and the results are output in real time, which plays a guiding role in the field. In this study, the mechanical properties of rock have not been considered, and the model is not refined enough. In order to obtain a refined model that is more generalized and more in line with engineering practice, future research will consider more input of tunneling parameters and rock mechanics parameters.

Author Contributions: Conceptualization, T.Y. and T.W.; methodology, T.Y., X.H., G.S. and B.L.; software, T.W.; validation, B.L., S.L., G.S., H.S. and T.W.; formal analysis, X.P. and X.H.; investigation, B.L., X.P., H.S. and S.L.; resources, T.W.; data curation, T.Y., H.S. and S.L.; writing—original draft preparation, T.Y., G.S., T.W. and X.H.; writing—review and editing, T.W.; visualization, B.L., G.S. and X.P.; supervision, X.H. and B.L.; project administration, B.L., G.S., H.S. and S.L.; funding acquisition, X.H. and S.L. All authors have read and agreed to the published version of the manuscript.

Funding: It was supported by the National Natural Science Foundation of China Regional Joint Key Project (Grant No. U22A20234), Wuhan City Knowledge Innovation Special Project(Grant No. 2022010801010162), the National Natural Science Foundation of China (Grant No. 52074258), Hubei Provincial Key Research and Development Program (Grant No. 2021BCA133), Hubei Provincial Natural Science Foundation Outstanding Youth Project (Grant No. 2022CFA084) and China State Construction Engineering Corporation Science and Technology Research and Development Program Support (Grant No. CSCEC-2019-Z-19).

Data Availability Statement: Data are contained within the article.

Conflicts of Interest: Author Hongbing Shi was employed by China Construction Civil Infrastructure Corp., Ltd. Author Shaoran Liu was employed by China Construction South Investment Co., Ltd. Author Guangzu Sheng was employed by Wuhan Urban Construction Group Construction Management Co., Ltd. The funding sponsors had no role in the design of the study; in the collection, analyses, or interpretation of data; in the writing of the manuscript, and in the decision to publish the results.

References

1. Song, T.; Liu, C.; Chen, F.; Xu, Y.; Wang, S.; He, C. Research on the Selection and Tunneling Adaptability of Dual-mode Shield/TBM in Metro Tunnel. *J. Rallway Eng. Soc.* **2022**, *39*, 17.
2. Editorial Department of China Highway and Transport. Review on China's Traffic Tunnel Engineering Research: 2022. *China Highw. Transp.* **2022**, *35*, 1–40. [CrossRef]
3. Chen, F.; Huang, Z.; He, C.; Meng, Q.; Lai, X.; Liu, C.; Wang, S. Selection of reasonable mode conversion point for earth pressure/slurry dual-mode shield in round graval-mudstone composite strata. *China Civ. Eng. J.* **2021**, *54*, 48–57. [CrossRef]
4. He, C.; Chen, F.; Huang, Z.; Meng, Q.; Liu, C.; Wang, S. Adaptability and Excavation Parameters of Compound Strata with Dual-Mode Shield Tunneling. *Chin. J. Geotech. Eng.* **2021**, *43*, 43–52.
5. Xie, X.; Yang, C.; Wang, Q.; Zeng, L.; Hou, J.; Zhou, B. Settlement Analysis and Control Research on Shield Tunnel Crossing the Yangtze River Embankment in Nanjing and Yan Road Overpass. *Chin. J. Rock Mech. Eng.* **2021**, *40*, 3313–3322.
6. Shen, X.; Yuan, D.J.; Jin, D.L. Influence of shield attitude change on shield–soil interaction. *Appl. Sci.* **2019**, *9*, 1812. [CrossRef]
7. Jin, D.L.; Shen, X.; Yuan, D.J. Theoretical analysis of three-dimensional ground displacements induced by shield tunneling. *Appl. Math. Model.* **2020**, *79*, 85–105.
8. Chen, C.; Shi, P.; Jia, P.; Dong, M. Correlation analysis of shield driving parameters and structural deformation prediction based on MK-LSTM algorithm. *J. Jilin Univ. (Eng. Technol. Ed.)* **2023**, 1–10. [CrossRef]
9. Zhao, B.; Zhou, J.; Tan, Z. Variation of shield boring parameters and correlation analysis in mixed ground. *China Civ. Eng. J.* **2017**, *50*, 140–144.
10. Qin, R.; Li, X. Influence of shield driving speed on ground surface settlement in clay layers. *China Civ. Eng. J.* **2020**, *53*, 1–6.
11. Zhao, J.; Gong, Q.M.; Eisensten, Z. Tunnelling through a frequently changing and mixed ground: A case history in Singapore. *Tunn. Undergr. Space Technol.* **2007**, *22*, 388–400. [CrossRef]
12. Farrokh, E.; Rostami, J. Correlation of tunnel convergence with TBM operational parameters and chip size in the Ghomroud tunnel, Iran. *Tunn. Undergr. Space Technol.* **2008**, *23*, 700–710. [CrossRef]
13. Lüth, S.; Giese, R.; Otto, P.; Krüger, K.; Mielitz, S.; Bohlen, T.; Dickmann, T. Seismic investigations of the Piora Basin using S-wave conversions at the tunnel face of the Piora adit (Gotthard Base Tunnel). *Int. J. Rock Mech. Min. Sci.* **2008**, *45*, 86–93. [CrossRef]
14. Sapigni, M.; Berti, M.; Bethaz, E.; Busillo, A.; Cardone, G. TBM performance estimation using rock mass classifications. *Int. J. Rock Mech. Min. Sci.* **2002**, *39*, 771–788. [CrossRef]
15. Kahraman, S.; Bilgin, N.; Feridunoglu, C. Dominant rock properties affecting the penetration rate of percussive drills. *Int. J. Rock Mech. Min. Sci.* **2003**, *40*, 711–723. [CrossRef]
16. Hassanpour, J.; Rostami, J.; Khamehchiyan, M.; Bruland, A. Developing new equations for TBM performance prediction in carbonate-argillaceous rocks: A case history of Nowsood water conveyance tunn. *Geomech. Geoengin. Int. J.* **2009**, *4*, 287–297. [CrossRef]
17. Wang, H.; Fu, D. Mathematical and physical model and relationship analysis among various parameters in earth pressure balance shield tunneling. *China Civ. Eng. J.* **2006**, *39*, 86–90.
18. Zhang, Z.; Li, T.; Han, A.; Su, M.; Huang, X. Prediction model of shield tunneling speed and cutterhead torque in complex strata and its study on strata adaptability. *Tunn. Constr.* **2016**, *36*, 1449–1455.
19. Xu, Q. Study on the Simulated Model Test of Shield Machine' Working Parameters Applicable to Different Stratums and Its Theoretical Investigation. Ph.D. Thesis, Tongji University, Shanghai, China, 2006.
20. Xu, H.; Zhou, J.; Asteris, P.G.; Jahed Armaghani, D.; Tahir, M.M. Supervised machine learning techniques to the prediction of tunnel boring machine penetration rate. *Appl. Sci.* **2019**, *9*, 3715. [CrossRef]
21. Zhang, Z.; Li, X.; Ji, J. Prediction model of TBM excavation parameters based on LS-SVM. *J. Hohai Univ. (Nat. Sci.)* **2021**, *49*, 373–379.
22. Li, C.; Li, T.; Li, Z.; Zhan, J. Prediction and Analysis of Shield Tunneling Parameters in Compound Strata Based on BP Neural Network. *China Civ. Eng. J.* **2017**, *50*, 145–150.
23. Hou, S.; Liu, Y.; Zhang, K. Prediction of TBM Excavation Parameters Based on IPSO-BP Hybrid Model. *Chin. J. Rock Mech. Eng.* **2020**, *39*, 1648–1657.
24. Qiu, D.; Fu, K.; Xue, Y.; Li, Z.; Li, G.; Kong, F. LSTM tile-series prediction model for TBM tunneling parameters of deep-buried tunnels and application research. *J. Cent. S. Univ. (Sci. Technol.)* **2021**, *52*, 2646–2660.
25. *GB 50307-2012;* Code for Geotechnical Investigations of Urban Rail Transit. China Planning Press: Beijing, China, 2012.
26. Wang, S.; Wang, Y.; Li, X.; Liu, L.; Yin, T. Study of Standardized Pre-processing Method of TBM Tunnelling Data. *Mod. Tunn. Technol.* **2022**, *59*, 38–44.
27. Liu, F.T.; Ting, K.M.; Zhou, Z.H. Isolation-based anomaly detection. In Proceedings of the 2008 Eighth IEEE International Conference on Data Mining, Pisa, Italy, 15–19 December 2008; pp. 413–422.
28. Xiao, H.H.; Yang, W.K.; Hu, J.; Zhang, Y.P.; Jing, L.J.; Chen, Z.Y. Significance and methodology: Preprocessing the big data for machine learning on TBM performance. *Undergr. Space* **2022**, *7*, 680–701. [CrossRef]
29. Zorlu, K.; Gokceoglu, C.; Ocakoglu, F.; Nefeslioglu, H.A.; Acikalin, S.J.E.G. Prediction of uniaxial compressive strength of sandstones using petrography—Based models. *Eng. Geol.* **2008**, *96*, 141–158. [CrossRef]

applied sciences

MDPI

Article

Prediction of Utility Tunnel Performance in a Soft Foundation during an Operation Period Based on Deep Learning

Wei Gao [1,*], Shuangshuang Ge [1], Yangqinchu Gao [2] and Shuo Yuan [1]

[1] Key Laboratory of Ministry of Education for Geomechanics and Embankment Engineering, College of Civil and Transportation Engineering, Hohai University, Nanjing 210024, China
[2] Alameda College, Peralta Community College District, Oakland, CA 94606, USA
* Correspondence: wgaowh@163.com; Tel.: +86-025-83772037

Abstract: The underground utility tunnel in a soft foundation is generally affected by the serious disturbance of the vehicle load during the operation period. Therefore, in this study, for the typical utility tunnel engineering in Suqian City of Jiangsu Province, China, field tests were conducted to monitor the performance of the utility tunnel structure in a soft foundation affected by the ground traffic loads during the operation period. Based on the test results, the datasets whose number is 15,376, composed of the five main disturbance factors (four vehicle operating load parameters and one operating time parameter), and the corresponding two main structure responses (displacement and stress) have been constructed. Based on the obtained datasets, using the proposed new deep learning model called WO-DBN, in which the seven hyperparameters of a deep belief network (DBN) are determined by the whale optimization algorithm (WOA), the safety responses of the utility tunnel structure have been predicted. The results show that for the prediction results, the average absolute error for the displacement is 0.1604, and for the stress, it is 12.3726, which are not significant and can meet the requirement of the real engineering. Therefore, the deep learning model can accurately predict the performance of the utility tunnel structure under a vehicle load and other disturbances, and the model has good applicability.

Keywords: field test; safety response; tunnel structure; vehicle load; WO-DBN

Citation: Gao, W.; Ge, S.; Gao, Y.; Yuan, S. Prediction of Utility Tunnel Performance in a Soft Foundation during an Operation Period Based on Deep Learning. *Appl. Sci.* **2024**, *14*, 2334. https://doi.org/10.3390/app14062334

Academic Editor: Douglas O'Shaughnessy

Received: 8 February 2024
Revised: 7 March 2024
Accepted: 8 March 2024
Published: 10 March 2024

1. Introduction

The underground utility tunnel is one kind of underground structure in which there are more than one utility pipe or cable. For the utility tunnel, it can not only reduce the excavation needs and costs, and consequently reduce traffic congestion caused by excavation, but also provide enough shallow underground space to avoid utility interference and enough space for new utilities, which meets the sustainable development requirements of underground space. Therefore, in light of urban development and the growing need for public facilities in big cities, the utility tunnel has been rapidly developed in the world. The first utility tunnel was built in France in 1850 [1]. However, the development of utility tunnels in China was very late, and the first one was built in Shanghai in 1994 [2]. Actually, the rapid development of the utility tunnel in China is from the 21st century, and recently, China has made big progress in utility tunnel construction as an essential urban infrastructure development. In China, the eastern coastal area is the economically developed region, and there are many big cities. In this area, the soft soil layers are widely distributed. Therefore, in China, many utility tunnels are constructed in the soft foundation. As a kind of urban underground engineering that is shallowly buried under the road, pavement and green areas, the utility tunnel in a soft foundation will be seriously affected by the ground traffic loads during the operation period. Because the mechanical properties of soft soils are very poor, under the loading disturbance, large deformation will be caused for the soft soils. Therefore, it is very hard work to determine the long-term mechanical status of soft soil after a disturbance, which will cause significant difficulties for the safety operation of the utility tunnels. Thus, it is a very important work to analyze the safety status of utility tunnels constructed in a soft foundation during the operation period.

Nowadays, there are many works on the safety analysis of utility tunnels during operation periods affected by external loads similar to ground traffic loads. These studies include two main types, which are experimental studies and numerical studies. For the experimental studies, based on the in-field explosion experiments, the protective performance of utility tunnels under a ground accidental explosion has been studied, which includes two types of utility tunnels (steel bars-reinforced concrete utility tunnel and basalt fiber reinforced polymer bars-reinforced utility tunnel) [3,4]. By a similar ground surface explosion experiment, the structural dynamic responses of a double-box-reinforced concrete utility tunnel buried in calcareous sand have been analyzed [5]. Moreover, based on a quasi-static experiment, the cross-sectional mechanical behavior of a prefabricated multi-cabin-reinforced concrete utility tunnel under free-field racking deformation by earthquake action has been studied [6]. Also, based on quasi-static load tests on the full-size model, the mechanical behavior of a prefabricated concrete utility tunnel and the seismic behavior of top joints for the hybrid precast utility tunnel consisting of the precast composite top slab and double-skin sidewalls with reserved rebar have been researched [7,8]. Based on the quasi-static cyclic tests for simulating earthquake action, the seismic performance of the precast concrete composite walls of utility tunnels with a grouting-sleeve connection out-of-plane has been studied [9]. By using the low-cyclic-loading tests conducted on the bottom joints, the seismic behavior of the prefabricated utility tunnel created by transferring reserved rebars from the double-skin concrete sidewalls to the bottom slab has been investigated [10]. Moreover, the shaking table model test is one widely used method for studying the dynamic response of utility tunnels, and there are many works involving this experimental method. For example, by employing a multi-shaking table array system for a 1/30-scale utility tunnel model, the transverse response of a reinforced concrete utility tunnel under near-fault ground motions with and without a velocity pulse has been studied [11]. And by a series of shaking table model tests, the dynamic responses of a prefabricated corrugated steel utility tunnel under earthquake action have been studied [12]. To ensure the sustainable and safe operation of a cast-in-place concrete utility tunnel over a design life of 100 years, the seismic response pattern of a utility tunnel in a layered liquefiable site by earthquake excitation has been studied by the shaking table tests [13]. By using the 1/20 scaled shaking table tests to simulate the earthquake excitation, the seismic performance of a prefabricated T-shaped cross micro-concrete utility tunnel has been studied [14]. And in a study [15], the dynamic response of a reinforced concrete utility tunnel under earthquake seismic action has been investigated through the shaking table tests. Moreover, in the study [16], the seismic failure mechanism of a prefabricated corrugated steel utility tunnel on liquefiable ground under earthquake excitation was investigated by the shaking table test, too. From the abovementioned studies, it can be found that two types of dynamic loads (ground surface explosion and earthquake excitation) have been simulated in these experimental studies. Most of them studied the dynamic response of utility tunnels by the influence of a short time disturbance. Therefore, there is a big gap between those experimental studies and the study on a utility tunnel affected by ground traffic loads during the operation period, which is a long-term disturbance.

For the numerical studies, by using the two-dimensional (2D) numerical model constructed by the general finite element software (ABAQUS 6.12), the response characteristics of a concrete rectangular tunnel that is similar to a utility tunnel in soft soil subjected to transversal ground shaking have been analyzed [17]. Using the 2D numerical model constructed by the same software, too, the dynamic behavior of the pipe-arched main compartment for the corrugated steel utility tunnel has been investigated [18]. By using the general finite element software (ABAQUS 6.10), a three-dimensional (3D) numerical model has been constructed for analyzing the mechanical performance of the perforated steel plate-reinforced utility tunnel with the use of ultra-high-performance concrete and an engineered cementitious composite under dynamic loads [19]. Also based on the 3D real-site numerical model constructed by the same software, the seismic responses of prefabricated utility tunnels with different incident angles of P-waves were analyzed [20]. Based on the 3D rigorous numerical model developed in the platform of Displacement Analyzer (DIANA), the cross-sectional mechanical behavior of a prefabricated multi-cabin-reinforced concrete utility tunnel under free-field racking deformation by earthquake action has been studied [6]. Using the 3D finite element model of the explicit dynamic software (LS-DYNA R8.0), the blast performance of a reinforced concrete utility tunnel subjected to a ground surface explosion has been investigated [21]. Moreover, by using the 2D discrete element numerical model of Particle Flow Code (PFC) software (PFC 2D 5.0), the dynamic response of the reinforced concrete utility tunnel in soft soil of a horizontal non-homogeneous site has been analyzed [15]. And, to investigate the seismic performance of prefabricated corrugated steel utility tunnels in liquefiable soil, the numerical plane model constructed by the finite difference software of Fast Lagrangian Analysis of Continua (FLAC) 3D (FLAC 3.0) has been used [16]. In most of the abovementioned studies, the effect of earthquake action on the mechanical behavior of utility tunnels has been investigated, except one in which the effect of a ground surface explosion has been considered. Therefore, no study is on the long-term effect of the ground traffic loads during the operation period on the performance of the utility tunnel.

Currently, with the development of artificial intelligence and information technology, some new methods have been used to analyze the safety operation of utility tunnels. For example, a building information modeling (BIM)-based framework for the operation and maintenance of utility tunnels has been constructed, which includes three modules (BIM model, database, and monitoring system) [22,23]. Based on the past monitoring data, a multi-layer long short-term memory and recurrent neural network architecture has been proposed for forecasting the temperature and relative humidity inside utility tunnels [24]. Moreover, based on the selected information, the risk assessment of utility tunnels can be conducted using different methods, such as an integrated model based on dynamic hazard scenario identification, Bayesian networks and risk analysis [25] and risk interaction-based deep learning [26]. Lastly, based on the collected data from previous studies, by using the traditional artificial neural network (ANN), the damage of a tunnel deeply buried in a mountain affected by an earthquake and landslide has been predicted [27]. It can be found that, in those studies, although the safety operation of a utility tunnel or a tunnel affected by some factors using monitored information has been analyzed, the used information is collected from other studies. Therefore, only the research framework or simple method have been proposed, and a prediction of the safety of a utility tunnel affected by ground traffic loads during the operation period has not been conducted.

In summarizing the previous studies, it can be found that the current studies are on the analysis of the dynamic mechanical behavior of utility tunnels affected by short time loads such as ground surface explosions or earthquake excitation, and the new methods are only used for the system analysis of the utility tunnel operation based on the collected information from other studies. Therefore, there are no studies on the prediction and evaluation of the safety performance of utility tunnels under long-term disturbance by ground vehicle loads during the operation period, especially the studies using the new methods, such as deep learning. To address this gap, here, field tests have been conducted for one real utility tunnel to obtain the big data information about the safety performance of a utility tunnel under disturbance by ground loads during the operation period. And based on a deep learning method–deep belief network (DBN) and the swarm intelligence optimization method–whale optimization algorithm (WOA), one new deep learning model–whale optimization deep belief network (WO-DBN) has been proposed. Lastly, based on the collected big data information, and by using the new deep learning model, the prediction of the performance of a utility tunnel under disturbance by ground vehicle loads during the operation period has been conducted.

The novelty and aim of this study are summarized as follows: (1) it is the first study on the prediction of the safety performance of a utility tunnel under long-term disturbance by ground vehicle loads during the operation period, (2) it is a new study on monitoring the long-term performance of utility tunnels by field tests, (3) it is the first study on the selection and analysis of big data to describe the safety of utility tunnels under long-term disturbances during the operation period and (4) it is the first study on an application deep learning method for predicting the safety performance of a utility tunnel under long-term disturbance during the operation period.

The rest of this paper is as follows. Section 2 gives the methodologies, including DBN, WOA, WO-DBN and field test methods. The results are provided in Section 3. Section 4 includes the discussions. Finally, the main conclusions of this study are summarized in Section 5.

2. Methodologies

2.1. Deep Belief Network (DBN)

The DBN has the special network structure represented by the multiple-layer restricted Boltzmann machines (RBMs) [28], as shown in Figure 1. For its excellent data analysis ability, DBN has become one of the most commonly used deep learning methods, and it is widely used in big data prediction, data mining, recognition and classification [28,29]. For DBN, the unsupervised pre-training method is applied to learn the input features by the multiple-layer RBMs. Multiple-layer RBMs will increase the upper bound of the log-likelihood, which can greatly enhance the data mining ability and improve the prediction accuracy. In comparison with ANN, in DBN, the initial weights are learned from the structure of the input data, which is closer to the global optimum. The learned weights are used as the initial values of other networks with the same structure, and thus, the drawbacks of initialization parameters falling into the local optima and a long training time can be avoided. Moreover, for its complex network structure, the DBN can treat big data problems. Nowadays, DBNs have already been used to solve the big data problems in geotechnical engineering [30]. Therefore, in this study, the DBN is applied to the performance prediction of a utility tunnel in a soft foundation during the operation period.

For one RBM in the DBN (Figure 2), there are only two layers, which are the visible layer ($\boldsymbol{v}$) and hidden layer ($\boldsymbol{h}$). Therefore, the visible layer of the first RBM is the first layer of the DBN, for which the original data are inputted. The hidden layer of the first RBM is the second layer of the DBN and is also the visible layer of the second RBM. The DBN including multiple hidden layers can be constructed. For multiple RBMs in the DBN, the DBN has a strong learning ability which can extract deep features of the complex data [31].

As a typical deep learning method, the training of a DBN is complex and generally includes two parts, unsupervised learning and supervised learning, corresponding to two stages, namely, forward pre-training and reverse fine-tuning. The training process of the DBN is shown in Figure 3.

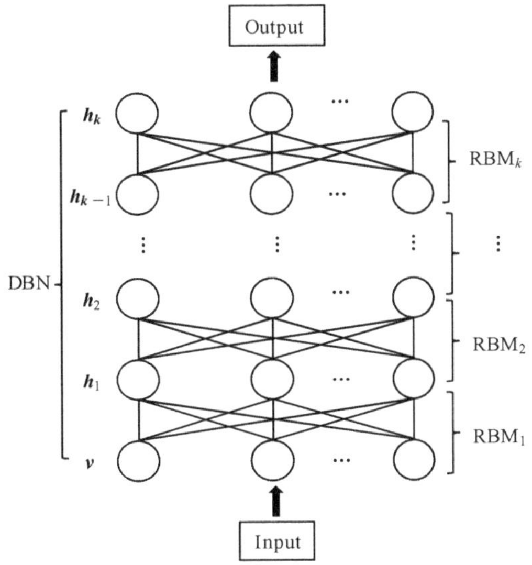

Figure 1. Structure of the DBN.

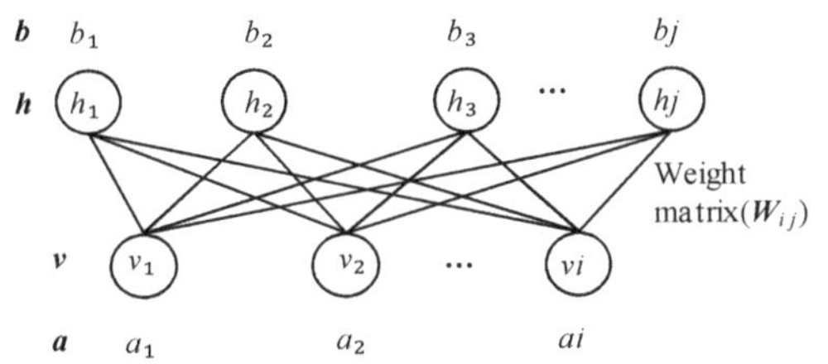

Figure 2. Structure of the RBM.

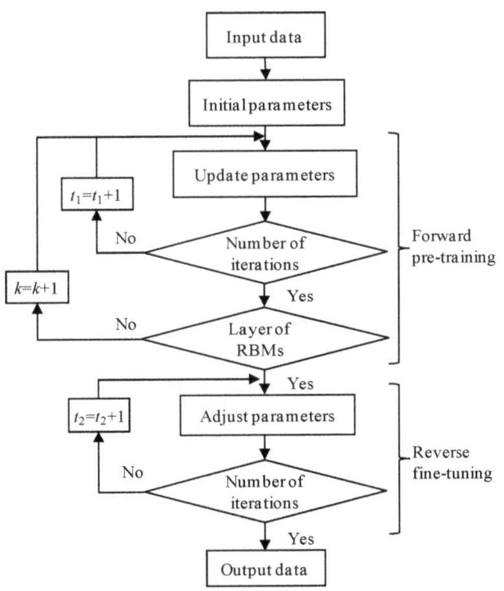

Figure 3. Training process of the DBN.

The detailed steps are as follows:

(1) Parameter initialization

First, for the first RBM, its model parameters $\boldsymbol{\theta}= (\boldsymbol{W}, \boldsymbol{a}, \boldsymbol{b})$ should be initialized randomly, which include the connection weights $\boldsymbol{W} = (w_{ij}) \in (0,1)$ between the visible layer v_i and the hidden layer $\boldsymbol{h}_j$ and the biases of visible and hidden layers ($\boldsymbol{a} = (a_1, a_2, \cdots, a_i)^T$ and $\boldsymbol{b} = (b_1, b_2, \cdots, b_j)^T$). The other initial parameters should be determined by the control variable method [29], including the number of RBM layers K, the numbers of neurons in hidden layers n_k (k = 1, K), the learning rate η and the maximum iteration numbers of pre-training and fine-tuning (T_1 and T_2).

(2) Forward pre-training

In this stage, the network is trained by the unsupervised learning method. Therefore, the input data have no labels, and the model parameters of DBN are updated by the greedy layer-by-layer learning algorithm [31]. That is to say, the first RBM is trained by the original input data, and thus, its model parameters can be obtained. Then, the model parameters of the first RBM remain unchanged, and its output data are used to train the second RBM until the last RBM is trained. The flow chart of the greedy layer-by-layer learning algorithm is shown in Figure 4, whose detail process is as follows:

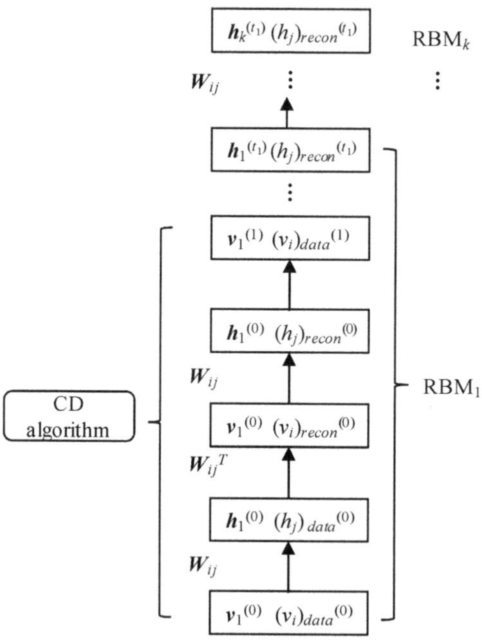

Figure 4. Flow chart of the greedy layer-by-layer learning algorithm.

First, for the first RBM, which is trained by the contrastive divergence (CD) algorithm, its outputs (the hidden layer, reconstructed visible layer and reconstructed hidden layer) can be obtained using the following equations:

$$(h_j)_{data}^{(t_1)} = p(h_j^{(t_1)} = 1 \big| \boldsymbol{v}^{(t_1)}) = \sigma(b_j + \sum_i (v_i)_{data}^{(t_1)} w_{ij}), \tag{1}$$

$$(v_i)_{recon}^{(t_1)} = p(v_i^{(t_1)} = 1 \big| \boldsymbol{h}^{(t_1)}) = \sigma(a_i + \sum_j (h_j)_{data}^{(t_1)} w_{ij}), \tag{2}$$

$$(h_j)_{recon}^{(t_1)} = p(h_j^{(t_1)} = 1 \big| \boldsymbol{v}^{(t_1)}) = \sigma(b_j + \sum_i (v_i)_{recon}^{(t_1)} w_{ij}), \tag{3}$$

where $(h_j)^{(t_1)}_{data}$ represents the calculated output of the hidden layer neuron (h_j) at the $t1$th update, $(v_i)^{(t_1)}_{data}$ represents the input of the visible layer neuron (v_i) at the $t1$th update. $(v_i)^{(t_1)}_{recon}$ and $(h_j)^{(t_1)}_{recon}$ represents the reconstructed output of the visible layer neuron (v_i) and hidden layer neuron (h_j) at the $t1$th update, respectively. $p(h_j^{(t_1)} = 1 \middle| v^{(t_1)})$ represents the probability that the hidden layer is activated when it is updated. $\sigma(x)$ is the logistic sigmoid function. The probability that the hidden layer, the reconstructed visible layer and the reconstructed hidden layer are activated when they are updated is the corresponding output value.

After the visible layer and hidden layer are reconstructed, the model parameters of this RBM are updated as

$$\Delta w_{ij} = \eta\left(\left\langle (v_i)^{(t_1)}_{data}(h_j)^{(t_1)}_{data}\right\rangle - \left\langle (v_i)^{(t_1)}_{recon}(h_j)^{(t_1)}_{recon}\right\rangle\right), \tag{4}$$

$$\Delta a_i = \eta\left((v_i)^{(t_1)}_{data} - (v_i)^{(t_1)}_{recon}\right), \tag{5}$$

$$\Delta b_j = \eta\left((h_j)^{(t_1)}_{data} - (h_j)^{(t_1)}_{recon}\right), \tag{6}$$

where Δw_{ij} is the update values of the weight, Δa_j is the bias of the visible layer, Δb_j is the bias of the hidden layer and η is the learning rate. Through the above formula, the initial weights of the model are learned from the structure of the input data in the pre-training stage, which can greatly improve the performance of the DBN.

Then, it is judged whether the number of pre-trainings (t_1) reaches the maximum number of iterations (T_1). If the number of pre-trainings (t_1) does not reach the maximum number of iterations (T_1), the updated model parameters are used as the initial value to continue to update the first RBM. Until the iteration number of pre-training (t_1) reaches the maximum number of iterations (T_1), the pre-training of the first RBM ends. The model parameters of this RBM are used for the visible layer of the next RBM.

Finally, the CD algorithm is used to train the high-level RBMs sequentially. Until the model parameters of all RBMs are updated, the pre-training of the DBN is over. The greedy layer-by-layer learning algorithm optimizes the weight of a DBN that is linear to network size and depth in time complexity. Moreover, since the approximation of the likelihood function only requires one step in this algorithm, the training time is significantly reduced.

It can be found that, using the greedy layer-by-layer learning algorithm, the training of the DBN can be simplified to the training of multiple RBMs. Therefore, for the DBN, the computing process can be simplified, and the training speed can be improved. Moreover, the data in this stage are not labeled, and thus, the data mining ability of the DBN has been improved too.

(3) Reverse fine-tuning

In this stage, the network is trained by the supervised learning method. Therefore, the data have labels. In the training, based on the errors between the computed output and the real output values, the weight and bias of the network have been updated by using the back propagation method. Therefore, in this stage, to further optimize the DBN, its parameters are fine-tuned after pre-training, whose details are as follows.

In this stage, the DBN is trained by a contrastive version of the wake–sleep algorithm called the updown algorithm [31]. Here, for the top layer, the mean square error (MSE) can be obtained as

$$\text{MSE} = \frac{1}{J}\sum_{j=1}^{J}(\rho_j - O_j)^2, \tag{7}$$

where J is the neuron number of the top layer. ρ_j and o_j represent the computed and real output values of the top layer of the jth neuron, respectively.

Then, by using the gradient descent method, the connection weights and bias are continuously updated based on the computed MSE, which is propagated backward layer by layer. Finally, when the fine-tuning iteration whose number is t_2 reaches the maximum iteration whose number is T_2, the reverse fine-tuning is over.

In summary, for the DBN composed of multiple RBMs, multiple-layer RBMs will increase the upper bound of the log-likelihood, which means a stronger learning ability. The training process is divided into two parts, which are an unsupervised pre-training procedure performed in a bottom-up manner and a supervised up-down fine-tuning process. The pre-training process can be regarded as feature learning, in which the better initial values of weights can be determined, and then, the updown algorithm is to adjust the whole network.

2.2. Whale Optimization Algorithm (WOA)

A WOA is a typical swarm intelligence optimization method [32] for simulating the efficient hunting behavior of humpback whale populations. Whales use spiral bubble nets to hunt their prey in order to achieve optimal results, and thus, whale herds possess extremely high swarm intelligence. In the WOA, the position of each whale represents one solution of the optimization problem. During whale hunting, whales exhibit two behaviors. One is the shrinking circle movement, in which all whales move towards other whales, and another is spiral bubble hunting, in which the whales swim in a circular motion and spray bubbles to drive their prey. The whales randomly choose these two behaviors for hunting. That is to say, whales will randomly choose whether to swim towards the whale in the optimal position or randomly choose a whale as their target and approach it. Therefore, there are three modes in the behavior patterns of whales, which are shrinking circle movement, spiral bubble hunting, and the exploration of prey. Using those three modes, the whales continuously update their position until they reach the optimal position.

For the above complex behavior patterns of whales, there are excellent exploration and exploitation abilities for the WOA. In the WOA, the current optimal solution is assumed to be the target prey. In the exploration phase, whales swim to a randomly selected individual, that is, a global search is performed. However, in the exploitation phase, whales swim around the target prey in a shrinking circle and along a spiral-shaped path simultaneously, that is, a local refinement search is performed. Moreover, a probability of 50% was applied to select a shrinking encircling mechanism or spiral model for updating the whale position in the WOA.

The flow chart of the WOA is shown in Figure 5.

The details of the WOA are as follows.

(1) The initial values of the parameters should be determined, which include the solution dimension D, whale population size N, maximum iteration number T_{max}, and spiral size constant b. Moreover, the initial positions of whales $\vec{Z_i} = (z_i^1, z_i^2, \cdots, z_i^D)$, $i = 1, 2, \cdots, N$ are assigned randomly.

(2) The optimal individual and its position are recorded, which is represented as $\vec{Z^*} = (z^1, z^2, \cdots, z^D)$. The adaptation of each individual to the environment is assessed.

(3) If the iteration number t reaches the maximum iteration number T_{max}, the process is over and the optimal solution should be outputted. Otherwise, the process continues.

(4) The probability p and coefficients vector $\vec{r}$ are assigned randomly. According to the probability p and the adaptive variation of the search vector $\vec{A}$, the individual position is updated as follows:

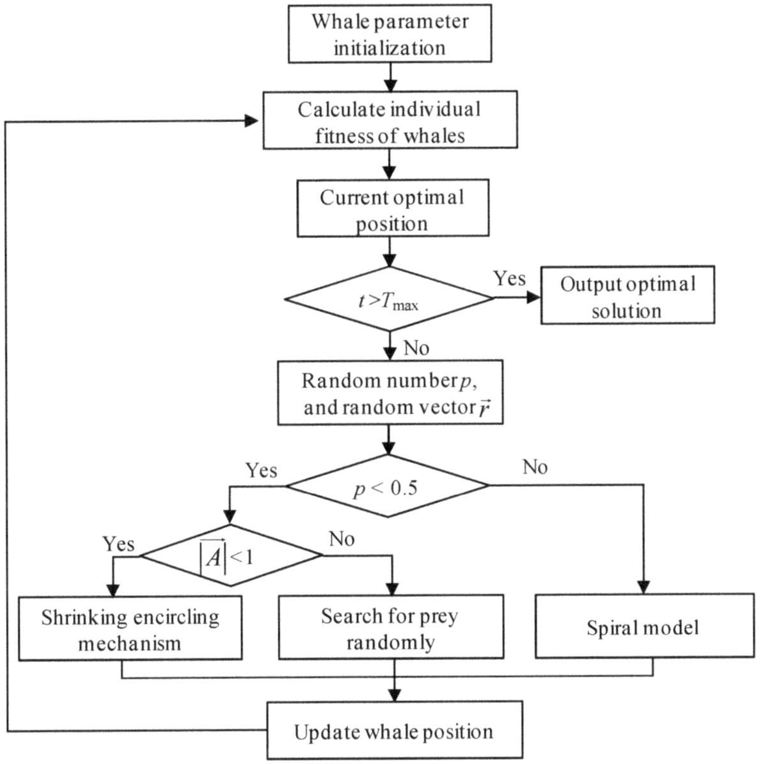

Figure 5. Flow chart of the WOA.

When $p < 50\%$ and $|\vec{A}| < 1$, the shrinking encircling mechanism is used to update the individual position, which is expressed as

$$\vec{Z}(t+1) = \vec{Z}^*(t) - \vec{A} \cdot \vec{B}, \quad (8)$$

where $\vec{Z}(t+1)$ is the next position vector and $\vec{Z}^*(t)$ is the current optimal position vector. It is worth noting that if there is a better solution, it needs to be updated in each iteration. t is the current iteration number, and

$$\vec{A} = 2\vec{a} \cdot \vec{r} - \vec{a}, \quad (9)$$

$$\vec{B} = \left| \vec{C} \cdot \vec{Z}^*(t) - \vec{Z}(t) \right|, \quad (10)$$

where $\vec{A}$ is one coefficient vector, $\vec{a}$ is a vector that decreases linearly from two to zero throughout the computing process, $\vec{r}$ is a random vector in the range of [0, 1], $\vec{Z}(t)$ is the current position vector and $\vec{C}$ is another coefficient vector which can be expressed as

$$\vec{C} = 2\vec{r}, \quad (11)$$

When $p < 50\%$ and $|\vec{A}| \geq 1$, the random search is used to update the position, which is expressed as

$$\vec{Z}(t+1) = \vec{Z}_{rand} - \vec{A} \cdot \left| \vec{C} \cdot \vec{Z}_{rand} - \vec{Z}(t) \right|, \tag{12}$$

where $\vec{Z}_{rand}$ is a random position vector. Here, $|\vec{A}| \geq 1$ emphasizes that the exploration shows that the WOA algorithm performs a global search.

Lastly, when $p \geq 50\%$, the spiral upward movement is used to update the position, which is expressed as

$$\vec{Z}(t+1) = \vec{Z^*}(t) + \vec{B'} \cdot e^{bl} \cdot \cos(2\pi l), \tag{13}$$

where b represents a constant spiral size, l is a random number in the range of $[-1, 1]$ and

$$\vec{B'} = \left| \vec{Z^*}(t) - \vec{Z}(t) \right|, \tag{14}$$

After the position updating operation, the process returns to step (2).

The previous study [32] shows that, because the exploitation and exploration phases have been conducted separately and in almost half of the iterations each, the WOA can solve the global optimization easily and with a high convergence speed. Moreover, the adaptive change in the $\vec{A}$ allows the WOA to make a smooth transition between exploitation and exploration. It is worth noting that the two main internal parameters, $\vec{A}$ and $\vec{C}$, need to be adjusted in the WOA.

2.3. *Whale Optimization Deep Belief Network (WO-DBN)*

For a complex structure of a DBN which is composed of multiple-layer RBMs, there are many hyperparameters in the DBN that should be determined beforehand. Generally, those hyperparameters are determined by the control variable method. However, using the control variable method, there are three main shortcomings [31], which are as follows: (1) For many search parameters, it is hard work to implement a control variable method. (2) For a small search range, the searched optimal value is only the result in this small range, rather than the real optimal one. (3) The theoretical basis of the control variable method is lacking. To solve those shortcomings, it is a very suitable way to use the optimization method to select the suitable hyperparameters of a DBN. As one good global optimization method, the WOA can be used to select the optimal hyperparameters of the machine learning model [33]. Therefore, in this study, the WOA is used to determine the suitable hyperparameters of the DBN, and a new deep learning model called a whale optimization deep belief network (WO-DBN) is proposed, whose flow chart is shown in Figure 6.

It must be noted that, in this deep learning model, based on the experience and our tests, there are four hidden layers. The ReLU activation function, which only requires a threshold to obtain the activation value and has a very fast convergence speed, is used [34].

The main steps of the new deep learning model are as follows:

(1) Based on the experience, the searched ranges of the DBN hyperparameters that should be determined are provided. The hyperparameters include the neuron numbers for the first, second, third and fourth hidden layers (n_1, n_2, n_3 and n_4), the learning rate η and the maximum iteration numbers for pre-training and fine-tuning (T_1 and T_2), which are taken as the whale position $\vec{Z}$, represented as,

$$\vec{Z} = [n_1, n_2, n_3, n_4, \eta, T_1, T_2], \tag{15}$$

(2) Based on the experiments and our tests, the initial parameters of the WOA are given, including the maximum iteration number T_{max}, the whale population size N and the spiral size constant b. According to the number of optimized hyperparameters, the dimension D is determined to be 7. Moreover, the initial whale population is generated randomly.

(3) According to the solved problem, the fitness function is defined as

$$f = \frac{\sum_{i=1}^{M} \left(y_i - y_i'\right)^2}{M}, \tag{16}$$

where y_i' and y_i are the computed and real values of the ith sample, respectively, and M is the number of samples.

By using this fitness function, the fitness values of whale individuals can be obtained. According to its fitness value, the most adaptive individual is taken as the prey, whose position is selected as the target position.

(4) Based on the computing operations in the fourth step of the WOA, the positions of whale individuals have been updated, and the new whale population is generated.

(5) If the iteration number reaches the maximum iteration number T_{max}, the process is over. Otherwise, the process returns to step (3).

(6) The whale individual with the best position is selected and outputted, whose position is the optimized DBN hyperparameters. And by using those hyperparameters, the suitable DBN model can be obtained.

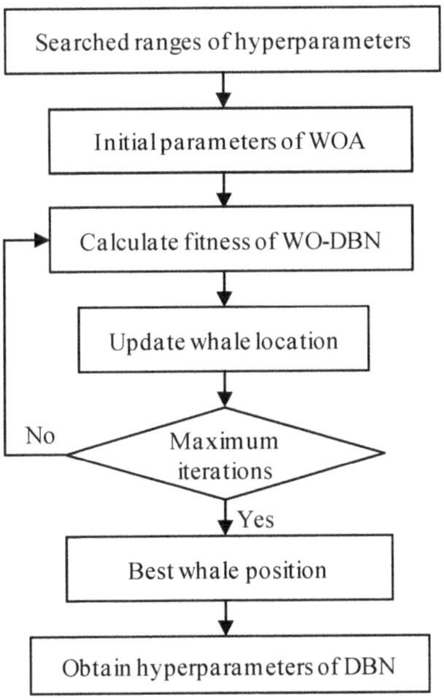

Figure 6. Flow chart of the WO-DBN.

2.4. Filed Test

The utility tunnel engineering is in the high-speed railway business district of Suqian City, Jiangsu Province of China. For this engineering, one section is along the Guangzhou Road, whose width is 40 m and length is 1851 m. The single cabin structure of this utility tunnel should meet the entering requirement of power, communication and water supply pipelines in the tunnel. The utility tunnel is under the central green isolation zone, as shown in Figure 7.

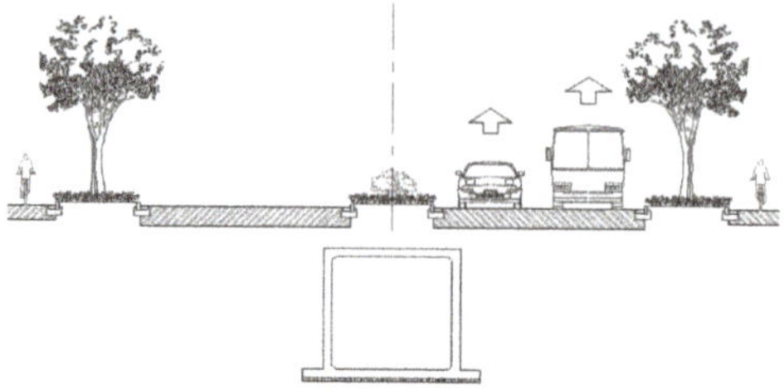

Figure 7. Layout of the utility tunnel.

For this utility tunnel engineering, the main tunnel structure is constructed by the segmented open excavation and cast-in-place and is one long strip single-cabin-reinforced concrete structure. The length of one segment is 25 m. The width and height of a standard section for this utility tunnel structure are 3 m and 3.5 m, and those for the directly buried outlet shaft are 6 m and 5.65 m. For a large size and complex structure for the section of the directly buried outlet shaft, in this study, the utility tunnel of the directly buried outlet shaft between the pile number of GZK 0 + 820 to GZK 0 + 900 is selected as the research object. For this utility tunnel of the directly buried outlet shaft, the covered depth is only about 2 m.

The site of this utility tunnel engineering belongs to the Xuhuai Yellow River alluvial plain geomorphic area, which is a unit of the abandoned ancient Yellow River channel geomorphology and has a flat terrain. By using the geological investigation method, such as the drilling method, in the report of engineering geological exploration, the engineering geological cross-section of this utility tunnel engineering can be obtained, which is shown in Figure 8.

From Figure 8 and according to the report of engineering geological exploration, there are six soil layers, for which the basic engineering geological conditions are summarized as follows.

For layer 1, it is the plain fill soil layer with the color of yellow or grayish yellow, it is loose and its main composition is silt, including the plant roots and stems, and locally contains crushed stones, bricks, etc. For its poor mechanical properties, this layer has been removed.

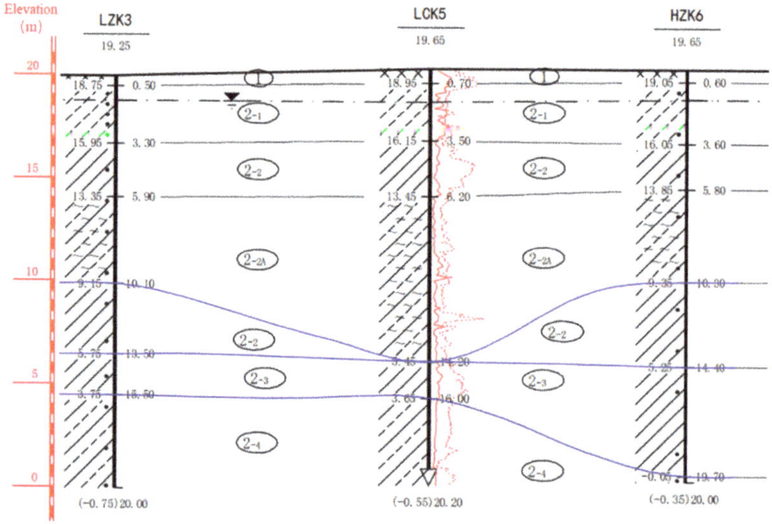

Figure 8. Cross-section of engineering geology.

For layer 2-1, it is the silt layer with the color of grayish yellow, and it is wet and slightly dense. For this layer, the thin layers of soft plastic clay are mixed in local layers, and there is a feeling of sand when rubbing hands with a rapid shaking response. This layer has low toughness and dry strength with moderate compressibility.

For layer 2-2, it is the clay layer, whose color is from grayish brown to yellowish gray with soft plastic and local plastic. For this layer, the thin layers of silt are partially mixed with the glossy cut surface. Moreover, this soil layer has high toughness and dry strength, with no shaking response and medium to high compressibility.

For layer 2-2A, it is the sullage silty clay layer, whose color is from grayish brown to yellowish gray, too. For this layer with a mechanical property from soft plastic to flow plastic, the thin layers of silt are also mixed with the slight glossy cut surface. Moreover, this soil layer has medium toughness and dry strength, with no shaking response and high compressibility.

For layer 2-3, which is the clay layer whose color is grayish brown or yellowish gray, it is from soft to plastic, with the glossy cut surface including a small number of iron manganese spots. This layer has high toughness and dry strength with no shaking response and medium compressibility.

For layer 2-4, it is the silty clay layer whose color is from gray yellow to brown yellow, and the mechanical property is from plastic to hard plastic. Moreover, the local layers of this layer are clay including iron manganese spots and a small number of calcareous nodules. As for layer 2-3, this layer has high toughness and dry strength, with no shaking response and medium compressibility, too.

From the above analysis, it can be found that, for the soil layers in which the utility tunnel is located, there are mainly two parts, which are soft soil in lower layers and filling soil in upper layers. The soft soil layers mainly include the sullage silty clay and the silty clay. Therefore, the soil layers in which the utility tunnel is located belongs to the typical soft soil.

Because the utility tunnel engineering, which is a typical soft foundation engineering, is shallowly buried under the green isolation areas, it will be seriously affected by the ground traffic load during the operation period, which will cause large deformation for the utility tunnel. Moreover, for locations near multiple ground transportation routes, this engineering will be repeatedly affected by the loads of incoming and outgoing ground vehicles. Therefore, for this utility tunnel engineering with its big size and complex structure during the operation period, the ground traffic load will significantly influence its safety. It is very important work to monitor and predict the safety for this utility tunnel during the operation period.

To monitor the response of the utility tunnel structure during the operation period in-site, the steel stress gauge and concrete strain gauge are placed at the middle of the floor, the side wall and the roof of the structure for the directly buried outlet shaft. Moreover, the contact pressure monitoring sensors are embedded in soil at different depths along the side wall of the utility tunnel and along the floor. The layout of field monitoring points is shown in Figure 9.

In this study, the used data acquisition instrument is the dynamic and static signal testing system of Chentu CT5808W, whose adjustable maximum frequency is 50 Hz. This instrument can capture the slight changes in sensors within a short period of time. In the field test, to simulate the vehicle load under normal two-way traffic and one-way traffic conditions, on the second lane from east to west and the opposite lane, vehicle load action tests have been conducted, with rear axle loads of 8.3 tons, 9 tons and 11 tons. The vehicle load, which is a one-time action, is applied by a dump truck.

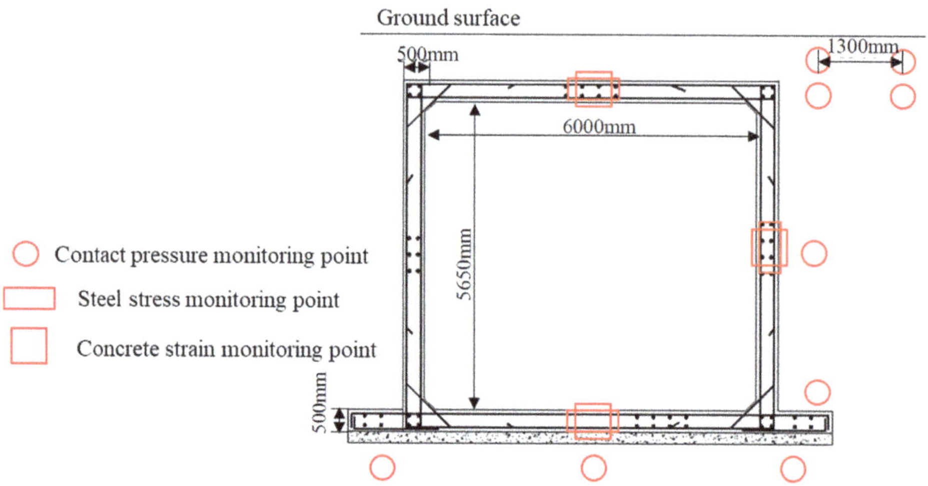

Figure 9. Layout of monitoring points.

3. Results

3.1. Construction Datasets from Field Tests Results

From the field test, it can be found that the sensors buried in the roof of the structure can obviously reflect the disturbance of vehicle loads during the operation period. Therefore, the results of those sensors are used in this study. From the comprehensive analysis of the engineering condition, site investigation and real measurement data, it can be found that the main load disturbance factors affecting the response of the utility tunnel structure include the vehicle driving speed, the magnitude of the vehicle load, the lateral distance between the vehicle wheel load center and the mid span of the structure roof, the symmetry of the vehicle load distribution and the operating time. The disturbance response of the utility tunnel structure is mainly represented by the settlement and the horizontal stress at the mid span of the structure roof. It must be noted that the settlement of the mid span of the structure roof can be computed by the obtained concrete strain results at the corresponding position, and the horizontal stress at the mid span of the structure roof is also computed by the obtained steel stress results at the corresponding position. Therefore, to predict the safety of the utility tunnel structure, the four vehicle operating load parameters and one operating time parameter are selected as the influence factors, which are the input variables of the prediction model, and the response factors of the utility tunnel structure are the settlement and the horizontal stress at the mid span of the structure roof, which are the output variables of the prediction model. According to the requirements for the input and output variables of the prediction model, field test data have been organized to establish the datasets for the training the model. For the large number of field test datasets, in order to illustrate the form of the dataset, only part of the data are provided here, as shown in Table 1.

It must be noted that, in Table 1, the "1" in the third column represents load symmetry, and "2" represents load asymmetry. For the data in each line, the first five ones represent the five disturbance factors, and the last two represent the two corresponding responses of the utility tunnel structure affected by the first five disturbance factors. The data in each line are one training sample of the prediction model, and the datasets of the prediction model are constructed by numerous lines of the training sample.

Finally, it must be noted that, because the load disturbance factors can be determined easily in the field tests at one engineering site, in this study, only the main load disturbance factors have been considered. Actually, in real engineering, there are other influence factors on the performance of the utility tunnel structure, such as the type of soft foundation soil.

However, those factors are unchanged for one particular utility tunnel during the operation period. Therefore, those influence factors cannot be taken as the disturbance factors for one particular form of engineering. Moreover, for one particular utility tunnel, the influence of the type of soft foundation soil has been represented in the implicit relationship between the load disturbance factors and structural responses. Therefore, for this prediction study, the influence of the type of soft foundation soil is not considered. That is to say, the type of soft foundation soil does not affect the predictive accuracy of the deep learning model.

Table 1. Part data of field test datasets.

Input Variables					Output Variables	
Operating Time /h	Vehicle Speed /km/h	Symmetry of Vehicle Load	Magnitude of Vehicle Load /kN	Load Position /m	Vertical Displacement /mm	Horizontal Stress /kPa
1	22.5	1	156.49	5.25	2.0053	720.342
1.6	41.25	1	169.17	5.25	2.2926	730.722
2.17	60	1	190.01	5.25	2.2596	725.702
1.26	60	1	195.81	5.25	2.4146	728.099
4.31	41.25	1	180.01	5.25	2.0452	720.963
4.63	80	2	90.01	6.39	1.8	391.573
1.17	41.25	2	81.25	7.34	1.17	303.95
4.06	80	2	100.1	6.39	2.24	383.41
1.25	80	2	96.58	6.39	2.76	340.47
1	41.25	2	96.58	7.34	2.16	288.28

3.2. Process of Prediction by WO-DBN

The prediction process of the settlement and the horizontal stress at the mid span of the structure roof based on WO-DBN is shown in Figure 10.

From Figure 10, the computing process of the prediction structure response of the utility tunnel is as follows.

(1) Based on the field test results, the datasets for the safety prediction model of the utility tunnel structure during the operation period have been constructed, that is, the field test datasets for the response of the utility tunnel structure during the operation period (including the structure deformation index data and structure stress index data) have been selected. Here, the field test data of a one-year simulated operation period have been used.

(2) The field test datasets have been preprocessed. The duplicate data and obvious abnormal data have been removed. The zero values in the datasets have been revised to the very small values (greater than zero). After preprocessing, a total of 15,376 sets of valid data have been obtained.

It should be noted that, if the amount of training data is very huge, to reduce the workload of processing the training data, the digital signal processing methods such as the wavelet analysis or filtration methods can be employed to optimize the selection and preprocessing of training data.

(3) Considering the total amount of data, the datasets have been divided. Approximately 80% of the data have been selected as the training sets, including a total of 12,110 sets, and the remaining data are as the testing sets. The used datasets are summarized in Table 2.

Table 2. The used datasets.

Name	Field Test Datasets	Training Sets	Testing Sets
Number of sets	15,376	12,110	3266

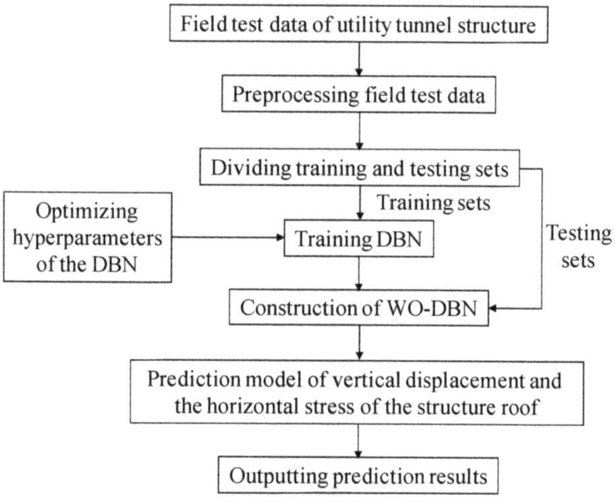

Figure 10. Flow chart of the prediction structure response of the utility tunnel based on WO-DBN.

(4) The hyperparameters of the DBN have been optimized by the WOA, and the prediction model by the DBN based on the optimization results has been established. The predication model has been trained by the training sets to determine the parameters of the predication model, and the WO-DBN model for the prediction of the vertical displacement and the horizontal stress at the mid span of the structure roof can be obtained.

(5) The data of the load and operating time of the testing sets have been substituted into the obtained WO-DBN model, and the corresponding structure response can be obtained, which are the predicted results of the vertical displacement and the horizontal stress at the mid span of the structure roof.

3.3. Evaluation Index of the Prediction Model

In this study, the square root mean square error (RMSE), mean absolute error (MAE) and correlation coefficient (R) are used to evaluate the prediction accuracy of the model, which are as follows:

$$\text{RMSE} = \sqrt{\frac{\sum\limits_{i=1}^{N} (y_i - y'_i)^2}{N}}, \tag{17}$$

$$\text{MAE} = \frac{\sum\limits_{i=1}^{N} |y_i - y'_i|}{N}, \tag{18}$$

$$\text{R} = \frac{\sum\limits_{i=1}^{N} (y_i - \overline{y})(y'_i - \overline{y'})}{\sqrt{\sum\limits_{i=1}^{N} (y_i - \overline{y})^2 \sum\limits_{i=1}^{N} (y'_i - \overline{y'})^2}}, \tag{19}$$

where y'_i represents the prediction value corresponding to the i-th input data, $\overline{y'}$ represents the average value of prediction values, y_i represents the real test value corresponding to the i-th input data, $\overline{y}$ represents the average value of real test values and N represents the number of prediction samples.

3.4. Analysis of Predication Results

The training sets are used to train the predication model, and the WO-DBN model can be obtained. The optimized hyperparameters of the DBN, which are the parameters of the WO-DBN, are summarized in Table 3.

Table 3. Parameters of the WO-DBN model.

Parameters	n_1	n_2	n_3	n_4	η	t_1	t_2
Values	44	47	34	18	0.0044	13	60

The testing sets have been substituted into the obtained WO-DBN model, and the prediction results of the structure response can be obtained, which are as shown in Figure 11. It must be noted that, for comparison, the real test results are also shown in this figure.

It must be noted that, in Figure 11, to show the computing errors between the prediction results and the real test ones clearly, the scattered points of vertical displacement and the horizontal stress of the structure roof corresponding to different samples were artificially connected to form variation curves. Therefore, the variation curves in Figure 11 have no practical meaning.

From Figure 11, it can be found that the prediction results of the WO-DBN model are in agreement with the real test results—that is, the performance of the constructed prediction model in this study is suitable.

To deeply analyze the prediction results, based on Equations (17)–(19), the quantitative evaluation indexes (RMSE, MAE and R) can be obtained, which are summarized in Table 4.

Table 4. Evaluation indexes of the WO-DBN prediction model.

Indexes	RMSE	MAE	R
Vertical Displacement	0.2312	0.1604	0.9742
Horizontal stress	22.0217	12.3726	0.6825

From Table 4, it can be found that, for the vertical displacement of the structure roof, the values of RMSE and MAE are all small, and the value of R is 0.9742, which is near 1. Therefore, the prediction results are very good. However, for the horizontal stress of the structure roof, the values of RMSE and MAE are large. The reason is that the computing values of RMSE and MAE are related with the magnitude of the data [27,35]. Because the magnitude of the data for the horizontal stress is much larger than that for the vertical displacement, there are big differences for the computing values of RMSE and MAE between them. Moreover, the computing value of R is not related with the magnitude of the data [27,35]. Therefore, its result is relatively effective. Actually, for the horizontal stress of the structure roof, its computing value of R is 0.6825, which is near 0.7, that is, the prediction result of WO-DBN is somewhat suitable too. Obviously, the prediction result of vertical displacement is better than that of horizontal stress, that is, the obtained WO-DBN prediction model has a better predictive effect on the structure deformation response. The reason is that the displacement response of the utility tunnel structure to the disturbances such as operation loads is more obvious than the stress response, which is represented by the more obvious change in results in Figure 11a than that in Figure 11b.

From the above analysis, it can be found that, based on the constructed WO-DBN model and considering the disturbance factors of the utility tunnel structure during the operation period, the corresponding disturbance response of the utility tunnel structure can be predicted well. Therefore, this study can provide an effective way to dynamically predict the safety response of the utility tunnel structure during the operation period based on the big data analysis by the artificial intelligent method.

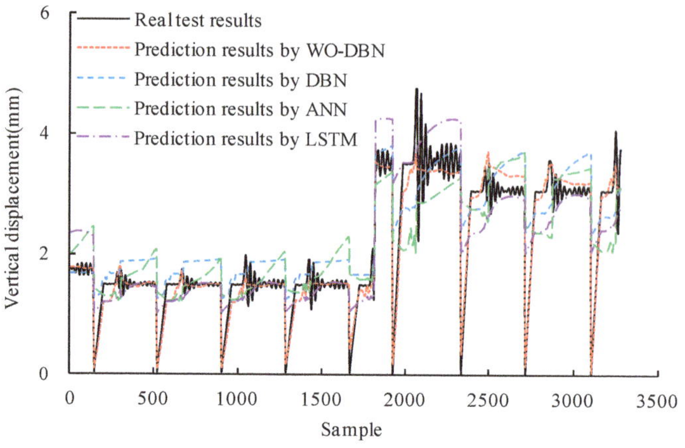

(a)

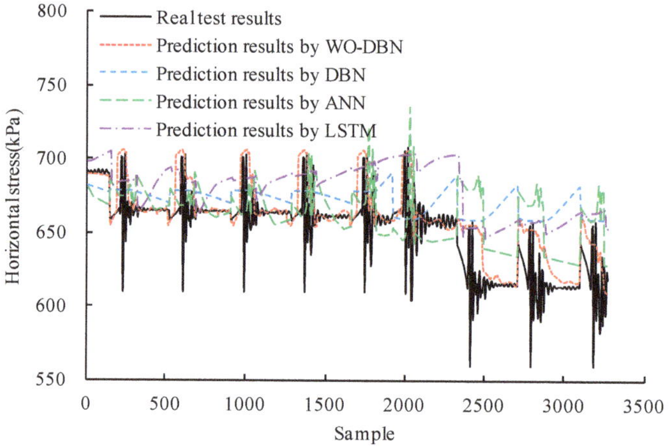

(b)

Figure 11. Prediction results of the structure response for the utility tunnel: (**a**) Vertical displacement at the mid span of the structure roof; (**b**) Horizontal stress at the mid span of the structure roof.

4. Discussion

To further analyze the performance of the WO-DBN model, a comparison study between WO-DBN, DBN, ANN and another deep learning method called long short-term memory (LSTM) has been conducted.

For the DBN model, the trial-and-error method is generally used to determine its hyperparameters, which can be described as follows.

(1) According to the experience, the possible ranges of hyperparameters have been determined.

(2) Several sets of hyperparameter combinations have been selected in their possible ranges by the experimental design or other methods, and these hyperparameter combinations have been used for prediction.

(3) The prediction errors of models by the different combinations of hyperparameters have been used for comparison, that is, the evaluation indexes of the prediction models have been compared. The parameter combination with the smallest prediction error has been selected as the final hyperparameters of the prediction model.

By using the trial-and-error method, the determined hyperparameters of the DBN model are summarized in Table 5.

Table 5. Parameters of the DBN model.

Parameters	n_1	n_2	n_3	n_4	η	t_1	t_2
Values	45	50	30	10	0.0014	15	65

As one typical artificial intelligence method, the ANN is generally used for prediction study [27,35–37]. Therefore, it is used here for comparison. By using the trial-and-error method, too, its hyperparameters can be determined, which are summarized in Table 6.

Table 6. Parameters of the ANN and LSTM models.

	Parameters	n_1	n_2	η	*Epoch*
ANN	Values	32	16	0.01	50
LSTM	Values	16	8	0.001	100

It should be noted that, in Table 6, n_1 and n_2 are neuron numbers for the first and second hidden layers, respectively. η is the learning rate, and *Epoch* is the maximum iteration number.

Moreover, for the LSTM which belongs to the recurrent neural network (RNN) [38], it is one generally used deep learning method for the prediction study [38,39]. Therefore, here, for a comparison with the new deep learning model (WO-DBN), the LSTM is applied. By using the trial-and-error method, the hyperparameters of LSTM can also be determined. Because the determined structure of LSTM is the same as that of ANN, the hyperparameters of LSTM are also summarized in Table 6.

By using the determined parameters in Tables 5 and 6, the prediction models for the response of the utility tunnel structure based on DBN, ANN and LSTM can be constructed. For comparison, the prediction results of the DBN, ANN and LSTM models are also shown in Figure 11. From Figure 11, it can be found that there are some differences between the prediction results of the three models (DBN, ANN and LSTM) and the real test results, and the results of WO-DBN approach the real data more. Therefore, the prediction results of the new WO-DBN model are the best, which are much better than those of the three models. Moreover, to compare the four models (WO-DBN, DBN, ANN and LSTM) more clearly, the quantitative evaluation indexes (RMSE, MAE and R) of the four prediction models are summarized in Table 7.

Table 7. Comparison of evaluation indexes for four prediction models (WO-DBN, DBN, ANN and LSTM).

		WO-DBN	**DBN**	**ANN**	**LSTM**
Vertical displacement	RMSE	0.2312	0.5697	0.6563	0.5595
	MAE	0.1604	0.4285	0.4852	0.3777
	R	0.9742	0.8642	0.7585	0.8435
Horizontal stress	RMSE	22.0217	31.2888	38.1835	34.5109
	MAE	12.3726	24.7234	30.4551	28.2891
	R	0.6825	0.5011	0.3411	0.5145

From Table 7, it can be found that, both for the vertical displacement and for the horizontal stress, the computing errors of the WO-DBN model are all much less than those of the three other models (DBN, ANN and LSTM). For example, the R for the vertical displacement by WO-DBN is 0.9742, which is much larger than those by DBN (0.8642), ANN (0.7585) and LSTM (0.8435). The R for the horizontal stress by WO-DBN is 0.6825, which is also much larger than those by DBN (0.5011), ANN (0.3411) and LSTM (0.5145).

Therefore, the performance of the prediction model by WO-DBN is significantly better than those by DBN, ANN and LSTM. The proposed WO-DBN model is a more suitable method for predicting the performance of the utility tunnel structure during the operation period. Moreover, in four prediction models, the performance of ANN is the poorest, whose computed R for the vertical displacement and the horizontal stress is the least. The computed R for the vertical displacement by DBN is larger than that by LSTM, but the computed R for the horizontal stress by DBN is slightly less than that by LSTM. However, with comprehensive consideration of the computing errors, the performance of DBN is superior. Therefore, for the prediction of the performance of the utility tunnel structure during the operation period, the order of the four models is as follows: WO-DBN, DBN, LSTM and ANN.

Although the proposed WO-DBN model can well predict the performance of the utility tunnel structure in a soft foundation during the operation period and its performance is much better than that of other models (DBN, ANN and LSTM), as a preliminary study, the amount of used data is not very large, and the considered structure disturbance factors are not very comprehensive. Moreover, in this deep learning model, the WOA is only used to select the hyperparameters of the DBN, and the DBN is still trained by the traditional greedy layer-by-layer learning algorithm. Finally, in this study, the offline data are used in the prediction model, which restricts the real-time prediction. Therefore, the future works can be summarized as follows: (1) a prediction study on the real big data of the measured disturbance of the utility tunnel structure during the operation period, (2) a prediction study considering more structure disturbance factors for the utility tunnel structure during the operation period, (3) the development of a new deep learning model, in which both the hyperparameters and algorithm parameters of DBN are all optimized by the WOA and (4) a prediction study based on the real-time monitoring data to enhance the predictive capabilities of the deep learning model.

5. Conclusions

In this study, to monitor the performance of the utility tunnel structure in a soft foundation affected by ground traffic loads during the operation period, the field tests have been conducted in the typical engineering site (one utility tunnel engineering in Suqian City of Jiangsu Province, China). From the field test results, the five main disturbance factors (four vehicle operating load parameters and one operating time parameter) on the utility tunnel structure and the two main structural safety responses (displacement and stress) have been determined to construct the big datasets whose number is 15,376 for the prediction model. To treat the datasets, one new deep learning model called WO-DBN has been proposed, in which the WOA is used to optimize the hyperparameters of the DBN. Finally, using the WO-DBN model, and based on the datasets, the main safety responses of the utility tunnel structure have been predicted. From the studies, the following conclusions can be drawn: (1) For the utility tunnel structure in a soft foundation during the operation period, the two main safety responses affected by the five main disturbance factors such as the load and time are the displacement and stress at the mid span of the structure roof. (2) For its special structure and good computing performance, the deep learning method (DBN) can analyze the big data from the field tests well; however, for the numerous hyperparameters of DBN, which cannot be determined easily, the WOA is used to optimize those hyperparameters, and a new deep learning model (WO-DBN) is proposed to treat the big data. (3) Based on the new WO-DBN model, the dynamic safety status of the utility tunnel structure in a soft foundation during the operation period can be predicted well, whose results show that the computing errors are not significant (the average absolute error for the displacement is 0.1604 and that for the stress is 12.3726) and can meet the requirement of real engineering.

Author Contributions: Conceptualization, W.G.; methodology, W.G. and S.G.; software, S.G. and Y.G.; validation, S.G. and S.Y.; formal analysis, S.G.; investigation, S.Y.; resources, W.G.; data curation, S.G. and S.Y.; writing—original draft preparation, W.G.; writing—review and editing, W.G. and Y.G.;

visualization, S.G.; supervision, W.G.; project administration, W.G.; funding acquisition, W.G. All authors have read and agreed to the published version of the manuscript.

Funding: This research received no external funding.

Institutional Review Board Statement: Not applicable.

Informed Consent Statement: Not applicable.

Data Availability Statement: Restrictions apply to the availability of these data. The data were obtained from Suqian High-speed Railway Construction and Development Co., Ltd. and are available from the authors with the permission of Suqian High-speed Railway Construction and Development Co., Ltd.

Conflicts of Interest: The authors declare no conflicts of interest.

References

1. Luo, Y.; Alaghbandrad, A.; Genger, T.; Hammad, A. History and recent development of multi-purpose utility tunnels. *Tunn. Undergr. Space Technol.* **2020**, *103*, 103511. [CrossRef]
2. Yang, C.; Peng, F.-L. Discussion on the Development of Underground Utility Tunnels in China. *Procedia Eng.* **2016**, *165*, 540–548. [CrossRef]
3. Zhou, Q.; He, H.; Liu, S.; Wang, P.; Zhou, Y.; Zhou, J.; Fan, H.; Jin, F. Evaluation of blast-resistant ability of shallow-buried rein-forced concrete urban utility tunnel. *Eng. Fail. Anal.* **2021**, *119*, 105003. [CrossRef]
4. Zhou, Q.; He, H.-G.; Liu, S.-F.; Chen, X.-S.; Tang, Z.-X.; Liu, Y.; Qiu, Z.-Y.; Li, S.-S.; Wang, H.; Zhou, Y.-Z.; et al. Blast resistance evaluation of urban utility tunnel reinforced with BFRP bars. *Def. Technol.* **2020**, *17*, 512–530. [CrossRef]
5. Pan, Y.; Zong, Z.; Li, J.; Qian, H.; Wu, C. Investigating the dynamic response of a double-box utility tunnel buried in calcareous sand against ground surface explosion. *Tunn. Undergr. Space Technol.* **2024**, *146*, 105636. [CrossRef]
6. Zhao, G.; Zhu, L.; Wu, S.; Liu, W.; Duan, S. Experimental and numerical investigation on the cross-sectional mechanical behavior of prefabricated multi-cabin RC utility tunnels. *Structures* **2022**, *42*, 466–479. [CrossRef]
7. Zhang, J.; Zhang, Y.; Peng, C.; Lei, Y.; Zhang, A.; Zuo, Z.; Chen, Z. Experimental and Numerical Investigation into Full-Scale Model of New Type Assembled Integral Utility Tunnel. *Buildings* **2023**, *13*, 1428. [CrossRef]
8. Xue, W.; Chen, S.; Wang, Q. Cyclic Loading Test Conducted on the Bottom Joints of a Hybrid Precast Utility Tunnel Composed of Double-Skin Sidewalls and a Precast Bottom Slab. *Buildings* **2024**, *14*, 341. [CrossRef]
9. Zhong, Z.; Li, G.; Li, J.; Shen, J.; Zhao, M.; Du, X. Experimental study on out-of-plane seismic performance of precast composite sidewalls of utility tunnel with grouting-sleeve joints. *Undergr. Space* **2024**, *16*, 1–17. [CrossRef]
10. Xue, W.; Chen, S.; Bai, H. Pseudo-Static Tests on Top Joints of Hybrid Precast Utility Tunnel. *Buildings* **2023**, *13*, 2567. [CrossRef]
11. Chen, H.; El Naggar, M.H.; Chu, J.; Li, X.; He, Q.; Wang, L.; Liu, X.; Zhou, L. Transverse response of utility tunnel under near fault ground motions: Multi-shake table array tests. *Soil Dyn. Earthq. Eng.* **2023**, *174*, 108135. [CrossRef]
12. Yue, F.; Liu, B.; Zhu, B.; Jiang, X.; Chen, S.; Jaisee, S.; Chen, L.; Lv, B. Shaking table investigations on seismic performance of prefabricated corrugated steel utility tunnels. *Tunn. Undergr. Space Technol.* **2020**, *105*, 103579. [CrossRef]
13. Yao, A.; Tian, T.; Gong, Y.; Li, H. Shaking Table Tests of Seismic Response of Multi-Segment Utility Tunnels in a Layered Liquefiable Site. *Sustainability* **2023**, *15*, 6030. [CrossRef]
14. Liang, J.; Zhang, J.; Dong, B.; Xu, A.; Ba, Z. Shaking table tests on the seismic performance of prefabricated T-shaped cross utility tunnel. *Structures* **2023**, *58*, 105516. [CrossRef]
15. Huang, D.-L.; Zong, Z.-L.; Tao, X.-X.; Liu, Q.; Huang, Z.-Y.; Tang, A.-P. Seismic response of utility tunnel in horizontal nonhomogeneous site based on improved discrete element method. *Structures* **2023**, *57*, 105179. [CrossRef]
16. Yue, F.; Liu, B.; Zhu, B.; Jiang, X.; Chen, L.; Liao, K. Shaking table test and numerical simulation on seismic performance of prefabricated corrugated steel utility tunnels on liquefiable ground. *Soil Dyn. Earthq. Eng.* **2020**, *141*, 106527. [CrossRef]
17. Tsinidis, G. Response characteristics of rectangular tunnels in soft soil subjected to transversal ground shaking. *Tunn. Undergr. Space Technol.* **2017**, *62*, 1–22. [CrossRef]
18. Zhang, Y.; Liu, B.; Meng, L. Structural behavior and soil arching state of underground corrugated steel utility tunnel. *J. Constr. Steel Res.* **2023**, *203*, 107798. [CrossRef]
19. Huang, B.-T.; Zhu, J.-X.; Weng, K.-F.; Huang, J.-Q.; Dai, J.-G. Prefabricated UHPC-concrete-ECC underground utility tunnel reinforced by perforated steel plate: Experimental and numerical investigations. *Case Stud. Constr. Mater.* **2022**, *16*, e00856. [CrossRef]
20. Huang, Z.; Feng, Y.; Tang, A.; Liu, Q. Influence of oblique incidence of P-waves on seismic response of prefabricated utility tunnels considering joints. *Soil Dyn. Earthq. Eng.* **2023**, *167*, 107797. [CrossRef]
21. Qian, H.; Zong, Z.; Wu, C.; Li, J.; Gan, L. Numerical study on the behavior of utility tunnel subjected to ground surface explosion. *Thin-Walled Struct.* **2021**, *161*, 107422. [CrossRef]
22. Yin, X.; Liu, H.; Chen, Y.; Wang, Y.; Al-Hussein, M. A BIM-based framework for operation and maintenance of utility tunnels. *Tunn. Undergr. Space Technol.* **2020**, *97*, 103252. [CrossRef]

23. Yang, L.; Zhang, F.; Yang, F.; Qian, P.; Wang, Q.; Wu, Y.; Wang, K. Generating Topologically Consistent BIM Models of Utility Tunnels from Point Clouds. *Sensors* **2023**, *23*, 6503. [CrossRef]
24. Peng, F.-L.; Qiao, Y.-K.; Yang, C. A LSTM-RNN based intelligent control approach for temperature and humidity environment of urban utility tunnels. *Heliyon* **2023**, *9*, e13182. [CrossRef]
25. Wu, J.; Bai, Y.; Fang, W.; Zhou, R.; Reniers, G.; Khakzad, N. An Integrated Quantitative Risk Assessment Method for Urban Underground Utility Tunnels. *Reliab. Eng. Syst. Saf.* **2021**, *213*, 107792. [CrossRef]
26. Xue, G.; Liu, S.; Ren, L.; Gong, D. Risk assessment of utility tunnels through risk interaction-based deep learning. *Reliab. Eng. Syst. Saf.* **2024**, *241*, 109626. [CrossRef]
27. Ansari, A.; Rao, K.; Jain, A.; Ansari, A. Formulation of multi-hazard damage prediction (MhDP) model for tunnelling projects in earthquake and landslide-prone regions: A novel approach with artificial neural networking (ANN). *J. Earth Syst. Sci.* **2023**, *132*, 164. [CrossRef]
28. Liu, W.; Wang, Z.; Liu, X.; Zeng, N.; Liu, Y.; Alsaadi, F.E. A survey of deep neural network architectures and their applications. *Neurocomputing* **2017**, *234*, 11–26. [CrossRef]
29. Gong, B.; Shu, C.; Han, S.; Cheng, S.-G. Mine Vegetation Identification via Ecological Monitoring and Deep Belief Network. *Plants* **2021**, *10*, 1099. [CrossRef]
30. Zhou, C.; Gao, W.; Cui, S.; Zhong, X.; Hu, C.; Chen, X. Ground Settlement of High-Permeability Sand Layer Induced by Shield Tunneling: A Case Study under the Guidance of DBN. *Geofluids* **2020**, *2020*, 6617468. [CrossRef]
31. Hinton, G.E.; Osindero, S.; Teh, Y.-W. A Fast Learning Algorithm for Deep Belief Nets. *Neural Comput.* **2006**, *18*, 1527–1554. [CrossRef]
32. Mirjalili, S.; Lewis, A. The whale optimization algorithm. *Adv. Eng. Softw.* **2016**, *95*, 51–67. [CrossRef]
33. Zhou, J.; Zhu, S.; Qiu, Y.; Armaghani, D.; Zhou, A.; Yong, W. Predicting tunnel squeezing using support vector machine opti-mized by whale optimization algorithm. *Acta Geotech.* **2022**, *17*, 1343–1366. [CrossRef]
34. Krizhevsky, A.; Sutskever, I.; Hinton, G.E. Imagenet classification with deep convolutional neural networks. *Commun. ACM* **2017**, *60*, 84–90. [CrossRef]
35. Ahmad, S.A.; Ahmed, H.U.; Rafiq, S.K.; Ahmad, D.A. Machine learning approach for predicting compressive strength in foam concrete under varying mix designs and curing periods. *Smart Constr. Sustain. Cities* **2023**, *1*, 16. [CrossRef]
36. Nguyen, L.Q.; Le, T.T.T.; Nguyen, T.G.; Tran, D.T. Prediction of underground mining-induced subsidence: Artificial neural network based approach. *Min. Miner. Depos.* **2023**, *17*, 45–52. [CrossRef]
37. Kim, Y.; Lee, S.S. Application of Artificial Neural Networks in Assessing Mining Subsidence Risk. *Appl. Sci.* **2020**, *10*, 1302. [CrossRef]
38. Yang, T.; Wen, T.; Huang, X.; Liu, B.; Shi, H.; Liu, S.; Peng, X.; Sheng, G. Predicting Model of Dual-Mode Shield Tunneling Parameters in Complex Ground Using Recurrent Neural Networks and Multiple Optimization Algorithms. *Appl. Sci.* **2024**, *14*, 581. [CrossRef]
39. He, Y.; Chen, Q. Construction and Application of LSTM-Based Prediction Model for Tunnel Surrounding Rock Deformation. *Sustainability* **2023**, *15*, 6877. [CrossRef]

MDPI AG
Grosspeteranlage 5
4052 Basel
Switzerland
Tel.: +41 61 683 77 34

Applied Sciences Editorial Office
E-mail: applsci@mdpi.com
www.mdpi.com/journal/applsci

www.ingramcontent.com/pod-product-compliance
Lightning Source LLC
Chambersburg PA
CBHW061600220926
51850CB00005B/383
9783725822478